# HANDBOOK
## OF
# Environmental
# Degradation Rates

**PHILIP H. HOWARD**
**ROBERT S. BOETHLING**
**WILLIAM F. JARVIS**
**WILLIAM M. MEYLAN**
**EDWARD M. MICHALENKO**

**HEATHER TAUB PRINTUP**
Editor

 LEWIS PUBLISHERS

**Library of Congress Cataloging-in-Publication Data**

Howard, Philip H.
     Handbook of environmental degradation rates / Philip H. Howard.
     Includes bibliographical references and indexes.

     1. Pollutants--Environmental aspects--Handbooks, manuals, etc.
     2. Decomposition (Chemistry)--Handbooks, manuals, etc.  I. Title.
     TD193.H73  1991
     628.5'2--dc20                                 90-47759

ISBN 0-87371-358-3

*Second Printing 1991*

LEWIS PUBLISHERS, INC.
121 South Main Street, Chelsea, Michigan 48118

PRINTED IN THE UNITED STATES OF AMERICA

## TABLE OF CONTENTS

# CHEMICAL NAME DIRECTORY

iv

# CHEMICAL NAME DIRECTORY

# CHEMICAL NAME DIRECTORY

# CHEMICAL NAME DIRECTORY

# CHEMICAL NAME DIRECTORY

# CHEMICAL NAME DIRECTORY

# CHEMICAL NAME DIRECTORY

# CHEMICAL NAME DIRECTORY

# CHEMICAL NAME DIRECTORY

# INTRODUCTION

This book is the result of work completed by the Syracuse Research Corporation in August 1989 for the Environmental Protection Agency (EPA). The task entailed the compilation of rate constants for individual abiotic and biotic degradation processes for chemicals of anthropogenic origin as they pertain to the environmental compartments of soil, water, and air. Estimates were performed for chemicals in which rate data could not be located in the available literature. A range of half-lives was established for both individual degradation processes and specific environmental media; the latter were based on the former. Half-lives for soil, water, and air do not account for the transport of a chemical between environmental compartments, and therefore, overall ranges are not necessarily representative of a chemical's actual persistence within a particular environmental medium.

Part I of the book outlines the approach employed in developing the Chemical Fate Rate Constants for the Superfund Amendments and Reauthorization Act (SARA) Section 313 chemicals and Superfund Health Evaluation Manual Chemicals. Altogether, the strategy was applied to 331 chemicals. Part II contains a test set of seven chemicals, which are reviewed in detail. These sample chemicals provide a better understanding of the approach that was used. Because of limitations placed upon time and available funds, the level of documentation for the examples in Part II was not maintained for the majority of chemicals contained in Part III.

# PART I: STRATEGY FOR DETERMINING ENVIRONMENTAL DEGRADATION RATE CONSTANTS

**Sources of Data.** The Syracuse Research Corporation's (SRC) Environmental Fate Data Bases (EFDB) were the primary sources of information; DATALOG, CHEMFATE, BIOLOG, and BIODEG files were searched for pertinent data. The review of data was greatly facilitated when CHEMFATE and/or BIODEG contained records on a chemical, since these files have actual experimental data. For chemicals not included in CHEMFATE or BIODEG, the reviewer used DATALOG for pointers to references pertaining to abiotic degradation processes and BIOLOG for references with information on biodegradation.

Physical/chemical properties, retrieved from SRC's CHEMFATE files, were helpful in the interpretation of degradation studies. For example, knowing the volatility and adsorption characteristics of a chemical are important for discerning the results of a semi-continuous or continuous activated sludge treatment test.

**Degradation Data Review and Analysis.** With the exception of removal efficiencies in wastewater treatment simulations, all individual process and media half-lives were for degradation processes and did not consider transport processes such as volatilization or adsorption. Thus, the degradation half-lives reported would have to be entered into some type of model that would consider transport processes in order to get the overall half-life in a given medium. Attempts were made to isolate data on individual processes within specific environmental media. For instance, the rate constant for the vapor phase reaction with photochemically produced hydroxyl radicals was considered independently from other photooxidation reactions in air such as reaction with ozone (only important for olefins). Biodegradation was separated into categories of aerobic and anaerobic. Removal efficiencies from wastewater treatment simulation studies were recorded for applicable chemicals. The abiotic degradation processes of hydrolysis, photolysis, and oxidation/reduction were also examined individually. Very little information regarding reduction reactions was located in the available literature. In all cases, experimental data were preferred over estimation techniques. The following sections outline the logic employed for each process as it pertains to air, water, or soil:

**Hydroxyl radical reaction in air:** For a considerable number of chemicals, experimental data were not found in the available literature. Therefore, rate constants were estimated according to the method of Atkinson, R (1987A). Based upon monitoring data for relatively pristine and polluted air, minimum and maximum atmospheric concentrations of $3X10^5$ and $3X10^6$ hydroxyl radicals/cm$^3$ (Atkinson, R, 1985; Hewitt, CN and Harrison, RM, 1985) were combined with either the experimental or estimated rate constant to give a range of half-lives.

**Ozone reaction in air:** The rate constant estimation method of Atkinson, R and Carter, WPL (1984) was used for those olefins which lacked experimental data. Representative ozone concentrations of $5.0X10^{11}$ to $3.0X10^{12}$ molecules/cm$^3$ were based on monitoring data of pristine and polluted atmospheres (Atkinson, R and Carter, WPL, 1984); and again, either experimental or estimated rate constants were applied to low and high ozone concentrations to yield a range of half-lives in air.

**Photolysis in air or water:** Half-lives for direct photolysis were based on sound experimental data in which sunlight or artificial light was used to irradiate a chemical in solution (preferably aqueous solution) at environmentally significant wavelengths (>290 nm). Photolysis rates in air were assumed to equal those in solution.

The Lambda max and whether or not absorption of UV light occurs at wavelengths greater than 290 nm was documented for chemicals with available UV absorption spectra in polar and non-polar solution (methanol, ethanol and hexane). This information may indicate a chemical has the potential to undergo photolysis in the environment, but an estimate of rate would not be prudent since a chemical that absorbs UV light at environmentally significant wavelengths does not necessarily photodegrade (there are many ways that a chemical can lose the absorbed energy without photodegrading).

**Photooxidation reactions with alkylperoxy ($RO_2 \cdot$), hydroxyl radicals ($HO \cdot$) or singlet oxygen in water ($^1O_2$):** For some chemical classes such as aromatic amines and phenols, reactions with photooxidants in sunlit waters can be important. As with photooxidation in air, the range of half-lives was calculated from reaction rate constants and monitored upper and lower levels of oxidants in relatively oligotrophic and eutrophic natural waters (Table 1). Class estimates were used for appropriate chemical class members that did not possess specific data.

**Table 1: Photooxidant concentrations in natural waters**

|  | Low | High | Reference |
|---|---|---|---|
| $HO \cdot$ | $5.0 \times 10^{-19}$ M | $2.0 \times 10^{-17}$ M | Mill, T et al. 1980<br>Mill, T 1989 |
| $RO_2 \cdot$ | $1.0 \times 10^{-11}$ M | $5.0 \times 10^{-10}$ M | Mill, T et al. 1980<br>Mill, T 1989 |
| $^1O_2$ | $1.0 \times 10^{-15}$ M | $1.0 \times 10^{-13}$ M | Scully, FE Jr., Hoigne, J 1987<br>Haag, WR, Hoigne, J 1986<br>Mill, T 1989 |

**Hydrolysis in water or soil:** Hydrolysis only applies to a limited number of chemicals that have hydrolyzable groups such as esters, aliphatic halogens, amides, carbamates, and phosphate esters. Relatively fast and slow half-lives were calculated from first or second order rate constants determined at pH 5, 7, and 9, and temperatures of 20 or 25°C. The PC-GEMS program HYDRO (U.S. EPA 1987) was used to estimate hydrolysis rates for esters that lacked experimental data. Despite the fact that humic materials can accelerate the hydrolysis rate for acid-catalyzed reactions or inhibit the hydrolysis rate for base-catalyzed reactions through adsorption to substrate (Perdue, EM 1983), hydrolysis half-lives in surface water, ground water, and soil were assumed to be identical, unless experimental data suggested otherwise.

**Aerobic biodegradation in water:** Data from surface water grab samples or river die-away studies were preferred over aqueous screening tests or properly controlled field studies. The low and high rate constants from grab sample data were used to establish the range of half-lives. If only one rate constant was available, the corresponding half-life was treated as a middle value. Additional information from screening studies aided in defining the estimated duration between low and high half-lives.

Our review of the literature did not generate grab sample data for a substantial number of chemicals. Aqueous field studies were considered when all other degradation or transport processes could be ruled out or taken into account (e.g., pond die-away studies of water soluble, non-volatile chemicals, or field studies that show a chemical is persistent).

In the absence of grab sample or reliable field data, aqueous screening studies were examined. A classification scheme was established for assigning high and low half-lives based upon the screening data. Chemicals that consistently degrade at a rapid, moderate, slow, and resistant rates were assigned to the appropriate range of biodegradation half-lives (Table 2). When little or conflicting data existed, these half-life ranges may have been adjusted based upon the available information and, where possible, data for related chemicals. When no biodegradation data were available, the categories were assigned by examining the functional groups in the chemical and using biodegradation "rules of thumb" (e.g., halogen or nitro groups slow biodegradation, and alcohol and carboxyl acid groups increase biodegradation) (Howard, P et al. 1987). If aerobic soil biodegradation rates were the only data available, the half-life for aerobic biodegradation in water was assumed to equal the biodegradation half-life in soil.

**Table 2: The classification and range of half-lives for the biodegradation of chemicals in environmental media**

| Class | Low $t^{1}/_{2}$ | High $t^{1}/_{2}$ |
|---|---|---|
| Fast | 1 day | 7 days |
| Moderately fast | 7 days | 4 weeks |
| Slow | 4 weeks | 6 months |
| Resistant | 6 months | 1 year |

**Aerobic biodegradation in soil:** The protocol for interpreting biodegradability tests of chemicals in soil was similar to that in water. Preferably, microbial degradation rates from one or more soil grab sample study were used to designate the range of half-lives. Field studies were acceptable if abiotic degradation and transport processes were not factors (e.g., a soil plot study of a non-volatile pesticide which does not hydrolyze, photodegrade, or leach). In the absence of credible grab sample or field rate data, screening tests were utilized to classify chemicals within

xvi

a standard category of half-lives (Table 2). Again, biodegradation rates in soil were assumed to equal those in natural surface waters.

**Anaerobic biodegradation:** Anaerobic biodegradation is generally much slower than aerobic biodegradation. The reverse is true only for certain chemicals such as chlorinated hydrocarbons. Nitroaromatics are also rapidly reduced in oxygen deficient environments, but it is questionable as to whether or not this is due to microbial action. Furthermore, biodegradation has been less studied under anaerobic conditions, and for many chemicals, information was not available. Unless experimental data were located, anaerobic biodegradation half-lives were conservatively estimated to be four times longer than the range for aerobic biodegradation.

**Biodegradation in biological treatment plants:** For some chemicals, percent removals from continuous and semi-continuous activated sludge treatment plant simulation studies or from influent/effluent monitoring data of conventional biological treatment facilities were obtained from the literature. If only one data point was found, it was listed as the high removal percentage.

**Determination of Overall Half-lives.** An overall range of half-lives was determined for each chemical within the environmental compartments of soil, surface water, ground water, and air. Half-lives were based upon the high and low degradation rates of the most important degradation process within a particular medium. In cases where every degradation rate was an estimate that did not differ drastically from the others, the final range of half-lives was selected from the process thought to possess the most reliable estimation technique.

**Soil:** Half-lives from the fastest reaction were usually transferred directly to the overall range of half-lives for soil. Biodegradation was the most common degradation process to dominate in soil, with the exception of those chemicals which undergo rapid hydrolysis. Soil and water hydrolysis half-lives were used interchangeably even though the higher organic matter concentrations in soil may increase the hydrolysis rate for acid-catalyzed reactions or slow the hydrolysis rate for base-catalyzed reactions (Perdue, EM 1983); actual hydrolysis data for soil were rarely found. Thus, most hydrolysis half-lives in soil were based upon rate data determined in water. For chemicals with overlapping biodegradation and hydrolysis half-lives, the respective fastest high and fastest low half-lives comprise the final range in soil. When hydrolysis was unimportant and biodegradation rate data in soil did not exist, the final range of half-lives in soil was assumed to equal the unacclimated aqueous aerobic biodegradation half-life. In many cases, the unacclimated aqueous aerobic biodegradation half-life was developed from screening studies and/or deduced using the "rules of thumb" previously mentioned.

**Surface water:** Half-lives for a chemical in surface water were derived in the same manner as those in soil. Generally, the most significant degradation process was the source of low and high half-lives. Biodegradation and then direct photolysis and photooxidation were the dominant degradation processes in surface water; hydrolysis was important for a small number of chemicals. Overlapping individual half-lives were merged to yield the most appropriate range of half-lives when more than one degradation process was environmentally relevant. For some pesticides in which only soil biodegradation rates were available, half-lives were assumed to be equal to those in aerobic soil.

**Ground water:** The range of half-lives for a chemical in ground water was also determined by the most important degradation process or processes. Biodegradation and

xvii

hydrolysis, to a lesser extent, were the principal means of degradation. Grab sample or field studies with dependable rate data were seldom found for chemicals in ground water. In general, biodegradation proceeds at a slow rate compared to surface waters because unacclimated ground water microbial populations are limited in terms of both numbers and enzymatic capability. Ground waters also maintain varying levels of oxygen and are more likely to be anaerobic. Hence, the rate of biodegradation in ground water was assumed to be one-half that in surface water and overall half-lives were conservatively estimated to be twice the unacclimated aqueous aerobic biodegradation half-lives, unless the data suggested otherwise. For chemicals that degrade rapidly under anaerobic conditions, the low half-life in ground water usually equalled the low aqueous anaerobic biodegradation half-life, and the high half-life was based upon the high aqueous aerobic biodegradation half life.

**Air:** The range of half-lives for a chemical in air was estimated using the fastest degradation process, usually reaction with hydroxyl radicals. For certain chemical classes, photooxidations with ozone and direct photolysis were important. Hydrolysis in air was important for only a few chemicals. When more than one reaction was environmentally important, a range of combined half-lives was calculated. For instance, the corresponding low and high half-lives for the gas-phase photooxidations of certain olefins with hydroxyl radicals and ozone were combined to yield a range of half-lives that was conditional upon simultaneous reactions.

# PART II: SAMPLE EVALUATIONS

The following section provides a fate rate table for each chemical, preceded by a brief review of the available rate data associated with each degradation process. Fate rate tables appear just as they are presented in Part III.

## 1,4-Dioxane (Dioxane)

**Physical Properties:** Based upon its physical properties, 1,4-Dioxane is not expected to significantly volatilize from water or to strongly adsorb to soil or sediments. It is expected to leach through soil rapidly, provided degradation is sufficiently slow.

### Abiotic Degradation

**Hydrolysis:** No data available; not expected to be significant under normal environmental conditions.

**Reduction:** No data available; not expected to be significant under normal environmental conditions.

**Photolysis:** No data available; not expected to be significant under normal environmental conditions.

**Photooxidation - Water:** Based upon experimental rate data, photooxidation with alkylperoxyl radicals ($RO_2$) is slow (low $t_{1/2} \approx 18$ years at an estimated high concentration of $5 \times 10^{-10}$ M $RO_2$), relative to reaction with hydroxyl radicals ($OH$) (high $t_{1/2} \approx 9.1$ years at an estimated low concentration of $5 \times 10^{-19}$ M $OH$).

**Photooxidation - Air:** Based upon estimated rate data, reaction with ozone ($O_3$) is very slow (low $t_{1/2} \approx 4.3$ years at an experimental high concentration of $3 \times 10^{12}$ $O_3/cm^3$) relative to reaction with $OH$ (high $t_{1/2} \approx 3.4$ days based upon measured rate constant for reaction of 1,3,5-trioxane with $OH$ and an experimental low concentration of $3 \times 10^5$ $OH/cm^3$).

(**Photooxidation - Polluted Air:** 50% degradation reported in 3.4 hours in air with 5 ppm NO, irradiated by sunlamps).

**Biodegradation:** The data located indicate that dioxane is resistant to biodegradation. The only grab sample experiments utilized trench leachate samples from a shallow-land waste burial site, but the study did not report incubation times for the experiments involving dioxane. Under aerobic conditions, 0% and 10% degradation were observed with and without an added N source, respectively; under anaerobic conditions, 4% and 13% degradation were observed, respectively.

Screening tests indicate little or no biodegradation of dioxane by sewage seed and activated sludge. Acclimation appears to have little effect on degradation rates. The MITI test confirms dioxane either is not degraded or is degraded slowly.

### Selection of Data/Determination of Rates and Half-lives:
### Half-lives:

**Soil:** No data concerning degradation in soil were located. Since hydrolysis is probably insignificant, biodegradation was selected as the most significant degradation mechanism. Based upon experiments in aqueous systems, biodegradation in soil is expected to be slow. An estimated range of half-lives of 1 to 6 months for dioxane in soil appears to be reasonable based upon scientific judgement.

**Air:** Photooxidation by $OH$ appears to be the most rapid degradation process in the atmosphere. High and low half-lives were calculated using estimations of high and low concentrations of ·OH expected in relatively polluted and pristine air, respectively. These ·OH concentrations were based upon reported experimental data.

<u>Surface Water</u>: The $t_{1/2}$ in surface water is based upon the estimated biodegradation rates because the rates for other degradation processes are either very slow (photooxidation) or are expected to be zero or negligible (direct photolysis, hydrolysis, and reduction).

<u>Ground Water</u>: The $t_{1/2}$ in ground water is based upon the estimated aerobic biodegradation rates (and assumed to be two times longer) because the rate of hydrolysis is negligible.

# 1,4-Dioxane

<u>CAS Registry Number:</u>  123-91-1

<u>Half-lives:</u>
  · **Soil:**                                      High:     4320 hours                (6 months)
                                                   Low:       672 hours                (4 weeks)
*Comment:* Scientific judgement based upon estimated unacclimated aqueous aerobic biodegradation half-life.

  · **Air:**                                       High:       81 hours                (3.4 days)
                                                   Low:       8.1 hours                (0.34 days)
*Comment:* Based upon photooxidation half-life in air.

  · **Surface Water:**                             High:     4320 hours                (6 months)
                                                   Low:       672 hours                (4 weeks)
*Comment:* Based upon estimated unacclimated aqueous aerobic biodegradation half-life.

  · **Ground Water:**                              High:     8640 hours                (12 months)
                                                   Low:      1344 hours                (8 weeks)
*Comment:* Scientific judgement based upon estimated unacclimated aqueous aerobic biodegradation half-life.

<u>Aqueous Biodegradation (unacclimated):</u>
  · **Aerobic half-life:**                         High:     4320 hours                (6 months)
                                                   Low:       672 hours                (4 weeks)
*Comment:* Scientific judgement based upon unacclimated aerobic aqueous screening test data which confirmed resistance to biodegradation (Kawasaki, M (1980); Sasaki, S (1978)).

  · **Anaerobic half-life:**                       High:   17280 hours                (24 months)
                                                   Low:      2688 hours                (16 weeks)
*Comment:* Scientific judgement based upon estimated aerobic biodegradation half-life.

  · **Removal/secondary treatment:**               High:       No data
                                                   Low:
*Comment:*

<u>Photolysis:</u>
  · **Atmos photol half-life:**                    High:  Will not directly photolyze
                                                   Low:
*Comment:*

  · **Max light absorption (nm):**                 No data

*Comment:*

**· Aq photol half-life:**                          High:
                                                    Low:
*Comment:*

**Photooxidation half-life:**
  **· Water:**                            High:  $8 \times 10^4$ hours      (9.1 years)
                                                    Low:    1608 hours        (67 days)
*Comment:* Based upon measured rates for reaction with hydroxyl radicals in water (Dorfman, LM and Adams, GE (1973); Anbar, M and Neta, P (1967)).

  **· Air:**                              High:    81 hours        (3.4 days)
                                                    Low:    8.1 hours      (0.34 days)
*Comment:* Based upon measured rate constant for reaction of 1,3,5-trioxane with hydroxyl radicals in air (Atkinson, R (1987A)).

**Reduction half-life:**                            High:  No reducible groups
                                                    Low:
*Comment:*

**Hydrolysis:**
  **· First-order hydr half-life:**
*Comment:*

  **· Acid rate const $(M(H+)\text{-}hr)^{-1}$:**    No hydrolyzable groups
*Comment:*

  **· Base rate const $(M(OH-)\text{-}hr)^{-1}$:**
*Comment:*

## Vinyl Chloride

Physical Properties:  Based upon its physical properties, vinyl chloride can be expected to rapidly volatilize from water, soil, and surfaces.  It is also expected to leach through soil due to its estimated moderate mobility in soil.

Abiotic Degradation
    Hydrolysis:  No data available; not expected to be significant under normal environmental conditions.
    Reduction:  No data available; not expected to be significant under normal environmental conditions.
    Photolysis:  No data available; not expected to be significant under normal environmental conditions.
    Photooxidation - Water:  No data available.
    Photooxidation - Air:  Based upon experimental rate data, reaction with ozone ($O_3$) is very slow relative to reaction with $\cdot OH$ (high $t_{1/2} \approx 97$ hours at an experimental low concentration of $3 \times 10^5$ $\cdot OH/cm^3$).

Biodegradation:  An anaerobic ground water field study of the biodegradation of tetrachloroethene, trichloroethene, cis- and trans-1,2-dichloroethene, and vinyl chloride yielded an estimate of 59 months for the half-life of vinyl chloride. The only other data available are from screening tests.  Under aerobic conditions, the amount of degradation observed ranges from none (0% BODT in 30 days using acclimated sewage seed) to significant (21.5% $CO_2$ evolution in 5 days using activated sludge).

Selection of Data/Determination of Rates and Half-lives:
Half-lives:
    Soil:  No data concerning degradation were located.  Since hydrolysis is probably insignificant, biodegradation was selected as the most significant degradation mechanism.  The range of half-lives was estimated using scientific judgement and data from aqueous screening experiments.

    Air:  Based upon experimental data, photooxidation by $\cdot OH$ appears to be the most rapid degradation process in the atmosphere.  High and low half-lives were calculated using an experimentally determined range of high and low concentrations of $\cdot OH$ expected in relatively polluted and pristine air, respectively.

    Surface Water:  The $t_{1/2}$ in water is based upon the estimated biodegradation rates because the rates for other degradation processes are either very slow (photooxidation) or are expected to be zero or negligible (direct photolysis, hydrolysis, and reduction).  The biodegradation rates were estimated using scientific judgement and data from the few aqueous screening experiments.

    Ground Water:  The $t_{1/2}$ in ground water is based upon the estimated biodegradation rates because the rate of hydrolysis is negligible. The low $t_{1/2}$ is based upon the estimated low unacclimated aqueous aerobic biodegradation half-life (and assumed to be two times longer) and the high $t_{1/2}$ is based upon the estimated half-life for anaerobic biodegradation from a ground water field study.

# Vinyl chloride

**CAS Registry Number:** 75-01-4

## Half-lives:
· **Soil:**                                          High:    4320 hours            (6 months)

Low:      672 hours            (4 weeks)

*Comment:* Scientific judgement based upon estimated unacclimated aqueous aerobic biodegradation half-life.

· **Air:**                                           High:       97 hours

Low:      9.7 hours

*Comment:* Based upon photooxidation half-life in air.

· **Surface Water:**                                 High:    4320 hours            (6 months)

Low:      672 hours            (4 weeks)

*Comment:* Scientific judgement based upon estimated unacclimated aqueous aerobic biodegradation half-life.

· **Ground Water:**                                  High:  69,000 hours            (95 months)

Low:     1344 hours            (8 weeks)

*Comment:* Scientific judgement based upon estimated unacclimated aqueous aerobic biodegradation half-life (low $t_{1/2}$) and an estimated half-life for anaerobic biodegradation of vinyl chloride from a ground water field study of chlorinated ethenes (Silka, LR and Wallen, DA (1988)).

## Aqueous Biodegradation (unacclimated):
· **Aerobic half-life:**                             High:    4320 hours            (6 months)

Low:      672 hours            (4 weeks)

*Comment:* Scientific judgement based upon aqueous screening test data (Freitag, D et al. (1984A); Helfgott, TB et al. (1977)).

· **Anaerobic half-life:**                           High:   17280 hours            (24 months)

Low:     2688 hours            (16 weeks)

*Comment:* Scientific judgement based upon estimated unacclimated aqueous aerobic biodegradation half-life.

· **Removal/secondary treatment:**                   High:       No data

Low:

*Comment:*

## Photolysis:
· **Atmos photol half-life:**                        High:  Not expected to directly photolyze

Low:

*Comment:* Scientific judgement based upon measured rate of photolysis in water (Callahan, MA et al. (1979A)).

· **Max light absorption (nm):**
*Comment:*

· **Aq photol half-life:**         High:  Not expected to directly photolyze
                                  Low:
*Comment:* Scientific judgement based upon measured rate of photolysis in water (Callahan, MA et al. (1979A)).

## Photooxidation half-life:
· **Water:**                       High:       No data
                                  Low:

*Comment:*

· **Air:**                         High:       97 hours
                                  Low:        9.7 hours
*Comment:* Based upon measured rate constant for reaction with hydroxyl radicals in air (Atkinson, R (1985)).

## Reduction half-life:                High:  Not expected to be significant
                                  Low:

*Comment:*

## Hydrolysis:
· **First-order hydr half-life:**
*Comment:*

· **Acid rate const $(M(H+)-hr)^{-1}$:**    No hydrolyzable groups
*Comment:*

· **Base rate const $(M(OH-)-hr)^{-1}$:**
*Comment:*

## Diallate

**Physical Properties:** Based upon its physical properties, diallate is not expected to significantly volatilize from water, soil or surfaces or to significantly adsorb to soils or sediments. Diallate is expected to be moderately mobile in soil.

### Abiotic Degradation

**Hydrolysis:** Experimental data are available for the base catalyzed and first order rates of hydrolysis. The half-lives estimated using these data range from 3.8 years at pH 9 to 6.6 years at pH 7.

**Reduction:** No data available; not expected to be significant under normal environmental conditions.

**Photolysis:** No data available; not expected to be significant under normal environmental conditions.

**Photooxidation - Water:** No experimental data available.

**Photooxidation - Air:** No experimental data available. The rate for reaction with ozone is expected to be slow relative to reaction with hydroxyl radicals. The rate for reaction with hydroxyl radicals was estimated using the method of Atkinson (1987A). This estimated rate combined with an experimentally determined range of hydroxyl radical concentrations results in a range of 0.58 to 5.8 hours for the half-life.

**Biodegradation:** No data were found concerning aqueous biodegradation. Adequate data were found concerning biodegradation in soils. Half-lives in these experiments, most of which included adequate sterile controls, ranged from 252 hours to 3 months.

### Selection of Data/Determination of Rates and Half-lives:
**Half-lives:**

**Soil:** Since hydrolysis is expected to be very slow based upon rates for aqueous hydrolysis, the range of half-lives reported in experiments for biodegradation in soil was selected.

**Air:** Photooxidation by ·OH appears to be the most rapid degradation process in the atmosphere. High and low half-lives were calculated using an estimated rate of reaction and an experimentally determined range of high and low concentrations of ·OH expected in relatively polluted and pristine air, respectively.

**Surface Water:** The $t_{1/2}$ in water is based upon the estimated biodegradation rates because the rates for other degradation processes are either very slow (photooxidation) or are expected to be zero or negligible (direct photolysis, hydrolysis, and reduction). Since no aqueous biodegradation rates were available, the rate of aqueous biodegradation was estimated using scientific judgement and the rates reported for biodegradation in soil.

**Ground Water:** The $t_{1/2}$ in ground water is based upon the estimated aerobic biodegradation rates (and assumed to be two times longer) because the rate of hydrolysis is negligible.

# Diallate

**CAS Registry Number:** 2303-16-4

**Structure:**

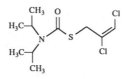

## Half-lives:
### · Soil:
|  |  |  |
|---|---|---|
| High: | 2160 hours | (3 months) |
| Low: | 252 hours | (0.35 months) |

*Comment:* Based upon aerobic soil die-away test data (Anderson, JPE and Domsch, KH (1976); Smith, AE (1970)).

### · Air:
|  |  |
|---|---|
| High: | 5.8 hours |
| Low: | 0.58 hours |

*Comment:* Based upon estimated photooxidation half-life in air.

### · Surface Water:
|  |  |  |
|---|---|---|
| High: | 2160 hours | (3 months) |
| Low: | 252 hours | (0.35 month) |

*Comment:* Scientific judgement based upon aerobic soil die-away test data (Anderson, JPE and Domsch, KH (1976); Smith, AE (1970)).

### · Ground Water:
|  |  |  |
|---|---|---|
| High: | 4320 hours | (6 months) |
| Low: | 504 hours | (0.70 month) |

*Comment:* Scientific judgement based upon aerobic soil die-away test data (Anderson, JPE and Domsch, KH (1976); Smith, AE (1970)).

## Aqueous Biodegradation (unacclimated):
### · Aerobic half-life:
|  |  |  |
|---|---|---|
| High: | 2160 hours | (3 months) |
| Low: | 252 hours | (0.35 month) |

*Comment:* Scientific judgement based upon aerobic soil die-away test data (Anderson, JPE and Domsch, KH (1976); Smith, AE (1970)).

### · Anaerobic half-life:
|  |  |  |
|---|---|---|
| High: | 8640 hours | (12 months) |
| Low: | 1008 hours | (2.1 months) |

*Comment:* Scientific judgement based upon aerobic soil die-away test data (Anderson, JPE and Domsch, KH (1976); Smith, AE (1970)).

### · Removal/secondary treatment:
|  |  |
|---|---|
| High: | No data |
| Low: |  |

*Comment:*

**Photolysis:**
   · **Atmos photol half-life:**              High:  Not expected to directly photolyze
                                                  Low:

   *Comment:*

   · **Max light absorption (nm):**
   *Comment:*

   · **Aq photol half-life:**                  High:
                                                  Low:

   *Comment:*

**Photooxidation half-life:**
   · **Water:**                             High:        No data
                                                    Low:

   *Comment:*

   · **Air:**                                 High:        5.8 hours
                                                 Low:        0.58 hours
   *Comment:* Based upon estimated rate constant for reaction with hydroxyl radicals in air (Atkinson, R (1987A)).

**Reduction half-life:**                        High:  Not expected to be significant
                                                  Low:

   *Comment:*

**Hydrolysis:**
   · **First-order hydr half-life:**       6.6 years                     ($t_{1/2}$ at pH 7)
   *Comment:* Based upon measured first-order and base catalyzed hydrolysis rate constant (Ellington, JJ et al. (1987)).

   · **Acid rate const (M(H+)-hr)$^{-1}$:**     No data         ($t_{1/2} \approx 6.6$ years at pH 5)
   *Comment:* Scientific judgement based upon measured first-order and base catalyzed hydrolysis rate constant (Ellington, JJ et al. (1987)).

   · **Base rate const (M(OH-)-hr)$^{-1}$:**    0.9 M$^{-1}$ hr$^{-1}$      ($t_{1/2} \approx 3.8$ years at pH 9)
   *Comment:* Based upon measured first-order and base catalyzed hydrolysis rate constant (Ellington, JJ et al. (1987)).

## Diaminotoluenes

<u>Physical Properties</u>: Based upon the physical properties, diaminotoluenes are not expected to significantly volatilize from water or soil, or strongly adsorb to soil or sediment. Based on their water solubility, diaminotoluenes are expected to be mobile in soil. However, aromatic amines have been shown to form covalent bonds with humic materials such as those present in soils. This process may limit the mobility of diaminotoluenes in soil.

<u>Abiotic Degradation</u>

<u>Hydrolysis</u>: No data available; not expected to be significant under normal environmental conditions.
<u>Reduction</u>: No data available; not expected to be significant under normal environmental conditions.
<u>Photolysis</u>: No data available; not expected to be significant under normal environmental conditions.
<u>Photooxidation - Water</u>: No experimental data were found. Based upon analogy to rates expected for aromatic amines as a class, a range for $t_{1/2}$ of 0.8 to 2.6 days was estimated.
<u>Photooxidation - Air</u>: No experimental data were found. The rate for reaction with ozone is expected to be slow relative to reaction with hydroxyl radicals. The rate for reaction with hydroxyl radicals was estimated using the method of Atkinson (1987A). The half-lives based upon the estimated rate constant ranged from 0.27 to 2.7 hours.

<u>Biodegradation</u>: The only biodegradation data available were data for aerobic degradation from a single screening test and a biological treatment simulation experiment. The screening test was an MITI test which confirmed that 2,4-diaminotoluene was either not biodegraded or slowly biodegraded under the conditions of the experiment. Although some removal of 2,4-diaminotoluene was observed for 2,4-diaminotoluene in the treatment simulation experiment (34% COD removal after 4 days using activated sludge with aeration), no control for volatilization was reported. These data suggest that biodegradation can be expected to be moderately slow to slow.

<u>Selection of Data/Determination of Rates and Half-lives</u>:
<u>Half-lives</u>:

<u>Soil</u>: No data concerning degradation in soil were located. Since hydrolysis is probably insignificant, biodegradation was selected as the most significant degradation mechanism. Based upon experiments in aqueous systems, biodegradation in soil is expected to be slow. An estimated range of half-lives of 1 to 6 months for diaminotoluenes in soil appears to be reasonable based upon scientific judgement.

<u>Air</u>: Photooxidation by ·OH appears to be the most rapid degradation process in the atmosphere. High and low half-lives were calculated using an estimated rate of reaction and estimations of high and low concentrations of ·OH expected in relatively polluted and pristine air, respectively. These ·OH concentrations were based upon reported experimental data.

<u>Surface Water</u>: The range of half-lives in water is based upon the estimated rate for photooxidation which leads to lower $t_{1/2}$ values than the estimated biodegradation rates.

<u>Ground Water</u>: The $t_{1/2}$ in ground water is based upon the estimated aerobic biodegradation rates (and

assumed to be two times longer) because the rate of hydrolysis is negligible.

# Diaminotoluenes

<u>CAS Registry Number:</u>  25376-45-8

<u>Half-lives:</u>
· **Soil:**                                    High:    4320 hours              (6 months)
                                               Low:      672 hours              (4 weeks)
*Comment:* Scientific judgement based upon estimated unacclimated aqueous aerobic biodegradation half-life.


· **Air:**                                     High:     2.7 hours
                                               Low:      0.27 hours
*Comment:* Based upon estimated photooxidation half-life in air.


· **Surface Water:**                           High:    1740 hours              (72 days)
                                               Low:       31 hours              (1.3 days)
*Comment:* Based upon estimated photooxidation half-life in water.


· **Ground Water:**                            High:    8640 hours              (12 months)
                                               Low:     1344 hours              (8 weeks)
*Comment:* Scientific judgement based upon estimated unacclimated aqueous aerobic biodegradation half-life.


<u>Aqueous Biodegradation (unacclimated):</u>
· **Aerobic half-life:**                       High:    4320 hours              (6 months)
                                               Low:      672 hours              (4 weeks)
*Comment:* Scientific judgement based upon unacclimated aerobic aqueous screening test data for 2,4-diaminotoluene which confirmed resistance to biodegradation (Sasaki, S (1978)).


· **Anaerobic half-life:**                     High:  17,280 hours              (24 months)
                                               Low:     2688 hours              (16 weeks)
*Comment:* Scientific judgement based upon estimated unacclimated aerobic aqueous biodegradation half-life.


· **Removal/secondary treatment:**             High:         34%
                                               Low:
*Comment:* Based upon activated sludge degradation results using a fill and draw method (Matsui, S et al. (1975)).


<u>Photolysis:</u>
· **Atmos photol half-life:**                  High:
                                               Low:
*Comment:*

**· Max light absorption (nm):**      No data
*Comment:*


**· Aq photol half-life:**            High:
                                      Low:

*Comment:*


## Photooxidation half-life:

**· Water:**                          High:    1740 hours         (72 days)
                                      Low:       31 hours        (1.3 days)
*Comment:* Scientific judgement based upon estimated half-life for reaction of aromatic amines with hydroxyl radicals in water (Mill, T (1989); Guesten, H et al. (1981)).   It is assumed that diaminotoluenes react twice as fast as aniline.


**· Air:**                            High:    2.7 hours
                                      Low:     0.27 hours
*Comment:* Based upon estimated rate constant for reaction with hydroxyl radicals in air (Atkinson, R (1987A)).


## Reduction half-life:                High:  Not expected to be significant
                                       Low:

*Comment:*


## Hydrolysis:
**· First-order hydr half-life:**
*Comment:*


**· Acid rate const $(M(H+)-hr)^{-1}$:**      No hydrolyzable groups
*Comment:*


**· Base rate const $(M(OH-)-hr)^{-1}$:**
*Comment:*

## 2,4-Dinitrotoluene

**Physical Properties:** Based upon its physical properties, 2,4-dinitrotoluene is not expected to significantly volatilize from water, soil, or surfaces, to strongly adsorb to soil or sediment. It can be expected to be moderately mobile in soil.

Abiotic Degradation

Hydrolysis: No data available; not expected to be significant under normal environmental conditions.

Photolysis: Experimental data for direct photolysis in distilled water are available ($t_{1/2} \approx 23$ to $72$ hours). No data were available for photolysis in air.

Photooxidation - Water: Based upon experimental rate data, photooxidation is fast in natural waters ($t_{1/2} \approx 3$ to $33$ hours).

Photooxidation - Air: No experimental data available. The rate for reaction with ozone is expected to be slow relative to reaction with hydroxyl radicals. The rate for reaction with hydroxyl radicals was estimated using the method of Atkinson (1987A). The half-lives based upon the estimated reaction rate constant ranged from 284 to 2840 hours.

**Biodegradation:** Aerobic die-away test data for experiments in fresh water from Waconda Bay and Searsville Pond give a range of half-lives ($t_{1/2} \approx 1$ to $4.1$ hours) but this is when considerable amounts of yeast extract were added. When no yeast extract was added, 0% degradation in 42 days was noted. The reduction of the nitro group observed in these studies may not be due to biodegradation. Therefore, it is concluded that 2,4-dinitrotoluene is fairly resistant to biodegradation under unacclimated aerobic conditions. Anaerobic die-away tests in Waconda Bay water lead to an estimate of a half-life on the order of 2 days.

Selection of Data/Determination of Rates and Half-lives:

Half-lives:

Soil: No data concerning degradation in soil were located. Since hydrolysis is probably insignificant biodegradation was selected as the most significant degradation mechanism. Based upon experiments in aqueous systems, 2,4-diaminotoluene is expected to be fairly resistant to biodegradation in soil An estimated range of half-lives of 1 to 6 months in soil appears to be reasonable based upon scientific judgement. Anaerobic conditions in soil may give increased rates of biodegradation based upon aqueous anaerobic experiments.

Air: Photolysis appears to be the most rapid degradation process in the atmosphere based upon photolysis rates in water.

Surface Water: The range of half-lives in water is based upon the estimated rate for photooxidation which leads to lower $t_{1/2}$ values than the estimated biodegradation rates.

Ground Water: The $t_{1/2}$ in ground water is based upon the estimated biodegradation rates because the rate of hydrolysis is negligible. The low $t_{1/2}$ is based upon the estimated low unacclimated aqueous anaerobic biodegradation half-life and the high $t_{1/2}$ is based upon the estimated high half-life for aerobic biodegradation (and assumed to be two times longer).

# 2,4-Dinitrotoluene

**CAS Registry Number:** 121-14-2

## Half-lives:

· **Soil:**    High:    4320 hours    (6 months)

Low:    672 hours    (4 weeks)

*Comment:* Scientific judgement based upon estimated unacclimated aqueous aerobic biodegradation half-life.

· **Air:**    High:    72 hours

Low:    23 hours

*Comment:* Based upon estimated photolysis half-life in air.

· **Surface Water:**    High:    33 hours

Low:    3 hours

*Comment:* Based upon photooxidation half-life in natural water.

· **Ground Water:**    High:    8640 hours    (12 months)

Low:    48 hours    (2 days)

*Comment:* Scientific judgement based upon estimated unacclimated aqueous anaerobic (low $t_{1/2}$) and aerobic (high $t_{1/2}$) biodegradation half-lives.

## Aqueous Biodegradation (unacclimated):

· **Aerobic half-life:**    High:    4320 hours    (6 months)

Low:    672 hours    (4 weeks)

*Comment:* Scientific judgement based upon aerobic natural water die-away test data (Spanggord, RJ et al. (1981)). Reduction of the nitro group observed in these studies may not be due to biodegradation.

· **Anaerobic half-life:**    High:    240 hours    (10 days)

Low:    48 hours    (2 days)

*Comment:* Scientific judgement based upon anaerobic natural water die-away test data (Spanggord, RJ et al. (1980)). Reduction of the nitro group observed in these studies may not be due to biodegradation.

· **Removal/secondary treatment:**    High:    No data

Low:

*Comment:*

## Photolysis:

· **Atmos photol half-life:**    High:    72 hours

Low:    23 hours

*Comment:* Scientific judgement based upon measured photolysis rates in water (Mill, T and Mabey, W (1985); Simmons, MS and Zepp, RG (1986)).

· **Max light absorption (nm):** No data
*Comment:*

· **Aq photol half-life:** High: 72 hours
Low: 23 hours
*Comment:* Based upon measured photolysis rates in water (Mill, T and Mabey, W (1985); Simmons, MS and Zepp, RG (1986)).

**Photooxidation half-life:**
· **Water:** High: 33 hours
Low: 3 hours
*Comment:* Based upon measured photooxidation rates in natural waters (Spanggord, RJ et al. (1980); Simmons, MS and Zepp, RG (1986)).

· **Air:** High: 2840 hours (118 days)
Low: 284 hours (11.8 days)
*Comment:* Based upon estimated rate constant for reaction with hydroxyl radicals in air (Atkinson, R (1987A)).

**Reduction half-life:** High: No data
Low:
*Comment:*

**Hydrolysis:**
· **First-order hydr half-life:**
*Comment:*

· **Acid rate const (M(H+)-hr)$^{-1}$:** No hydrolyzable groups
*Comment:*

· **Base rate const (M(OH-)-hr)$^{-1}$:**
*Comment:*

## 2,6-Dinitrotoluene

**Physical Properties:** Based upon its physical properties, 2,6-dinitrotoluene is not expected to significantly volatilize from water, soil, or surfaces or strongly adsorb to soil or sediment. It can be expected to be moderately mobile in soil.

Abiotic Degradation

Hydrolysis: No data available; not expected to be significant under normal environmental conditions.

Reduction: No data available; not expected to be significant under normal environmental conditions.

Photolysis: Experimental data for direct photolysis in distilled water are available ($t_{1/2} \approx 17$ to $25$ hours). No data were available for photolysis in air.

Photooxidation - Water: Based upon experimental rate data, photooxidation is fast in natural waters ($t_{1/2} \approx 2$ to $17$ hours).

Photooxidation - Air: No experimental data available. The rate for reaction with ozone is expected to be slow relative to reaction with hydroxyl radicals. The rate for reaction with hydroxyl radicals was estimated using the method of Atkinson (1987A). The half-lives based upon this estimated rate range from 284 to 2840 hours.

**Biodegradation:** Aerobic die-away test data for experiments in fresh water from Waconda Bay and Searsville Pond give a range of half-lives ($t_{1/2} \approx 5$ to $87$ hours) but this is when considerable amounts of yeast extract were added. When no yeast extract was added, 0% degradation in 42 days was noted. The reduction of the nitro group observed in these studies may not be due to biodegradation. Therefore, it is concluded that 2,6-dinitrotoluene is fairly resistant to biodegradation under unacclimated aerobic conditions. No anaerobic die-away test data were available. Data from a screening test utilizing raw municipal sewage seed suggest that biodegradation occurs under anaerobic conditions, but no rates were reported. A half-life for anaerobic biodegradation of 2,6-dinitrotoluene can be estimated from available data for degradation of 2,4-dinitrotoluene. For 2,4-dinitrotoluene, anaerobic die-away tests in Waconda Bay water lead to an estimate of a half-life on the order of 2 days.

Selection of Data/Determination of Rates and Half-lives:

Half-lives:

Soil: No data concerning degradation in soil were located. Since hydrolysis is probably insignificant, biodegradation was selected as the most significant degradation mechanism. Based upon experiments in aqueous systems, 2,6-dinitrotoluene is expected to be fairly resistant to biodegradation in soil. An estimated range of half-lives of 1 to 6 months in soil appears to be reasonable based upon scientific judgement. Anaerobic conditions in soil may give increased rates of biodegradation based upon aqueous anaerobic experiments.

Air: Photolysis appears to be the most rapid degradation process in the atmosphere based upon photolysis rates in water.

Surface Water: The range of half-lives in water is based upon the estimated rate for photooxidation which leads to lower $t_{1/2}$ values than the estimated biodegradation rates.

Ground Water: The $t_{1/2}$ in ground water is based upon the estimated biodegradation rates because the rate of hydrolysis is negligible. The low $t_{1/2}$ is based upon the estimated low unacclimated aqueous anaerobic biodegradation half-life and the high $t_{1/2}$ is based upon the estimated high half-life for aerobic biodegradation (and assumed to be two times longer).

# 2,6-Dinitrotoluene

**CAS Registry Number:** 606-20-2

**Half-lives:**
  · **Soil:**                           High:     4320 hours           (6 months)
                                         Low:      672 hours           (4 weeks)

*Comment:* Scientific judgement based upon estimated unacclimated aqueous aerobic biodegradation half-life.

  · **Air:**                             High:     25 hours
                                           Low:     17 hours

*Comment:* Based upon estimated photolysis half-life in air.

  · **Surface Water:**               High:     17 hours
                                         Low:      2 hours

*Comment:* Based upon photooxidation half-life in natural water.

  · **Ground Water:**              High:     8640 hours          (12 months)
                                         Low:      48 hours            (2 days)

*Comment:* Scientific judgement based upon estimated unacclimated aqueous anaerobic biodegradation half-life for 2,4-dinitrotoluene (low $t_{1/2}$) and estimated unacclimated aqueous aerobic biodegradation half-life for 2,6-dinitrotoluene (high $t_{1/2}$).

**Aqueous Biodegradation (unacclimated):**
  · **Aerobic half-life:**            High:     4320 hours          (6 months)
                                         Low:      672 hours          (4 weeks)

*Comment:* Scientific judgement based upon aerobic natural water die-away test data (Spanggord, RJ et al. (1981)). Reduction of the nitro group observed in these studies may not be due to biodegradation.

  · **Anaerobic half-life:**         High:     300 hours           (13 days)
                                         Low:      48 hours            (2 days)

*Comment:* Scientific judgement based upon anaerobic natural water die-away test data for 2,4-dinitrotoluene (Spanggord, RJ et al. (1980)). Reduction of the nitro group observed in these studies may not be due to biodegradation.

  · **Removal/secondary treatment:**     High:
                                           Low:

*Comment:*

**Photolysis:**
  · **Atmos photol half-life:**        High:     25 hours

                                        Low:        17 hours
*Comment:* Scientific judgement based upon measured photolysis rates in water (Simmons, MS and Zepp, RG (1986); Mill, T and Mabey, W (1985)).

· **Max light absorption (nm):**          No data
*Comment:*

· **Aq photol half-life:**               High:       25 hours
                                         Low:        17 hours
*Comment:* Based upon measured photolysis rates in water (Simmons, MS and Zepp, RG (1986); Mill, T and Mabey, W (1985)).

## Photooxidation half-life:
· **Water:**                             High:       17 hours
                                         Low:        2 hours
*Comment:* Based upon measured photooxidation rates in natural waters (Simmons, MS and Zepp, RG (1986)).

· **Air:**                       High:    2840 hours        (118 days)
                                 Low:     284 hours         (11.8 days)
*Comment:* Based upon estimated rate constant for reaction with hydroxyl radicals in air (Atkinson, R (1987A)).

## Reduction half-life:                  High:       No data
                                         Low:
*Comment:*

## Hydrolysis:
· **First-order hydr half-life:**
*Comment:*

· **Acid rate const $(M(H+)-hr)^{-1}$:**      No hydrolyzable groups
*Comment:*

· **Base rate const $(M(OH-)-hr)^{-1}$:**
*Comment:*

## Bis(2-ethylhexyl) Phthalate

**Physical Properties:** Based upon its physical properties, bis(2-ethylhexyl) phthalate is expected to slowly volatilize from water and adsorb very strongly to soil and sediment. It is not expected to significantly leach through soil.

**Abiotic Degradation**

**Hydrolysis:** A base-catalyzed hydrolysis rate constant has been reported ($t_{1/2} \approx 20$ years at pH 9), but no rate data for first-order or acid catalyzed hydrolysis are available.

**Reduction:** No data available; not expected to be significant under normal environmental conditions.

**Photolysis:** No data available; however, an estimate of the half-life can be made using experimental data available for photolysis of dimethyl phthalate in water which corresponds to a low half-life of 144 days. Since no data for photolysis in air were available, a range of half-lives for this process can be estimated from the estimated half-lives for photolysis in water.

**Photooxidation - Water:** Based upon experimental rate data, photooxidation by alkylperoxyl radicals is moderately slow to very slow ($t_{1/2} \approx 44$ days to 1.6 years).

**Photooxidation - Air:** No experimental data was found. The rate for reaction with ozone is expected to be slow relative to reaction with hydroxyl radicals. The rate for reaction with hydroxyl radicals was estimated using the method of Atkinson (1987A). The half-lives based upon this estimated rate range from 2.9 to 29 hours.

**Biodegradation:** Aerobic die-away test data are available from studies which utilized grab samples of water from a variety of sources. Half-lives for aerobic degradation range from 5 to 23 days. Two anaerobic experiments gave conflicting results. Significant degradation was observed ($t_{1/2} \approx 42$ days) in a die-away test which utilized alluvial garden soil under flooded, anaerobic conditions. In a microcosm experiment which included freshwater sediment, no degradation was observed in the sediment under anaerobic conditions. Three of four anaerobic screening studies utilizing digester sludge observed no degradation after incubation for up to 70 days; the remaining study observed slight degradation (9% theoretical methane production) after 70 days.

**Selection of Data/Determination of Rates and Half-lives:**
**Half-lives:**

**Soil:** No data concerning degradation in soil were located. Since hydrolysis is probably insignificant, biodegradation was selected as the most significant degradation mechanism. Based upon experiments in aqueous systems, biodegradation in soil is expected to be rapid to moderately rapid.

**Air:** Photooxidation by ·OH appears to be the most rapid degradation process in the atmosphere. High and low half-lives were calculated using an estimated rate of reaction and estimations of high and low concentrations of ·OH expected in relatively polluted and pristine air, respectively. These ·OH concentrations were based upon reported experimental data.

**Surface Water:** The $t_{1/2}$ in water is based upon the estimated biodegradation rates because the rates for other degradation processes are either very slow (photooxidation) or are expected to be zero or negligible by comparison (direct photolysis, hydrolysis, and reduction).

xli

Ground Water: The $t_{1/2}$ in ground water is based upon the estimated biodegradation rates because the rate of hydrolysis is negligible. The low $t_{1/2}$ is based upon the estimated low unacclimated aqueous aerobic biodegradation half-life (and assumed to be two times longer) and the high $t_{1/2}$ is based upon the estimated high half-life for anaerobic biodegradation.

# Bis-(2-ethylhexyl) Phthalate

**CAS Registry Number:** 117-81-7

## Half-lives:

· **Soil:**            High:    550 hours       (23 days)
                                     Low:     120 hours       (5 days)
*Comment:* Scientific judgement based upon unacclimated aqueous aerobic biodegradation half-life.

· **Air:**             High:    29 hours
                                     Low:     2.9 hours
*Comment:* Based upon estimated photooxidation half-life in air.

· **Surface Water:**      High:    550 hours       (23 days)
                                     Low:     120 hours       (5 days)
*Comment:* Based upon unacclimated aqueous aerobic biodegradation half-life.

· **Ground Water:**      High:    9336 hours      (389 days)
                                     Low:     240 hours       (10 days)
*Comment:* Scientific judgement based upon estimated unacclimated aqueous anaerobic (high $t_{1/2}$) and aerobic (low $t_{1/2}$) biodegradation half-life.

## Aqueous Biodegradation (unacclimated):

· **Aerobic half-life:**      High:    550 hours       (23 days)
                                     Low:     120 hours       (5 days)
*Comment:* Based upon grab sample die-away test data and scientific judgement (Schouten, MJ et al. (1979); Johnson, BT and Lulves, W (1975)).

· **Anaerobic half-life:**      High:    9336 hours      (389 days)
                                     Low:     980 hours       (42 days)
*Comment:* Scientific judgement based upon anaerobic die-away test data (low $t_{1/2}$) (Shanker, R et al. (1985) and anaerobic aqueous screening studies in which 0-9% theoretical methane production was observed in up to 70 days (Horowitz, A et al. (1982); Shelton, DR et al. (1984)).

· **Removal/secondary treatment:**      High:        91%
                                               Low:        70%
*Comment:* high: Graham, PR (1973); low: Saeger, VW and Tucker, ES (1976).

## Photolysis:

· **Atmos photol half-life:**      High:    4800 hours      (200 days)
                                     Low:     3500 hours      (144 days)
*Comment:* Scientific judgement based upon measured rate of aqueous photolysis for dimethyl phthalate (Wolfe, NL et al. (1980A)).

· **Max light absorption (nm):** No data
*Comment:*

· **Aq photol half-life:**  High:  4800 hours      (200 days)
                           Low:   3500 hours      (144 days)
*Comment:*  Scientific judgement based upon measured rate of aqueous photolysis for dimethyl phthalate (Wolfe, NL et al. (1980A)).

## Photooxidation half-life:
· **Water:**  High: $1.4 \times 10^4$ hours      (584 days)
             Low:  1056 hours        (44 days)
*Comment:*  Scientific judgement based upon measured rate data for reaction with peroxyl radicals in water (Wolfe, NL et al. (1980A)).

· **Air:**  High:  29 hours
           Low:   2.9 hours
*Comment:*  Based upon estimated rate constant for reaction with hydroxyl radicals in air (Atkinson, R (1987A)).

## Reduction half-life:  High:  No data
                        Low:
*Comment:*

## Hydrolysis:
· **First-order hydr half-life:**  No data ($t_{1/2} \approx 2,000$ years at pH 7)
*Comment:*  Scientific judgement based upon measured base catalyzed hydrolysis rate constant (Wolfe, NL et al. (1980B)).

· **Acid rate const $(M(H+)-hr)^{-1}$:**  No data

· **Base rate const $(M(OH-)-hr)^{-1}$:**  $1.1 \times 10^{-4}$ l/mole-sec ($t_{1/2} \approx 20$ years at pH 9)
*Comment:*  (Wolfe, NL et al. (1980B)).

PART III: ENVIRONMENTAL DEGRADATION RATE CONSTANTS

# Formaldehyde

**CAS Registry Number:** 50-00-0

**Half-lives:**
  · **Soil:**              High:    168 hours        (7 days)
                          Low:     24 hours         (1 day)
  *Comment:* Scientific judgement based upon unacclimated aqueous aerobic biodegradation half-life.

  · **Air:**               High:    6 hours
                          Low:     1.25 hours
  *Comment:* Based upon photolysis half-life in the gas phase.

  · **Surface Water:**     High:    168 hours        (7 days)
                          Low:     24 hours         (1 day)
  *Comment:* Scientific judgement based upon unacclimated aqueous aerobic biodegradation half-life.

  · **Ground Water:**      High:    336 hours        (14 days)
                          Low:     48 hours         (2 days)
  *Comment:* Scientific judgement based upon unacclimated aqueous aerobic biodegradation half-life.

**Aqueous Biodegradation (unacclimated):**
  · **Aerobic half-life:**    High:    168 hours        (7 days)
                             Low:     24 hours         (1 day)
  *Comment:* Scientific judgement based upon unacclimated aqueous aerobic biodegradation screening test data (Heukelekian, H and Rand, MC (1955); Gellman, I and Heukelekian, H (1950)).

  · **Anaerobic half-life:**   High:    672 hours        (28 days)
                             Low:     96 hours         (4 days)
  *Comment:* Scientific judgement based upon unacclimated aqueous aerobic biodegradation half-life.

  · **Removal/secondary treatment:**   High:    99%
                                      Low:     57%
  *Comment:* Removal percentages based upon data from a semicontinuous sewage (low $t_{1/2}$) and continuous activated sludge (high $t_{1/2}$) biological treatment simulator (low $t_{1/2}$: Hatfield, R (1957); high $t_{1/2}$: Canalstuca, J (1983)).

**Photolysis:**
  · **Atmos photol half-life:**        High:    6 hours

Low:    1.25 hours

*Comment:* Based upon measured photolysis for gas phase formaldehyde irradiated with simulated sunlight; $t_{1/2}$ calculated for sunlight photolysis (Calvert, JG et al. (1972); Su, F et al. (1979)).

· **Max light absorption (nm):**       lambda max approximately 300 nm (gas phase)
*Comment:* Significant absorption extends to approximately 360 nm in the gas phase (Calvert, JG et al. (1972)).

· **Aq photol half-life:**         High:    No data
                                       Low:
*Comment:* Not expected to be significant since formaldehyde exists almost entirely as a hydrate (which does not absorb sunlight) in water solution (Chameides, WL and Davis, DD (1983)).

## Photooxidation half-life:

· **Water:**                     High: $1.9 \times 10^5$ hours        (22 years)
                                      Low:    4813 hours        (201 days)
*Comment:* Based upon measured rate constant for reaction with hydroxyl radical in water (Dorfman, LM and Adams, GE (1973)).

· **Air:**                         High:    71.3 hours        (2.97 days)
                                      Low:    7.13 hours
*Comment:* Based upon measured rate constant for reaction with hydroxyl radical in air (Atkinson, R (1985)).

## Reduction half-life:

                                          High:    No data
                                          Low:
      *Comment:*

## Hydrolysis:

· **First-order hydr half-life:**
*Comment:*

· **Acid rate const $(M(H+)\text{-hr})^{-1}$:**  No hydrolyzable groups
*Comment:*

· **Base rate const $(M(OH-)\text{-hr})^{-1}$:**
*Comment:*

2

# Phenobarbital

<u>**CAS Registry Number:**</u> 50-06-6

<u>**Structure:**</u>

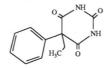

<u>**Half-lives:**</u>
- **· Soil:**              High:      672 hours           (4 weeks)
                          Low:       168 hours           (1 week)
  *Comment:* Scientific judgement based upon estimated aqueous aerobic biodegradation half-life.

- **· Air:**               High:      3.9 hours
                          Low:       0.4 hours           (23 minutes)
  *Comment:* Scientific judgement based upon estimated photooxidation half-life in air.

- **· Surface Water:**     High:      672 hours           (4 weeks)
                          Low:       168 hours           (1 week)
  *Comment:* Scientific judgement based upon estimated aqueous aerobic biodegradation half-life.

- **· Ground Water:**      High:      8640 hours          (8 weeks)
                          Low:       336 hours           (2 weeks)
  *Comment:* Scientific judgement based upon estimated aqueous aerobic biodegradation half-life.

<u>**Aqueous Biodegradation (unacclimated):**</u>
- **· Aerobic half-life:**   High:      672 hours           (4 weeks)
                          Low:       168 hours           (1 week)
  *Comment:* Scientific judgement.

- **· Anaerobic half-life:** High:      2688 hours          (16 weeks)
                          Low:       672 hours           (4 weeks)
  *Comment:* Scientific judgement based upon estimated aqueous aerobic biodegradation half-life.

- **· Removal/secondary treatment:**   High:      No data
                          Low:
  *Comment:*

<u>**Photolysis:**</u>
- **· Atmos photol half-life:**        High:
                          Low:
  *Comment:*

· **Max light absorption (nm):** No data
*Comment:*

· **Aq photol half-life:** High:
Low:

*Comment:*

**Photooxidation half-life:**
· **Water:** High: No data
Low:

*Comment:*

· **Air:** High: 3.9 hours
Low: 0.4 hours (23 minutes)
*Comment:* Scientific judgement based upon estimated rate data for hydroxyl radicals in air (Atkinson, R (1987A)).

**Reduction half-life:** High: No data
Low:

*Comment:*

**Hydrolysis:**
· **First-order hydr half-life:** No data
*Comment:*

· **Acid rate const (M(H+)-hr)$^{-1}$:**
*Comment:*

· **Base rate const (M(OH-)-hr)$^{-1}$:**
*Comment:*

# Mitomycin C

**CAS Registry Number:** 50-07-7

**Structure:**

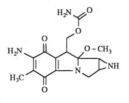

**Half-lives:**
- **Soil:** High: 672 hours (4 weeks)

  Low: 168 hours (1 week)

  *Comment:* Scientific judgement based upon estimated aqueous aerobic biodegradation half-life.

- **Air:** High: 2.1 hours

  Low: 0.2 hours (13 minutes)

  *Comment:* Scientific judgement based upon estimated photooxidation half-life in air.

- **Surface Water:** High: 672 hours (4 weeks)

  Low: 168 hours (1 week)

  *Comment:* Scientific judgement based upon estimated aqueous aerobic biodegradation half-life.

- **Ground Water:** High: 8640 hours (8 weeks)

  Low: 336 hours (2 weeks)

  *Comment:* Scientific judgement based upon estimated aqueous aerobic biodegradation half-life.

**Aqueous Biodegradation (unacclimated):**
- **Aerobic half-life:** High: 672 hours (4 weeks)

  Low: 168 hours (1 week)

  *Comment:* Scientific judgement.

- **Anaerobic half-life:** High: 2688 hours (16 weeks)

  Low: 672 hours (4 weeks)

  *Comment:* Scientific judgement based upon estimated aqueous aerobic biodegradation half-life.

- **Removal/secondary treatment:** High: No data

  Low:

  *Comment:*

**Photolysis:**
- **Atmos photol half-life:** High:

Low:

*Comment:*

**· Max light absorption (nm):**     No data
*Comment:*

**· Aq photol half-life:**     High:
                         Low:

*Comment:*

## Photooxidation half-life:
**· Water:**     High:     No data
            Low:

*Comment:*

**· Air:**     High:     2.1 hours
          Low:     0.2 hours     (13 minutes)
*Comment:* Scientific judgement based upon estimated rate data for hydroxyl radicals in air (Atkinson, R (1987A)).

## Reduction half-life:
     High:     No data
     Low:
*Comment:*

## Hydrolysis:
**· First-order hydr half-life:**     1872 hours     (78 days)
*Comment:* Based upon a neutral rate constant ($k_N = 3.7 \times 10^{-3}$ hr$^{-1}$) at pH 7 and 25 °C (Ellington, JJ et al. (1988A)).

**· Acid rate const (M(H+)-hr)$^{-1}$:**
*Comment:*

**· Base rate const (M(OH-)-hr)$^{-1}$:**     3.0
*Comment:* ($t_{1/2}$ = 72 days) Based upon neutral ($k_N = 3.7 \times 10^{-3}$ hr$^{-1}$) and base rate constants (3.0 M$^{-1}$ hr$^{-1}$) at pH 9 and 25 °C (Ellington, JJ et al. (1988A)).

# Cyclophosphamide

<u>**CAS Registry Number:**</u>  50-18-0

<u>**Structure:**</u>

<u>**Half-lives:**</u>
  · **Soil:**                            High:     672 hours          (4 weeks)
                                      Low:      168 hours          (7 days)
*Comment:* Scientific judgement based upon estimated unacclimated aqueous aerobic biodegradation half-life.

  · **Air:**                             High:     3.91 hours
                                        Low:      0.391 hours
*Comment:* Scientific judgement based upon estimated photooxidation half-life in air.

  · **Surface Water:**              High:     672 hours          (4 weeks)
                                      Low:      168 hours          (7 days)
*Comment:* Scientific judgement based upon estimated unacclimated aqueous aerobic biodegradation half-life.

  · **Ground Water:**             High:     976 hours          (41 days)
                                        Low:      336 hours          (14 days)
*Comment:* Scientific judgement based upon estimated aqueous aerobic biodegradation half-life (low $t_{1/2}$) and hydrolysis half-life (high $t_{1/2}$).

<u>**Aqueous Biodegradation (unacclimated):**</u>
  · **Aerobic half-life:**          High:     672 hours          (4 weeks)
                                        Low:      168 hours          (7 days)
*Comment:* Scientific judgement.

  · **Anaerobic half-life:**        High:     2688 hours       (16 weeks)
                                        Low:      672 hours          (28 days)
*Comment:* Scientific judgement based upon estimated aqueous aerobic biodegradation half-lives.

  · **Removal/secondary treatment:**    High:     No data
                                          Low:
*Comment:*

7

## Photolysis:
· **Atmos photol half-life:**          High:
                                              Low:

*Comment:*

· **Max light absorption (nm):**     No data
*Comment:*

· **Aq photol half-life:**            High:
                                              Low:

*Comment:*

## Photooxidation half-life:
· **Water:**                     High:       No data
                                              Low:

*Comment:*

· **Air:**                       High:       3.91 hours
                                              Low:       0.391 hours
*Comment:* Scientific judgement based upon estimated rate constant for reaction with hydroxyl radicals in air (Atkinson, R (1987A)).

## Reduction half-life:
                                            High:       No data
                                            Low:
*Comment:*

## Hydrolysis:
· **First-order hydr half-life:**                   976 hours             (41 days)
*Comment:* ($t_{1/2}$ at pH 7 and 25°C) Based upon measured neutral hydrolysis rate constant (Ellington, JJ et al. (1987)).

· **Acid rate const $(M(H+)-hr)^{-1}$:**     No data
*Comment:*

· **Base rate const $(M(OH-)-hr)^{-1}$:**     No data
*Comment:*

# DDT

**CAS Registry Number:** 50-29-3

**Structure:**

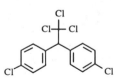

**Half-lives:**

· **Soil:** High: $1.4 \times 10^5$ hours      (15.6 years)

Low:   17520 hours      (2 years)

*Comment:* Based upon observed rates of biodegradation in aerobic soils under field conditions (low $t_{1/2}$: Lichtenstein, EP and Schultz, KR (1959); high $t_{1/2}$: Stewart, DKR and Chisholm, D (1971)).

· **Air:** High:   177 hours      (7.4 days)

Low:   17.7 hours

*Comment:* Scientific judgement based upon estimated photooxidation half-life in air.

· **Surface Water:** High:   8400 hours      (350 days)

Low:   168 hours      (7 days)

*Comment:* Scientific judgement based upon estimated photooxidation half-life in water.

· **Ground Water:** High: $2.7 \times 10^5$ hours      (31.3 years)

Low:   384 hours      (16 days)

*Comment:* Based upon anaerobic flooded soil die-away study data for DDT for two flooded soils (low $t_{1/2}$) (Castro, TF and Yoshida, T (1971)) and observed rates of biodegradation of DDT in aerobic soils under field conditions (low $t_{1/2}$: Lichtenstein, EP and Schultz, KR (1959); high $t_{1/2}$: Stewart, DKR and Chisholm, D (1971)).

**Aqueous Biodegradation (unacclimated):**

· **Aerobic half-life:** High: $1.37 \times 10^5$ hours      (15.6 years)

Low:   17520 hours      (2 years)

*Comment:* Based upon observed rates of biodegradation in aerobic soils under field conditions (low $t_{1/2}$: Lichtenstein, EP and Schultz, KR (1959); high $t_{1/2}$: Stewart, DKR and Chisholm, D (1971).

· **Anaerobic half-life:** High:   2400 hours      (100 days)

Low:   384 hours      (16 days)

*Comment:* Based upon anaerobic flooded soil die-away study data for two flooded soils (Castro, TF and Yoshida, T (1971)).

**· Removal/secondary treatment:**　　　High:　　　100%
　　　　　　　　　　　　　　　　　　　　　　　Low:

*Comment:* Removal percentage based upon data form a continuous activated sludge biological treatment simulator (Patterson, JW and Kodukala, PS (1981)).

## Photolysis:
**· Atmos photol half-life:**　　　　　　High: Not expected to be significant
　　　　　　　　　　　　　　　　　　　　　Low:

*Comment:* Scientific judgement based upon 0% degradation of DDT observed in distilled water solutions under sunlight for 7 days (Callahan, MA et al. (1979)) and measured rate of photolysis in distilled water solution under artificial light at >290 nm (Zepp, RG et al. (1977A)).

**· Max light absorption (nm):**　　　　lambda max = 238, 268, 277 nm (hexane)
*Comment:* Absorption extends to approximately 282 nm (Gore, RC et al. (1971)).

**· Aq photol half-life:**　　　　　　　High: Not expected to be significant
　　　　　　　　　　　　　　　　　　　　Low:

*Comment:* Scientific judgement based upon 0% degradation of DDT observed in distilled water solutions under sunlight for 7 days (Callahan, MA et al. (1979)) and measured rate of photolysis in distilled water solution under artificial light at >290 nm (Zepp, RG et al. (1977A)).

## Photooxidation half-life:
**· Water:**　　　　　　　　　　　　　High:　　8400 hours　　　　　(350 days)
　　　　　　　　　　　　　　　　　　　Low:　　168 hours　　　　　(7 days)
*Comment:* Scientific judgement based upon measured rate of photooxidation in two natural waters under sunlight for 7 days (low $t_{1/2}$) and 56 days (high $t_{1/2}$) (Callahan, MA et al. (1979)).

**· Air:**　　　　　　　　　　　　　　High:　　177 hours　　　　　(7.4 days)
　　　　　　　　　　　　　　　　　　　Low:　　17.7 hours
*Comment:* Scientific judgement based upon estimated rate constant for reaction with hydroxyl radical in air (Atkinson, R (1987A)).

## Reduction half-life:
　　　　　　　　　　　　　　　　　　　High: No data
　　　　　　　　　　　　　　　　　　　Low:

*Comment:*

## Hydrolysis:
**· First-order hydr half-life:**　　　　$1.94 \times 10^5$ hours (22 years)
*Comment:* ($t_{1/2}$ at pH 7 and 27°C) Calculated from measured base catalyzed hydrolysis constant (Wolfe, NL et al. (1977)).

**· Acid rate const $(M(H+)-hr)^{-1}$:**　　No data
*Comment:* ($t_{1/2}$ = $1.94 \times 10^7$ hours (2,200 years) at pH 5 and 27°C) Calculated from measured base catalyzed hydrolysis constant (Wolfe, NL et al. (1977)).

**· Base rate const (M(OH-)-hr)$^{-1}$:**

*Comment:* ($t_{1/2}$ = 1.94X10$^5$ hours (22 years) at pH 9 and 27°C) Calculated from measured base catalyzed hydrolysis constant (Wolfe, NL et al. (1977)).

# Benzo(a)pyrene

**CAS Registry Number:** 50-32-8

**Structure:**

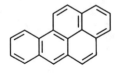

**Half-lives:**
- **Soil:**
  - High: 12720 hours       (1.45 years)
  - Low: 1368 hours       (57 days)

  *Comment:* Based upon aerobic soil die-away test data at 10-30°C (Coover, MP and Sims, RCC (1987); Groenewegen, D and Stolp, H (1976)).

- **Air:**
  - High: 1.1 hours
  - Low: 0.37 hours

  *Comment:* Scientific judgement based upon estimated photolysis half-life in air.

- **Surface Water:**
  - High: 1.1 hours
  - Low: 0.37 hours

  *Comment:* Scientific judgement based upon estimated photolysis half-life in water.

- **Ground Water:**
  - High: 25440 hours       (2.90 years)
  - Low: 2736 hours       (114 days)

  *Comment:* Scientific judgement based upon estimated unacclimated aqueous aerobic biodegradation half-life.

**Aqueous Biodegradation (unacclimated):**
- **Aerobic half-life:**
  - High: 12720 hours       (1.45 years)
  - Low: 1368 hours       (57 days)

  *Comment:* Based upon aerobic soil die-away test data at 10-30°C (Coover, MP and Sims, RCC (1987); Groenewegen, D and Stolp, H (1976)).

- **Anaerobic half-life:**
  - High: 50880 hours       (5.8 years)
  - Low: 5472 hours       (228 days)

  *Comment:* Scientific judgement based upon estimated unacclimated aqueous aerobic biodegradation half-life.

- **Removal/secondary treatment:**
  - High: No data
  - Low:

  *Comment:*

## Photolysis:
· **Atmos photol half-life:**         High:     1.1 hours

                                             Low:     0.37 hours

*Comment:* Based upon measured photolysis rate constant for midday winter sunlight at 35°N latitude in 20% aqueous acetonitrile (high $t_{1/2}$) (Smith, JH et al. (1978)) and adjusted for approximate summer sunlight intensity (low $t_{1/2}$) (Lyman, WJ et al. (1982)).

· **Max light absorption (nm):**       lambda max = 347, 364, 384 nm (ethanol); 219, 226, 254, 265, 272, 283, 291, 330, 345, 363, 379, 383, 393, 402 nm (cyclohexane)

*Comment:* Ethanol spectrum: Radding, SB et al. (1976); cyclohexane spectrum: (IARC (1983A)).

· **Aq photol half-life:**            High:     1.1 hours

                                             Low:     0.37 hours

*Comment:* Based upon measured photolysis rate constant for midday winter sunlight at 35°N latitude in 20% aqueous acetonitrile (high $t_{1/2}$) (Smith, JH et al. (1978)) and adjusted for approximate summer sunlight intensity (low $t_{1/2}$) (Lyman, WJ et al. (1982)).

## Photooxidation half-life:
· **Water:**                         High:    10349 hours            (431 days)

                                             Low:      207 hours               (8.6 days)

*Comment:* Based upon measured rate constant for reaction with alkylperoxyl radical in water (Smith, JH et al. (1978)).

· **Air:**                              High:     4.28 hours

                                             Low:     0.428 hours

*Comment:* Scientific judgement based upon estimated rate constant for reaction with hydroxyl radical in air (Atkinson, R (1987A)).

## Reduction half-life:
                                             High:  No data

                                             Low:

*Comment:*

## Hydrolysis:
· **First-order hydr half-life:**

*Comment:*

· **Acid rate const (M(H+)-hr)$^{-1}$:**     No hydrolyzable groups

*Comment:*

· **Base rate const (M(OH-)-hr)$^{-1}$:**

*Comment:*

# 2,4-Dinitrophenol

**CAS Registry Number:** 51-28-5

**Half-lives:**
   · **Soil:**
                                        High:    6312 hours      (8.77 months)
                                          Low:     1622 hours      (2.25 months)

*Comment:* Based upon data from aerobic soil column studies (Kincannon, DF and Lin, YS (1985)) and aerobic soil die-away test data (Sudhakar-Barik, R and Sethunathan, N (1978A)).

   · **Air:**
                                          High:    1114 hours      (46.4 days)
                                          Low:     111 hours       (4.6 days)

*Comment:* Scientific judgement based upon estimated photooxidation half-life in air.

   · **Surface Water:**
                                  High:    3840 hours      (160 days)
                                  Low:      77 hours       (3.2 days)

*Comment:* Scientific judgement based upon reported reaction rate constants for $RO_2\cdot$ with the phenol class (Mill, T and Mabey, W (1985)).

   · **Ground Water:**
                               High:   12624 hours    (17.5 months)
                               Low:      68 hours       (2.8 days)

*Comment:* Scientific judgement based upon estimated aqueous aerobic biodegradation half-life (high $t_{1/2}$) and estimated aqueous anaerobic biodegradation half-life (low $t_{1/2}$).

**Aqueous Biodegradation (unacclimated):**
   · **Aerobic half-life:**
                              High:    6312 hours      (8.77 months)
                              Low:     1622 hours      (2.25 months)

*Comment:* Based upon data from aerobic soil column studies (Kincannon, DF and Lin, YS (1985)) and aerobic soil die-away test data (Sudhakar-Barik, R and Sethunathan, N (1978A)).

   · **Anaerobic half-life:**
                           High:    170 hours      (7.1 days)
                           Low:      68 hours      (2.8 days)

*Comment:* Based upon anaerobic flooded soil die-away tests (Sudhakar-Barik, R and Sethunathan, N (1978A)).

   · **Removal/secondary treatment:**    High:    >99.9%
                                            Low:

*Comment:* Removal percentage based upon data from a continuous activated sludge biological treatment simulator (Kincannon, DF et al. (1983B)).

**Photolysis:**
   · **Atmos photol half-life:**
                            High:  No data
                            Low:

*Comment:*

· **Max light absorption (nm):** lambda max = 365 nm (extends to = 480 nm) at pH 8; lambda max = 265 nm (extends to = 390 nm) at pH 1
*Comment:* (Overcash, MR et al. (1982)).

· **Aq photol half-life:**       High:  No data
                                 Low:

*Comment:*

## Photooxidation half-life:

· **Water:**                     High:    3840 hours       (160 days)
                                 Low:       77 hours       (3.2 days)
*Comment:* Scientific judgement based upon reported reaction rate constants for $RO_2 \cdot$ with the phenol class (Mill, T and Mabey, W (1985)).

· **Air:**                       High:    1114 hours       (46.4 days)
                                 Low:      111 hours       (4.6 days)
*Comment:* Scientific judgement based upon estimated rate constant for reaction with hydroxyl radical in air (Atkinson, R (1987A)).

## Reduction half-life:        High:  No data
                               Low:

*Comment:*

## Hydrolysis:
· **First-order hydr half-life:**
*Comment:*

· **Acid rate const $(M(H+)-hr)^{-1}$:**       No hydrolyzable groups
*Comment:*

· **Base rate const $(M(OH-)-hr)^{-1}$:**
*Comment:*

# Nitrogen mustard

<u>CAS Registry Number:</u>  51-75-2

<u>Structure:</u>

<u>Half-lives:</u>
   · **Soil:**                       High:      24 hours
                                      Low:      30 minutes
   *Comment:*  Based upon aqueous hydrolysis half-lives.

   · **Air:**                        High:       8.9 hours
                                      Low:       0.89 hours
   *Comment:*  Based upon estimated photooxidation half-lives in air.

   · **Surface Water:**              High:      24 hours
                                      Low:      30 minutes
   *Comment:*  Based upon aqueous hydrolysis half-lives.

   · **Ground Water:**               High:      24 hours
                                      Low:      30 minutes
   *Comment:*  Based upon aqueous hydrolysis half-lives.

<u>Aqueous Biodegradation (unacclimated):</u>
   · **Aerobic half-life:**          High:    4320 hours        (6 months)
                                      Low:      672 hour        (4 weeks)
   *Comment:* Scientific judgement based upon a general resistance of tertiary amines to biodegrade.

   · **Anaerobic half-life:**        High:   17280 hours        (24 months)
                                      Low:     2688 hours       (16 weeks)
   *Comment:*  Scientific judgement based upon estimated aqueous aerobic biodegradation half-lives.

   · **Removal/secondary treatment:**   High: No data
                                        Low:
   *Comment:*

<u>Photolysis:</u>
   · **Atmos photol half-life:**        High: No data

16

Low:

*Comment:*

· **Max light absorption (nm):**
*Comment:*

· **Aq photol half-life:**  High:
Low:

*Comment:*

**Photooxidation half-life:**
· **Water:**  High:  No data
Low:

*Comment:*

· **Air:**  High:  8.9 hours
Low:  0.89 hours
*Comment:* Based upon estimated reaction rate constant with ·OH (Atkinson, R (1987A)).

**Reduction half-life:**  High:  No data
Low:

*Comment:*

**Hydrolysis:**
· **First-order hydr half-life:**  High:  24 hours
Low:  30 minutes
*Comment:* High and low values are based upon scientific judgement of experimentally reported reaction kinetics of nitrogen mustard and similar homologs in dilute aqueous solutions at various temperatures (Cohen, B et al. (1948); Bartlett, PD et al. (1947)).

· **Acid rate const $(M(H+)-hr)^{-1}$:**
*Comment:*

· **Base rate const $(M(OH-)-hr)^{-1}$:**
*Comment:*

# Ethyl carbamate

**CAS Registry Number:** 51-79-6

**Structure:**

$$H_2N \overset{O}{\underset{||}{C}} O \diagdown CH_3$$

**Half-lives:**

- **Soil:**
  High: 168 hours (7 days)
  Low: 24 hours (1 day)
  *Comment:* Scientific judgement based upon estimated unacclimated aerobic aqueous biodegradation half-life.

- **Air:**
  High: 30 hours (1.25 days)
  Low: 3.0 hours
  *Comment:* Scientific judgement based upon estimated photooxidation half-life in air.

- **Surface Water:**
  High: 168 hours (7 days)
  Low: 24 hours (1 day)
  *Comment:* Scientific judgement based upon estimated unacclimated aerobic aqueous biodegradation half-life.

- **Ground Water:**
  High: 336 hours (14 days)
  Low: 48 hours (2 days)
  *Comment:* Scientific judgement based upon estimated unacclimated aerobic aqueous biodegradation half-life.

**Aqueous Biodegradation (unacclimated):**

- **Aerobic half-life:**
  High: 168 hours (7 days)
  Low: 24 hours (1 day)
  *Comment:* Scientific judgement.

- **Anaerobic half-life:**
  High: 672 hours (28 days)
  Low: 96 hours (4 days)
  *Comment:* Scientific judgement based upon estimated unacclimated aerobic aqueous biodegradation half-life.

- **Removal/secondary treatment:**
  High: No data
  Low:
  *Comment:*

18

## Photolysis:
   **· Atmos photol half-life:**               High:
                                               Low:

   *Comment:*

   **· Max light absorption (nm):**            No data
   *Comment:*

   **· Aq photol half-life:**                  High:
                                               Low:

   *Comment:*

## Photooxidation half-life:
   **· Water:**                                High:       No data
                                               Low:

   *Comment:*

   **· Air:**                                  High:       30 hours          (1.25 days)
                                               Low:        3.0 hours
   *Comment:* Scientific judgement based upon estimated rate constant for reaction with hydroxyl radicals in air (Atkinson, R (1987A)).

## Reduction half-life:                        High:       No data
                                               Low:

   *Comment:*

## Hydrolysis:
   **· First-order hydr half-life:**           $2.56 \times 10^6$ hours          (292 years)
   *Comment:* $t_{1/2}$ at pH 7 - Based upon neutral and base catalyzed hydrolysis rate constants (Ellington,                    JJ et al. (1987A)).

   **· Acid rate const $(M(H+)-hr)^{-1}$:**              No data
   *Comment:* $(t_{1/2} = 2.66 \times 10^6$ hours (304 years) at pH 5) Based upon neutral and base catalyzed hydrolysis rate constants (Ellington, JJ et al. (1987A)).

   **· Base rate const $(M(OH-)-hr)^{-1}$:**             $0.11 \ M^{-1} \ s^{-1}$
   *Comment:* $(t_{1/2} = 5.10 \times 10^5$ hours (58.2 years) at pH 9) Based upon neutral and base catalyzed hydrolysis rate constants (Ellington, JJ et al. (1987A)).

19

# Trichlorofon

**CAS Registry Number:** 52-68-6

**Structure:**

**Half-lives:**
- **Soil:** High: 1080 hours (45 days)

  Low: 24 hours (1 day)

  *Comment:* Scientific judgement based upon unacclimated soil grab sample data (low $t_{1/2}$: Guirguis, MW and Shafik, MT (1975); high $t_{1/2}$: Kostovetskii, YI et al. (1976)).

- **Air:** High: 101 hours

  Low: 1 hour

  *Comment:* Scientific judgement based upon estimated photooxidation half-life in air.

- **Surface Water:** High: 588 hours (3.5 weeks)

  Low: 22 hours

  *Comment:* Scientific judgement based upon aqueous hydrolysis half-lives for pH 6 (high $t_{1/2}$) and 8 (low $t_{1/2}$) at 25 °C (Chapman, RA and Cole, CM (1982)).

- **Ground Water:** High: 588 hours (3.5 weeks)

  Low: 22 hours

  *Comment:* Scientific judgement based upon aqueous hydrolysis half-lives for pH 6 (high $t_{1/2}$) and 8 (low $t_{1/2}$) at 25 °C (Chapman, RA and Cole, CM (1982)).

**Aqueous Biodegradation (unacclimated):**
- **Aerobic half-life:** High: 1080 hours (45 days)

  Low: 24 hours (1 day)

  *Comment:* Scientific judgement based upon unacclimated soil grab sample data (low $t_{1/2}$: Guirguis, MW and Shafik, MT (1975); high $t_{1/2}$: Kostovetskii, YI et al. (1976)).

- **Anaerobic half-life:** High: 4320 hours (180 days)

  Low: 96 hours (4 days)

  *Comment:* Scientific judgement based upon unacclimated aerobic biodegradation half-life.

- **Removal/secondary treatment:** High: No data

  Low:

  *Comment:*

20

## Photolysis:
· **Atmos photol half-life:**    High:    No data
                                 Low:

*Comment:*

· **Max light absorption (nm):**    lambda max <200
*Comment:* No absorbance of UV light at wavelengths >240 nm (Gore, RC et al. (1971)).

· **Aq photol half-life:**    High:    No data
                              Low:

*Comment:*

## Photooxidation half-life:
· **Water:**    High:    No data
                Low:

*Comment:*

· **Air:**    High:    101 hours    (4.2 days)
              Low:    1 hour
*Comment:* Scientific judgement based upon an estimated rate constant for the vapor phase reaction with hydroxyl radicals in air (Atkinson, R (1987A)).

## Reduction half-life:
                High:    No data
                Low:

*Comment:*

## Hydrolysis:
· **First-order hydr half-life:**    68 hours    (2.8 days)
*Comment:* Scientific judgement based upon first order rate constant at pH 7 and 25 °C (Chapman, RA and Cole, CM (1982)).

· **Acid rate const (M(H+)-hr)$^{-1}$:**    $1.18 \times 10^{-3}$ hr$^{-1}$
*Comment:* ($t_{1/2}$ = 3.5 weeks) Based upon first order rate constant at pH 6 and 25 °C (Chapman, RA and Cole, CM (1982)).

· **Base rate const (M(OH-)-hr)$^{-1}$:**    $3.15 \times 10^{-2}$ hr$^{-1}$
*Comment:* ($t_{1/2}$ = 22 hours) Based upon first order rate constant at pH 8 and 25 °C (Chapman, RA and Cole, CM (1982)).

# Dibenz(a,h)anthracene

**CAS Registry Number:** 53-70-3

**Structure:**

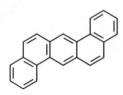

**Half-lives:**

· **Soil:**                                    High:    22560 hours              (2.58 years)
                                                      Low:      8664 hours                (361 days)
*Comment:* Based upon aerobic soil die-away test data (Coover, MP and Sims, RC (1987); Sims, RC (1990)).

· **Air:**                                     High:    4.28 hours
                                                      Low:     0.428 hours
*Comment:* Scientific judgement based upon estimated photooxidation half-life in air.

· **Surface Water:**                  High:       782 hours              (32.6 days)
                                                      Low:          6 hours
*Comment:* Scientific judgement based upon sunlight photolysis half-life in water.

· **Ground Water:**                   High:    45120 hours              (5.15 years)
                                                      Low:    17328 hours              (1.98 years)
*Comment:* Scientific judgement based upon estimated unacclimated aqueous aerobic biodegradation half-life.

**Aqueous Biodegradation (unacclimated):**

· **Aerobic half-life:**              High:    22560 hours              (2.58 years)
                                                      Low:      8664 hours                (361 days)
*Comment:* Based upon aerobic soil die-away test data (Coover, MP and Sims, RCC (1987); Sims, RC (1990)).

· **Anaerobic half-life:**           High:    90240 hours              (10.3 years)
                                                      Low:    34656 hours              (3.96 years)
*Comment:* Scientific judgement based upon estimated unacclimated aqueous aerobic biodegradation half-life.

· **Removal/secondary treatment:**    High:       No data
                                                                   Low:
*Comment:*

## Photolysis:

**· Atmos photol half-life:**        High:    782 hours       (32.6 days)

                                               Low:      6 hours

*Comment:* Scientific judgement based upon measured rate of photolysis in heptane under November sunlight (high $t_{1/2}$) (Muel, B and Saguem, S (1985)) and the above data adjusted by ratio of sunlight photolysis half-lives for benz(a)anthracene in water vs. heptane (low $t_{1/2}$) (Smith, JH et al. (1978); Muel, B and Saguem, S (1985)).

**· Max light absorption (nm):** lambda max = 215, 221, 231, 272, 277, 285, 297, 319, 331, 347 nm (no solvent reported)
*Comment:* IARC (1983A).

**· Aq photol half-life:**        High:    782 hours       (32.6 days)

                                     Low:      6 hours

*Comment:* Scientific judgement based upon measured rate of photolysis in heptane under November sunlight (high $t_{1/2}$) (Muel, B and Saguem, S (1985)) and the above data adjusted by ratio of sunlight photolysis half-lives for benz(a)anthracene in water vs. heptane (low $t_{1/2}$) (Smith, JH et al. (1978); Muel, B and Saguem, S (1985)).

## Photooxidation half-life:

**· Water:**        High:    No data

                 Low:

*Comment:*

**· Air:**        High:    4.28 hours

                 Low:      0.428 hours

*Comment:* Scientific judgement based upon estimated rate constant for reaction with hydroxyl radical in air (Atkinson, R (1987A)).

## Reduction half-life:

                 High:    No data

                 Low:

*Comment:*

## Hydrolysis:

**· First-order hydr half-life:**
*Comment:*

**· Acid rate const $(M(H+)\text{-}hr)^{-1}$:**       No hydrolyzable groups
*Comment:*

**· Base rate const $(M(OH\text{-})\text{-}hr)^{-1}$:**
*Comment:*

# 2-Acetylaminofluorene

**CAS Registry Number:** 53-96-3

**Structure:**

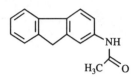

**Half-lives:**

  · **Soil:**
          High:    4320 hours       (6 months)
          Low:     672 hours        (4 weeks)

*Comment:* Scientific judgement based upon estimated unacclimated aqueous aerobic biodegradation half-life.

  · **Air:**
          High:    7.2 hours
          Low:     0.72 hours

*Comment:* Based upon estimated photooxidation half-life in air (Atkinson, R (1987A)).

  · **Surface Water:**
          High:    4320 hours       (6 months)
          Low:     672 hours        (4 weeks)

*Comment:* Scientific judgement based upon estimated unacclimated aqueous aerobic biodegradation half-life.

  · **Ground Water:**
          High:    8640 hours      (12 months)
          Low:     1344 hours       (8 weeks)

*Comment:* Scientific judgement based upon estimated unacclimated aqueous aerobic biodegradation half-life.

**Aqueous Biodegradation (unacclimated):**

  · **Aerobic half-life:**
          High:    4320 hours       (6 months)
          Low:     672 hours        (4 weeks)

*Comment:* Scientific judgement based upon unacclimated aerobic aqueous screening test data (Fochtman, EG (1981); Tabak, HH et al. (1981A); Lutin, PA et al. (1965)).

  · **Anaerobic half-life:**
          High:    17280 hours     (24 months)
          Low:     2880 hours      (16 weeks)

*Comment:* Scientific judgement based upon estimated unacclimated aqueous aerobic biodegradation half-life.

  · **Removal/secondary treatment:**
          High:    No data
          Low:

*Comment:*

## Photolysis:
· **Atmos photol half-life:**                              High:      No data
                                                           Low:
*Comment:*

· **Max light absorption (nm):**          lambda max = 288 nm (methanol)
*Comment:*  Sadtler UV No. 4150.

· **Aq photol half-life:**                                 High:      No data
                                                           Low:

*Comment:*

## Photooxidation half-life:
· **Water:**                                               High:      No data
                                                           Low:

*Comment:*

· **Air:**                                                 High:      7.2 hours
                                                           Low:       0.72 hours
*Comment:*  Based upon estimated rate constant for reaction with hydroxyl radicals in air
(Atkinson, R (1987A)).

## Reduction half-life:                                    High:  Not expected to be significant
                                                           Low:
*Comment:*

## Hydrolysis:
· **First-order hydr half-life:**          $2.3 \times 10^{-6}$ hr$^{-1}$
*Comment:*  Based upon measured first order and base catalyzed hydrolysis rate constants ($t_{1/2} = 34$
years at pH 7.0) (Ellington, JJ et al. (1987)).

· **Acid rate const (M(H+)-hr)$^{-1}$:**                  No data
*Comment:*  Scientific judgement based upon measured first order and base catalyzed hydrolysis
rate constants ($t_{1/2} = 34$ years at pH 5.0) (Ellington, JJ et al. (1987)).

· **Base rate const (M(OH-)-hr)$^{-1}$:**     $6 \times 10^{-3}$ M$^{-1}$hr$^{-1}$
*Comment:*  Based upon measured first order and base catalyzed hydrolysis rate constants ($t_{1/2} = 33$
years at pH 9.0) (Ellington, JJ et al. (1987)).

# N-Nitrosodiethylamine

<u>CAS Registry Number:</u>  55-18-5

<u>Half-lives:</u>
  · **Soil:**                              High:    4320 hours        (6 months)
                                        Low:     480 hours         (20 days)
*Comment:* Based upon aerobic soil die-away data (high $t_{1/2}$: Tate, RL III and Alexander, M (1975); low $t_{1/2}$: Oliver, JE et al. (1979)).

  · **Air:**                               High:    8 hours
                                        Low:     4 hours
*Comment:* Based upon measured rate for sunlight photolysis in water (high $t_{1/2}$: Oliver, JE (1981); low $t_{1/2}$: Zhang, Z et al. (1983)).

  · **Surface Water:**                High:    8 hours
                                        Low:     4 hours
*Comment:* Based upon measured rates for sunlight photolysis in water (high $t_{1/2}$: Oliver, JE (1981); low $t_{1/2}$: Zhang, Z et al. (1983)).

  · **Ground Water:**               High:    8640 hours       (12 months)
                                        Low:     960 hours        (40 days)
*Comment:* Scientific judgement based upon estimated unacclimated aqueous aerobic biodegradation half-lives.

<u>Aqueous Biodegradation (unacclimated):</u>
  · **Aerobic half-life:**           High:    4320 hours       (6 months)
                                        Low:     480 hours        (20 days)
*Comment:* Based upon aerobic soil die-away data (high $t_{1/2}$: Tate, RL III and Alexander, M (1975); low $t_{1/2}$: Oliver, JE et al. (1979)).

  · **Anaerobic half-life:**       High:    17280 hours     (24 months)
                                        Low:     1920 hours      (80 days)
*Comment:* Scientific judgement based upon estimated unacclimated aqueous aerobic biodegradation half-lives.

  · **Removal/secondary treatment:**    High:    No data
                                        Low:
*Comment:*

<u>Photolysis:</u>
  · **Atmos photol half-life:**       High:    8 hours
                                        Low:     4 hours

*Comment:* Based upon measured rates for sunlight photolysis in water (high $t_{1/2}$: Oliver, JE (1981); low $t_{1/2}$: Zhang, Z et al. (1983)).

· **Max light absorption (nm):**         lambda max = 234, 345 nm (water)
*Comment:* IARC (1978).

· **Aq photol half-life:**         High:      8 hours
                                         Low:       4 hours
*Comment:* Based upon measured rates for sunlight photolysis in water (high $t_{1/2}$: Oliver, JE (1981); low $t_{1/2}$: Zhang, Z et al. (1983)).

## Photooxidation half-life:
· **Water:**         High:      No data
                             Low:

*Comment:*

· **Air:**         High:      37 hours
                     Low:       3.7 hours
*Comment:* Scientific judgement based upon estimated rate constant for reaction with hydroxyl radical in air (Atkinson, R (1987A)).

## Reduction half-life:
                                       High:      No data
                                       Low:
*Comment:*

## Hydrolysis:
· **First-order hydr half-life:**         Not expected to be significant
*Comment:* IARC (1978).

· **Acid rate const $(M(H+)-hr)^{-1}$:**         No data
*Comment:*

· **Base rate const $(M(OH-)-hr)^{-1}$:**         No data
*Comment:*

# Benzamide

<u>CAS Registry Number:</u>  55-21-0

<u>Half-lives:</u>

· **Soil:**                                      High:      360 hours              (15 days)
                                                 Low:       48 hours               (2 days)
*Comment:*  Scientific judgement based upon grab sample aerobic soil column test data (Fournier, JC and Salle, J (1974)).

· **Air:**                                       High:      31 hours
                                                 Low:       3.1 hours
*Comment:*  Based upon estimated photooxidation half-life in air.

· **Surface Water:**                             High:      360 hours              (15 days)
                                                 Low:       48 hours               (2 days)
*Comment:*  Scientific judgement based upon grab sample aerobic soil column test data (Fournier, JC and Salle, J (1974)).

· **Ground Water:**                              High:      720 hours              (30 days)
                                                 Low:       96 hours               (4 days)
*Comment:*  Scientific judgement based upon grab sample aerobic soil column test data (Fournier, JC and Salle, J (1974)).

<u>Aqueous Biodegradation (unacclimated):</u>

· **Aerobic half-life:**                         High:      360 hours              (15 days)
                                                 Low:       48 hours               (2 days)
*Comment:*  Scientific judgement based upon grab sample aerobic soil column test data (Fournier, JC and Salle, J (1974)).

· **Anaerobic half-life:**                       High:      1440 hours             (60 days)
                                                 Low:       192 hours              (8 days)
*Comment:*  Scientific judgement based upon grab sample aerobic soil column test data (Fournier, JC and Salle, J (1974)).

· **Removal/secondary treatment:**               High:      No data
                                                 Low:
*Comment:*

<u>Photolysis:</u>

· **Atmos photol half-life:**                    High:      No data
                                                 Low:
*Comment:*

28

· **Max light absorption (nm):**         No data
*Comment:*

· **Aq photol half-life:**         High:    No data
                                   Low:
*Comment:*

## Photooxidation half-life:
   · **Water:**                     High: 7.4X104 hours          (8.4 years)
                                    Low:     960 hours          (40 days)
*Comment:* Based upon measured rates for reaction with hydroxyl radicals in water (high $t_{1/2}$: Anbar, M et al. (1966); low $t_{1/2}$: Dorfman, LM and Adams, GE (1973)).

   · **Air:**                       High:    31 hours
                                    Low:     3.1 hours
*Comment:* Based upon estimated rate constant for reaction with hydroxyl radicals in air (Atkinson, R (1987A)).

## Reduction half-life:                High:    No data
                                       Low:
*Comment:*

## Hydrolysis:
*Comment:* Not expected to be significant based upon estimated half-lives for hydrolysis of acetamide of 261, 3950, and 46 years at pH 5, 7, and 9, respectively, which were calculated using experimental acid and base hydrolysis rate constants for acetamide (Mabey, W and Mill, T (1978)).

· **First-order hydr half-life:**         No data
*Comment:*

· **Acid rate const (M(H+)-hr)$^{-1}$:**         No data
*Comment:*

· **Base rate const (M(OH-)-hr)$^{-1}$:**         No data
*Comment:*

# Nitroglycerin

CAS Registry Number: 55-63-0

## Half-lives:

**· Soil:**
| | | |
|---|---|---|
| High: | 168 hours | (7 days) |
| Low: | 48 hours | (2 days) |

*Comment:* Scientific judgement based upon estimated unacclimated aerobic aqueous biodegradation half-life.

**· Air:**
| | |
|---|---|
| High: | 17.6 hours |
| Low: | 1.76 hours |

*Comment:* Scientific judgement based upon estimated photooxidation half-life in air.

**· Surface Water:**
| | | |
|---|---|---|
| High: | 168 hours | (7 days) |
| Low: | 48 hours | (2 days) |

*Comment:* Scientific judgement based upon estimated unacclimated aerobic aqueous biodegradation half-life.

**· Ground Water:**
| | | |
|---|---|---|
| High: | 336 hours | (14 days) |
| Low: | 96 hours | (4 days) |

*Comment:* Scientific judgement based upon estimated unacclimated aerobic aqueous biodegradation half-life.

## Aqueous Biodegradation (unacclimated):

**· Aerobic half-life:**
| | | |
|---|---|---|
| High: | 168 hours | (7 days) |
| Low: | 48 hours | (2 days) |

*Comment:* Scientific judgement based upon munitions chemical-contaminated river water die-away test data (low $t_{1/2}$) (Wendt, TM et al. (1978)) and activated sludge screening test data (high $t_{1/2}$) (Spanggord, RJ et al. (1980A)).

**· Anaerobic half-life:**
| | | |
|---|---|---|
| High: | 672 hours | (28 days) |
| Low: | 192 hours | (8 days) |

*Comment:* Scientific judgement based upon estimated unacclimated aerobic aqueous biodegradation half-life.

**· Removal/secondary treatment:**
| | |
|---|---|
| High: | 100% |
| Low: | |

*Comment:* Removal percentage based upon continuous activated sludge treatment simulator data (Wendt, TM et al. (1978)).

## Photolysis:

**· Atmos photol half-life:**
| | | |
|---|---|---|
| High: | 2784 hours | (116 days) |

Low:    928 hours        (38.7 days)

*Comment:* Scientific judgement based upon measured photolysis rate in distilled water under winter sunlight (high $t_{1/2}$) and adjusted for approximate summer sunlight intensity (low $t_{1/2}$) (Spanggord, RJ et al. (1980A)).

· **Max light absorption (nm):**        No lambda max reported

*Comment:* Significant absorption up to 323 nm in distilled water solution (Spanggord, RJ et al. (1980A)).

· **Aq photol half-life:**        High:    2784 hours        (116 days)
                                      Low:    928 hours        (38.7 days)

*Comment:* Scientific judgement based upon measured photolysis rate in distilled water under winter sunlight directly (high $t_{1/2}$) and adjusted for approximate summer sunlight intensity (low $t_{1/2}$) (Spanggord, RJ et al. (1980A)).

## Photooxidation half-life:

· **Water:**        High:    4695 hours        (196 days)
                          Low:    2711 hours        (113 days)

*Comment:* Scientific judgement based upon measured sunlight photooxidation rates in naturally eutrophic pond water uncontaminated by munitions chemicals (high $t_{1/2}$) and in river water contaminated by munitions chemicals (low $t_{1/2}$); the rates are corrected for direct sunlight photolysis rate in distilled water and for the rate of an unidentified degradation process which occurred in the dark (Spanggord, RJ et al. (1980A)).

· **Air:**        High:    17.6 hours
                Low:    1.76 hours

*Comment:* Scientific judgement based upon estimated rate constant for reaction with hydroxyl radicals in air (Atkinson, R (1987A)).

## Reduction half-life:        High:    No data
                                    Low:

*Comment:*

## Hydrolysis:

· **First-order hydr half-life:**        81600 hours (9.3 years) at pH 7

*Comment:* Scientific judgement based upon base catalyzed hydrolysis rate constant (Ellington, JJ et al. (1987A)).

· **Acid rate const $(M(H+)-hr)^{-1}$:**        No data

*Comment:*

· **Base rate const $(M(OH-)-hr)^{-1}$ :**    $2.36 \times 1^{-2}$ $M^{-1}$ $s^{-1}$

*Comment:* ($t_{1/2}$ = 816 hours (34 days) at pH 9) Scientific judgement based upon base catalyzed hydrolysis rate constant (Ellington, JJ et al. (1987A)).

# Methylthiouracil

CAS Registry Number: 56-04-2

Structure:

H₃C — NH — S
NH
O

(structure as drawn: methyl-substituted pyrimidine ring with S and O substituents)

## Half-lives:
  · **Soil:**                                     High:     672 hours           (4 weeks)
                                              Low:     168 hours           (1 week)
    *Comment:* Scientific judgement based upon estimated aqueous aerobic biodegradation half-life.

  · **Air:**                                       High:     3.4 hours
                                              Low:     0.3 hours       (22 minutes)
    *Comment:* Scientific judgement based upon estimated photooxidation half-life in air.

  · **Surface Water:**                     High:     672 hours           (4 weeks)
                                              Low:     168 hours           (1 week)
    *Comment:* Scientific judgement based upon estimated aqueous aerobic biodegradation half-life.

  · **Ground Water:**                    High:     8640 hours        (8 weeks)
                                              Low:     336 hours           (2 weeks)
    *Comment:* Scientific judgement based upon estimated aqueous aerobic biodegradation half-life.

## Aqueous Biodegradation (unacclimated):
  · **Aerobic half-life:**                 High:     672 hours           (4 weeks)
                                              Low:     168 hours           (1 week)
    *Comment:* Scientific judgement.

  · **Anaerobic half-life:**            High:     2688 hours        (16 weeks)
                                              Low:     672 hours           (4 weeks)
    *Comment:* Scientific judgement based upon estimated aqueous aerobic biodegradation half-life.

  · **Removal/secondary treatment:**     High:     No data
                                              Low:
    *Comment:*

## Photolysis:
  · **Atmos photol half-life:**         High:
                                              Low:
    *Comment:*

· **Max light absorption (nm):**  No data
*Comment:*

· **Aq photol half-life:**  High:
  Low:
*Comment:*

**Photooxidation half-life:**
  · **Water:**  High:  No data
    Low:
*Comment:*

  · **Air:**  High:  3.4 hours
    Low:  0.3 hours  (22 minutes)
*Comment:* Scientific judgement based upon estimated rate data for hydroxyl radicals and ozone in air (Atkinson, R and Carter, WPL (1984)).

**Reduction half-life:**  High:  No data
  Low:
*Comment:*

**Hydrolysis:**
  · **First-order hydr half-life:**  71832 hours  (8.2 years)
*Comment:* Based upon a neutral rate constant ($k_N = 9.7 \times 10^{-6}$ $hr^{-1}$) at pH 7 and 25 °C (Ellington, JJ et al. (1988A)).

  · **Acid rate const $(M(H+)\text{-}hr)^{-1}$:**
*Comment:*

  · **Base rate const $(M(OH\text{-})\text{-}hr)^{-1}$:**
*Comment:*

# Carbon tetrachloride

**CAS Registry Number:** 56-23-5

**Half-lives:**
  · **Soil:** High: 8640 hours (1 year)
                                       Low: 4320 hours (6 months)
  *Comment:* Scientific judgement based upon estimated aqueous aerobic biodegradation half-life.

  · **Air:** High: $1.6 \times 10^5$ hours (18.3 years)
                                     Low: $1.6 \times 10^4$ hours (1.8 years)
  *Comment:* Based upon photooxidation half-life in air.

  · **Surface Water:** High: 8640 hours (1 year)
                                     Low: 4032 hours (6 months)
  *Comment:* Scientific judgement based upon estimated aqueous aerobic biodegradation half-life.

  · **Ground Water:** High: 8640 hours (1 year)
                                     Low: 168 hours (7 days)
  *Comment:* Scientific judgement based upon estimated aqueous aerobic biodegradation half-life and acclimated anaerobic sediment/aquifer grab sample data (low $t_{1/2}$: Parsons, F et al. (1985)).

**Aqueous Biodegradation (unacclimated):**
  · **Aerobic half-life:** High: 8640 hours (1 year)
                                     Low: 4032 hours (6 months)
  *Comment:* Scientific judgement based upon acclimated aerobic screening test data (Tabak, HH et al. (1981)).

  · **Anaerobic half-life:** High: 672 hours (28 days)
                                     Low: 168 hours (7 days)
  *Comment:* Scientific judgement based upon unacclimated anaerobic screening test data (Bouwer, EJ and McCarty, PL (1983)) and acclimated anaerobic sediment/aquifer grab sample data (Parsons, F et al. (1985)).

  · **Removal/secondary treatment:** High: 99%
                                     Low:
  *Comment:* Based upon % degraded under anaerobic continuous flow conditions (Bouwer, EJ and McCarty, PL (1983)).

**Photolysis:**
  · **Atmos photol half-life:** High: Not expected to be important
                                     Low:
  *Comment:* Scientific judgement based upon aqueous photolysis data.

· **Max light absorption (nm):** Lambda max = 275, 250, 220
*Comment:* Absorption maxima (Hubrich, C and Stuhl, F (1980)). No absorbance above 290 nm (Hustert, K et al.(1981)).

· **Aq photol half-life:** High: Not expected to be important
Low:
*Comment:* Stable in water exposed to sunlight (EPA Photochemical Degradation Test) (Hustert, K et al.(1981)).

## Photooxidation half-life:
· **Water:** High: No data
Low:

*Comment:*

· **Air:** High: $1.6 \times 10^5$ hours (18.3 years)
Low: $1.6 \times 10^4$ hours (1.8 years)
*Comment:* Based upon measured rate data for the vapor phase reaction with hydroxyl radicals in air (Atkinson, R (1985)).

## Reduction half-life:
High: No data
Low:

*Comment:*

## Hydrolysis:
· **First-order hydr half-life:** 7000 years
*Comment:* Scientific judgement based upon reported rate constant ($4.8 \times 10^{-7}$ $mol^{-1}$ $s^{-1}$) at pH 7 and 25 °C (Mabey, W and Mill, T (1978)).

· **Acid rate const $(M(H+)-hr)^{-1}$:**
*Comment:*

· **Base rate const $(M(OH-)-hr)^{-1}$:**
*Comment:*

# 3-Methylcholanthrene

<u>CAS Registry Number:</u>  56-49-5

<u>Structure:</u>

<u>Half-lives:</u>
· **Soil:**                                  High:    33600 hours                    (3.84 years)
                                            Low:    14616 hours                     (1.67 years)
*Comment:*  Scientific judgement based upon mineralization half-life in freshwater and estuarine ecosystems (Heitkamp, MA (1988)).

· **Air:**                                   High:    3.17 hours
                                            Low:    0.317 hours
*Comment:*  Scientific judgement based upon estimated photooxidation half-life in air.

· **Surface Water:**                        High:    33600 hours                    (3.84 years)
                                            Low:    14616 hours                     (1.67 years)
*Comment:*  Scientific judgement based upon mineralization half-life in freshwater and estuarine ecosystems (Heitkamp, MA (1988)).

· **Ground Water:**                         High:    67200 hours                    (7.67 years)
                                            Low:    29232 hours                     (3.34 years)
*Comment:*  Scientific judgement based upon estimated unacclimated aqueous aerobic biodegradation half-life.

<u>Aqueous Biodegradation (unacclimated):</u>
· **Aerobic half-life**:                     High:    33600 hours                    (3.84 years)
                                            Low:    14616 hours                     (1.67 years)
*Comment:*  Scientific judgement based upon mineralization half-life in freshwater and estuarine ecosystems (Heitkamp, MA (1988)).

· **Anaerobic half-life**:                   High: 134400 hours                     (15.3 years)
                                            Low:    58464 hours                     (6.67 years)
*Comment:*  Scientific judgement based upon estimated unacclimated aqueous aerobic biodegradation half-life.

· **Removal/secondary treatment:**          High:        No data
                                            Low:

36

*Comment:*

## Photolysis:
· **Atmos photol half-life:**   High:   No data
                                Low:
*Comment:*

· **Max light absorption (nm):**   No data
*Comment:* No absorption above 300 nm (no solvent reported) (Radding, SB et al. (1976)).

· **Aq photol half-life:**   High:   No data
                             Low:
*Comment:*

## Photooxidation half-life:
· **Water:**   High:   No data
               Low:
*Comment:*

· **Air:**   High:   3.17 hours
             Low:    0.317 hours
*Comment:* Scientific judgement based upon estimated rate constant for reaction with hydroxyl radical in air (Atkinson, R (1987A)).

## Reduction half-life:   High:   No data
                          Low:
*Comment:*

## Hydrolysis:
· **First-order hydr half-life:**
*Comment:*

· **Acid rate const $(M(H+)-hr)^{-1}$:**   No hydrolyzable groups
*Comment:*

· **Base rate const $(M(OH-)-hr)^{-1}$:**
*Comment:*

# Diethylstilbestrol

**CAS Registry Number:** 56-53-1

**Structure:**

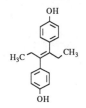

**Half-lives:**

**· Soil:**
High: 4320 hours (6 months)
Low: 672 hours (4 weeks)
*Comment:* Scientific judgement based upon estimated unacclimated aqueous aerobic biodegradation half-life.

**· Air:**
High: 0.0358 hours
Low: 0.00593 hours
*Comment:* Scientific judgement based upon estimated photooxidation half-life in air.

**· Surface Water:**
High: 3840 hours (160 days)
Low: 66 hours (2.75 days)
*Comment:* Scientific judgement based upon estimated photooxidation half-life in water.

**· Ground Water:**
High: 8640 hours (12 months)
Low: 1344 hours (8 weeks)
*Comment:* Scientific judgement based upon estimated unacclimated aqueous aerobic biodegradation half-life.

**Aqueous Biodegradation (unacclimated):**

**· Aerobic half-life:**
High: 4320 hours (6 months)
Low: 672 hours (4 weeks)
*Comment:* Scientific judgement based upon aqueous aerobic biodegradation screening test data (Lutin, PA et al. (1965); Malaney, GW et al. (1967)).

**· Anaerobic half-life:**
High: 17280 hours (24 months)
Low: 2688 hours (16 weeks)
*Comment:* Scientific judgement based upon estimated unacclimated aqueous aerobic biodegradation half-life.

**· Removal/secondary treatment:**
High: No data
Low:

*Comment:*

**Photolysis:**
  · **Atmos photol half-life:**               High:
                                              Low:
  *Comment:*

  · **Max light absorption (nm):**            No data
  *Comment:*

  · **Aq photol half-life:**                  High:
                                              Low:

  *Comment:*

**Photooxidation half-life:**
  · **Water:**                    High:     3840 hours            (160 days)
                                  Low:       66 hours            (2.75 days)
  *Comment:* Scientific judgement based upon reported reaction rate constants for ·OH and $RO_2$·
  (low $t_{1/2}$) and $RO_2$ (high $t_{1/2}$) with the phenol class (Mill, T and Mabey, W (1985); Guesten, H et
  al. (1981)).

  · **Air:**                      High:  0.0358 hours
                                  Low: 0.00593 hours
  *Comment:* Scientific judgement based upon estimated rate constants for reaction with hydroxyl
  radical (Atkinson, R (1987A)) and ozone (Atkinson, R and Carter, WPL (1984)) in air.

**Reduction half-life:**                      High:       No data
                                              Low:

  *Comment:*

**Hydrolysis:**
  · **First-order hydr half-life:**
  *Comment:*

  · **Acid rate const $(M(H+)\text{-hr})^{-1}$:**      No hydrolyzable groups
  *Comment:*

  · **Base rate const $(M(OH\text{-})\text{-hr})^{-1}$:**
  *Comment:*

# Benz(a)anthracene

**CAS Registry Number:** 56-55-3

**Structure:**

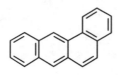

## Half-lives:

**· Soil:**                    High:   16320 hours          (1.86 years)
                               Low:    2448 hours           (102 days)
*Comment:* Based upon aerobic soil die-away test data at 10-30°C (Coover, MP and Sims, RCC (1987); Groenewegen, D and Stolp, H (1976)).

**· Air:**                     High:    3 hours
                               Low:     1 hour
*Comment:* Scientific judgement based upon estimated photolysis half-life in air.

**· Surface Water:**           High:    3 hours
                               Low:     1 hour
*Comment:* Scientific judgement based upon estimated photolysis half-life in water.

**· Ground Water:**            High:   32640 hours          (3.73 years)
                               Low:    4896 hours           (204 days)
*Comment:* Scientific judgement based upon estimated unacclimated aqueous aerobic biodegradation half-life.

## Aqueous Biodegradation (unacclimated):

**· Aerobic half-life:**       High:   16320 hours          (1.86 years)
                               Low:    2448 hours           (102 days)
*Comment:* Based upon aerobic soil die-away test data at 10-30°C (Coover, MP and Sims, RCC (1987); Groenewegen, D and Stolp, H (1976)).

**· Anaerobic half-life:**     High:   65280 hours          (7.45 years)
                               Low:    9792 hours           (1.12 years)
*Comment:* Scientific judgement based upon estimated unacclimated aqueous aerobic biodegradation half-life.

**· Removal/secondary treatment:**   High:    No data
                                     Low:
*Comment:*

## Photolysis:
· **Atmos photol half-life:**        High:        3 hours
                                     Low:         1 hour
*Comment:* Scientific judgement based upon measured photolysis rate constant for midday March sunlight on a cloudy day (Smith, JH et al. (1978)) and adjusted for approximate summer and winter sunlight intensity (Lyman, WJ et al. (1982)).

· **Max light absorption (nm):** lambda max = 314, 327, 341, 359, 376, 386 nm (ethanol); 222, 227, 254, 267 nm (ethanol).
*Comment:* Spectrum #1: Radding, SB et al. (1976); spectrum #2: IARC (1983A).

· **Aq photol half-life:**          High:        3 hours
                                    Low:         1 hour
*Comment:* Scientific judgement based upon measured photolysis rate constant for midday March sunlight on a cloudy day (Smith, JH et al. (1978)) and adjusted for approximate summer and winter sunlight intensity (Lyman, WJ et al. (1982)).

## Photooxidation half-life:
· **Water:**        High:        3850 hours        (160 days)
                    Low:          77 hours         (3.2 days)
*Comment:* Based upon measured rate constant for reaction with alkylperoxyl radical in water (Radding, SB et al. (1976)).

· **Air:**          High:        8.01 hours
                    Low:         0.801 hours
*Comment:* Scientific judgement based upon estimated rate constant for reaction with hydroxyl radical in air (Atkinson, R (1987A)).

## Reduction half-life:                High:        No data
                                       Low:
*Comment:*

## Hydrolysis:
· **First-order hydr half-life:**
*Comment:*

· **Acid rate const (M(H+)-hr)$^{-1}$:**        No hydrolyzable groups
*Comment:*

· **Base rate const (M(OH-)-hr)$^{-1}$:**
*Comment:*

41

# 1,1-Dimethyl hydrazine

CAS Registry Number: 57-14-7

Half-lives:
· Soil:                                High:    528 hours              (22 days)
                                       Low:     192 hours              (8 days)
Comment: Scientific judgement based upon estimated unacclimated aqueous aerobic biodegradation half-life.

· Air:                                 High:    7.7 hours
                                       Low:     0.8 hours              (46 minutes)
Comment: Scientific judgement based upon estimated photooxidation half-life in air.

· Surface Water:                       High:    528 hours              (22 days)
                                       Low:     192 hours              (8 days)
Comment: Scientific judgement based upon unacclimated freshwater grab sample data (Braun, BA and Zirrolli, JA (1983)).

· Ground Water:                        High:    1056 hours             (44 days)
                                       Low:     384 hours              (16 days)
Comment: Scientific judgement based upon estimated unacclimated aqueous aerobic biodegradation half-life.

Aqueous Biodegradation (unacclimated):
· Aerobic half-life:                   High:    528 hours              (22 days)
                                       Low:     192 hours              (8 days)
Comment: Scientific judgement based upon unacclimated freshwater grab sample data (Braun, BA and Zirrolli, JA (1983)).

· Anaerobic half-life:                 High:    2112 hours             (8 days)
                                       Low:     768 hours              (32 days)
Comment: Scientific judgement based upon unacclimated aerobic biodegradation half-life.

· Removal/secondary treatment:         High:    No data
                                       Low:
Comment:

Photolysis:
· Atmos photol half-life:              High:    No data
                                       Low:
Comment:

· **Max light absorption (nm):**　　　　　　　No data
*Comment:*

· **Aq photol half-life:**　　　　　　High:　　No data
　　　　　　　　　　　　　　　　　　　Low:
*Comment:*

**Photooxidation half-life:**
· **Water:**　　　　　　　　　　　　　High:　　No data
　　　　　　　　　　　　　　　　　　　Low:
*Comment:*

· **Air:**　　　　　　　　　　　　　　High:　　7.7 hours
　　　　　　　　　　　　　　　　　　　Low:　　0.8 hours　　　　　(46 minutes)
*Comment:* Scientific judgement based upon an estimated rate constant for the vapor phase reaction with hydroxyl radicals in air (Atkinson, R (1987A)).

**Reduction half-life:**　　　　　　　High:　　No data
　　　　　　　　　　　　　　　　　　　Low:
*Comment:*

**Hydrolysis:**
· **First-order hydr half-life:**　　　　Not expected to be important
*Comment:* Stable under aqueous hydrolysis conditions (Braun, BA and Zirrolli, JA (1983)).

· **Acid rate const (M(H+)-hr)$^{-1}$:**
*Comment:*

· **Base rate const (M(OH-)-hr)$^{-1}$:**
*Comment:*

# Strychnine

CAS Registry Number: 57-24-9

Structure:

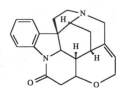

Half-lives:
   · **Soil:**                          High:   672 hours          (4 weeks)
                                        Low:    168 hours          (7 days)
   *Comment:* Scientific judgement based upon unacclimated aqueous aerobic biodegradation half-life.

   · **Air:**                           High:   1.96 hours
                                        Low:    0.196 hours
   *Comment:* Scientific judgement based upon estimated photooxidation half-life in air.

   · **Surface Water:**                 High:   672 hours          (4 weeks)
                                        Low:    168 hours          (7 days)
   *Comment:* Scientific judgement based upon unacclimated aqueous aerobic biodegradation half-life.

   · **Ground Water:**                  High:   1344 hours         (8 weeks)
                                        Low:    336 hours          (14 days)
   *Comment:* Scientific judgement based upon unacclimated aqueous aerobic biodegradation half-life.

Aqueous Biodegradation (unacclimated):
   · **Aerobic half-life:**             High:   672 hours          (4 weeks)
                                        Low:    168 hours          (7 days)
   *Comment:* Scientific judgement.

   · **Anaerobic half-life:**           High:   2688 hours         (16 weeks)
                                        Low:    672 hours          (28 days)
   *Comment:* Scientific judgement based upon unacclimated aqueous aerobic biodegradation half-life.

   · **Removal/secondary treatment:**   High:   No data
                                        Low:
   *Comment:*

<u>Photolysis:</u>
- **Atmos photol half-life:**        High:
                                         Low:

*Comment:*

- **Max light absorption (nm):**    No data
*Comment:*

- **Aq photol half-life:**          High:
                                         Low:

*Comment:*

<u>Photooxidation half-life:</u>
- **Water:**                     High:        No data
                                         Low:

*Comment:*

- **Air:**                         High:     1.96 hours
                                       Low:     0.196 hours
*Comment:* Scientific judgement based upon estimated rate constant for reaction with hydroxyl radical in air (Atkinson, R (1987A)).

<u>Reduction half-life:</u>           High:        No data
                                         Low:

*Comment:*

<u>Hydrolysis:</u>
- **First-order hydr half-life:**        No data
*Comment:*

- **Acid rate const $(M(H+)-hr)^{-1}$:**
*Comment:*

- **Base rate const $(M(OH-)-hr)^{-1}$:**
*Comment:*

# beta-Propiolactone

CAS Registry Number: 57-57-8

## Half-lives:

· **Soil:**  High: 3.4 hours

Low: 0.058 hours (3.5 minutes)

*Comment:* Based on measured hydrolysis rate constant in water at 25°C (high $t_{1/2}$: Butler, AR and Gold, V (1962A); low $t_{1/2}$: Mabey, W and Mill, T (1978)).

· **Air:**  High: 1800 hours (75 days)

Low: 180 hours (7.5 days)

*Comment:* Based upon estimated photooxidation half-life in air.

· **Surface Water:**  High: 3.4 hours

Low: 0.058 hours (3.5 minutes)

*Comment:* Based on measured hydrolysis rate constant in water at 25°C (high $t_{1/2}$: Butler, AR and Gold, V (1962A); low $t_{1/2}$: Mabey, W and Mill, T (1978)).

· **Ground Water:**  High: 3.4 hours

Low: 0.058 hours (3.5 minutes)

*Comment:* Based on measured hydrolysis rate constant in water at 25°C (high $t_{1/2}$: Butler, AR and Gold, V (1962A); low $t_{1/2}$: Mabey, W and Mill, T (1978)).

## Aqueous Biodegradation (unacclimated):

· **Aerobic half-life:**  High: 168 hours (7 days)

Low: 24 hours (1 day)

*Comment:* Scientific judgement based upon unacclimated aerobic aqueous screening test data (Malaney, GW et al. (1967)).

· **Anaerobic half-life:**  High: 672 hours (28 days)

Low: 96 hours (4 days)

*Comment:* Scientific judgement based upon estimated aqueous aerobic biodegradation half-life.

· **Removal/secondary treatment:**  High: No data

Low:

*Comment:*

## Photolysis:

· **Atmos photol half-life:**  High: No data

Low:

*Comment:*

· **Max light absorption (nm):** No data
*Comment:*

· **Aq photol half-life:** High: No data
Low:
*Comment:*

**Photooxidation half-life:**
· **Water:** High: No data
Low:
*Comment:*

· **Air:** High: 1800 hours (75 days)
Low: 180 hours (7.5 days)
*Comment:* Based upon estimated rate constant for reaction with hydroxyl radicals in air (Atkinson, R (1985)).

**Reduction half-life:** High: No data
Low:
*Comment:*

**Hydrolysis:**
· **First-order hydr half-life:** High: 3.4 hours
Low: 0.058 hours (3.5 minutes)
*Comment:* Based on measured hydrolysis rate constant in water at 25°C (high $t_{1/2}$: Butler, AR and Gold, V (1962A); low $t_{1/2}$: Mabey, W and Mill, T (1978)).

· **Acid rate const (M(H+)-hr)$^{-1}$:** No data
*Comment:*

· **Base rate const (M(OH-)-hr)$^{-1}$:** No data
*Comment:*

# Chlordane

CAS Registry Number: 57-74-9

Structure:

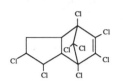

Half-lives:
   · **Soil:**                          High:   33264 hours        (3.8 years)
                                        Low:    5712 hours         (283 days)
   *Comment:* Scientific judgement based upon unacclimated aerobic river die-away test data (low $t_{1/2}$: Eichelberger, JW and Lichtenberg, JJ (1971)) and reported soil grab sample data (high $t_{1/2}$: Castro, TF and Yoshida, T (1971)).

   · **Air:**                           High:   51.7 hours         (2.2 days)
                                        Low:    5.2 hours
   *Comment:* Scientific judgement based upon estimated photooxidation half-life in air.

   · **Surface Water:**                 High:   33264 hours        (3.8 years)
                                        Low:    5712 hours         (283 days)
   *Comment:* Scientific judgement based upon unacclimated aerobic river die-away test data (low $t_{1/2}$: Eichelberger, JW and Lichtenberg, JJ (1971)) and reported soil grab sample data (high $t_{1/2}$: Castro, TF and Yoshida, T (1971)).

   · **Ground Water:**                  High:   66528 hours        (7.6 years)
                                        Low:    11424 hours        (566 days)
   *Comment:* Scientific judgement based upon estimated aqueous aerobic biodegradation half-life.

Aqueous Biodegradation (unacclimated):
   · **Aerobic half-life:**             High:   33264 hours        (3.8 years)
                                        Low:    5712 hours         (283 days)
   *Comment:* Scientific judgement based upon unacclimated aerobic river die-away test data (low $t_{1/2}$: Eichelberger, JW and Lichtenberg, JJ (1971)) and reported soil grab sample data (high $t_{1/2}$: Castro, TF and Yoshida, T (1971)).

   · **Anaerobic half-life:**           High:   168 hours          (7 days)
                                        Low:    24 hours           (1 day)
   *Comment:* Scientific judgement based upon soil and freshwater mud grab sample data for aldrin, dieldrin, endrin and heptachlor epoxide (Maule, A et al. (1987)).

   · **Removal/secondary treatment:**   High:   No data

Low:

*Comment:*

## Photolysis:
· **Atmos photol half-life:**          High:     No data
                                         Low:

*Comment:*

· **Max light absorption (nm):**  lambda max approximately 210 nm
*Comment:*  In hexane, chlordane does not absorb UV light at wavelengths greater than 280 nm (Gore, RC et al. (1971)).

· **Aq photol half-life:**           High:     No data
                                         Low:

*Comment:*

## Photooxidation half-life:
· **Water:**                      High:     No data
                                         Low:

*Comment:*

· **Air:**                         High:     51.7 hours        (2.2 days)
                                         Low:      5.2 hours
*Comment:*  Scientific judgement based upon an estimated rate constant for vapor phase reaction with hydroxyl radicals in air (Atkinson, R (1987A)).

## Reduction half-life:
                                         High:     No data
                                         Low:

*Comment:*

## Hydrolysis:
· **First-order hydr half-life:**        >197000 years
*Comment:*  Scientific judgement based upon base rate constant ($4.3 \times 10^{-3}$ $M^{-1}$ $hr^{-1}$) at pH 7 and 25 °C (Ellington, JJ et al. (1987)).

· **Acid rate const $(M(H+)-hr)^{-1}$:**
*Comment:*

· **Base rate const $(M(OH-)-hr)^{-1}$:**     $4.3 \times 10^{-3}$ $M^{-1}$ $hr^{-1}$
*Comment:*  Based upon a rate constant at pH 9 and 25 °C (Ellington, JJ et al. (1987)).

49

# 7,12-Dimethylbenz(a)anthracene

**CAS Registry Number:** 57-97-6

**Structure:**

**Half-lives:**

| · **Soil:** | High: | 672 hours | (28 days) |
| | Low: | 480 hours | (20 days) |

*Comment:* Based upon aerobic soil die-away test data (Sims, RC (1990)).

| · **Air:** | High: | 3.20 hours |
| | Low: | 0.32 hours |

*Comment:* Scientific judgement based upon estimated photooxidation half-life in air.

| · **Surface Water:** | High: | 672 hours | (28 days) |
| | Low: | 480 hours | (20 days) |

*Comment:* Based upon aerobic soil die-away test data (Sims, RC (1990)).

| · **Ground Water:** | High: | 1344 hours | (56 days) |
| | Low: | 960 hours | (40 days) |

*Comment:* Scientific judgement based upon estimated unacclimated aqueous aerobic biodegradation half-life.

**Aqueous Biodegradation (unacclimated):**

| · **Aerobic half-life:** | High: | 672 hours | (28 days) |
| | Low: | 480 hours | (20 days) |

*Comment:* Based upon aerobic soil die-away test data (Sims, RC (1990)).

| · **Anaerobic half-life:** | High: | 2688 hours | (112 days) |
| | Low: | 1920 hours | (80 days) |

*Comment:* Scientific judgement based upon estimated unacclimated aqueous aerobic biodegradation half-life.

| · **Removal/secondary treatment:** | High: | No data |
| | Low: | |

*Comment:*

**Photolysis:**

**· Atmos photol half-life:**        High:        No data
                                     Low:

*Comment:*

**· Max light absorption (nm):** lambda max = 234.6, 263, 275, 285, 295.5, 346 nm (methanol)
*Comment:* Sadtler UV No. 2413.

**· Aq photol half-life:**        High:        No data
                                   Low:

*Comment:*

## Photooxidation half-life:

**· Water:**        High: $1.375 \times 10^6$ hours        (157 years)
                    Low:    13750 hours        (1.57 years)

*Comment:* Based upon measured rate constant for reaction with singlet oxygen in benzene (Stevens, B et al. (1974)).

**· Air:**        High:        3.20 hours
                  Low:        0.32 hours

*Comment:* Scientific judgement based upon estimated rate constant for reaction with hydroxyl radical in air (Atkinson, R (1987A)).

## Reduction half-life:        High:        No data
                                Low:

*Comment:*

## Hydrolysis:

**· First-order hydr half-life:**
*Comment:*

**· Acid rate const $(M(H+)\text{-}hr)^{-1}$:**        No hydrolyzable groups
*Comment:*

**· Base rate const $(M(OH\text{-})\text{-}hr)^{-1}$:**
*Comment:*

# gamma-Hexachlorocyclohexane (Lindane)

## Half-lives:
- **Soil:**  High:  5765 hours  (240 days)
  Low:  330 hours  (13.8 days)
  *Comment:* Based upon hydrolysis half-life (Ellington, JJ et al. (1987)).

- **Air:**  High:  92.4 hours  (3.85 days)
  Low:  9.24 hours
  *Comment:* Scientific judgement based upon estimated photooxidation half-life in air.

- **Surface Water:**  High:  5765 hours  (240 days)
  Low:  330 hours  (13.8 days)
  *Comment:* Based upon hydrolysis half-life (Ellington, JJ et al. (1987)).

- **Ground Water:**  High:  5765 hours  (240 days)
  Low:  142 hours  (5.9 days)
  *Comment:* Based upon hydrolysis half-life (Ellington, JJ et al. (1987)).

## Aqueous Biodegradation (unacclimated):
- **Aerobic half-life:**  High:  9912 hours  (413 days)
  Low:  744 hours  (31 days)
  *Comment:* Based upon aerobic soil die-away study data (low $t_{1/2}$: Brahmaprakash, GP et al. (1985); high $t_{1/2}$: Kohnen, R et al. (1975)).

- **Anaerobic half-life:**  High:  734 hours  (30 days)
  Low:  142 hours  (5.9 days)
  *Comment:* Based upon anaerobic flooded soil die-away study data (low $t_{1/2}$: Brahmaprakash, GP et al. (1985); high $t_{1/2}$: Zhang, S et al. (1982)).

- **Removal/secondary treatment:**  High:  25%
  Low:
  *Comment:* Removal percentage based upon data from a continuous activated sludge biological treatment simulator (Petrasek, AC et al. (1983)).

## Photolysis:
- **Atmos photol half-life:**  High:
  Low:
  *Comment:*

- **Max light absorption (nm):**  No data

52

*Comment:*

· **Aq photol half-life:**               High:
                                         Low:

*Comment:*

## Photooxidation half-life:
· **Water:**                        High:       No data
                                           Low:

*Comment:*

· **Air:**                           High:     92.4 hours          (3.85 days)
                                         Low:      9.24 hours
*Comment:* Scientific judgement based upon estimated rate constant for reaction with hydroxyl radical in air (Atkinson, R (1987A)).

## Reduction half-life:                     High:     No data
                                         Low:

*Comment:*

## Hydrolysis:
· **First-order hydr half-life:**            4957 hours         (207 days)
*Comment:* ($t_{1/2}$ at pH 7 and 25°C) Based upon measured neutral and base catalyzed hydrolysis rate constants (Ellington, JJ et al. (1987)).

· **Acid rate const $(M(H+)-hr)^{-1}$:**     No data
*Comment:* ($t_{1/2}$ = 5765 hours (240 days) at pH 5 and 25°C) Based upon measured neutral and base catalyzed hydrolysis rate constants (Ellington, JJ et al. (1987)).

· **Base rate const $(M(OH-)-hr)^{-1}$:**     198
*Comment:* ($t_{1/2}$ = 330 hours (13.8 days) at pH 9 and 25°C) Based upon measured neutral and base catalyzed hydrolysis rate constants (Ellington, JJ et al. (1987)).

# 2,3,4,6-Tetrachlorophenol

**CAS Registry Number:** 58-90-2

**Half-lives:**

· **Soil:**                        High:    4320 hours              (6 months)

                                        Low:     672 hours               (4 weeks)

*Comment:* Scientific judgement based upon estimated unacclimated aqueous aerobic biodegradation half-life.

· **Air:**                         High:    3644 hours              (151.9 days)

                                        Low:     364.4 hours            (15.2 days)

*Comment:* Scientific judgement based upon estimated photooxidation half-life in air.

· **Surface Water:**            High:    336 hours               (14 days)

                                        Low:     1 hour

*Comment:* Scientific judgement based upon aqueous photolysis data for 2,4,5-, 2,4,6-trichlorophenol and pentachlorophenol.

· **Ground Water:**             High:    8640 hours              (12 months)

                                        Low:     1344 hours             (8 weeks)

*Comment:* Scientific judgement based upon estimated unacclimated aqueous aerobic biodegradation half-life.

**Aqueous Biodegradation (unacclimated):**

· **Aerobic half-life:**        High:    4032 hours              (6 months)

                                        Low:     672 hours               (4 weeks)

*Comment:* Scientific judgement based upon acclimated aerobic screening test data (Alexander, M and Aleem, MIH (1961)).

· **Anaerobic half-life:**      High:    16128 hours            (24 months)

                                        Low:     2688 hours             (16 weeks)

*Comment:* Scientific judgement based upon unacclimated aerobic biodegradation half-lives.

· **Removal/secondary treatment:**    High:    91%

                                        Low:

*Comment:* Based upon % degraded under aerobic continuous flow conditions (Salkinoja-Salonen, MS et al. (1984)).

**Photolysis:**

· **Atmos photol half-life:**   High:    No data

                                        Low:

*Comment:*

· **Max light absorption (nm):**  lambda max = 301
*Comment:*  In methanol (Sadtler UV No. 23265).


· **Aq photol half-life:**              High:        336 hours                (14 days)
                                                         Low:          1 hour
*Comment:*  Scientific judgement based upon aqueous photolysis data for 2,4,5-, 2,4,6-
trichlorophenol and pentachlorophenol.

**Photooxidation half-life:**
  · **Water:**                              High:      3480 hours              (145 days)
                                                        Low:        66.0 hours              (2.75 days)
*Comment:*  Scientific judgement based upon reported reaction rate constants for ·OH and $RO_2$·
with the phenol class (Mill, T and Mabey, W (1985); Guesten, H et al. (1981)).

  · **Air:**                                  High:      3644 hours              (151.9 days)
                                                        Low:      364.4 hours              (15.2 days)
*Comment:*  Scientific judgement based upon an estimated rate constant for the vapor phase
reaction with hydroxyl radicals in air (Atkinson, R (1987A)).

**Reduction half-life:**                High:       No data
                                                        Low:
  *Comment:*

**Hydrolysis:**
  · **First-order hydr half-life:**        No hydrolyzable groups
  Comment:  Rate constant at neutral pH is zero (Kollig, HP et al. (1987A)).

  · **Acid rate const (M(H+)-hr)$^{-1}$:**        0.0
  *Comment:*  Based upon measured rate data at acid pH (Kollig, HP et al. (1987A)).

  · **Base rate const (M(OH-)-hr)$^{-1}$:**        0.0
  *Comment:*  Based upon measured rate data at basic pH (Kollig, HP et al. (1987A)).

# N-Nitrosomorpholine

<u>CAS Registry Number:</u>  59-89-2

<u>Structure:</u>

<u>Half-lives:</u>
&middot; **Soil:**

| | High: | 4320 hours | (6 months) |
|---|---|---|---|
| | Low: | 672 hours | (4 weeks) |

*Comment:* Scientific judgement based upon estimated unacclimated aqueous aerobic biodegradation half-life.

&middot; **Air:**

| | High: | 18.0 hours |
|---|---|---|
| | Low: | 0.9 hours |

*Comment:* Based upon combined estimated half-life ranges for photolysis and photooxidation.

&middot; **Surface Water:**

| | High: | 170 hours |
|---|---|---|
| | Low: | 1.7 hours |

*Comment:* Based upon estimated half-lives for photolysis.

&middot; **Ground Water:**

| | High: | 8640 hours | (12 months) |
|---|---|---|---|
| | Low: | 1344 hours | (8 weeks) |

*Comment:* Scientific judgement based upon estimated unacclimated aqueous aerobic biodegradation half-life.

<u>Aqueous Biodegradation (unacclimated):</u>
&middot; **Aerobic half-life:**

| | High: | 4320 hours | (6 months) |
|---|---|---|---|
| | Low: | 672 hours | (4 weeks) |

*Comment:* Scientific judgement based upon estimated unacclimated aqueous rates predicted for the nitrosamine and cyclic ether classes.

&middot; **Anaerobic half-life:**

| | High: | 17280 hours | (24 months) |
|---|---|---|---|
| | Low: | 2688 hours | (16 weeks) |

*Comment:* Scientific judgement based upon estimated unacclimated aqueous aerobic biodegradation half-life.

&middot; **Removal/secondary treatment:**

| | High: | No data |
|---|---|---|
| | Low: | |

*Comment:*

**Photolysis:**
  · **Atmos photol half-life:**        High:    170 hours
                                         Low:     1.7 hours
  *Comment:* Scientific judgement based upon estimated aqueous photolysis half-lives.

  · **Max light absorption (nm):** lambda max = 346 nm
  *Comment:* Absorption maxima (peak) in water (IARC, 1978)

  · **Aq photol half-life:**           High:    170 hours
                                         Low:     1.7 hours
  *Comment:* Scientific judgement based upon experimental photolysis studies in water using artificial light source (Challis, BC and Li, BFL (1982)).

**Photooxidation half-life:**
  · **Water:**                      High:    No data
                                         Low:
  *Comment:*

  · **Air:**                         High:    18.4 hours
                                       Low:    1.84 hours
  *Comment:* Based upon estimated rate constant for reaction with hydroxyl radicals in air (Atkinson, R (1987A)).

**Reduction half-life:**           High:    No data
                                         Low:
  *Comment:*

**Hydrolysis:**
  · **First-order hydr half-life:**    No data
  *Comment:*

  · **Acid rate const (M(H+)-hr)$^{-1}$:**
  *Comment:*

  · **Base rate const (M(OH-)-hr)$^{-1}$:**
  *Comment:*

# 4-Aminoazobenzene

CAS Registry Number: 60-09-3

Half-lives:
  · Soil:                          High:     672 hours        (4 weeks)
                                     Low:      168 hours        (7 days)
*Comment:* Scientific judgement based upon estimated unacclimated aqueous aerobic biodegradation half-life.

  · Air:                         High:     9.7 hours
                                     Low:      0.97 hours
*Comment:* Based upon estimated photooxidation half-life in air.

  · Surface Water:           High:     672 hours        (4 weeks)
                                     Low:      62.4 hours
*Comment:* High value based upon estimated unacclimated aqueous aerobic biodegradation half-life.
Low value based upon estimated photooxidation half-life in water.

  · Ground Water:            High:    1344 hours      (8 weeks)
                                     Low:      336 hours      (14 days)
*Comment:* Scientific judgement based upon estimated unacclimated aqueous aerobic biodegradation half-life.

Aqueous Biodegradation (unacclimated):
  · Aerobic half-life:         High:     672 hours        (4 weeks)
                                     Low:      168 hours        (7 days)
*Comment:* Scientific judgement based upon aerobic aqueous screening test data (Urushigawa, Y and Yonezawa, Y (1977)).

  · Anaerobic half-life:       High:   2688 hours     (16 weeks)
                                     Low:      672 hours      (28 days)
*Comment:* Scientific judgement based upon estimated aqueous aerobic biodegradation half-life.

  · Removal/secondary treatment:    High:     No data
                                     Low:
*Comment:*

Photolysis:
  · Atmos photol half-life:      High: Insufficient data
                                     Low:
*Comment:*

58

· **Max light absorption (nm):** lambda max = 384 nm
*Comment:* Maxima in methanol solvent (Sadtler 142 UV).

· **Aq photol half-life:**  High: Insufficient data
                            Low:
*Comment:*

## Photooxidation half-life:
· **Water:**  High:   3840 hours        (160 days)
              Low:    62.4 hours        (2.6 days)
*Comment:* Scientific judgement based upon estimated rate constants (·OH and RO$_2$·) for the aromatic amine chemical class (Mill, T and Mabey, W, (1985)).

· **Air:**  High:   9.7 hours
            Low:    0.97 hours
*Comment:* Based upon estimated rate constant for reaction with hydroxyl radicals in air (Atkinson, R (1987A)).

## Reduction half-life:  High:    No data
                         Low:
*Comment:*

## Hydrolysis:
· **First-order hydr half-life:**  No hydrolyzable groups
*Comment:*

· **Acid rate const (M(H+)-hr)$^{-1}$:**
*Comment:*

· **Base rate const (M(OH-)-hr)$^{-1}$:**
*Comment:*

# Dimethylaminoazobenzene

**CAS Registry Number:** 60-11-7

**Half-lives:**

- **Soil:**  
  High: 672 hours  (4 weeks)  
  Low: 168 hours  (7 days)  
  *Comment:* Scientific judgement based upon unacclimated aqueous aerobic biodegradation half-life.

- **Air:**  
  High: 2.89 hours  
  Low: 0.289 hours  
  *Comment:* Scientific judgement based upon estimated photooxidation half-life in air.

- **Surface Water:**  
  High: 672 hours  (4 weeks)  
  Low: 31 hours  (1.3 days)  
  *Comment:* Scientific judgement based upon unacclimated aqueous aerobic biodegradation half-life (high $t_{1/2}$) and estimated photooxidation half-life in water (low $t_{1/2}$).

- **Ground Water:**  
  High: 1344 hours  (8 weeks)  
  Low: 336 hours  (14 days)  
  *Comment:* Scientific judgement based upon unacclimated aqueous aerobic biodegradation half-life.

**Aqueous Biodegradation (unacclimated):**

- **Aerobic half-life:**  
  High: 672 hours  (4 weeks)  
  Low: 168 hours  (7 days)  
  *Comment:* Scientific judgement based upon unacclimated aqueous aerobic biodegradation screening test data (Fochtman, EG (1981)).

- **Anaerobic half-life:**  
  High: 2688 hours  (16 weeks)  
  Low: 672 hours  (28 days)  
  *Comment:* Scientific judgement based upon unacclimated aqueous aerobic biodegradation half-life.

- **Removal/secondary treatment:**  
  High: No data  
  Low:  
  *Comment:*

**Photolysis:**

- **Atmos photol half-life:**  
  High:  
  Low:  
  *Comment:*

· **Max light absorption (nm):**                    No data
*Comment:*


· **Aq photol half-life:**                 High:
                                            Low:

*Comment:*


## Photooxidation half-life:
· **Water:**                               High:      1740 hours              (72 days)
                                           Low:       31 hours               (1.3 days)
*Comment:* Scientific judgement based upon photooxidation rate constants with ·OH and $RO_2$·
for he aromatic amine class (Mill, T and Mabey, W (1985); Guesten, H et al. (1981)).


· **Air:**                                 High:      2.89 hours
                                           Low:       0.289 hours
*Comment:* Scientific judgement based upon estimated rate constant for reaction with hydroxyl
radical in air (Atkinson, R (1987A)).


## Reduction half-life:                    High:      No data
                                           Low:
*Comment:*


## Hydrolysis:
· **First-order hydr half-life:**          No data
*Comment:*


· **Acid rate const $(M(H+)-hr)^{-1}$:**
*Comment:*


· **Base rate const $(M(OH-)-hr)^{-1}$:**
*Comment:*

# Methylhydrazine

<u>CAS Registry Number:</u>  60-34-4

<u>Half-lives:</u>

· **Soil:**                                    High:     576 hours              (24 days)

                                                Low:      312 hours              (13 days)

*Comment:*  Scientific judgement based upon estimated unacclimated aqueous aerobic biodegradation half-life.

· **Air:**                                     High:     0.37 hours

                                                Low:      0.060 hours

*Comment:*  Based upon photooxidation half-life in air.

· **Surface Water:**                           High:     576 hours              (24 days)

                                                Low:      312 hours              (13 days)

*Comment:*  Scientific judgement based upon estimated unacclimated aqueous aerobic biodegradation half-life.

· **Ground Water:**                            High:     1152 hours             (48 days)

                                                Low:      624 hours              (26 days)

*Comment:*  Scientific judgement based upon estimated unacclimated aqueous aerobic biodegradation half-life.

<u>Aqueous Biodegradation (unacclimated):</u>

· **Aerobic half-life**:                       High:     576 hours              (24 days)

                                                Low:      312 hours              (13 days)

*Comment:*  Based upon unacclimated grab sample experiments for fresh pond water (low $t_{1/2}$) and sea  water (high $t_{1/2}$) (Braun, BA and Zirrolli, JA (1983)).

· **Anaerobic half-life**:                     High:     2304 days              (98 days)

                                                Low:      1248 hours             (52 days)

*Comment:*  Scientific judgement based upon estimated unacclimated aqueous aerobic biodegradation half-life.

· **Removal/secondary treatment:**             High:     No data

                                                Low:

*Comment:*

<u>Photolysis:</u>

· **Atmos photol half-life:**                  High:

                                                Low:

*Comment:*

· **Max light absorption (nm):**                    No data
*Comment:*

· **Aq photol half-life:**                    High:
                                              Low:

*Comment:*

**Photooxidation half-life:**
  · **Water:**                                High:    No data
                                              Low:

*Comment:*

  · **Air:**                                  High:    0.37 hours
                                              Low:    0.060 hours
*Comment:*  Based upon measured rate constants for reaction in water with ozone (Tuazon, EC et al. (1982)) and hydroxyl radical (Harris, GW et al. (1979)).

**Reduction half-life:**                      High:    No data
                                              Low:

*Comment:*

**Hydrolysis:**
  · **First-order hydr half-life:**                   No data
*Comment:*

  · **Acid rate const $(M(H+)-hr)^{-1}$:**
*Comment:*

  · **Base rate const $(M(OH-)-hr)^{-1}$:**
*Comment:*

# Acetamide

CAS Registry Number: 60-35-5

Half-lives:
  · Soil:                       High:    168 hours        (7 days)
                                Low:     24 hours         (1 day)
  Comment: Scientific judgement based upon estimated unacclimated aqueous aerobic
  biodegradation half-life.

  · Air:                        High:    32 hours
                                Low:     3.2 hours
  Comment: Based upon estimated photooxidation half-life in air.

  · Surface Water:              High:    168 hours        (7 days)
                                Low:     24 hours         (1 day)
  Comment: Scientific judgement based upon estimated unacclimated aqueous aerobic
  biodegradation half-life.

  · Ground Water:               High:    336 hours        (14 days)
                                Low:     48 hours         (2 days)
  Comment: Scientific judgement based upon estimated unacclimated aqueous aerobic
  biodegradation half-life.

Aqueous Biodegradation (unacclimated):
  · Aerobic half-life:          High:    168 hours        (7 days)
                                Low:     24 hours         (1 day)
  Comment: Scientific judgement based upon aerobic aqueous screening test data (Malaney, GW
  and Gerhold, RW (1962, 1969); Urano, K and Kato, Z (1986)).

  · Anaerobic half-life:        High:    672 hours        (28 days)
                                Low:     96 hours         (4 days)
  Comment: Scientific judgement based upon estimated unacclimated aqueous aerobic
  biodegradation half-life.

  · Removal/secondary treatment: High:   No data
                                 Low:
  Comment:

Photolysis:
  · Atmos photol half-life:     High:    No data
                                Low:
  Comment:

· **Max light absorption (nm):**
*Comment:*

· **Aq photol half-life:**                     High:          No data
                                               Low:
*Comment:*

**Photooxidation half-life:**
· **Water:**                                   High:          No data
                                               Low:
*Comment:*

· **Air:**                                      High:          32 hours
                                               Low:          3.2 hours
*Comment:* Based upon estimated rate constant for reaction with hydroxyl radicals in air
(Atkinson, R (1987A)).

**Reduction half-life:**                       High:          No data
                                               Low:
*Comment:*

**Hydrolysis:**
· **First-order hydr half-life:**              34,625,700 hours               (3950 year)
*Comment:* pH 7, 25 °C (Mabey, W and Mill, T (1978))

· **Acid rate const (M(H+)-hr)$^{-1}$:**          0.03
*Comment:* 25 °C (Mabey, W and Mill, T (1978)); at pH 5: $t_{1/2} = 0.6931/0.03 \times 10^{-5} = 2{,}310{,}000$
hours  (264 years)

· **Base rate const (M(OH-)-hr)$^{-1}$:**          0.17
*Comment:* 25 °C (Mabey, W and Mill, T (1978)); at pH 9: $t_{1/2} = 0.6931/0.17 \times 10^{-5} = 408{,}000$
hours  (46.5 years)

# Dimethoate

**CAS Registry Number:** 60-51-5

**Structure:**

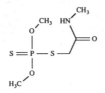

**Half-lives:**
  · **Soil:**                        High:      888 hours            (37 days)
                                      Low:       264 hours            (11 days)
  *Comment:* Based upon soil die-away test data for two soils (Bro-Rasmussen, F et al. (1970)).

  · **Air:**                         High:      4.69 hours
                                      Low:       0.469 hours
  *Comment:* Scientific judgement based upon estimated photooxidation half-life in air.

  · **Surface Water:**               High:      1344 hours           (56 days)
                                      Low:       264 hours            (11 days)
  *Comment:* Scientific judgement based upon estimated unacclimated aqueous aerobic biodegradation half-life.

  · **Ground Water:**                High:      2688 hours           (112 days)
                                      Low:       528 hours            (22 days)
  *Comment:* Scientific judgement based upon estimated unacclimated aqueous aerobic biodegradation half-life.

**Aqueous Biodegradation (unacclimated):**
  · **Aerobic half-life:**           High:      1344 hours           (56 days)
                                      Low:       264 hours            (11 days)
  *Comment:* Based upon river die-away test data (high $t_{1/2}$) (Eichelberger, JW and Lichtenberg, JJ (1971)) and soil die-away test data (low $t_{1/2}$) (Bro-Rasmussen, F et al. (1970)).

  · **Anaerobic half-life:**         High·      5376 hours           (224 days)
                                      Low:       1056 hours           (44 days)
  *Comment:* Scientific judgement based upon estimated unacclimated aqueous aerobic biodegradation half-life.

  · **Removal/secondary treatment:**  High:      No data
                                      Low:
  *Comment:*

66

## Photolysis:
**· Atmos photol half-life:**　　　　　　High:　　No data
　　　　　　　　　　　　　　　　　　　Low:

*Comment:*

**· Max light absorption (nm):**　　　lambda max <200 nm (water)
*Comment:* Absorption extends to approximately 265 nm (Gore, RC et al. (1971)).

**· Aq photol half-life:**　　　　　　High:　　No data
　　　　　　　　　　　　　　　　　Low:

*Comment:*

## Photooxidation half-life:
**· Water:**　　　　　　　　　　　High:　　No data
　　　　　　　　　　　　　　　　Low:

*Comment:*

**· Air:**　　　　　　　　　　　　High:　　4.69 hours
　　　　　　　　　　　　　　　　Low:　　0.469 hours
*Comment:* Scientific judgement based upon estimated rate constant for reaction with hydroxyl radicals in air (Atkinson, R (1987A)).

## Reduction half-life:　　　　　　High:　　No data
　　　　　　　　　　　　　　　　Low:

*Comment:*

## Hydrolysis:
**· First-order hydr half-life:**　　　2822 hours (118 days)
*Comment:* ($t_{1/2}$ at pH 5-7 and 25°C) Based upon measured neutral and base catalyzed hydrolysis rate constants (Ellington, JJ et al. (1987)).

**· Acid rate const $(M(H+)-hr)^{-1}$:**　　No data
*Comment:*

**· Base rate const $(M(OH-)-hr)^{-1}$:**　　756
*Comment:* ($t_{1/2}$ = 90.8 hours (3.78 days) at pH 9 and 25°C) Based upon measured neutral and base catalyzed hydrolysis rate constants (Ellington, JJ et al. (1987)).

# Dieldrin

CAS Registry Number: 60-57-1

**Structure:**

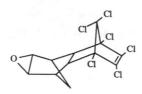

**Half-lives:**
**· Soil:**

High: 25920 hours (3 years)
Low: 4200 hours (175 days)

*Comment:* Scientific judgement based upon unacclimated aerobic soil grab sample data (low $t_{1/2}$: Castro and TF, Yoshida, T (1971)) and reported half-life in soil based on field data (high $t_{1/2}$: Kearney, PC et al. (1969C)).

**· Air:**

High: 40.5 hours (1.7 days)
Low: 4 hours

*Comment:* Scientific judgement based upon estimated photooxidation half-life in air.

**· Surface Water:**

High: 25920 hours (3 years)
Low: 4200 hours (175 days)

*Comment:* Scientific judgement based upon estimated aqueous aerobic biodegradation half-life.

**· Ground Water:**

High: 51840 hours (6 years)
Low: 24 hours (1 day)

*Comment:* Scientific judgement based upon estimated aqueous anaerobic and aerobic biodegradation half-lives.

**Aqueous Biodegradation (unacclimated):**
**· Aerobic half-life:**

High: 25920 hours (3 years)
Low: 4200 hours (175 days)

*Comment:* Scientific judgement based upon unacclimated aerobic soil grab sample data (low $t_{1/2}$: Castro, TF and Yoshida, T (1971)) and reported half-life in soil based on field data (high $t_{1/2}$: Kearney, PC et al. (1969C)).

**· Anaerobic half-life:**

High: 168 hours (7 days)
Low: 24 hours (1 day)

*Comment:* Scientific judgement based upon soil and freshwater mud grab sample data (Maule, A et al. (1987)).

· **Removal/secondary treatment:**     High:     No data

                                                  Low:

*Comment:*

## Photolysis:
· **Atmos photol half-life:**     High:     No data

                                                 Low:

*Comment:*

· **Max light absorption (nm):**     lambda max approximately 218 nm
*Comment:* In hexane, dieldrin does not absorb UV light at wavelengths greater than 290 nm (Gore, RC et al. (1971)).

· **Aq photol half-life:**     High:     No data

                                                 Low:

*Comment:*

## Photooxidation half-life:
· **Water:**     High:     No data

                                                 Low:

*Comment:*

· **Air:**     High:     40.5 hours     (1.7 days)

                                                 Low:     4 hours
*Comment:* Scientific judgement based upon an estimated rate constant for vapor phase reaction with hydroxyl radicals in air (Atkinson, R (1987A)).

## Reduction half-life:
                                                 High:     No data

                                                 Low:

*Comment:*

## Hydrolysis:
· **First-order hydr half-life:**     10.5 years
*Comment:* Based upon a first order rate constant ($7.5 \times 10^{-6}$ hr$^{-1}$) at pH 7 and 25 °C (Ellington, JJ et al. (1987A)).

· **Acid rate const (M(H+)-hr)$^{-1}$:**
*Comment:*

· **Base rate const (M(OH-)-hr)$^{-1}$:**
*Comment:*

# Amitrole

CAS Registry Number: 61-82-5

Structure:

Half-lives:
 · Soil:                                     High:    4320 hours            (6 months)
                                             Low:      672 hours            (4 weeks)
   Comment: Scientific judgement based upon estimated aqueous aerobic biodegradation half-life.

 · Air:                                      High:       32 hours           (1.3 days)
                                             Low:        3.2 hours
   Comment: Scientific judgement based upon estimated photooxidation half-life in air.

 · Surface Water:                            High:    4320 hours            (6 months)
                                             Low:      672 hours            (4 weeks)
   Comment: Scientific judgement based upon estimated aqueous aerobic biodegradation half-life.

 · Ground Water:                             High:    8640 hours            (12 months)
                                             Low:     1344 hours            (8 weeks)
   Comment: Scientific judgement based upon estimated aqueous aerobic biodegradation half-life.

Aqueous Biodegradation (unacclimated):
 · Aerobic half-life:                        High:    4032 hours            (6 months)
                                             Low:      672 hours            (4 weeks)
   Comment: Scientific judgement based upon reported half-lives in soil and water (Freed, VH and
   Haque, R (1973); Reinert, KH and Rodgers, JH (1987)).

 · Anaerobic half-life:                      High:   16128 hours            (24 months)
                                             Low:     2688 hours            (16 weeks)
   Comment: Scientific judgement based upon estimated aqueous aerobic biodegradation half-life.

 · Removal/secondary treatment:              High:     No data
                                             Low:
   Comment:

Photolysis:
 · Atmos photol half-life:                   High:     No data
                                             Low:

70

*Comment:*

· **Max light absorption (nm):**         lambda max  is approximately 195 nm
*Comment:* In water, amitrole does not absorb UV light at wavelengths greater than 350 nm
(Gore, RC et al. (1971)).

· **Aq photol half-life:**        High:       No data
                                     Low:
*Comment:*

## Photooxidation half-life:
· **Water:**                 High:       No data
                                     Low:
*Comment:*

· **Air:**                    High:       32 hours                (1.3 days)
                                     Low:       3.2 hours
*Comment:* Scientific judgement based upon an estimated rate constant for vapor phase reaction
with hydroxyl radicals in air (Atkinson, R (1987A)).

## Reduction half-life:                   High:       No data
                                     Low:
*Comment:*

## Hydrolysis:
· **First-order hydr half-life:**        Not expected to be  important
*Comment:*

· **Acid rate const (M(H+)-hr)$^{-1}$:**
*Comment:*

· **Base rate const (M(OH-)-hr)$^{-1}$:**
*Comment:*

# Phenacetin

<u>CAS Registry Number:</u> 62-44-2

<u>Structure:</u>

<u>Half-lives:</u>

· **Soil:**
|       |       |           |
|-------|-------|-----------|
| High: | 672 hours | (4 weeks) |
| Low:  | 168 hours | (7 days)  |

*Comment:* Scientific judgement based upon unacclimated aqueous aerobic biodegradation half-life.

· **Air:**
|       |       |
|-------|-------|
| High: | 8.30 hours |
| Low:  | 0.830 hours |

*Comment:* Scientific judgement based upon estimated photooxidation half-life in air.

· **Surface Water:**
|       |       |           |
|-------|-------|-----------|
| High: | 672 hours | (4 weeks) |
| Low:  | 168 hours | (7 days)  |

*Comment:* Scientific judgement based upon unacclimated aqueous aerobic biodegradation half-life.

· **Ground Water:**
|       |       |           |
|-------|-------|-----------|
| High: | 1344 hours | (8 weeks)  |
| Low:  | 336 hours  | (14 days)  |

*Comment:* Scientific judgement based upon unacclimated aqueous aerobic biodegradation half-life.

<u>Aqueous Biodegradation (unacclimated):</u>

· **Aerobic half-life:**
|       |       |           |
|-------|-------|-----------|
| High: | 672 hours | (4 weeks) |
| Low:  | 168 hours | (7 days)  |

*Comment:* Scientific judgement.

· **Anaerobic half-life:**
|       |       |           |
|-------|-------|-----------|
| High: | 2688 hours | (16 weeks) |
| Low:  | 672 hours  | (28 days)  |

*Comment:* Scientific judgement based upon unacclimated aqueous aerobic biodegradation half-life.

· **Removal/secondary treatment:**
|       |       |
|-------|-------|
| High: | No data |
| Low:  | |

*Comment:*

## Photolysis:
· **Atmos photol half-life:**        High:
                                           Low:

*Comment:*

· **Max light absorption (nm):**    No data
*Comment:*

· **Aq photol half-life:**         High:
                                           Low:

*Comment:*

## Photooxidation half-life:
· **Water:**                     High:        No data
                                           Low:

*Comment:*

· **Air:**                        High:        8.30 hours
                                         Low:       0.830 hours
*Comment:* Scientific judgement based upon estimated rate constant for reaction with hydroxyl radical in air (Atkinson, R (1987A)).

## Reduction half-life:
                                        High:        No data
                                         Low:
*Comment:*

## Hydrolysis:
· **First-order hydr half-life:**
*Comment:*

· **Acid rate const (M(H+)-hr)$^{-1}$:**    No data
*Comment:*

· **Base rate const (M(OH-)-hr)$^{-1}$:**
*Comment:*

# Ethyl methanesulfonate

<u>CAS Registry Number:</u>  62-50-0

<u>Half-lives:</u>
  · **Soil:**                                                       High:      77 hours                 (3.2 days)
                                                                         Low:       46 hours                 (1.9 days)
  *Comment:*  Scientific judgement based upon hydrolysis half-life in water.


  · **Air:**                                                        High:      74.7 hours              (3.1 days)
                                                                         Low:       7.47 hours
  *Comment:*  Scientific judgement based upon estimated photooxidation half-life in air.


  · **Surface Water:**                               High:      77 hours                 (3.2 days)
                                                                         Low:       46 hours                 (1.9 days)
  *Comment:*  Scientific judgement based upon hydrolysis half-life in water.


  · **Ground Water:**                                  High:      77 hours                 (3.2 days)
                                                                         Low:       46 hours                 (1.9 days)
  *Comment:*  Scientific judgement based upon hydrolysis half-life in water.

<u>Aqueous Biodegradation (unacclimated):</u>
  · **Aerobic half-life:**                           High:      672 hours               (4 weeks)
                                                                         Low:       168 hours               (7 days)
  *Comment:*  Scientific judgement.


  · **Anaerobic half-life:**                        High:      2688 hours             (16 weeks)
                                                                         Low:       672 hours               (28 days)
  *Comment:*  Scientific judgement based upon unacclimated aqueous aerobic biodegradation half-life.


  · **Removal/secondary treatment:**     High:          No data
                                                                         Low:
  *Comment:*


<u>Photolysis:</u>
  · **Atmos photol half-life:**                  High:
                                                                         Low:
  *Comment:*


  · **Max light absorption (nm):**          No data
  *Comment:*

**· Aq photol half-life:**   High:
                             Low:

*Comment:*

## Photooxidation half-life:
**· Water:**                 High:     No data
                             Low:

*Comment:*

**· Air:**                   High:     74.7 hours          (3.1 days)
                             Low:      7.47 hours
*Comment:* Scientific judgement based upon estimated rate constant for reaction of the sulfur-deoxy analog of ethyl methanesulfonate with hydroxyl radical in air (Atkinson, R (1987A)).

## Reduction half-life:       High:     No data
                             Low:

*Comment:*

## Hydrolysis:
**· First-order hydr half-life:**   High:     77 hours          (3.2 days)
                                    Low:      46 hours          (1.9 days)
*Comment:* ($t_{1/2}$ at pH 7 and 25°C (low $t_{1/2}$) and 20°C (high $t_{1/2}$)). Based upon measured neutral hydrolysis rate constants (high $t_{1/2}$: Ehrenberg, L et al. (1974); low $t_{1/2}$: Ellington, JJ et al. (1987)).

**· Acid rate const $(M(H+)-hr)^{-1}$:**     No data
*Comment:*

**· Base rate const $(M(OH-)-hr)^{-1}$:**    No data
*Comment:*

# Thioacetamide

CAS Registry Number: 62-55-5

Half-lives:
· Soil:                              High:      168 hours            (7 days)
                                     Low:        24 hours            (1 day)
Comment: Scientific judgement based upon estimated aqueous aerobic biodegradation half-life.

· Air:                               High:      31.7 hours
                                     Low:        3.2 hours
Comment: Scientific judgement based upon estimated photooxidation half-life in air.

· Surface Water:                     High:      168 hours            (7 days)
                                     Low:        24 hours            (1 day)
Comment: Scientific judgement based upon estimated aqueous aerobic biodegradation half-life.

· Ground Water:                      High:      336 hours            (14 days)
                                     Low:        48 hours            (2 days)
Comment: Scientific judgement based upon estimated unacclimated aqueous aerobic biodegradation half-life.

Aqueous Biodegradation (unacclimated):
· Aerobic half-life:                 High:      168 hours            (7 days)
                                     Low:        24 hours            (1 day)
Comment: Scientific judgement based upon aqueous aerobic screening test data for acetamide (Malaney, GW and Gerhold, RW (1962, 1969), Urano, K and Kato, Z (1986)).

· Anaerobic half-life:               High:      672 hours            (28 days)
                                     Low:        96 hours            (4 days)
Comment: Scientific judgement based upon estimated aqueous aerobic biodegradation half-life.

· Removal/secondary treatment:       High:      No data
                                     Low:
Comment:

Photolysis:
· Atmos photol half-life:            High:
                                     Low:
Comment:

· Max light absorption (nm):         No data
Comment:

76

**· Aq photol half-life:**　　　　　　High:
　　　　　　　　　　　　　　　　　　　Low:

*Comment:*

## Photooxidation half-life:
　**· Water:**　　　　　　　　　　　High:　　　No data
　　　　　　　　　　　　　　　　　　Low:

*Comment:*

　**· Air:**　　　　　　　　　　　　High:　　　31.7 hours
　　　　　　　　　　　　　　　　　　Low:　　　3.2 hours
*Comment:* Scientific judgement based upon estimated rate data for hydroxyl radicals in air
(Atkinson, R (1987A)).

## Reduction half-life:　　　　　　　High:　　　No data
　　　　　　　　　　　　　　　　　　Low:

*Comment:*

## Hydrolysis:
　**· First-order hydr half-life:**　　　8064 hours　(336 days)
　*Comment:* Based upon a neutral rate constant ($k_N$ = 8.6X10$^{-5}$ hr$^{-1}$) at pH 7 and 25 °C (Ellington,
JJ et al. (1987)).

　**· Acid rate const (M(H+)-hr)$^{-1}$:**　　　6.0X10$^{-2}$ M$^{-1}$ hr$^{-1}$
*Comment:* ($t_{1/2}$ = 333 days) Based upon a neutral ($k_N$ = 8.6X10$^{-5}$ hr$^{-1}$) and acid rate constants
(6.0X10$^{-2}$ M$^{-1}$ hr$^{-1}$) at pH 5 and 25 °C (Ellington, JJ et al. (1987)).

　**· Base rate const (M(OH-)-hr)$^{-1}$:**　　　1.4 M$^{-1}$ hr$^{-1}$
*Comment:* ($t_{1/2}$ = 289 days) Based upon neutral ($k_N$ = 8.6X10$^{-5}$ hr$^{-1}$) and base rate constants (1.4
M$^{-1}$ hr$^{-1}$) at pH 9 and 25 °C (Ellington, JJ et al. (1987)).

# Thiourea

<u>CAS Registry Number:</u>  62-56-6

<u>Half-lives:</u>
   · **Soil:**                                      High:      168 hours              (7 days)
                                                    Low:        24 hours              (1 day)
   *Comment:*  Scientific judgement based upon estimated aqueous aerobic biodegradation half-life.


   · **Air:**                                       High:       16 hours
                                                    Low:        1.6 hours
   *Comment:*  Scientific judgement based upon estimated photooxidation half-life in air.


   · **Surface Water:**                             High:      168 hours              (7 days)
                                                    Low:        24 hours              (1 day)
   *Comment:*  Scientific judgement based upon estimated aqueous aerobic biodegradation half-life.


   · **Ground Water:**                              High:      336 hours              (14 days)
                                                    Low:        48 hours              (2 days)
   *Comment:*  Scientific judgement based upon estimated unacclimated aqueous aerobic biodegradation half-life.

<u>Aqueous Biodegradation (unacclimated):</u>
   · **Aerobic half-life:**                         High:      168 hours              (7 days)
                                                    Low:        24 hours              (1 day)
   *Comment:*  Scientific judgement based upon aqueous aerobic screening test data (Freitag, D et al. (1985), Kolyada, TI (1969)).


   · **Anaerobic half-life:**                       High:      672 hours              (28 days)
                                                    Low:        96 hours              (4 days)
   *Comment:*  Scientific judgement based upon estimated aqueous aerobic biodegradation half-life.


   · **Removal/secondary treatment:**               High:      No data
                                                    Low:
   *Comment:*


<u>Photolysis:</u>
   · **Atmos photol half-life:**                    High:      No data
                                                    Low:
   *Comment:*


   · **Max light absorption (nm):**          lambda max = 241
   *Comment:*  No absorbance occurs above 310 nm in methanol (Sadtler UV No. 3292).

**· Aq photol half-life:**  High: No data
                            Low:

*Comment:*

## Photooxidation half-life:
**· Water:**  High: 81927 hours          (9.3 years)
             Low:  2048 hours           (85.3 days)
*Comment:* Based upon measured rate data for hydroxyl radicals in aqueous solution (Anbar, M = Neta, P (1967)).

**· Air:**  High: 16 hours
           Low:  1.6 hours
*Comment:* Scientific judgement based upon an estimated rate constant for the vapor phase reaction with hydroxyl radicals in air (Atkinson, R (1987A)).

## Reduction half-life:  High: No data
                         Low:

*Comment:*

## Hydrolysis:
**· First-order hydr half-life:**      No hydrolyzable groups
*Comment:* Stable under hydrolysis conditions for 6, 10, and 6 days at pHs of 3, 7, and 11, respectively, and 70 °C (Ellington, JJ et al. (1987A)).

**· Acid rate const $(M(H+)\text{-}hr)^{-1}$:**
*Comment:*

**· Base rate const $(M(OH-)\text{-}hr)^{-1}$:**
*Comment:*

# N-Nitrosodimethylamine

<u>CAS Registry Number:</u>  62-75-9

<u>Half-lives:</u>
  · **Soil:**                              High:     4320 hours              (6 months)
                                          Low:      504 hours               (21 days)
  *Comment:* Based upon aerobic soil die-away data (high $t_{1/2}$: Tate, RL III and Alexander, M
  (1975); low $t_{1/2}$: Oliver, JE et al. (1979)).

  · **Air:**                               High:     1 hour
                                          Low:      0.5 hours
  *Comment:* Based upon measured rate of photolysis in the vapor phase under sunlight (Hanst, PL
  et al. (1977)).

  · **Surface Water:**                     High:     1 hour
                                          Low:      0.5 hours
  *Comment:* Based upon measured rate of photolysis in the vapor phase under sunlight (Hanst, PL
  et al. (1977)).

  · **Ground Water:**                      High:     8640 hours              (12 months)
                                          Low:      1008 hours              (42 days)
  *Comment:* Scientific judgement based upon estimated unacclimated aqueous aerobic
  biodegradation half-lives.

<u>Aqueous Biodegradation (unacclimated):</u>
  · **Aerobic half-life:**                 High:     4320 hours              (6 months)
                                          Low:      504 hours               (21 days)
  *Comment:* Based upon aerobic soil die-away data (high $t_{1/2}$: Tate, RL III and Alexander, M
  (1975); low $t_{1/2}$: Oliver, JE et al. (1979)).

  · **Anaerobic half-life:**               High:     17280 hours             (24 months)
                                          Low:      2016 hours              (84 days)
  *Comment:* Scientific judgement based upon estimated unacclimated aqueous aerobic
  biodegradation half-lives.

  · **Removal/secondary treatment:**       High:     100%
                                          Low:      71.4%
  *Comment:* Removal percentages based upon data from a continuous sewage biological treatment
  simulator (Fochtman, EG and Eisenberg, W (1979)).

<u>Photolysis:</u>
  · **Atmos photol half-life:**            High:     1 hour

Low:      0.5 hours

*Comment:* Based upon measured rate of photolysis in the vapor phase under sunlight (Hanst, PL et al. (1977)).

· **Max light absorption (nm):**      lambda max = 230 and 332 nm (water)
*Comment:* IARC (1978).

· **Aq photol half-life:**      High:      1 hour
                                Low:      0.5 hours
*Comment:* Based upon measured rate of photolysis in the vapor phase under sunlight (Hanst, PL et al. (1977)).

## Photooxidation half-life:

· **Water:**      High:      No data
                  Low:

*Comment:*

· **Air:**      High:      254 hours            (10.6 days)
                Low:      25.4 hours
*Comment:* Scientific judgement based upon measured rate constant for reaction with hydroxyl radical in air (Atkinson, R (1985)).

## Reduction half-life:      High:      No data
                            Low:

*Comment:*

## Hydrolysis:

· **First-order hydr half-life:**      Not expected to be significant
*Comment:* IARC (1978).

· **Acid rate const $(M(H+)-hr)^{-1}$:**
*Comment:*

· **Base rate const $(M(OH-)-hr)^{-1}$:**
*Comment:*

# Carbaryl

**CAS Registry Number:** 63-25-2

**Structure:**

**Half-lives:**

- **Soil:**　　　　　　　　　　　　　　High:　　720 hours　　　　　(30 days)
　　　　　　　　　　　　　　　　　　　Low:　　 3.2 hours

  *Comment:* Scientific judgement based upon aqueous hydrolysis half-life for pH 9 and 28 °C (low $t_{1/2}$: Wolfe, NL et al. (1976)) and unacclimated aerobic biodegradation half-life.

- **Air:**　　　　　　　　　　　　　　 High:　　7.4 hours
　　　　　　　　　　　　　　　　　　　Low:　　4.5 minutes

  *Comment:* Scientific judgement based upon estimated photooxidation half-life in air.

- **Surface Water:**　　　　　　　　　 High:　　200 hours　　　　　(8.3 days)
　　　　　　　　　　　　　　　　　　　Low:　　 3.2 hours

  *Comment:* Scientific judgement based upon aqueous hydrolysis half-life for pH 9 and 28 °C (low $t_{1/2}$) and photolysis half-life for winter sunlight at 40 ° N (Wolfe, NL et al. (1976)).

- **Ground Water:**　　　　　　　　　 High:　　1440 hours　　　　(60 days)
　　　　　　　　　　　　　　　　　　　Low:　　 3.2 hours

  *Comment:* Scientific judgement based upon aqueous hydrolysis half-life for pH 9 and 28 °C (low $t_{1/2}$: Wolfe, NL et al. (1976)) and unacclimated aerobic biodegradation half-life.

**Aqueous Biodegradation (unacclimated):**

- **Aerobic half-life:**　　　　　　　 High:　　720 hours　　　　　(30 days)
　　　　　　　　　　　　　　　　　　　Low:　　 40 hours　　　　　 (1.7 days)

  *Comment:* Scientific judgement based upon unacclimated aerobic river die-away test data (low $t_{1/2}$: Eichelberger, JW and Lichtenberg, JJ (1971)) and freshwater grab sample data (high $t_{1/2}$: Wolfe, NL et al. (1978)).

- **Anaerobic half-life:**　　　　　　 High:　　2880 hours　　　　(120 days)
　　　　　　　　　　　　　　　　　　　Low:　　 160 hours　　　　 (6.7 days)

  *Comment:* Scientific judgement based upon unacclimated aerobic biodegradation half-life.

- **Removal/secondary treatment:**　　High:　　No data
　　　　　　　　　　　　　　　　　　　Low:

*Comment:*

## Photolysis:
  · **Atmos photol half-life:**         High:     200 hours         (8.3 days)
                                            Low:      52 hours
*Comment:* Scientific judgement based upon aqueous photolysis data (Wolfe, NL et al. (1976)).

  · **Max light absorption (nm):**      lambda max = 312.5, 280.0, 270.5, 221.5
*Comment:* No absorbance of UV light at wavelengths >340 nm in methanol (Sadtler UV No. 20579).

  · **Aq photol half-life:**            High:     200 hours         (8.3 days)
                                            Low:      52 hours
*Comment:* Based upon reported photolysis half-life for summer and winter sunlight at 40 ° N (Wolfe, NL et al. (1976)).

## Photooxidation half-life:
  · **Water:**                          High:     No data
                                            Low:
*Comment:*

  · **Air:**                              High:     7.4 hours
                                            Low:     4.5 minutes
*Comment:* Scientific judgement based upon an estimated rate constant for the vapor phase reaction with hydroxyl radicals in air (Atkinson, R (1987A)).

## Reduction half-life:                       High:     No data
                                            Low:
*Comment:*

## Hydrolysis:
  · **First-order hydr half-life:**      312 hours   (13 days)
*Comment:* Scientific judgement based upon base rate constant ($3.61 \times 10$ $M^{-1}$ $min^{-1}$) at pH 7 and 25 °C (Wolfe, NL et al. (1976)).

  · **First-order hydr half-life:**      31999 hours   (3.6 years)
*Comment:* Scientific judgement based upon base rate constant ($3.61 \times 10$ $M^{-1}$ $min^{-1}$) at pH 5 and 25 °C (Wolfe, NL et al. (1976)).

  · **Acid rate const $(M(H+)-hr)^{-1}$:**
*Comment:*

  · **Base rate const $(M(OH-)-hr)^{-1}$:**     $3.61 \times 10^2$ $M^{-1}$ $min^{-1}$
*Comment:* ($t_{1/2}$ = 3.2 hours) Based upon rate constant at pH 9 and 27 °C (Wolfe, NL et al.

(1976)).

# Ethanol

**CAS Registry Number:**  64-17-5

**Half-lives:**
- **Soil:**                          High:      24 hours
                                    Low:      2.6 hours

  *Comment:* Based upon soil die-away test data (Griebel, GE and Owens, LD (1972)).

- **Air:**                           High:      122 hours                (5.1 days)
                                    Low:      12.2 hours

  *Comment:* Based upon photooxidation half-life in air.

- **Surface Water:**            High:      26 hours
                                    Low:      6.5 hours

  *Comment:* Scientific judgement based upon estimated unacclimated aqueous aerobic biodegradation half-life.

- **Ground Water:**             High:      52 hours                (2.2 days)
                                    Low:      13 hours

  *Comment:* Scientific judgement based upon estimated unacclimated aqueous aerobic biodegradation half-life.

**Aqueous Biodegradation (unacclimated):**
- **Aerobic half-life:**         High:      26 hours
                                    Low:      6.5 hours

  *Comment:* Based upon river die-away test data for one sample of water from one river (Apoteker, A and Thevenot, DR (1983)).

- **Anaerobic half-life:**       High:      104 hours            (4.3 days)
                                    Low:      26 hours

  *Comment:* Scientific judgement based upon estimated unacclimated aqueous aerobic biodegradation half-life.

- **Removal/secondary treatment:**      High:      67%
                                    Low:

  *Comment:* Based upon biological oxygen demand results from an activated sludge dispersed seed aeration treatment simulator (Mills, EJ Jr and Stack, VT Jr (1954)).

**Photolysis:**
- **Atmos photol half-life:**      High:
                                    Low:

  *Comment:*

· **Max light absorption (nm):**       No data
*Comment:*

· **Aq photol half-life:**       High:
                                                    Low:

*Comment:*

**Photooxidation half-life:**
   · **Water:**       High: $3.2 \times 10^5$ hours      (36.6 years)
                                             Low:     8020 hours        (334 days)
*Comment:* Based upon measured rate constant for reaction with hydroxyl radical in water (Anbar, M and Neta, P (1967)).

   · **Air:**       High:     122 hours        (5.1 days)
                                             Low:     12.2 hours
*Comment:* Based upon measured rate constant for reaction with hydroxyl radical in air (Atkinson, R (1987)).

**Reduction half-life:**       High:     No data
                                                   Low:
*Comment:*

**Hydrolysis:**
   · **First-order hydr half-life:**
*Comment:*

   · **Acid rate const (M(H+)-hr)$^{-1}$:**       No hydrolyzable groups
*Comment:*

   · **Base rate const (M(OH-)-hr)$^{-1}$:**
*Comment:*

# Formic acid

<u>CAS Registry Number:</u>  64-18-6

<u>Half-lives:</u>
  · **Soil:**                                    High:        168 hours                  (7 days)
                                                         Low:          24 hours                   (1 day)
  *Comment:*  Scientific judgement based upon unacclimated aqueous aerobic biodegradation half-
  life.

  · **Air:**                                      High:        1331 hours                (55.5 days)
                                                         Low:          133 hours                 (5.55 days)
  *Comment:*  Based upon photooxidation half-life in air.

  · **Surface Water:**                     High:        168 hours                  (7 days)
                                                         Low:          24 hours                   (1 day)
  *Comment:*  Scientific judgement based upon unacclimated aqueous aerobic biodegradation half-
  life.

  · **Ground Water:**                      High:        336 hours                  (14 days)
                                                         Low:          48 hours                   (2 days)
  *Comment:*  Scientific judgement based upon unacclimated aqueous aerobic biodegradation half-
  life.

<u>Aqueous Biodegradation (unacclimated):</u>
  · **Aerobic half-life:**                   High:        168 hours                  (7 days)
                                                         Low:          24 hours                   (1 day)
  *Comment:*  Scientific judgement based upon unacclimated aqueous aerobic biodegradation
  screening test data (Heukelekian, H and Rand, MC (1955); Malaney, GW and Gerhold, RM
  (1969)).

  · **Anaerobic half-life:**                High:        672 hours                  (28 days)
                                                         Low:          96 hours                   (4 days)
  *Comment:*  Scientific judgement based upon unacclimated aqueous aerobic biodegradation half-
  life.

  · **Removal/secondary treatment:**     High:        No data
                                                                 Low:
  *Comment:*

<u>Photolysis:</u>
  · **Atmos photol half-life:**            High:
                                                         Low:

*Comment:*

· **Max light absorption (nm):**      No data
*Comment:*

· **Aq photol half-life:**      High:
                                        Low:
*Comment:*

## Photooxidation half-life:
· **Water:**      High: $1.4 \times 10^5$ hours      (15.7 years)
                       Low:     3438 hours          (143 days)
*Comment:* Based upon measured rate constant for reaction with hydroxyl radical in water (Dorfman, LM and Adams, GE (1973)).

· **Air:**      High:    1331 hours      (55.5 days)
                     Low:     133 hours       (5.55 days)
*Comment:* Based upon measured rate constant for reaction with hydroxyl radical in air (Atkinson, R (1985)).

## Reduction half-life:
                                    High:      No data
                                    Low:
*Comment:*

## Hydrolysis:
· **First-order hydr half-life:**
*Comment:*

· **Acid rate const $(M(H+)\text{-}hr)^{-1}$:**      No hydrolyzable groups
*Comment:*

· **Base rate const $(M(OH\text{-})\text{-}hr)^{-1}$:**
*Comment:*

# Diethyl sulfate

CAS Registry Number: 64-67-5

## Half-lives:
· **Soil:**                           High:      12 hours
                                      Low:       1.7 hours

*Comment:* Low $t_{1/2}$ based upon measured overall hydrolysis rate constant for diethyl sulfate (Robertson, RE and Sugamori, SE (1966)); high $t_{1/2}$ based upon observed maximum time for complete hydrolysis of dimethyl sulfate in neutral and slightly basic and acidic aqueous solutions (Lee, ML et al. (1980)).

· **Air:**                            High:      36 hours
                                      Low:       3.6 hours

*Comment:* Based upon estimated photooxidation half-life in air.

· **Surface Water:**                  High:      12 hours
                                      Low:       1.7 hours

*Comment:* Low $t_{1/2}$ based upon measured overall hydrolysis rate constant for diethyl sulfate (Robertson, RE and Sugamori, SE (1966)); high $t_{1/2}$ based upon observed maximum time for complete hydrolysis of dimethyl sulfate in neutral and slightly basic and acidic aqueous solutions (Lee, ML et al. (1980)).

· **Ground Water:**                   High:      12 hours
                                      Low:       1.7 hours

*Comment:* Low $t_{1/2}$ based upon measured overall hydrolysis rate constant for diethyl sulfate (Robertson, RE and Sugamori, SE (1966)); high $t_{1/2}$ based upon observed maximum time for complete hydrolysis of dimethyl sulfate in neutral and slightly basic and acidic aqueous solutions (Lee, ML et al. (1980)).

## Aqueous Biodegradation (unacclimated):
· **Aerobic half-life:**              High:      672 hours       (4 weeks)
                                      Low:       168 hours       (7 days)

*Comment:* Scientific judgement.

· **Anaerobic half-life:**            High:      2688 hours      (16 weeks)
                                      Low:       672 hours       (28 days)

*Comment:* Scientific judgement.

· **Removal/secondary treatment:**    High:      No data
                                      Low:

*Comment:*

## Photolysis:
- **Atmos photol half-life:**           High:     No data
                                              Low:

  *Comment:*

- **Max light absorption (nm):**     No data
  *Comment:*

- **Aq photol half-life:**             High:     No data
                                              Low:

  *Comment:*

## Photooxidation half-life:
- **Water:**                       High:  Not expected to be significant
                                              Low:

  *Comment:*

- **Air:**                           High:     36 hours
                                              Low:      3.6 hours

  *Comment:* Scientific judgement based upon estimated rate constant for reaction with hydroxyl radicals in air (Atkinson, R (1987A)).

## Reduction half-life:
                                            High:     No data
                                            Low:

    *Comment:*

## Hydrolysis:
- **First-order hydr half-life:**     1.7 hours
  *Comment:* ($t_{1/2}$ at pH 7) Based on measured overall rate constant for hydrolysis in water (Robertson, RE and Sugamori, SE (1966)).

- **Acid rate const $(M(H+)\text{-}hr)^{-1}$:**     No data
  *Comment:*

- **Base rate const $(M(OH\text{-})\text{-}hr)^{-1}$:**     No data
  *Comment:*

# Uracil mustard

**CAS Registry Number:** 66-75-1

**Structure:**

**Half-lives:**
- **Soil:**                    High:     24 hours
                               Low:      30 minutes
  *Comment:* Based upon aqueous hydrolysis half-lives.

- **Air:**                     High:     2.9 hours
                               Low:      0.29 hours          (18 minutes)
  *Comment:* Scientific judgement based upon estimated photooxidation half-life in air.

- **Surface Water:**           High:     24 hours
                               Low:      30 minutes
  *Comment:* Based upon aqueous hydrolysis half-lives.

- **Ground Water:**            High:     24 hours
                               Low:      30 minutes
  *Comment:* Based upon aqueous hydrolysis half-lives.

**Aqueous Biodegradation (unacclimated):**
- **Aerobic half-life:**       High:     672 hours         (4 weeks)
                               Low:      168 hours         (1 week)
  *Comment:* Scientific judgement.

- **Anaerobic half-life:**     High:     17280 hours       (24 months)
                               Low:      2688 hours        (16 weeks)
  *Comment:* Scientific judgement based upon estimated aqueous aerobic biodegradation half-lives.

- **Removal/secondary treatment:**   High:     No data
                                     Low:
  *Comment:*

**Photolysis:**
- **Atmos photol half-life:**  High:     No data
                               Low:

91

*Comment:*

**· Max light absorption (nm):**       lambda max = 257 nm
*Comment:* In ethanol. (IARC (1975A)).

  **· Aq photol half-life:**         High:      No data
                                        Low:
*Comment:*

**Photooxidation half-life:**
  **· Water:**                     High:      No data
                                        Low:
*Comment:*

  **· Air:**                         High:      2.9 hours
                                        Low:      0.29 hours         (18 minutes)
*Comment:* Scientific judgement based upon an estimated rate constant for the vapor phase reaction with hydroxyl radicals in air (Atkinson, R (1987A)).

**Reduction half-life:**               High:      No data
                                       Low:
*Comment:*

**Hydrolysis:**
                                High:      24 hours
                                Low:
*Comment:* Scientific judgement based upon reported experimental data on reaction kinetics of nitrogen mustard and similar homologs in dilute aqueous solutions at various temperatures (Cohen, B et al. (1948); Bartlett, PD et al. (1947)).

  **· Acid rate const (M(H+)-hr)$^{-1}$:**
*Comment:*

  **· Base rate const (M(OH-)-hr)$^{-1}$:**
*Comment:*

# Methanol

**CAS Registry Number:** 67-56-1

**Half-lives:**

· **Soil:**  High: 168 hours (7 days)

Low: 24 hours (1 day)

*Comment:* Scientific judgement based upon unacclimated grab sample of aerobic soil/water suspensions from ground water aquifers (low $t_{1/2}$: Scheunert, I et al.(1987); high $t_{1/2}$: Novak, JT et al. (1985)).

· **Air:**  High: 713 hours (29.7 days)

Low: 71 hours (3 days)

*Comment:* Based upon photooxidation half-life in air.

· **Surface Water:**  High: 168 hours (7 days)

Low: 24 hours (1 day)

*Comment:* Scientific judgement based upon estimated unacclimated aqueous aerobic biodegradation half-life.

· **Ground Water:**  High: 168 hours (7 days)

Low: 24 hours (1 day)

*Comment:* Scientific judgement based upon unacclimated grab sample of aerobic soil/water suspensions from ground water aquifers (low $t_{1/2}$: Scheunert, I et al.(1987); high $t_{1/2}$: Novak, JT et al. (1985)).

**Aqueous Biodegradation (unacclimated):**

· **Aerobic half-life:**  High: 168 hours (7 days)

Low: 24 hours (1 day)

*Comment:* Scientific judgement based upon unacclimated grab sample of aerobic soil/water suspensions from ground water aquifers (low $t_{1/2}$: Scheunert, I et al.(1987); high $t_{1/2}$: Novak, JT et al. (1985)).

· **Anaerobic half-life:**  High: 120 hours (5 days)

Low: 24 hours (1 day)

*Comment:* Scientific judgement based upon unacclimated grab sample of anaerobic marine water/sediment and soil/water suspensions (low $t_{1/2}$: Oremland, RS et al. (1982); high $t_{1/2}$: Scheunert, I et al.(1987)).

· **Removal/secondary treatment:**  High: 99%

Low: 86%

*Comment:* Based upon % degraded under acclimated aerobic semi-continuous flow conditions (high %: Hatfield, R (1957); low %: Swain, HM and Somerville, HJ (1978)).

<u>Photolysis:</u>
· **Atmos photol half-life:**  High:  No data
                             Low:
*Comment:*


· **Max light absorption (nm):**  No data
*Comment:*


· **Aq photol half-life:**  High: Not expected to be important
                          Low:
*Comment:* Stable under aqueous photolysis conditions (EPA test) (Hustert, K et al. (1981)).

<u>Photooxidation half-life:</u>
· **Water:**  High:  44774 hours      (5.1 years)
            Low:    1119 hours      (46.6 days)
*Comment:* Based upon measured rate data for hydroxyl radicals in aqueous solution (Dorfman, LM and Adams, GE (1973)).


· **Air:**  High:  713 hours      (29.7 days)
          Low:   71 hours       (3 days)
*Comment:* Based upon measured rate data for the vapor phase reaction with hydroxyl radicals in air (Atkinson, R (1985)).

<u>Reduction half-life:</u>  High:  No data
                         Low:
*Comment:*

<u>Hydrolysis:</u>
· **First-order hydr half-life:**  No hydrolyzable groups
*Comment:*

· **Acid rate const $(M(H+)-hr)^{-1}$:**
*Comment:*

· **Base rate const $(M(OH-)-hr)^{-1}$:**
*Comment:*

# Isopropanol

**CAS Registry Number:** 67-63-0

**Half-lives:**

| · **Soil:** | | High: | 168 hours | (7 days) |
| | | Low: | 24 hours | (1 day) |

*Comment:* Scientific judgement based upon estimated unacclimated aerobic aqueous biodegradation half-life.

| · **Air:** | | High: | 72 hours | (3 days) |
| | | Low: | 6.2 hours | (0.26 days) |

*Comment:* Based upon photooxidation half-life in air.

| · **Surface Water:** | | High: | 168 hours | (7 days) |
| | | Low: | 24 hours | (1 day) |

*Comment:* Scientific judgement based upon estimated unacclimated aerobic aqueous biodegradation half-life.

| · **Ground Water:** | | High: | 336 hours | (14 days) |
| | | Low: | 48 hours | (2 days) |

*Comment:* Scientific judgement based upon estimated unacclimated aerobic aqueous biodegradation half-life.

**Aqueous Biodegradation (unacclimated):**

| · **Aerobic half-life:** | | High: | 168 hours | (7 days) |
| | | Low: | 24 hours | (1 day) |

*Comment:* Scientific judgement based upon unacclimated aerobic aqueous screening test data (Gellman, I and Heukelekian, H (1955); Heukelekian, H and Rand, MC (1955); Price, KS et al. (1974); Takemoto, S et al. (1981); Wagner, R (1976)).

| · **Anaerobic half-life:** | | High: | 672 hours | (28 days) |
| | | Low: | 96 hours | (4 days) |

*Comment:* Scientific judgement based upon estimated aqueous aerobic biodegradation half-life and unacclimated anaerobic aqueous screening test data (Hou, CT et al. (1983A); Sonoda, Y and Seiko, Y (1968); Speece, RE (1983)).

| · **Removal/secondary treatment:** | | High: | No data |
| | | Low: | |

*Comment:*

**Photolysis:**

| · **Atmos photol half-life:** | | High: Not expected to be significant |

Low:

*Comment:*

· **Max light absorption (nm):**
*Comment:*

· **Aq photol half-life:**          High:
                                    Low:
*Comment:*

**Photooxidation half-life:**
    · **Water:**                    High: $1.9 \times 10^5$ hours          (22 years)
                                    Low:    4728 hours          (197 days)
*Comment:* Based upon measured rates for reaction with hydroxyl radicals in water (Dorfman, LM and Adams, GE (1973)).

    · **Air:**                      High:    72 hours          (3 days)
                                    Low:    6.2 hours          (0.26 days)
*Comment:* Based upon measured rates for reaction with hydroxyl radicals in air (Atkinson, R (1985); Atkinson, R (1987A)).

**Reduction half-life:**            High:    No data
                                    Low:
*Comment:*

**Hydrolysis:**
    · **First-order hydr half-life:**    No hydrolyzable groups
*Comment:*

    · **Acid rate const $(M(H+)-hr)^{-1}$:**
*Comment:*

    · **Base rate const $(M(OH-)-hr)^{-1}$:**
*Comment:*

# Acetone

<u>CAS Registry Number:</u>  67-64-1

## Half-lives:

| · Soil: | High: | 168 hours | (7 days) |
| | Low: | 24 hours | (1 day) |

*Comment:*  Scientific judgement based upon estimated unacclimated aqueous aerobic biodegradation half-life.

| · **Air:** | High: | 2790 hours | (116 days) |
| | Low: | 279 hours | (11.6 days) |

*Comment:*  Based upon photooxidation half-life in air.

| · **Surface Water:** | High: | 168 hours | (7 days) |
| | Low: | 24 hours | (1 day) |

*Comment:*  Scientific judgement based upon estimated unacclimated aqueous aerobic biodegradation half-life.

| · **Ground Water:** | High: | 336 hours | (14 days) |
| | Low: | 48 hours | (2 days) |

*Comment:*  Scientific judgement based upon estimated unacclimated aqueous aerobic biodegradation half-life.

## Aqueous Biodegradation (unacclimated):

| · **Aerobic half-life:** | High: | 168 hours | (7 days) |
| | Low: | 24 hours | (1 day) |

*Comment:*  Scientific judgement based upon unacclimated aqueous screening test data (Bridie, AL et al. (1979); Dore, M et al. (1975)).

| · **Anaerobic half-life:** | High: | 672 hours | (28 days) |
| | Low: | 96 hours | (4 days) |

*Comment:*  Scientific judgement based upon unacclimated aqueous aerobic biodegradation half-life.

| · **Removal/secondary treatment:** | High: | 75% |
| | Low: | 54% |

*Comment:*  Based upon % degraded under acclimated aerobic semi-continuous and continuous flow conditions (high %: Hatfield, R (1957); low %: Mills, EJ Jr and Stack, VT Jr (1954)).

## Photolysis:

| · **Atmos photol half-life:** | High: | No data |
| | Low: | |

*Comment:*

· **Max light absorption (nm):**     lambda max = 270
*Comment:* No absorbance occurs above 330 nm in methanol (Sadtler UV No. 89).

· **Aq photol half-life:**           High:     No data
                                     Low:
*Comment:*

**Photooxidation half-life:**
  · **Water:**                       High:3.97X10$^6$ hours          (453 years)
                                     Low: 9.92X10$^4$ hours          (11.3 years)
*Comment:* Based upon measured rate data for hydroxyl radicals in aqueous solution (Dorfman, LM and Adams, GE (1973)).

  · **Air:**                         High:    2790 hours             (116 days)
                                     Low:     279 hours              (11.6 days)
*Comment:* Based upon an measured rate data for the vapor phase reaction with hydroxyl radicals in air (Atkinson, R (1985)).

**Reduction half-life:**             High:     No data
                                     Low:
*Comment:*

**Hydrolysis:**
  · **First-order hydr half-life:**  No hydrolyzable groups
*Comment:*

  · **Acid rate const (M(H+)-hr)$^{-1}$:**
*Comment:*

  · **Base rate const (M(OH-)-hr)$^{-1}$:**
*Comment:*

# Chloroform

**CAS Registry Number:** 67-66-3

**Half-lives:**

· **Soil:**                          High:    4320 hours          (6 months)
                                      Low:      672 hours          (4 weeks)
*Comment:* Scientific judgement based upon estimated aqueous aerobic biodegradation half-life.

· **Air:**                           High:    6231 hours          (260 days)
                                      Low:      623 hours          (26 days)
*Comment:* Based upon photooxidation half-life in air.

· **Surface Water:**                 High:    4320 hours          (6 months)
                                      Low:      672 hours          (4 weeks)
*Comment:* Scientific judgement based upon estimated aqueous aerobic biodegradation half-life.

· **Ground Water:**                  High:   43200 hours          (5 years)
                                      Low:     1344 hours          (8 weeks)
*Comment:* Scientific judgement based upon estimated unacclimated aqueous aerobic biodegradation half-life and grab sample data of aerobic soil from a ground water aquifer (high $t_{1/2}$: Wilson, JT et al. (1983)).

**Aqueous Biodegradation (unacclimated):**

· **Aerobic half-life:**             High:    4320 hours          (6 months)
                                      Low:      672 hours          (4 weeks)
*Comment:* Scientific judgement based upon unacclimated aerobic screening test data (Kawasaki, M (1980); Flathman, PE and Dahlgran, JR (1982)).

· **Anaerobic half-life:**           High:     672 hours          (4 weeks)
                                      Low:      168 hours          (1 week)
*Comment:* Scientific judgement based upon unacclimated anaerobic screening test data (Bouwer, EJ and McCarty, PL (1983); Bouwer, EJ, et al. (1981)).

· **Removal/secondary treatment:**   High:         96%
                                      Low:
*Comment:* Based upon % degraded under aerobic continuous flow conditions (Bouwer, EJ and McCarty, PL (1983)).

**Photolysis:**

· **Atmos photol half-life:**        High:  Not expected to be important
                                      Low:
*Comment:* Scientific judgement based upon aqueous photolysis data.

· **Max light absorption (nm):**   Lambda max = 296.3, 275.8, 254.7, 220.9
*Comment:* Reported absorption maxima (Crutzen, PJ et al. (1978)).

· **Aq photol half-life:**    High: Not expected to be important
            Low:
*Comment:* Stable in water exposed to sunlight for one year (Dilling, WL et al.(1975)).

## Photooxidation half-life:
· **Water:**        High: $2.8 \times 10^7$ hours   (3140 years)
          Low: $6.9 \times 10^5$ hours   (78.5 years)
*Comment:* Based upon measured rate data for hydroxyl radicals in aqueous solution (Dorfman, LM and Adams, GE (1973)).

· **Air:**         High: 6231 hours   (260 days)
          Low:  623 hours   (26 days)
*Comment:* Based upon measured rate data for the vapor phase reaction with hydroxyl radicals in air (Atkinson, R (1985)).

## Reduction half-life:      High:  No data
               Low:
*Comment:*

## Hydrolysis:
· **First-order hydr half-life:**  3500 years
*Comment:* Scientific judgement based upon reported rate constant ($6.9 \times 10^{-12}$ s$^{-1}$) at pH 7 and 25 °C (Mabey, W and Mill, T (1978)).

· **Acid rate const (M(H+)-hr)$^{-1}$:**
*Comment:*

· **Base rate const (M(OH-)-hr)$^{-1}$:**
*Comment:*

# Hexachloroethane

**CAS Registry Number:** 67-72-1

**Half-lives:**

· **Soil:**            High:     4320 hours         (6 months)

Low:      672 hours         (4 weeks)

*Comment:* Scientific judgement based upon estimated aqueous aerobic biodegradation half-life.

· **Air:**            High: $>6X10^5$ hours         (>73 years)

Low: $>6X10^4$ hours         (>7.3 years)

*Comment:* Scientific judgement based upon an estimated maximum rate constant for reaction with hydroxyl radical (Singh, HB et al. (1980)).

· **Surface Water:**       High:     4320 hours         (6 months)

Low:      672 hours         (4 weeks)

*Comment:* Scientific judgement based upon estimated aqueous aerobic biodegradation half-life.

· **Ground Water:**       High:     8640 hours        (12 months)

Low:    1344 hours         (8 weeks)

*Comment:* Scientific judgement based upon estimated aqueous aerobic biodegradation half-life.

**Aqueous Biodegradation (unacclimated):**

· **Aerobic half-life:**      High:     4320 hours         (6 months)

Low:      672 hours         (4 weeks)

*Comment:* Scientific judgement based upon aqueous screening studies (Abrams, EF et al. (1975); Kawasaki, M (1980); Tabak, HH et al. (1981))

· **Anaerobic half-life:**    High:    17280 hours       (24 months)

Low:    2688 hours         (16 weeks)

*Comment:* Scientific judgement based upon estimated aqueous aerobic biodegradation half-life.

· **Removal/secondary treatment:**    High:       No data

Low:

*Comment:*

**Photolysis:**

· **Atmos photol half-life:**     High:       No data

Low:

*Comment:*

· **Max light absorption (nm):**     No data

*Comment:*

**· Aq photol half-life:**            High:     No data
                                         Low:

*Comment:*

## Photooxidation half-life:

**· Water:**                       High:  Not significant
                                         Low:

*Comment:* No abstractable hydrogens

**· Air:**                         High: $>6 \times 10^5$ hours        (>73 years)
                                         Low: $>6 \times 10^4$ hours        (>7.3 years)
*Comment:* Scientific judgement based upon an estimated maximum rate constant for reaction with hydroxyl radical (Singh, HB et al. (1980)).

## Reduction half-life:

                                         High:     No data
                                         Low:

*Comment:*

## Hydrolysis:

**· First-order hydr half-life:**       Not significant
*Comment:* No hydrolysis observed after 13 days at 85°C and pH 7 (Ellington, JJ et al. (1987)).

**· Acid rate const (M(H+)-hr)$^{-1}$:**       Not significant
*Comment:* No hydrolysis observed after 13 days at 85°C and pH 3 (Ellington, JJ et al. (1987)).

**· Base rate const (M(OH-)-hr)$^{-1}$:**       Not significant
*Comment:* No hydrolysis observed after 13 days at 85°C and pH 11 (Ellington, JJ et al. (1987)).

# Triaziquone

CAS Registry Number: 68-76-8

**Structure:**

**Half-lives:**

  · **Soil:**

| | High: | 744 hours | (31 days) |
|---|---|---|---|
| | Low: | 168 hours | (7 days) |

*Comment:* Scientific judgement based upon estimated aqueous hydrolysis half-life.

  · **Air:**

| | High: | 6.8 hours |
|---|---|---|
| | Low: | 0.68 hours |

*Comment:* Based upon estimated photooxidation half-life in air.

  · **Surface Water:**

| | High: | 744 hours | (31 days) |
|---|---|---|---|
| | Low: | 168 hours | (7 days) |

*Comment:* Scientific judgement based upon estimated aqueous hydrolysis half-life.

  · **Ground Water:**

| | High: | 744 hours | (31 days) |
|---|---|---|---|
| | Low: | 336 hours | (14 days) |

*Comment:* Scientific judgement based upon estimated aqueous hydrolysis half-life.

**Aqueous Biodegradation (unacclimated):**

  · **Aerobic half-life:**

| | High: | 672 hours | (4 weeks) |
|---|---|---|---|
| | Low: | 168 hours | (7 days) |

*Comment:* Scientific judgement based upon estimated biodegradation rates for the benzoquinone and aziridine chemical classes.

  · **Anaerobic half-life:**

| | High: | 2688 hours | (16 weeks) |
|---|---|---|---|
| | Low: | 672 hours | (28 days) |

*Comment:* Scientific judgement based upon estimated aerobic biodegradation half-life.

  · **Removal/secondary treatment:**

| | High: | No data |
|---|---|---|
| | Low: | |

*Comment:*

**Photolysis:**

**· Atmos photol half-life:**                     High:       No data
                                                   Low:

*Comment:*

**· Max light absorption (nm):**                  No data
*Comment:*

**· Aq photol half-life:**                        High:       No data
                                                   Low:

*Comment:*

**Photooxidation half-life:**
   **· Water:**                     High:       No data
                                                   Low:

*Comment:*

   **· Air:**                       High:       6.8 hours
                                                   Low:        0.68 hours
*Comment:* Based upon estimated rate constant for reaction with hydroxyl radicals in air (Atkinson, R (1987A)).

**Reduction half-life:**                          High:       No data
                                                   Low:

*Comment:*

**Hydrolysis:**
   **· First-order hydr half-life:**          372 hours     (15.5 days)
*Comment:* Scientific judgement based upon hydrolysis of aziridine at 25°C and neutral pH (Earley, JE et al. (1958)).

   **· Acid rate const (M(H+)-hr)$^{-1}$:**
*Comment:*

   **· Base rate const (M(OH-)-hr)$^{-1}$:**
*Comment:*

# N-Nitroso-N-methyl-N'-nitroguanidine

<u>CAS Registry Number:</u> 70-25-7

<u>Structure:</u>

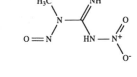

<u>Half-lives:</u>
  · **Soil:**                     High:        26 hours
                                  Low:         0.71 hours
  *Comment:* Based upon hydrolysis half-lives at pH 9 (low $t_{1/2}$) and pH 5 (high $t_{1/2}$) (Ellington, JJ et al. (1987)).

  · **Air:**                      High:        10.4 hours
                                  Low:         1.04 hours
  *Comment:* Scientific judgement based upon estimated photooxidation half-life in air.

  · **Surface Water:**            High:        26 hours
                                  Low:         0.71 hours
  *Comment:* Based upon hydrolysis half-lives at pH 9 (low $t_{1/2}$) and pH 5 (high $t_{1/2}$) (Ellington, JJ et al. (1987)).

  · **Ground Water:**             High:        26 hours
                                  Low:         0.71 hours
  *Comment:* Based upon hydrolysis half-lives at pH 9 (low $t_{1/2}$) and pH 5 (high $t_{1/2}$) (Ellington, JJ et al. (1987)).

<u>Aqueous Biodegradation (unacclimated):</u>
  · **Aerobic half-life:**        High:        672 hours          (4 weeks)
                                  Low:         168 hours          (7 days)
  *Comment:* Scientific judgement.

  · **Anaerobic half-life:**      High:        2688 hours         (16 weeks)
                                  Low:         672 hours          (28 days)
  *Comment:* Scientific judgement.

  · **Removal/secondary treatment:**   High:   No data
                                       Low:
  *Comment:*

<u>Photolysis:</u>

**· Atmos photol half-life:**　　　　　　　High:
　　　　　　　　　　　　　　　　　　　　　Low:
*Comment:*

**· Max light absorption (nm):**　　　　　No data
*Comment:*

**· Aq photol half-life:**　　　　　　　　High:
　　　　　　　　　　　　　　　　　　　　　Low:

*Comment:*

## Photooxidation half-life:
　　**· Water:**　　　　　　　　　　　　　High:　　　No data
　　　　　　　　　　　　　　　　　　　　　Low:

*Comment:*

　　**· Air:**　　　　　　　　　　　　　　High:　　　10.4 hours
　　　　　　　　　　　　　　　　　　　　　Low:　　　1.04 hours
*Comment:* Scientific judgement based upon estimated rate constant for reaction with hydroxyl radical in air (Atkinson, R (1987A)).

## Reduction half-life:　　　　　　　　　High:　　　No data
　　　　　　　　　　　　　　　　　　　　　Low:

*Comment:*

## Hydrolysis:
　　**· First-order hydr half-life:**　　　　19 hours
*Comment:* ($t_{1/2}$ at pH 7 and 25°C) Based upon measured acid and base catalyzed and neutral hydrolysis rate constants (Ellington, JJ et al. (1987)).

　　**· Acid rate const (M(H+)-hr)$^{-1}$:**　　4.9
*Comment:* ($t_{1/2}$ = 26 hours at pH 5 and 25°C) Based upon measured acid and base catalyzed and neutral hydrolysis rate constants (Ellington, JJ et al. (1987)).

　　**· Base rate const (M(OH-)-hr)$^{-1}$:**　　$9.5 \times 10^{-4}$
*Comment:* ($t_{1/2}$ = 0.71 hours at pH 9 and 25°C) Based upon measured acid and base catalyzed and neutral hydrolysis rate constants (Ellington, JJ et al. (1987)).

# Hexachlorophene

**CAS Registry Number:** 70-30-4

**Structure:**

**Half-lives:**

· **Soil:**
High: 7872 hours      (10.9 months)
Low: 6000 hours      (8.33 months)
*Comment:* Scientific judgement based upon unacclimated aqueous aerobic biodegradation half-life.

· **Air:**
High: 336 hours      (14 days)
Low: 3.36 hours      (1.4 days)
*Comment:* Scientific judgement based upon estimated photooxidation half-life in air.

· **Surface Water:**
High: 7872 hours      (10.9 months)
Low: 6000 hours.      (8.33 months)
*Comment:* Scientific judgement based upon unacclimated aqueous aerobic biodegradation half-life.

· **Ground Water:**
High: 15744 hours      (1.80 years)
Low: 12000 hours      (1.37 years)
*Comment:* Scientific judgement based upon unacclimated aqueous aerobic biodegradation half-life.

**Aqueous Biodegradation (unacclimated):**

· **Aerobic half-life:**
High: 7872 hours      (10.9 months)
Low: 6000 hours      (8.33 months)
*Comment:* Based upon aerobic estuarine river die-away test data (Lee, RF and Ryan, C (1979)).

· **Anaerobic half-life:**
High: 31488 hours      (3.59 years)
Low: 24000 hours      (2.74 years)
*Comment:* Scientific judgement based upon unacclimated aqueous aerobic biodegradation half-life.

· **Removal/secondary treatment:**
High: No data
Low:
*Comment:*

<u>Photolysis:</u>
- **· Atmos photol half-life:**         High:
                                               Low:

   *Comment:*

- **· Max light absorption (nm):**     No data
  *Comment:*

- **· Aq photol half-life:**            High:
                                             Low:

   *Comment:*

<u>Photooxidation half-life:</u>
- **· Water:**                   High:      No data
                                             Low:

   *Comment:*

- **· Air:**                       High:     336 hours            (14 days)
                                             Low:     33.6 hours       (1.4 days)
  *Comment:* Scientific judgement based upon estimated rate constant for reaction with hydroxyl radical in air (Atkinson, R (1987A)).

<u>Reduction half-life:</u>              High:     No data
                                             Low:
   *Comment:*

<u>Hydrolysis:</u>
- **· First-order hydr half-life:**
   *Comment:*

- **· Acid rate const $(M(H+)\text{-}hr)^{-1}$:**     No data
  *Comment:*

- **· Base rate const $(M(OH\text{-})\text{-}hr)^{-1}$:**
  *Comment:*

# 1-Butanol

<u>CAS Registry Number:</u> 71-36-3

<u>Half-lives:</u>
  · **Soil:**                     High:     168 hours         (7 days)
                                Low:      24 hours         (1 day)
*Comment:* Scientific judgement based upon estimated unacclimated aqueous aerobic biodegradation half-life.

  · **Air:**                      High:     87.7 hours       (3.7 days)
                                Low:      8.8 hours
*Comment:* Based upon photooxidation half-life in air.

  · **Surface Water:**        High:     168 hours         (7 days)
                                Low:      24 hours         (1 day)
*Comment:* Scientific judgement based upon unacclimated freshwater grab sample data (Hammerton, C (1955)) and aqueous screening test data (Bridie, AL et al. (1979)).

  · **Ground Water:**        High:    1296 hours     (54 days)
                                Low:      48 hours         (2 days)
*Comment:* Scientific judgement based upon estimated aqueous anaerobic biodegradation half-life.

<u>Aqueous Biodegradation (unacclimated):</u>
  · **Aerobic half-life**:     High:     168 hours         (7 days)
                                Low:      24 hours         (1 day)
*Comment:* Scientific judgement based upon unacclimated freshwater grab sample data (Hammerton, C (1955)) and aqueous screening test data (Bridie, AL et al. (1979)).

  · **Anaerobic half-life**:    High:    1296 hours     (54 days)
                                Low:      96 hours         (4 days)
*Comment:* Scientific judgement based upon acclimated anaerobic screening test data (high $t_{1/2}$) (Chou, WL et al. (1979)) and aqueous aerobic biodegradation half-life (low $t_{1/2}$).

  · **Removal/secondary treatment:**  High:      99%
                                Low:      31%
*Comment:* Based upon % degraded under acclimated aerobic semi-continuous and continuous flow conditions (high %: Hatfield, R (1957); low %: Matsui, S et al. (1975)).

<u>Photolysis:</u>
  · **Atmos photol half-life:**    High:
                                Low:
*Comment:*

· **Max light absorption (nm):**        No data
*Comment:*

· **Aq photol half-life:**        High:
                                          Low:

*Comment:*

## Photooxidation half-life:
· **Water:**        High: 104000 hours        (11.9 years)
                                Low:    2602 hours         (108 days)
*Comment:* Based upon measured rate data for hydroxyl radicals in aqueous solution (Dorfman, LM and Adams, GE (1973)).

· **Air:**        High:    87.7 hours        (3.7 days)
                              Low:     8.8 hours
*Comment:* Based upon measured rate data for the vapor phase reaction with hydroxyl radicals in air (Atkinson, R (1985)).

## Reduction half-life:
                                  High:      No data
                                  Low:

*Comment:*

## Hydrolysis:
· **First-order hydr half-life:**        No hydrolyzable groups
*Comment:*

· **Acid rate const $(M(H+)-hr)^{-1}$:**
*Comment:*

· **Base rate const $(M(OH-)-hr)^{-1}$:**
*Comment:*

# Benzene

**CAS Registry Number:** 71-43-2

**Half-lives:**
- **Soil:**                        High:    384 hours          (16 days)
                                    Low:     120 hours          (5 days)
  *Comment:* Scientific judgement based upon unacclimated aqueous aerobic biodegradation half-life.

- **Air:**                         High:    501 hours          (20.9 days)
                                    Low:     50.1 hours         (2.09 days)
  *Comment:* Based upon photooxidation half-life in air.

- **Surface Water:**               High:    384 hours          (16 days)
                                    Low:     120 hours          (5 days)
  *Comment:* Scientific judgement based upon unacclimated aqueous aerobic biodegradation half-life.

- **Ground Water:**                High:    17280 hours        (24 months)
                                    Low:     240 hours          (10 days)
  *Comment:* Scientific judgement based upon unacclimated aqueous aerobic (low $t_{1/2}$) and anaerobic (high $t_{1/2}$) biodegradation half-life.

**Aqueous Biodegradation (unacclimated):**
- **Aerobic half-life:**           High:    384 hours          (16 days)
                                    Low:     120 hours          (5 days)
  *Comment:* Based upon river die-away data (high $t_{1/2}$) (Vaishnav, DD and Babeu, L (1987)) and upon sea water die-away test data (low $t_{1/2}$) (Van der Linden, AC (1978)).

- **Anaerobic half-life:**         High:    17280 hours        (24 months)
                                    Low:     2688 hours         (16 weeks)
  *Comment:* Scientific judgement based upon unacclimated aqueous anaerobic biodegradation screening test data (Horowitz, A et al. (1982)).

- **Removal/secondary treatment:**  High:        100%
                                     Low:         44%
  *Comment:* Removal percentages based upon data from continuous activated sludge biological treatment simulators (Stover, EL and Kincannon, DF (1983); Feiler, HD et al. (1979)).

**Photolysis:**
- **Atmos photol half-life:**      High:    16152 hours        (673 days)
                                    Low:     2808 hours         (117 days)

111

*Comment:* Scientific judgement based upon measured photolysis half-lives in deionized water (Hustert, K et al. (1981)).

· **Max light absorption (nm):**          lambda max is approximately 239, 244, 249, 255, 261, 268 nm                                            (cyclohexane).
*Comment:* Absorption in cyclohexane extends to approximately 285 nm; absorption in the gas phase extends to approximately 275 nm (Howard, PH and Durkin, PR (1975)).

· **Aq photol half-life:**          High:   16152 hours          (673 days)
                                    Low:     2808 hours          (117 days)
*Comment:* Scientific judgement based upon measured photolysis half-lives in deionized water (Hustert, K et al. (1981)).

## Photooxidation half-life:
· **Water:**          High:$3.21 \times 10^5$ hours          (36.6 years)
                      Low:     8021 hours          (334 days)
*Comment:* Based upon measured rate constant for reaction with hydroxyl radical in water (Guesten, H et al. (1981)).

· **Air:**          High:     501 hours          (20.9 days)
                    Low:     50.1 hours          (2.09 days)
*Comment:* Based upon measured rate constant for reaction with hydroxyl radical in air (Atkinson, R (1985)).

## Reduction half-life:          High:     No data
                                 Low:
*Comment:*

## Hydrolysis:
· **First-order hydr half-life:**
*Comment:*

· **Acid rate const $(M(H+)-hr)^{-1}$:**          No hydrolyzable groups
*Comment:*

· **Base rate const $(M(OH-)-hr)^{-1}$:**
*Comment:*

# 1,1,1-Trichloroethane

<u>CAS Registry Number:</u> 71-55-6

<u>Half-lives:</u>
- **Soil:**                                         High:    6552 hours              (39 weeks)
                                                     Low:     3360 hours              (20 weeks)

    *Comment:* Scientific judgement based upon estimated unacclimated aqueous aerobic biodegradation half-life.

- **Air:**                                          High:    53929 hours             (6.2 years)
                                                     Low:     5393 hours              (225 days)

    *Comment:* Based upon photooxidation half-life in air.

- **Surface Water:**                                High:    6552 hours              (39 weeks)
                                                     Low:     3360 hours              (20 weeks)

    *Comment:* Scientific judgement based upon estimated unacclimated aqueous aerobic biodegradation half-life.

- **Ground Water:**                                 High:    13104 hours             (78 weeks)
                                                     Low:     3360 hours              (20 weeks)

    *Comment:* Scientific judgement based upon estimated unacclimated aqueous aerobic biodegradation half-life and sub-soil grab sample data from a ground water aquifer (low $t_{1/2}$: Wilson, JT et al. (1983)).

<u>Aqueous Biodegradation (unacclimated):</u>
- **Aerobic half-life:**                            High:    6552 hours              (39 weeks)
                                                     Low:     3360 hours              (20 weeks)

    *Comment:* Scientific judgement based upon unacclimated aerobic sea water grab sample data (high $t_{1/2}$: Pearson, CR and McConnell, G (1975)) and sub-soil grab sample data from a ground water aquifer (low $t_{1/2}$: Wilson, JT et al. (1983)).

- **Anaerobic half-life:**                          High:    26208 hours             (156 weeks)
                                                     Low:     13440 hours             (80 weeks)

    *Comment:* Scientific judgement based upon unacclimated aerobic biodegradation half-life.

- **Removal/secondary treatment:**                  High:    No data
                                                     Low:

    *Comment:*

<u>Photolysis:</u>
- **Atmos photol half-life:**                       High:  Not expected to be important
                                                     Low:

*Comment:* Scientific judgement based upon aqueous photolysis data.

· **Max light absorption (nm):**         Lambda max = 250, 220
*Comment:* Absorption maxima (Hubrich, C and Stuhl, F (1980)).

· **Aq photol half-life:**         High:  Not expected to be important
                                   Low:
*Comment:* Stable in water exposed to sunlight for one year (Dilling, WL et al. (1975)).

**Photooxidation half-life:**
· **Water:**                       High:      No data
                                   Low:

*Comment:*

· **Air:**                         High:   53929 hours         (6.2 years)
                                   Low:    5393 hours          (225 days)
*Comment:* Based upon measured rate data for the vapor phase reaction with hydroxyl radicals in air (Atkinson, R (1985)).

**Reduction half-life:**           High:      No data
                                   Low:
*Comment:*

**Hydrolysis:**
· **First-order hydr half-life:**         0.73 years
*Comment:* Scientific judgement based upon reported rate constant (0.96 $yr^{-1}$) at pH 7 and 25 °C (Kollig, HP et al. (1987A)).

· **Acid rate const $(M(H+)-hr)^{-1}$:**
*Comment:*

· **Base rate const $(M(OH-)-hr)^{-1}$:**
*Comment:*

# Methoxychlor

**CAS Registry Number:** 72-43-5

**Structure:**

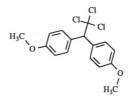

**Half-lives:**

· **Soil:** High: 8760 hours (1 year)

Low: 4320 hours (6 months)

*Comment:* Scientific judgement based upon very slow biodegradation observed in an aerobic soil die-away test study (Fogel, S et al. (1982)).

· **Air:** High: 11.2 hours

Low: 1.12 hours

*Comment:* Scientific judgement based upon estimated photooxidation half-life in air (Atkinson, R (1987A)).

· **Surface Water:** High: 5.4 hours

Low: 2.2 hours

*Comment:* Based upon measured photooxidation in river water exposed to midday May sunlight (Zepp, RG et al. (1976)).

· **Ground Water:** High: 8760 hours (1 year)

Low: 1200 hours (50 days)

*Comment:* Scientific judgement based upon aerobic (high $t_{1/2}$) and anaerobic (low $t_{1/2}$) soil die-away test study data (Fogel, S et al. (1982)).

**Aqueous Biodegradation (unacclimated):**

· **Aerobic half-life:** High: 8760 hours (1 year)

Low: 4320 hours (6 months)

*Comment:* Scientific judgement based upon very slow biodegradation observed in an aerobic soil die-away test study (Fogel, S et al. (1982)).

· **Anaerobic half-life:** High: 4320 hours (6 months)

Low: 1200 hours (50 days)

*Comment:* Scientific judgement based upon anaerobic soil die-away test study data (Fogel, S et al. (1982)).

· **Removal/secondary treatment:**        High:        No data
                                          Low:

*Comment:*

## Photolysis:
· **Atmos photol half-life:**        High:        2070 hours        (86.3 days)
                                      Low:        300 hours        (12.5 days)
*Comment:* Scientific judgement based upon measured photolysis rates in distilled water under midday summer sunlight (low $t_{1/2}$) (Wolfe, NL et al. (1976)); high $t_{1/2}$ based upon measured photolysis rates in distilled water under midday summer sunlight (Zepp, RG et al. (1976)) and adjusted for approximate winter sunlight intensity (Lyman, WJ et al. (1982)).

· **Max light absorption (nm):**        lambda max = 229, 246, 276, and 283 nm (methanol)
*Comment:* Significant absorption to 294 nm (Sadtler UV No. 5497).

· **Aq photol half-life:**        High:        2070 hours        (86.3 days)
                                  Low:        300 hours        (12.5 days)
*Comment:* Scientific judgement based upon measured photolysis rates in distilled water under midday summer sunlight (low $t_{1/2}$) (Wolfe, NL et al. (1976)); high $t_{1/2}$ based upon measured photolysis rates in distilled water under midday summer sunlight (Zepp, RG et al. (1976)) and adjusted for approximate winter sunlight intensity (Lyman, WJ et al. (1982)).

## Photooxidation half-life:
· **Water:**        High:        5.4 hours
                    Low:        2.2 hours
*Comment:* Based upon measured photooxidation in river water exposed to midday May sunlight (Zepp, RG et al. (1976)).

· **Air:**        High:        11.2 hours
                  Low:        1.12 hours
*Comment:* Scientific judgement based upon estimated rate constant for reaction with hydroxyl radical (Atkinson, R (1987A)).

## Reduction half-life:        High:        No data
                               Low:

*Comment:*

## Hydrolysis:
· **First-order hydr half-life:**        9198 hours        (1.05 years)
*Comment:* ($t_{1/2}$ at pH 7 and 25°C) Scientific judgement based upon neutral and base catalyzed hydrolysis rate constants (Kollig, HP et al. (1987A)).

· **Acid rate const (M(H+)-hr)$^{-1}$:**        0.00
*Comment:* ($t_{1/2}$ = 9240 hours (1.06 years) at pH 5 and 25°C) Scientific judgement based upon

116

neutral and base catalyzed hydrolysis rate constants (Kollig, HP et al. (1987A)).

· **Base rate const (M(OH-)-hr)$^{-1}$:**        1.23X10$^4$ M$^{-1}$ year$^{-1}$
*Comment:* (t$_{1/2}$ = 7779 hours (0.888 years) at pH 9 and 25°C) Scientific judgement based upon neutral and base catalyzed hydrolysis rate constants (Kollig, HP et al. (1987A)).

# DDD

<u>CAS Registry Number:</u>  72-54-8

<u>Structure:</u>

<u>Half-lives:</u>
 · **Soil:**                                     High: 1.4X10$^5$ hours           (15.6 years)
                                                 Low:    17520 hours             (2 years)
*Comment:*  Based upon observed rates of biodegradation of DDT in aerobic soils under field conditions (low $t_{1/2}$:  Lichtenstein, EP and Schultz, KR (1959); high $t_{1/2}$:  Stewart, DKR and Chisholm, D (1971)).

 · **Air:**                                      High:      177 hours            (7.4 days)
                                                 Low:     17.7 hours
*Comment:*  Scientific judgement based upon estimated photooxidation half-life in air.

 · **Surface Water:**                            High: 1.4X10$^5$ hours           (15.6 years)
                                                 Low:    17520 hours             (2 years)
*Comment:*  Based upon observed rates of biodegradation of DDT in aerobic soils under field conditions (low $t_{1/2}$:  Lichtenstein, EP and Schultz, KR (1959); high $t_{1/2}$:  Stewart, DKR and Chisholm, D (1971)).

 · **Ground Water:**                             High: 2.7X10$^5$ hours           (31.3 years)
                                                 Low:    1680 hours              (70 days)
*Comment:*  Based upon anaerobic flooded soil die-away study data for DDD for two flooded soils (low $t_{1/2}$) (Castro, TF and Yoshida, T (1971)) and observed rates of biodegradation of DDT in aerobic soils under field conditions (low $t_{1/2}$:  Lichtenstein, EP and Schultz, KR (1959); high $t_{1/2}$:  Stewart, DKR and Chisholm, D (1971)).

<u>Aqueous Biodegradation (unacclimated):</u>
 · **Aerobic half-life:**                        High: 1.4X10$^5$ hours           (15.6 years)
                                                 Low:    17520 hours             (2 years)
*Comment:*  Based upon observed rates of biodegradation of DDT in aerobic soils under field conditions (low $t_{1/2}$:  Lichtenstein, EP and Schultz, KR (1959); high $t_{1/2}$:  Stewart, DKR and Chisholm, D (1971)).

 · **Anaerobic half-life:**                      High:     7056 hours            (294 days)

Low:     1680 hours               (70 days)

*Comment:* Based upon anaerobic flooded soil die-away study data for DDD for two flooded soils (Castro, TF and Yoshida, T (1971)).

· **Removal/secondary treatment:**     High:     100%

Low:     17%

*Comment:* Removal percentage based upon data from a continuous activated sludge biological treatment simulator (Patterson, JW and Kodukala, PS (1981)).

## Photolysis:

· **Atmos photol half-life:**          High:     No data

Low:

*Comment:*

· **Max light absorption (nm):**       lambda max = 230, 270, 278 nm (hexane)

*Comment:* Absorption extends to approximately 284 nm (Gore, RC et al. (1971)).

· **Aq photol half-life:**             High:     No data

Low:

*Comment:*

## Photooxidation half-life:

· **Water:** ·                         High:     No data

Low:

*Comment:*

· **Air:**                             High:     133 hours          (5.54 days)

Low:     13.3 hours

*Comment:* Scientific judgement based upon estimated rate constant for reaction with hydroxyl radical in air (Atkinson, R (1987A)).

## Reduction half-life:                 High:     No data

Low:

*Comment:*

## Hydrolysis:

· **First-order hydr half-life:**      $2.09 \times 10^5$ hours (28 years)

*Comment:* ($t_{1/2}$ at pH 7 and 25°C) Calculated from measured neutral and base catalyzed hydrolysis constants (Ellington, JJ et al. (1987)).

· **Acid rate const $(M(H+)-hr)^{-1}$:**     No data

*Comment:* ($t_{1/2}$ = $2.47 \times 10^5$ (28 years) at pH 5 and 25°C) Calculated from measured neutral and base catalyzed hydrolysis constants (Ellington, JJ et al. (1987)).

119

· **Base rate const (M(OH-)-hr)$^{-1}$:**        5.2

*Comment:* ($t_{1/2}$ = 12646 hours (1.4 years) at pH 9 and 25°C) Calculated from measured neutral and base catalyzed hydrolysis constants (Ellington, JJ et al. (1987)).

# DDE

**CAS Registry Number:** 72-55-9

**Structure:**

**Half-lives:**
 · **Soil:**                                          High: 1.4X10$^5$ hours                (15.6 years)
                                                          Low:   17520 hours                     (2 years)
*Comment:*  Based upon observed rates of biodegradation of DDT in aerobic soils under field conditions (low t$_{1/2}$: Lichtenstein, EP and Schultz, KR (1959); high t$_{1/2}$: Stewart, DKR and Chisholm, D (1971)).

 · **Air:**                                           High:    177 hours                       (7.4 days)
                                                          Low:    17.7 hours
*Comment:*  Scientific judgement based upon estimated photooxidation half-life in air.

 · **Surface Water:**                       High:    146 hours                       (6.1 days)
                                                          Low:     15 hours
*Comment:*  Scientific judgment based upon estimated photolysis half-life in water.

 · **Ground Water:**                        High: 2.7X10$^5$ hours                (31.3 years)
                                                          Low:     384 hours                      (16 days)
*Comment:*  Based upon anaerobic flooded soil die-away study data for DDT for two flooded soils (low t$_{1/2}$) (Castro, TF and Yoshida, T (1971)) and observed rates of biodegradation of DDT in aerobic soils under field conditions (low t$_{1/2}$: Lichtenstein, EP and Schultz, KR (1959); high t$_{1/2}$: Stewart, DKR and Chisholm, D (1971)).

**Aqueous Biodegradation (unacclimated):**
 · **Aerobic half-life:**                   High: 1.4X10$^5$ hours                (15.6 years)
                                                          Low:   17520 hours                     (2 years)
*Comment:*  Based upon observed rates of biodegradation of DDT in aerobic soils under field conditions (low t$_{1/2}$: Lichtenstein, EP and Schultz, KR (1959); high t$_{1/2}$: Stewart, DKR and Chisholm, D (1971)).

 · **Anaerobic half-life:**               High:    2400 hours                      (100 days)
                                                          Low:     384 hours                      (16 days)
*Comment:*  Based upon anaerobic flooded soil die-away study data for DDT for two flooded soils

(Castro, TF and Yoshida, T (1971)).

· **Removal/secondary treatment:**      High:       39%
                                                 Low:
*Comment:* Removal percentage based upon data from a continuous activated sludge biological treatment simulator (Patterson, JW and Kodukala, PS (1981)).

## Photolysis:
· **Atmos photol half-life:**      High:    146 hours        (6.1 days)
                                                 Low:      15 hours
*Comment:* Scientific judgement based upon measured photolysis rate constants for DDE irradiated at >290 nm with artificial light in water solution (high $t_{1/2}$: Zepp, RG and Cline, DM (1977); low $t_{1/2}$: Draper, WM (1985)).

· **Max light absorption (nm):**      lambda max = 248 nm (hexane)
*Comment:* Absorption extends to approximately 303 nm (Gore, RC et al. (1971)).

· **Aq photol half-life:**      High:    146 hours        (6.1 days)
                                          Low:      15 hours
*Comment:* Scientific judgement based upon measured photolysis rate constants for DDE irradiated at >290 nm with artificial light in water solution (high $t_{1/2}$: Zepp, RG and Cline, DM (1977); low $t_{1/2}$: Draper, WM (1985)).

## Photooxidation half-life:
· **Water:**      High:      No data
                            Low:
*Comment:*

· **Air:**      High:      40.9 hours
                  Low:      5.25 hours
*Comment:* Scientific judgement based upon estimated rate constants for reaction with hydroxyl radical (Atkinson, R (1987A)) and ozone (Atkinson, R and Carter, WPL (1984)) in air.

## Reduction half-life:
                               High:      No data
                               Low:
*Comment:*

## Hydrolysis:
· **First-order hydr half-life:**
*Comment:*

· **Acid rate const (M(H+)-hr)$^{-1}$:**      No hydrolyzable groups
*Comment:*

· **Base rate const (M(OH-)-hr)$^{-1}$:**
*Comment:*

# Methyl bromide

**CAS Registry Number:** 74-83-9

**Half-lives:**

· **Soil:**
  | | High: | 672 hours | (4 weeks) |
  | | Low: | 168 hours | (7 days) |

*Comment:* Scientific judgement based upon unacclimated aqueous aerobic biodegradation half-life.

· **Air:**
  | | High: | 16327 hours | (680 days) |
  | | Low: | 1633 hours | (68.0 days) |

*Comment:* Based upon photooxidation half-life in air.

· **Surface Water:**
  | | High: | 672 hours | (4 weeks) |
  | | Low: | 168 hours | (7 days) |

*Comment:* Scientific judgement based upon unacclimated aqueous aerobic biodegradation half-life.

· **Ground Water:**
  | | High: | 912 hours | (38 days) |
  | | Low: | 336 hours | (14 days) |

*Comment:* Scientific judgement based upon unacclimated aqueous aerobic biodegradation half-life (low $t_{1/2}$) and hydrolysis half-life (high $t_{1/2}$).

**Aqueous Biodegradation (unacclimated):**

· **Aerobic half-life:**
  | | High: | 672 hours | (4 weeks) |
  | | Low: | 168 hours | (7 days) |

*Comment:* Scientific judgement based upon unacclimated aerobic aqueous screening test data for bromoform from experiments utilizing settled domestic wastewater inoculum (Tabak, HH et al. (1981)).

· **Anaerobic half-life:**
  | | High: | 2688 hours | (16 weeks) |
  | | Low: | 672 hours | (28 days) |

*Comment:* Scientific judgement based upon unacclimated aqueous aerobic biodegradation half-life.

· **Removal/secondary treatment:**
  | | High: | No data |
  | | Low: | |

*Comment:*

**Photolysis:**

· **Atmos photol half-life:**
  | | High: | No data |
  | | Low: | |

124

*Comment:*

· **Max light absorption (nm):**          lambda max = 202 nm (gas phase)
*Comment:*  Absorption up to 262 nm (Robbins, DE (1976)).

· **Aq photol half-life:**          High:          No data
                                    Low:
*Comment:*

**Photooxidation half-life:**
· **Water:**          High:          No data
                      Low:
*Comment:*

· **Air:**          High:    16327 hours          (680 days)
                    Low:      1633 hours          (68.0 days)
*Comment:*  Based upon measured rates for reaction with hydroxyl radical in air (Atkinson, R (1985)).

**Reduction half-life:**          High:          No data
                                  Low:
*Comment:*

**Hydrolysis:**
· **First-order hydr half-life:**          High:    912 hours          (38 days)
                                           Low:     470 hours          (19.6 days)
*Comment:*  ($t_{1/2}$ at pH 7 and 20°C (high $t_{1/2}$) and 25°C (low $t_{1/2}$))
Based upon measured first order hydrolysis rate constant (high $t_{1/2}$: Ehrenberg, L et al. (1974); low $t_{1/2}$: Mabey, W and Mill, T (1978)).

· **Acid rate const $(M(H+)-hr)^{-1}$:**          No data
*Comment:*

· **Base rate const $(M(OH-)-hr)^{-1}$:**          No data
*Comment:*

125

# Ethylene

CAS Registry Number: 74-85-1

## Half-lives:
· **Soil:**           High:    672 hours        (4 weeks)
                      Low:     24 hours         (1 day)
*Comment:* Scientific judgement based upon biological screening tests in soil (Sawada, S and Totsuka, T (1986); Sawada S et al. (1986); Debont, JAM (1976); Nakano R and Kuwatsuka, S (1979)).

· **Air:**            High:    56 hours
                      Low:     6.2 hours
*Comment:* Based upon combined, measured photooxidation rate constants for ·OH and ozone (Atkinson, R (1985); Atkinson, R and Carter, WPL (1984)).

· **Surface Water:**  High:    672 hours        (4 weeks)
                      Low:     24 hours         (1 day)
*Comment:* Scientific judgement based upon soil biodegradation half-lives.

· **Ground Water:**   High:    1344 hours       (8 weeks)
                      Low:     48 hours         (2 days)
*Comment:* Scientific judgement based upon soil biodegradation half-lives.

## Aqueous Biodegradation (unacclimated):
· **Aerobic half-life:**    High:    672 hours        (4 weeks)
                            Low:     24 hours         (1 day)
*Comment:* Scientific judgement based upon soil biodegradation half-lives.

· **Anaerobic half-life:**  High:    2688 hours       (16 weeks)
                            Low:     96 hours         (4 days)
*Comment:* Scientific judgement based upon soil biodegradation half-lives.

· **Removal/secondary treatment:**   High:    No data
                                     Low:
*Comment:*

## Photolysis:
· **Atmos photol half-life:**   High: No photolyzable functions
                                Low:
*Comment:*

· **Max light absorption (nm):**

*Comment:*

**· Aq photol half-life:**  High:
Low:
*Comment:*

**Photooxidation half-life:**
  **Water:**  High:  76800 hours  (3200 days)
Low:  1920 hours  (80 days)
*Comment:* Based upon a measured rate constant for reaction with ·OH in aqueous solution
(Guesten H et al. (1981)).

**· Air:**  High:  56 hours
Low:  6.2 hours
*Comment:*  Based upon combined, measured photooxidation rate constants for ·OH and ozone
(Atkinson, R (1985); Atkinson, R and Carter, WPL (1984)).

**Reduction half-life:**  High:  No data
Low:
*Comment:*

**Hydrolysis:**
  **· First-order hydr half-life:**  No hydrolyzable functions
*Comment:*

  **· Acid rate const $(M(H+)-hr)^{-1}$:**
*Comment:*

  **· Base rate const $(M(OH-)-hr)^{-1}$:**
*Comment:*

# Methyl chloride

CAS Registry Number: 74-87-3

Half-lives:
 · Soil:                              High:    672 hours          (4 weeks)
                                      Low:     168 hours          (7 days)
 Comment: Scientific judgement based upon estimated aerobic biodegradation half-life.

 · Air:                               High:    14717 hours        (613 days)
                                      Low:     1472 hours         (61.3 days)
 Comment: Based upon photooxidation half-life in air.

 · Surface Water:                     High:    672 hours          (4 weeks)
                                      Low:     168 hours          (7 days)
 Comment: Scientific judgement based upon estimated aerobic biodegradation half-life.

 · Ground Water:                      High:    1344 hours         (8 weeks)
                                      Low:     336 hours          (14 days)
 Comment: Scientific judgement based upon estimated aerobic biodegradation half-life.

Aqueous Biodegradation (unacclimated):
 · Aerobic half-life:                 High:    672 hours          (4 weeks)
                                      Low:     168 hours          (7 days)
 Comment: Scientific judgement based upon unacclimated aerobic aqueous screening test data for
 dichloromethane from experiments utilizing settled domestic wastewater inoculum (Tabak, HH et
 al. (1981)) and activated sludge inoculum (Klecka, GM (1982)).

 · Anaerobic half-life:               High:    2688 hours         (16 weeks)
                                      Low:     672 hours          (28 days)
 Comment: Scientific judgement based upon estimated aerobic biodegradation half-life.

 · Removal/secondary treatment:       High:    No data
                                      Low:
 Comment:

Photolysis:
 · Atmos photol half-life:            High:    No data
                                      Low:
 Comment:

 · Max light absorption (nm):         lambda max <174 nm (gas)
 Comment: No absorption at >230 nm in the gas phase (Robbins, DE (1976)).

128

**· Aq photol half-life:**          High:       No data
                                    Low:

*Comment:*

## Photooxidation half-life:
**· Water:**                        High:       No data
                                    Low:

*Comment:*

**· Air:**                          High:    14717 hours          (613 days)
                                    Low:      1472 hours          (61.3 days)
*Comment:* Based upon measured rate constants for reaction with hydroxyl radical in air
(Atkinson, R (1985)).

## Reduction half-life:              High:       No data
                                    Low:

*Comment:*

## Hydrolysis:
**· First-order hydr half-life:**       7000 hours (292 days) at pH 7
*Comment:* Scientific judgement based upon neutral and base catalyzed hydrolysis rate constants
at 25°C extrapolated from data obtained at higher temperatures (Mabey, W and Mill, T (1978)).

**· Acid rate const $(M(H+)-hr)^{-1}$:**       $2.75 \times 10^{-8}$ $M^{-1}$ $s^{-1}$ ($t_{1/2}$ = 7000 hours at pH 5)
*Comment:* Scientific judgement based upon neutral and base catalyzed hydrolysis rate constants
at 25°C extrapolated from data obtained at higher temperatures (Mabey, W and Mill, T (1978)).

**· Base rate const $(M(OH-)-hr)^{-1}$:**       $2.76 \times 10^{-8}$ $M^{-1}$ $s^{-1}$ ($t_{1/2}$ = 6975 hours at pH 9)
*Comment:* Scientific judgement based upon neutral and base catalyzed hydrolysis rate constants
at 25°C extrapolated from data obtained at higher temperatures (Mabey, W and Mill, T (1978)).

# Methyl iodide

CAS Registry Number: <u>CAS Registry Number:</u> 74-88-4

## Half-lives:
· **Soil:**                                    High:        672 hours              (4 weeks)
                                               Low:         168 hours              (7 days)
*Comment:* Scientific judgement based upon aerobic aqueous biodegradation half-life.

· **Air:**                                     High:        5348 hours             (223 days)
                                               Low:         535 hours              (22.3 days)
*Comment:* Based upon photooxidation half-life in air.

· **Surface Water:**                           High:        672 hours              (4 weeks)
                                               Low:         168 hours              (7 days)
*Comment:* Scientific judgement based upon aerobic aqueous biodegradation half-life.

· **Ground Water:**                            High:        1344 hours             (8 weeks)
                                               Low:         336 hours              (2 weeks)
*Comment:* Scientific judgement based upon aerobic aqueous biodegradation half-life.

## Aqueous Biodegradation (unacclimated):
· **Aerobic half-life:**                       High:        672 hours              (4 weeks)
                                               Low:         168 hours              (7 days)
*Comment:* Scientific judgement.

· **Anaerobic half-life:**                     High:        2688 hours             (16 weeks)
                                               Low:         672 hours              (4 weeks)
*Comment:* Scientific judgement based upon aerobic aqueous biodegradation half-life.

· **Removal/secondary treatment:**             High:        No data
                                               Low:
*Comment:*

## Photolysis:
· **Atmos photol half-life:**                  High:        No data
                                               Low:
*Comment:*

· **Max light absorption (nm):**               lambda max = 260
*Comment:* Methyl iodide absorbs UV light continuously up to approximately 360 nm (Tsao, CW and Root, JW (1972)).

**· Aq photol half-life:**   High:   No data
                            Low:

*Comment:*

## Photooxidation half-life:
**· Water:**   High:   1440 hours   (60 days)
              Low:    480 hours    (20 days)
*Comment:* Scientific judgement based upon a measured rate constant ($k = 1.7 \times 10^{10}$ $M^{-1}$ $s^{-1}$) for reaction with hydrated electrons (average concn of $1.2 \times 10^{-17}$ M) in natural waters at 355 nm and 25 °C with a pH of 6.2 (Zepp, RG et al. (1987)).

**· Air:**   High:   5348 hours   (223 days)
            Low:    535 hours    (22.3 days)
*Comment:* Based upon a measured rate constant for vapor phase reaction with hydroxyl radicals in air (Garraway, J and Donovan, J (1979)).

## Reduction half-life:
                            High:   No data
                            Low:

*Comment:*

## Hydrolysis:
**· First-order hydr half-life:**   2640 hours   (110 days)
*Comment:* Scientific judgement based upon measured rate data at pH of 7 ($k_N = 7.28 \times 10^{-8}$ $s^{-1}$) (not based catalyzed) extrapolated to 25 °C (Mabey, W and Mill, T (1978)).

**· Acid rate const $(M(H+)-hr)^{-1}$:**
*Comment:*

**· Base rate const $(M(OH-)-hr)^{-1}$:**
*Comment:*

# Hydrocyanic acid

<u>CAS Registry Number:</u>  74-90-8

<u>Structure:</u>                               HC≡N

<u>Half-lives:</u>
· **Soil:**                          High:     4320 hours              (6 months)
                                 Low:       672 hours              (4 weeks)
   *Comment:*  Scientific judgement based upon estimated aqueous aerobic biodegradation half-life.

· **Air:**                           High:   21392 hours              (2.4 years)
                                 Low:      2139 hours              (89.1 days)
   *Comment:*  Based upon photooxidation half-life in air.

· **Surface Water:**                 High:     4320 hours              (6 months)
                                 Low:       672 hours              (4 weeks)
   *Comment:*  Scientific judgement based upon estimated aqueous aerobic biodegradation half-life.

· **Ground Water:**                  High:     8640 hours              (12 months)
                                 Low:      1344 hours              (8 weeks)
   *Comment:*  Scientific judgement based upon estimated aqueous aerobic biodegradation half-life.

<u>Aqueous Biodegradation (unacclimated):</u>
· **Aerobic half-life:**             High:     4032 hours              (6 months)
                                 Low:       672 hours              (4 weeks)
   *Comment:*  Scientific judgement based upon acclimated aerobic screening test data (Pettet, AJ and Mills, EV (1954)).

· **Anaerobic half-life:**           High:   16128 hours              (24 months)
                                 Low:      2688 hours              (16 weeks)
   *Comment:*  Scientific judgement based upon estimated aqueous aerobic biodegradation half-life.

· **Removal/secondary treatment:**   High:       No data
                                 Low:
   *Comment:*

<u>Photolysis:</u>
· **Atmos photol half-life:**        High:       No data
                                 Low:
   *Comment:*

· **Max light absorption (nm):**     No data

132

*Comment:*

**· Aq photol half-life:**                    High:        No data
                                              Low:

*Comment:*

**Photooxidation half-life:**
    **· Water:**                              High:        No data
                                              Low:

*Comment:*

**· Air:**                                    High:    21392 hours           (2.4 years)
                                              Low:     2139 hours            (89.1 days)

*Comment:* Scientific judgement based upon measured rate data for the vapor phase reaction with hydroxyl radicals in air (Atkinson, R (1985)).

**Reduction half-life:**                      High:        No data
                                              Low:

*Comment:*

**Hydrolysis:**
    **· First-order hydr half-life:**         No data
    *Comment:*

    **· Acid rate const $(M(H+)-hr)^{-1}$:**
    *Comment:*

    **· Base rate const $(M(OH-)-hr)^{-1}$:**
    *Comment:*

# Methylene bromide

**CAS Registry Number:** 74-95-3

**Half-lives:**
  · **Soil:**                             High:    672 hours      (4 weeks)
                                          Low:     168 hours      (7 days)
*Comment:* Scientific judgement based upon unacclimated aqueous aerobic biodegradation half-life.

  · **Air:**    High: 8510 hours (355 days)  Low: 851 hours (35.5 days)
*Comment:* Scientific judgement based upon estimated photooxidation half-life in air.

  · **Surface Water:**  High: 672 hours (4 weeks)  Low: 168 hours (7 days)
*Comment:* Scientific judgement based upon unacclimated aqueous aerobic biodegradation half-life.

  · **Ground Water:**  High: 1344 hours (8 weeks)  Low: 336 hours (14 days)
*Comment:* Scientific judgement based upon unacclimated aqueous aerobic biodegradation half-life.

**Aqueous Biodegradation (unacclimated):**
  · **Aerobic half-life:**  High: 672 hours (4 weeks)  Low: 168 hours (7 days)
*Comment:* Scientific judgement based upon unacclimated aerobic aqueous screening test data for bromoform from experiments utilizing settled domestic wastewater inoculum (Tabak, HH et al. (1981).

  · **Anaerobic half-life:**  High: 2688 hours (16 weeks)  Low: 672 hours (28 days)
*Comment:* Scientific judgement based upon unacclimated aqueous aerobic biodegradation half-life.

  · **Removal/secondary treatment:**  High: No data  Low:
*Comment:*

**Photolysis:**
  · **Atmos photol half-life:**  High:  Low:

134

*Comment:*

· **Max light absorption (nm):**     No data
*Comment:*

· **Aq photol half-life:**     High:
     Low:

*Comment:*

**Photooxidation half-life:**
· **Water:**     High:     No data
     Low:
*Comment:*

· **Air:**     High:     8510 hours     (355 days)
     Low:     851 hours     (35.5 days)
*Comment:* Scientific judgement based upon estimated rate constant for reaction with hydroxyl radical in air (Atkinson, R (1987A)).

**Reduction half-life:**     High:     No data
     Low:
*Comment:*

**Hydrolysis:**
· **First-order hydr half-life:**     $1.6 \times 10^6$ hours (183 years)
*Comment:* (half-life at pH 7 and 25°C) Based upon overall hydrolysis rate constant (Mill,T et al. (1982)).

· **Acid rate const $(M(H+)-hr)^{-1}$:**
*Comment:*

· **Base rate const $(M(OH-)-hr)^{-1}$:**
*Comment:*

# Chloroethane

CAS Registry Number: 75-00-3

Half-lives:
 · Soil:                                High:    672 hours          (4 weeks)
                                        Low:     168 hours          (7 days)
Comment: Scientific judgement based upon estimated unacclimated aerobic aqueous
biodegradation half-life.


 · Air:                                 High:    1604 hours         (66.8 days)
                                        Low:     160 hours          (6.67 days)
Comment: Based upon photooxidation half-life in air.


 · Surface Water:                       High:    672 hours          (4 weeks)
                                        Low:     168 hours          (7 days)
Comment: Scientific judgement based upon estimated unacclimated aerobic aqueous
biodegradation half-life.


 · Ground Water:                        High:    1344 hours         (8 weeks)
                                        Low:     336 hours          (14 days)
Comment: Scientific judgement based upon estimated unacclimated aerobic aqueous
biodegradation half-life.


Aqueous Biodegradation (unacclimated):
 · Aerobic half-life:                   High:    672 hours          (4 weeks)
                                        Low:     168 hours          (7 days)
Comment: Scientific judgement based upon aqueous aerobic screening test data for 1-
chloropropane and 1-chlorobutane (Gerhold, RM and Malaney, GW (1966)).


 · Anaerobic half-life:                 High:    2688 hours         (16 weeks)
                                        Low:     672 hours          (28 days)
Comment: Scientific judgement based upon estimated unacclimated aerobic aqueous
biodegradation half-life.


 · Removal/secondary treatment:         High:    No data
                                        Low:
Comment:


Photolysis:
 · Atmos photol half-life:              High:    No data
                                        Low:
Comment:

**· Max light absorption (nm):**      lambda max <160 nm (gas phase)
*Comment:* Absorption extends up to approximately 235 nm (Hubrich, C and Stuhl, F (1980)).

**· Aq photol half-life:**           High:      No data
                                     Low:

*Comment:*

**Photooxidation half-life:**
   **· Water:**                      High:      No data
                                     Low:

*Comment:*

   **· Air:**                        High:   1604 hours      (66.8 days)
                                     Low:    160 hours       (6.67 days)
*Comment:* Based upon measured rate constants for reaction with hydroxyl radical in air
(Atkinson, R (1985)).

**Reduction half-life:**             High:      No data
                                     Low:

*Comment:*

**Hydrolysis:**
   **· First-order hydr half-life:**      912 hours        (38 days) at 25°C
*Comment:* Based upon neutral hydrolysis rate constant which was extrapolated from data at
higher temperatures (Mabey, W and Mill, T (1978)).

   **· Acid rate const (M(H+)-hr)$^{-1}$:**      No data
   *Comment:*

   **· Base rate const (M(OH-)-hr)$^{-1}$:**     No data
   *Comment:*

# Vinyl chloride

**CAS Registry Number:** 75-01-4

**Half-lives:**

   · **Soil:**                         High:    4320 hours      (6 months)
                                        Low:     672 hours       (4 weeks)
*Comment:* Scientific judgement based upon estimated unacclimated aqueous aerobic biodegradation half-life.

   · **Air:**                         High:    97 hours
                                        Low:     9.7 hours
*Comment:* Based upon photooxidation half-life in air.

   · **Surface Water:**          High:    4320 hours      (6 months)
                                        Low:     672 hours       (4 weeks)
*Comment:* Scientific judgement based upon estimated unacclimated aqueous aerobic biodegradation half-life.

   · **Ground Water:**          High:  69,000 hours    (95 months)
                                        Low:    1344 hours       (8 weeks)
*Comment:* Scientific judgement based upon estimated unacclimated aqueous aerobic biodegradation half-life (low $t_{1/2}$) and an estimated half-life for anaerobic biodegradation of vinyl chloride from a ground water field study of chlorinated ethenes (Silka, LR and Wallen, DA (1988)).

**Aqueous Biodegradation (unacclimated):**

   · **Aerobic half-life:**        High:    4320 hours      (6 months)
                                        Low:     672 hours       (4 weeks)
*Comment:* Scientific judgement based upon aqueous screening test data (Freitag, D et al. (1984A); Helfgott, TB et al. (1977)).

   · **Anaerobic half-life:**     High:  17280 hours    (24 months)
                                        Low:    2688 hours      (16 weeks)
*Comment:* Scientific judgement based upon estimated unacclimated aqueous aerobic biodegradation half-life.

   · **Removal/secondary treatment:**    High:    No data
                                        Low:
*Comment:*

**Photolysis:**

   · **Atmos photol half-life:**    High:  Not expected to directly photolyze

Low:

*Comment:* Scientific judgement based upon measured rate of photolysis in water (Callahan, MA et al. (1979A)).

· **Max light absorption (nm):**
*Comment:*

| · **Aq photol half-life:** | High: | Not expected to directly photolyze |
| | Low: | |

*Comment:* Scientific judgement based upon measured rate of photolysis in water (Callahan, MA et al. (1979A)).

## Photooxidation half-life:

| · **Water:** | High: | No data |
| | Low: | |

*Comment:*

| · **Air:** | High: | 97 hours |
| | Low: | 9.7 hours |

*Comment:* Based upon measured rate constant for reaction with hydroxyl radicals in air (Atkinson, R (1985)).

| **Reduction half-life:** | High: | Not expected to be significant |
| | Low: | |

*Comment:*

## Hydrolysis:

· **First-order hydr half-life:**
*Comment:*

· **Acid rate const $(M(H+)\text{-}hr)^{-1}$:**     No hydrolyzable groups
*Comment:*

· **Base rate const $(M(OH\text{-})\text{-}hr)^{-1}$:**
*Comment:*

# Acetonitrile

**CAS Registry Number:** 75-05-8

**Half-lives:**
   · **Soil:**                          High:    672 hours            (4 weeks)
                                        Low:     168 hours            (1 week)
   *Comment:* Scientific judgement based upon estimated aqueous aerobic biodegradation half-life.

   · **Air:**                           High:    12991 hours          (1.5 years)
                                        Low:     1299 hours           (54 days)
   *Comment:* Based upon photooxidation half-life in air.

   · **Surface Water:**                 High:    672 hours            (4 weeks)
                                        Low:     168 hours            (1 week)
   *Comment:* Scientific judgement based upon aerobic river die-away test data (Ludzack, FJ et al. (1958)).

   · **Ground Water:**                  High:    8640 hours           (8 weeks)
                                        Low:     336 hours            (2 weeks)
   *Comment:* Scientific judgement based upon estimated aqueous aerobic biodegradation half-life.

**Aqueous Biodegradation (unacclimated):**
   · **Aerobic half-life:**             High:    672 hours            (4 weeks)
                                        Low:     168 hours            (1 week)
   *Comment:* Scientific judgement based upon aerobic river die-away test data (Ludzack, FJ et al. (1958)).

   · **Anaerobic half-life:**           High:    2688 hours           (16 weeks)
                                        Low:     672 hours            (4 weeks)
   *Comment:* Scientific judgement based upon estimated aqueous aerobic biodegradation half-life.

   · **Removal/secondary treatment:**   High:        94%
                                        Low:         27%
   *Comment:* Based upon % degraded under aerobic (high %) and anaerobic (low %) continuous flow conditions (Ludzack, FJ et al. (1961)).

**Photolysis:**
   · **Atmos photol half-life:**        High:  Not expected to be important
                                        Low:
   *Comment:* Stable under photolysis conditions in air (Kagiya, T et al. (1975)).

   · **Max light absorption (nm):**     lambda max not reported.

*Comment:* No absorbance occurs above 200 nm (Silverstein, RM and Bassler, GC (1967)).

· **Aq photol half-life:**       High:  Not expected to be important
                                 Low:
*Comment:* Scientific judgement based upon photolysis data in air.

**Photooxidation half-life:**
  · **Water:**                   High: $1.1X10^8$ hours        (12559 years)
                                 Low: $2.8X10^6$ hours         (314 years)
*Comment:* Based upon measured rate data for hydroxyl radicals in aqueous solution (Dorfman, LM and Adams, GE (1973)).

  · **Air:**                     High:   12991 hours           (1.5 years)
                                 Low:    1299 hours            (54 days)
*Comment:* Based upon measured rate data for the vapor phase reaction with hydroxyl radicals in air (Atkinson, R (1985)).

**Reduction half-life:**         High:      No data
                                 Low:
*Comment:*

**Hydrolysis:**
  · **First-order hydr half-life:**      >150,000 yr
*Comment:* Based upon base rate constant ($5.8X10^{-3}$ $M^{-1}$ $hr^{-1}$) at pH of 7 and 25 °C (Ellington, JJ et al. (1987)).

  · **Acid rate const $(M(H+)-hr)^{-1}$:**
*Comment:*

  · **Base rate const $(M(OH-)-hr)^{-1}$:**       $5.8X10^{-3}$
*Comment:* Reported rate constant at pH 11 and 25 °C (Ellington, JJ et al. (1987)).

# Dichloromethane

CAS Registry Number: 75-09-2

## Half-lives:
· **Soil:**　　　　　　　　　　　　High:　672 hours　　　　(4 weeks)

Low:　168 hours　　　　(7 days)

*Comment:* Scientific judgement based upon estimated unacclimated aqueous aerobic biodegradation half-life.

· **Air:**　　　　　　　　　　　　High:　4584 hours　　　(191 days)

Low:　458 hours　　　(19.1 days)

*Comment:* Based upon photooxidation half-life in air.

· **Surface Water:**　　　　　　　High:　672 hours　　　　(4 weeks)

Low:　168 hours　　　　(7 days)

*Comment:* Scientific judgement based upon estimated unacclimated aqueous aerobic biodegradation half-life.

· **Ground Water:**　　　　　　　High:　1344 hours　　　(8 weeks)

Low:　336 hours　　　(14 days)

*Comment:* Scientific judgement based upon estimated unacclimated aqueous aerobic biodegradation half-life.

## Aqueous Biodegradation (unacclimated):
· **Aerobic half-life:**　　　　　High:　672 hours　　　　(4 weeks)

Low:　168 hours　　　　(7 days)

*Comment:* Scientific judgement based upon unacclimated aerobic screening test data (Tabak, HH et al. (1981), Kawasaki, M (1980)).

· **Anaerobic half-life:**　　　　High:　2688 hours　　　(16 weeks)

Low:　672 hours　　　(28 days)

*Comment:* Scientific judgement based upon unacclimated aerobic biodegradation half-life.

· **Removal/secondary treatment:**　High:　　94.5%

Low:

*Comment:* Based upon % degraded under aerobic continuous flow conditions (Kincannon, DF et al. (1983)).

## Photolysis:
· **Atmos photol half-life:**　　High:　Not expected to be important

Low:

*Comment:* Scientific judgement based upon aqueous photolysis data.

· **Max light absorption (nm):**      Lambda max = 250, 220
*Comment:* Absorption maxima (Hubrich, C, Stuhl, F (1980)).

· **Aq photol half-life:**      High:  Not expected to be important
                                Low:
*Comment:* Stable in water exposed to sunlight for one year (Dilling, WL et al. (1975)).

## Photooxidation half-life:
· **Water:**                  High:      No data
                              Low:

*Comment:*

· **Air:**                    High:    4584 hours          (191 days)
                              Low:     458 hours           (19.1 days)
*Comment:* Based upon measured rate data for the vapor phase reaction with hydroxyl radicals in air (Atkinson, R (1985)).

## Reduction half-life:                High:      No data
                                        Low:
*Comment:*

## Hydrolysis:
· **First-order hydr half-life:**      704 years
*Comment:* Scientific judgement based upon reported rate constant ($3.2 \times 10^{-11}$ s$^{-1}$) at pH 7 and 25 °C (Mabey, W and Mill, T (1978)).

· **Acid rate const (M(H+)-hr)$^{-1}$:**
*Comment:*

· **Base rate const (M(OH-)-hr)$^{-1}$:**
*Comment:*

# Ethylene oxide

CAS Registry Number: 75-21-8

Half-lives:
· Soil:                          High:    285 hours              (11.9 days)
                                 Low:     251 hours              (10.5 days)
Comment: Scientific judgement based upon hydrolysis half-lives at pH 7 and 9 (high $t_{1/2}$) and pH 5 (low $t_{1/2}$) (Mabey, W and Mill, T (1978)).

· Air:                           High:    9167 hours             (382 days)
                                 Low:     917 hours              (38.2 days)
Comment: Based upon photooxidation half-life in air.

· Surface Water:                 High:    285 hours              (11.9 days)
                                 Low:     251 hours              (10.5 days)
Comment: Scientific judgement based upon hydrolysis half-lives at pH 7 and 9 (high $t_{1/2}$) and pH 5 (low $t_{1/2}$) (Mabey, W and Mill, T (1978)).

· Ground Water:                  High:    285 hours              (11.9 days)
                                 Low:     251 hours              (10.5 days)
Comment: Scientific judgement based upon hydrolysis half-lives at pH 7 and 9 (high $t_{1/2}$) and pH 5 (low $t_{1/2}$) (Mabey, W and Mill, T (1978)).

Aqueous Biodegradation (unacclimated):
· Aerobic half-life:             High:    4320 hours             (6 months)
                                 Low:     672 hours              (4 weeks)
Comment: Scientific judgement based upon unacclimated aerobic aqueous screening test data (Bridie, AL et al. (1979); Conway, RA et al. (1983)).

· Anaerobic half-life:           High:    17280 hours            (24 months)
                                 Low:     2688 hours             (16 weeks)
Comment: Scientific judgement based upon estimated aerobic biodegradation half-life.

· Removal/secondary treatment:   High:    No data
                                 Low:
Comment:

Photolysis:
· Atmos photol half-life:        High:
                                 Low:
Comment:

· **Max light absorption (nm):**      No data
*Comment:*

· **Aq photol half-life:**      High:
                                Low:
*Comment:*

**Photooxidation half-life:**
· **Water:**      High:  Expected to be extremely slow
                  Low:
*Comment:* Anbar, M and Neta, P (1967); Hendry, DG et al. (1974); Brown, SL et al. (1975C).

· **Air:**      High:   9167 hours           (382 days)
                Low:    917 hours            (38.2 days)
*Comment:* Based upon measured rate constant for reaction with hydroxyl radical (Atkinson, R (1987A)).

**Reduction half-life:**      High:   No data
                              Low:
*Comment:*

**Hydrolysis:**
· **First-order hydr half-life:**      285 hours at 25°C   (11.9 days)
*Comment:* ($t_{1/2}$ at pH 7) Based upon measured first-order and base and acid catalyzed hydrolysis rate constants (Mabey, W and Mill, T (1978)).

· **Acid rate const (M(H+)-hr)$^{-1}$:**      $9.3X1^{-3}$ $M^{-1}$ $s^{-1}$ at 25°C
*Comment:* ($t_{1/2}$ = 251 hours (10.5 days) at pH 5) Based upon measured first-order and base and acid catalyzed hydrolysis rate constants (Mabey, W and Mill, T (1978)).

· **Base rate const (M(OH-)-hr)$^{-1}$:**      $1.0X10^{-4}$ $M^{-1}$ $s^{-1}$ at 25°C
*Comment:* ($t_{1/2}$ = 285 hours (11.9 days) at pH 9) Based upon measured first-order and base and acid catalyzed hydrolysis rate constants (Mabey, W and Mill, T (1978)).

# Bromoform

**CAS Registry Number:** 75-25-2

**Half-lives:**
  · **Soil:**                        High:    4320 hours         (6 months)
                                     Low:     672 hours          (4 weeks)
*Comment:* Scientific judgement based upon unacclimated aqueous aerobic biodegradation half-life.

  · **Air:**                         High:    12989 hours       (541 days)
                                       Low:     1299 hours        (54.1 days)
*Comment:* Scientific judgement based upon estimated photooxidation half-life in air.

  · **Soil:**                        High:    4320 hours         (6 months)
                                     Low:     672 hours          (4 weeks)
*Comment:* Scientific judgement based upon unacclimated aqueous aerobic biodegradation half-life.

  · **Ground Water:**            High:    8640 hours         (12 months)
                                     Low:     1344 hours        (8 weeks)
*Comment:* Scientific judgement based upon unacclimated aqueous aerobic biodegradation half-life.

**Aqueous Biodegradation (unacclimated):**
  · **Soil:**                        High:    4320 hours         (6 months)
                                     Low:     672 hours          (4 weeks)
*Comment:* Scientific judgement based upon unacclimated aerobic aqueous screening test data from experiments utilizing settled domestic wastewater inoculum (Bouwer, EJ et al. (1984)).

  · **Anaerobic half-life**:          High:    17280 hours       (24 months)
                                     Low:     2688 hours        (16 weeks)
*Comment:* Scientific judgement based upon unacclimated aqueous aerobic biodegradation half-life.

  · **Removal/secondary treatment:**    High:            68%
                                     Low:
*Comment:* Removal percentage based upon data from a continuous activated sludge biological treatment simulator (Hannah, SA et al. (1986)).

**Photolysis:**
  · **Atmos photol half-life:**         High:
                                     Low:

*Comment:*

· **Max light absorption (nm):**     No data
*Comment:*

· **Aq photol half-life:**     High:
                                      Low:

*Comment:*

## Photooxidation half-life:
  · **Water:**     High:     No data
                          Low:

*Comment:*

  · **Air:**     High:    12989 hours        (541 days)
                        Low:     1299 hours        (54.1 days)
*Comment:* Scientific judgement based upon estimated rate constant for reaction with hydroxyl radical in air (Atkinson, R (1987A)).

## Reduction half-life:     High:     No data
                                Low:
*Comment:*

## Hydrolysis:
  · **First-order hydr half-life:**     $6.02 \times 10^6$ hours (687 years)
*Comment:* ($t_{1/2}$ at pH 7 and 25°C)
Based upon measured base catalyzed hydrolysis rate constant (Mabey, W and Mill, T (1978)).

  · **Acid rate const $(M(H+)\text{-}hr)^{-1}$:**     No data
*Comment:*

  · **Base rate const $(M(OH\text{-})\text{-}hr)^{-1}$:**     1.15 $M^{-1}$ $s^{-1}$
*Comment:* ($t_{1/2} = 6.02 \times 10^4$ hours (6.9 years) at pH 9 and 25°C)
Based upon measured base catalyzed hydrolysis rate constant (Mabey, W and Mill, T (1978)).

# 1,1-Dichloroethane

CAS Registry Number:  75-34-3

## Half-lives:
**· Soil:**                          High:    3696 hours              (22 weeks)
                                     Low:     768 hours               (32 days)
*Comment:* Scientific judgement based upon methane acclimated soil grab sample data (low $t_{1/2}$: Henson, JM et al. (1989)) and sub-soil grab sample data from a ground water aquifer (high $t_{1/2}$: Wilson, JT et al. (1983)).

**· Air:**                           High:    2468 hours              (103 days)
                                     Low:     247 hours               (10.3 days)
*Comment:* Based upon photooxidation half-life in air.

**· Surface Water:**                 High:    3696 hours              (22 weeks)
                                     Low:     768 hours               (32 days)
*Comment:* Scientific judgement based upon estimated aqueous aerobic biodegradation half-life.

**· Ground Water:**                  High:    8640 hours              (22 weeks)
                                     Low:     1344 hours              (64 days)
*Comment:* Scientific judgement based upon estimated aqueous aerobic biodegradation half-life and sub-soil grab sample data from a ground water aquifer (high $t_{1/2}$: Wilson, JT et al. (1983)).

## Aqueous Biodegradation (unacclimated):
**· Aerobic half-life:**             High:    3696 hours              (22 weeks)
                                     Low:     768 hours               (32 days)
*Comment:* Scientific judgement based upon estimated methane acclimated soil grab sample data (low $t_{1/2}$: Henson, JM et al. (1989)) and sub-soil grab sample data from a ground water aquifer (high $t_{1/2}$: Wilson, JT et al. (1983)).

**· Anaerobic half-life:**           High:    14784 hours             (88 weeks)
                                     Low:     3072 hours              (128 days)
*Comment:* Scientific judgement based upon estimated aqueous aerobic biodegradation half-life.

**· Removal/secondary treatment:**   High:    No data
                                     Low:
*Comment:*

## Photolysis:
**· Atmos photol half-life:**        High:    No data
                                     Low:
*Comment:*

· **Max light absorption (nm):**          No data
*Comment:*

· **Aq photol half-life:**          High:      No data
                                    Low:
*Comment:*

**Photooxidation half-life:**
· **Water:**          High:      No data
                                    Low:
*Comment:*

· **Air:**          High:    2468 hours        (103 days)
                            Low:     247 hours        (10.3 days)
*Comment:* Based upon measured rate data for the vapor phase reaction with hydroxyl radicals in air (Atkinson, R (1985)).

**Reduction half-life:**          High:      No data
                                    Low:
*Comment:*

**Hydrolysis:**
· **First-order hydr half-life:**      No data
*Comment:*

· **Acid rate const $(M(H+)\text{-}hr)^{-1}$:**
*Comment:*

· **Base rate const $(M(OH\text{-})\text{-}hr)^{-1}$:**
*Comment:*

# 1,1-Dichloroethylene

<u>CAS Registry Number:</u> 75-35-4

<u>Half-lives:</u>

· **Soil:**                           High:    4320 hours          (6 months)
                                      Low:      672 hours          (4 weeks)
*Comment:* Scientific judgement based upon estimated aqueous aerobic biodegradation half-life.

· **Air:**                            High:    98.7 hours          (4.1 days)
                                      Low:       9.9 hours
*Comment:* Based upon photooxidation half-life in air.

· **Surface Water:**                  High:    4320 hours          (6 months)
                                      Low:      672 hours          (4 weeks)
*Comment:* Scientific judgement based upon estimated aqueous aerobic biodegradation half-life.

· **Ground Water:**                   High:    3168 hours          (132 days)
                                      Low:     1344 hours          (8 weeks)
*Comment:* Scientific judgement based upon estimated aqueous aerobic biodegradation half-life and anaerobic grab sample data for soil from a ground water aquifer receiving landfill leachate (Wilson, BH et al. (1986)).

<u>Aqueous Biodegradation (unacclimated):</u>

· **Aerobic half-life:**              High:    4320 hours          (6 months)
                                      Low:      672 hours          (4 weeks)
*Comment:* Scientific judgement based upon acclimated aerobic soil screening test data (Tabak, HH et al. (1981)).

· **Anaerobic half-life:**            High:    4152 hours          (173 days)
                                      Low:     1944 hours          (81 days)
*Comment:* Scientific judgement based upon anaerobic sediment grab sample data (Barrio-Lage, G et al. (1986)).

· **Removal/secondary treatment:**    High:        92%
                                      Low:        58%
*Comment:* Based upon % degraded under aerobic continuous flow conditions (Hannah, SA et al. (1986)).

<u>Photolysis:</u>

· **Atmos photol half-life:**         High:    No data
                                      Low:
*Comment:*

· **Max light absorption (nm):**          No data
*Comment:*

· **Aq photol half-life:**          High:          No data
                                    Low:
*Comment:*

**Photooxidation half-life:**
  · **Water:**          High:          No data
                        Low:
*Comment:*

  · **Air:**          High:          98.7 hours          (4.1 days)
                      Low:          9.9 hours
*Comment:* Based upon measured rate data for the vapor phase reaction with hydroxyl radicals in air (Goodman, MA et al. (1986)).

**Reduction half-life:**          High:          No data
                                  Low:
*Comment:*

**Hydrolysis:**
  · **First-order hydr half-life:**          No hydrolyzable groups
*Comment:* Rate constant at pH 3, 7, and 11 is zero (Kollig, HP et al. (1987A)).

  · **Acid rate const $(M(H+)-hr)^{-1}$:**
*Comment:*

  · **Base rate const $(M(OH-)-hr)^{-1}$:**
*Comment:*

# Phosgene

<u>CAS Registry Number:</u>  75-44-5

<u>Half-lives:</u>
  · **Soil:**                        High:        1 hour
                                     Low:    0.05 hours              (3 minutes)
  *Comment:*  Scientific judgement based upon hydrolysis half-life in 89% acetone soln at -20 °C
  (Ugi, I and Beck, F (1961)).

  · **Air:**                         High:$7.2X10^{14}$ hours        ($8.2X10^{10}$ years)
                                     Low:  989880 hours             (113 years)
  *Comment:*  Scientific judgement based upon hydrolysis half-lives in air.

  · **Surface Water:**               High:        1 hour
                                     Low:    0.05 hours              (3 minutes)
  *Comment:*  Scientific judgement based upon hydrolysis half-life in 89% acetone soln at -20 °C
  (Ugi, I and Beck, F (1961)).

  · **Ground Water:**                High:        1 hour
                                     Low:    0.05 hours              (3 minutes)
  *Comment:*  Scientific judgement based upon hydrolysis half-life in 89% acetone soln at -20 °C
  (Ugi, I and Beck, F (1961)).

<u>Aqueous Biodegradation (unacclimated):</u>
  · **Aerobic half-life:**           High:     672 hours            (4 weeks)
                                     Low:      168 hours            (7 days)
  *Comment:*  Scientific judgement.

  · **Anaerobic half-life:**         High:     2688 hours           (16 weeks)
                                     Low:      672 hours            (28 days)
  *Comment:*  Scientific judgement based upon aerobic aqueous biodegradation half-life.

  · **Removal/secondary treatment:**  High:        No data
                                      Low:
  *Comment:*

<u>Photolysis:</u>
  · **Atmos photol half-life:**      High:        No data
                                     Low:
  *Comment:*

  · **Max light absorption (nm):**   lambda max = 274.5, 264.5, 254.3, 234.2

*Comment:* No absorbance of UV light at wavelengths >290 nm (Montgomery, CW and Rollefson GK (1933)).

· **Aq photol half-life:**  High: No data
Low:

*Comment:*

**Photooxidation half-life:**
· **Water:**  High: No data
Low:

*Comment:*

· **Air:**  High: Not expected to be important
Low:

*Comment:* Based upon rate data for vapor phase reaction with hydroxyl radicals in air (Cupitt, LT (1980)).

**Reduction half-life:**  High: No data
Low:

*Comment:*

**Hydrolysis:**
· **First-order hydr half-life:**  1.9 hours
*Comment:* Scientific judgement based upon measured rate data ($k_N = 1.02 \times 10^{-4}$ s$^{-1}$) in 89% acetone solution at -20 °C (Ugi, I and Beck, F (1961)).

· **Acid rate const (M(H+)-hr)$^{-1}$:**  No data
*Comment:*

· **Base rate const (M(OH-)-hr)$^{-1}$:**  No data
*Comment:*

**Hydrolysis in Air:**  High: $7.2 \times 10^{14}$ hours  ($8.2 \times 10^{10}$ years)
Low: 989880 hours  (113 years)
*Comment:* Scientific judgement based upon hydrolysis rate constant ($3.67 \times 10^{-7}$ mol-1 s$^{-1}$) in air at sea level and the stratosphere (Snelson, A et al. (1978)).

# 2-Methylaziridine

**CAS Registry Number:** 75-55-8

**Structure:**

NH
$\triangleright$— CH$_3$

**Half-lives:**
  **· Soil:**                            High:      870 hours        (36 days)
                                         Low:       87 hours         (3.6 days)
*Comment:* Scientific judgement based upon hydrolysis half-lives for pH 5 to 7 at 25 and 5 °C (Ellington, JJ et al. (1987)).

  **· Air:**                             High:      10.6 hours
                                         Low:       1.1 hours
*Comment:* Scientific judgement based upon estimated photooxidation half-life in air.

  **· Surface Water:**                   High:      870 hours        (36 days)
                                         Low:       87 hours         (3.6 days)
*Comment:* Scientific judgement based upon hydrolysis half-lives for pH 5 to 7 at 25 and 5 °C (Ellington, JJ et al. (1987)).

  **· Ground Water:**                    High:      870 hours        (36 days)
                                         Low:       87 hours         (3.6 days)
*Comment:* Scientific judgement based upon hydrolysis half-lives for pH 5 to 7 at 25 and 5 °C (Ellington, JJ et al. (1987)).

**Aqueous Biodegradation (unacclimated):**
  **· Aerobic half-life:**               High:      672 hours        (4 weeks)
                                         Low:       168 hours        (1 week)
*Comment:* Scientific judgement.

  **· Anaerobic half-life:**             High:      2688 hours       (16 weeks)
                                         Low:       672 hours        (4 weeks)
*Comment:* Scientific judgement based upon estimated aqueous aerobic biodegradation half-life.

  **· Removal/secondary treatment:**     High:      No data
                                         Low:
*Comment:*

**Photolysis:**

154

**· Atmos photol half-life:**        High:
                                       Low:

*Comment:*

**· Max light absorption (nm):**     No data
*Comment:*

**· Aq photol half-life:**         High:
                                       Low:

*Comment:*

**Photooxidation half-life:**
    **· Water:**                   High:      No data
                                       Low:

*Comment:*

    **· Air:**                     High:    10.6 hours
                                       Low:     1.1 hours
*Comment:* Scientific judgement based upon estimated rate data for hydroxyl radicals in air (Atkinson, R (1987A)).

**Reduction half-life:**           High:      No data
                                       Low:

*Comment:*

**Hydrolysis:**
    **· First-order hydr half-life:**     87 hours  (3.6 days)
*Comment:* Based upon neutral rate constant ($k_N = 8.0 \times 10^{-3}$ hr$^{-1}$) at pH 7 and 25 °C (Ellington, JJ et al. (1987)).

    **· First-order hydr half-life:**     870 hours  (36 days)
*Comment:* Based upon a neutral rate constant ($k_N = 8.0 \times 10^{-4}$ hr$^{-1}$) at pH 7 and 5 °C (Ellington, JJ et al. (1987)).

    **· Acid rate const (M(H+)-hr)$^{-1}$:**     $4.0 \times 10^{-3}$
*Comment:* ($t_{1/2}$ = 87 days) Based upon a neutral ($k_N = 8.0 \times 10^{-3}$ hr$^{-1}$) and acid rate constants at pH 5 and 25 °C (Ellington, JJ et al. (1987)).

    **· Base rate const (M(OH-)-hr)$^{-1}$:**
*Comment:*

# tert-Butyl alcohol

CAS Registry Number:  75-65-0

## Half-lives:
· **Soil:**                                         High:    4800 hours                    (200 days)
                                                    Low:      360 hours                    (15 days)
*Comment:*  Based upon soil microcosm studies (Novak, JT et al. (1985); Wilson, WG et al. (1986)).

· **Air:**                                          High:      590 hours                   (24.5 days)
                                                    Low:         59 hours                  (2.45 days)
*Comment:*  Based upon measured photooxidation half-life in air.

· **Surface Water:**                                High:    4320 hours                    (6 months)
                                                    Low:      672 hours                    (4 weeks)
*Comment:*  Scientific judgement based upon unacclimated aqueous aerobic biodegradation half-life.

· **Ground Water:**                                 High:    8640 hours                    (12 months)
                                                    Low:     1344 hours                    (8 weeks)
*Comment:*  Scientific judgement based upon unacclimated aqueous aerobic biodegradation half-lives.

## Aqueous Biodegradation (unacclimated):
· **Aerobic half-life:**                            High:    4320 hours                    (6 months)
                                                    Low:      677 hours                    (4 weeks)
*Comment:*  Scientific judgement based upon river die-away studies (Hammerton, C (1955); Sokolova, LP and Kaplin, VT (1969)).

· **Anaerobic half-life:**                          High:   12000 hours                    (500 days)
                                                    Low:     2400 hours                    (100 days)
*Comment:*  Based upon observed degradation rates in microcosm studies simulating anaerobic aquifers (Novak JT et al. (1985)).

· **Removal/secondary treatment:**                  High:       No data
                                                    Low:
*Comment:*

## Photolysis:
· **Atmos photol half-life:**                       High: No photolyzable functions
                                                    Low:
*Comment:*

156

· **Max light absorption (nm):**
*Comment:*

· **Aq photol half-life:**                                    High:
                                                            Low:

*Comment:*

**Photooxidation half-life:**
  · **Water:**                                    High: $5.7 \times 10^8$ hours        (64500 years)
                                                            Low:    18480 hours            (771 days)
*Comment:* Based upon measured reaction with ·OH radicals in water (Anbar M and Neta P (1967); Dorfman LM and Adams GE (1973)).

  · **Air:**                                        High:        590 hours        (24.5 days)
                                                            Low:          59 hours        (2.45 days)
*Comment:*  Based upon measured photooxidation half-life in air via reaction with ·OH radicals (Atkinson, R (1985)).

**Reduction half-life:**                                    High:        No data
                                                            Low:
*Comment:*

**Hydrolysis:**
  · **First-order hydr half-life:**                No hydrolyzable functions
*Comment:*

  · **Acid rate const (M(H+)-hr)$^{-1}$:**
*Comment:*

  · **Base rate const (M(OH-)-hr)$^{-1}$:**
*Comment:*

# Trichlorofluoromethane

CAS Registry Number: 75-69-4

## Half-lives:
· **Soil:**                                         High:    8640 hours              (1 year)
                                                   Low:     4320 hours              (6 months)
*Comment:* Scientific judgement based upon estimated aqueous aerobic biodegradation half-life.

· **Air:**                                          High: 1.3X10$^6$ hours            (147 years)
                                                   Low: 1.3X10$^5$ hours             (14.7 years)
*Comment:* Based upon photooxidation half-life in air.

· **Surface Water:**                                High:    8640 hours              (1 year)
                                                   Low:     4032 hours              (6 months)
*Comment:* Scientific judgement based upon estimated aqueous aerobic biodegradation half-life.

· **Ground Water:**                                 High:   17280 hours              (2 years)
                                                   Low:     8640 hours              (12 months)
*Comment:* Scientific judgement based upon estimated aqueous aerobic biodegradation half-life.

## Aqueous Biodegradation (unacclimated):
· **Aerobic half-life:**                            High:    8640 hours              (1 year)
                                                   Low:     4032 hours              (6 months)
*Comment:* Scientific judgement based upon acclimated aerobic screening test data (Tabak, HH et al. (1981)).

· **Anaerobic half-life:**                          High:   34560 hours              (4 years)
                                                   Low:    16128 hours              (24 months)
*Comment:* Scientific judgement based upon estimated aqueous aerobic biodegradation half-life.

· **Removal/secondary treatment:**                  High:        No data
                                                   Low:
*Comment:*

## Photolysis:
· **Atmos photol half-life:**                       High:        No data
                                                   Low:
*Comment:*

· **Max light absorption (nm):** Lambda max = 260, 240, 230
*Comment:* Absorption maxima (Hubrich, C and Stuhl, F (1980)). No absorbance above 290 nm (Hustert, K et al.(1981)).

**· Aq photol half-life:**     High:  No data
                Low:

*Comment:*

## Photooxidation half-life:
 **· Water:**        High:  No data
                Low:

*Comment:*

 **· Air:**         High: $1.3 \times 10^6$ hours   (147 years)
                Low: $1.3 \times 10^5$ hours   (14.7 years)

*Comment:* Based upon an measured rate data for the vapor phase reaction with hydroxyl radicals in air (Atkinson, R (1985)).

## Reduction half-life:
                High:  No data
                Low:

*Comment:*

## Hydrolysis:
 **· First-order hydr half-life:**  No data
 *Comment:*

 **· Acid rate const $(M(H+)-hr)^{-1}$:**
 *Comment:*

 **· Base rate const $(M(OH-)-hr)^{-1}$:**
 *Comment:*

# Dichlorodifluoromethane

**CAS Registry Number:** 75-71-8

**Half-lives:**

· **Soil:**                                             High:    4320 hours              (6 months)
                                                          Low:      672 hours              (4 weeks)
  *Comment:* Scientific judgement based upon estimated aqueous aerobic biodegradation half-life.

· **Air:**                                              High:    21180 hours             (2.4 years)
                                                          Low:      2118 hours             (88.3 days)
  *Comment:* Based upon photooxidation half-life in air.

· **Surface Water:**                                    High:    4320 hours              (6 months)
                                                          Low:      672 hours              (4 weeks)
  *Comment:* Scientific judgement based upon estimated aqueous aerobic biodegradation half-life.

· **Ground Water:**                                     High:    8640 hours              (12 months)
                                                          Low:      1344 hours             (8 weeks)
  *Comment:* Scientific judgement based upon estimated aqueous aerobic biodegradation half-life.

**Aqueous Biodegradation (unacclimated):**

· **Aerobic half-life:**                                High:    4032 hours              (6 months)
                                                          Low:      672 hours              (4 weeks)
  *Comment:* Scientific judgement based upon acclimated aerobic screening test data for trichlorofluoromethane (Tabak, HH et al. (1981)).

· **Anaerobic half-life:**                              High:    16128 hours             (24 months)
                                                          Low:      2688 hours             (16 weeks)
  *Comment:* Scientific judgement based upon estimated aqueous aerobic biodegradation half-life.

· **Removal/secondary treatment:**                      High:    No data
                                                          Low:
  *Comment:*

**Photolysis:**

· **Atmos photol half-life:**                           High:    No data
                                                          Low:
  *Comment:*

· **Max light absorption (nm):** Lambda max <200
  *Comment:* Absorption maxima (Weast, RC et al. (1962)).

**· Aq photol half-life:**         High:      No data
                                   Low:

*Comment:*

## Photooxidation half-life:
**· Water:**                       High:      No data
                                   Low:

*Comment:*

**· Air:**                         High:    21180 hours         (2.4 years)
                                   Low:     2118 hours          (88.3 days)

*Comment:* Based upon measured rate data for the vapor phase reaction with hydroxyl radicals in air (Atkinson, R (1985)).

## Reduction half-life:           High:      No data
                                   Low:

*Comment:*

## Hydrolysis:
**· First-order hydr half-life:**  No data

*Comment:*

**· Acid rate const (M(H+)-hr)$^{-1}$:**

*Comment:*

**· Base rate const (M(OH-)-hr)$^{-1}$:**

*Comment:*

# Dalapon

CAS Registry Number: 75-99-0

Structure:

$$O$$
$$Cl$$
$$\\ _{OH}$$
$$H_3C \quad Cl$$

Half-lives:
· Soil:                               High:    1440 hours              (60 days)
                                      Low:     336 hours              (14 days)
Comment: Scientific judgement based upon unacclimated aerobic soil grab sample data (low $t_{1/2}$:
Corbin, FT and Upchurch, RP (1967); high $t_{1/2}$: Kaufman, DD and Doyle, RD (1977)).

· Air:                                High:    2893 hours              (120.5 days)
                                      Low:     289 hours              (12.1 days)
Comment: Scientific judgement based upon estimated photooxidation half-life in air.

· Surface Water:                      High:    1440 hours              (60 days)
                                      Low:     336 hours              (14 days)
Comment: Scientific judgement based upon estimated aqueous aerobic biodegradation half-life.

· Ground Water:                       High:    2880 hours              (120 days)
                                      Low:     672 hours              (28 days)
Comment: Scientific judgement based upon estimated aqueous aerobic biodegradation half-life.

Aqueous Biodegradation (unacclimated):
· Aerobic half-life:                  High:    1440 hours              (60 days)
                                      Low:     336 hours              (14 days)
Comment: Scientific judgement based upon unacclimated aerobic soil grab sample data (low $t_{1/2}$:
Corbin, FT and Upchurch, RP (1967); high $t_{1/2}$: Kaufman, DD and Doyle, RD (1977)).

· Anaerobic half-life:                High:    5760 hours              (240 days)
                                      Low:     1344 hours             (56 days)
Comment: Scientific judgement based upon estimated aqueous aerobic biodegradation half-life.

· Removal/secondary treatment:        High:    No data
                                      Low:
Comment:

Photolysis:
· Atmos photol half-life:             High:    No data

Low:

*Comment:*

· **Max light absorption (nm):**          No data
*Comment:*

· **Aq photol half-life:**          High:          No data
                                     Low:

*Comment:*

**Photooxidation half-life:**
  · **Water:**                        High:          No data
                                      Low:

*Comment:*

  · **Air:**                          High:     2893 hours          (120.5 days)
                                      Low:      289 hours           (12.1 days)
*Comment:* Scientific judgement based upon an estimated rate constant for vapor phase reaction with hydroxyl radicals in air (Atkinson, R (1987A)).

**Reduction half-life:**               High:          No data
                                       Low:

*Comment:*

**Hydrolysis:**
  · **First-order hydr half-life:**    No data
*Comment:*

  · **Acid rate const (M(H+)-hr)$^{-1}$:**
*Comment:*

  · **Base rate const (M(OH-)-hr)$^{-1}$:**
*Comment:*

# 1,1,2-Trichloro-1,2,2-trifluoroethane

**CAS Registry Number:** 76-13-1

## Half-lives:

· **Soil:**
    High:   8640 hours      (1 year)
    Low:    4320 hours      (6 months)
*Comment:* Scientific judgement based upon estimated aqueous aerobic biodegradation half-life.

· **Air:**
    High: $8.8 \times 10^6$ hours    (1000 years)
    Low: $3.5 \times 10^5$ hours    (40 years)
*Comment:* Tropospheric rate based upon measured reaction rates with O(1D) radicals (Davidson, JA et al. (1978); Pitts, JN Jr et al. (1974)).

· **Surface Water:**
    High:   8640 hours      (1 year)
    Low:    4320 hours      (6 months)
*Comment:* Scientific judgement based upon estimated aqueous aerobic biodegradation half-life.

· **Ground Water:**
    High:  17280 hours     (2 years)
    Low:   1440 hours      (1 year)
*Comment:* Scientific judgement based upon estimated aqueous aerobic biodegradation half-life.

## Aqueous Biodegradation (unacclimated):

· **Aerobic half-life:**
    High:   8640 hours      (1 year)
    Low:    4320 hours      (6 months)
*Comment:* Scientific judgement based upon a relative resistance of completely halogenated aliphatics to biodegrade.

· **Anaerobic half-life:**
    High:  34560 hours     (4 years)
    Low:  17280 hours     (2 years)
*Comment:* Scientific judgement based upon estimated aqueous aerobic biodegradation half-life.

· **Removal/secondary treatment:**
    High:    No data
    Low:
*Comment:*

## Photolysis:

· **Atmos photol half-life:**
    High:    Infinite
    Low:
*Comment:* Does not directly photolyze in the troposphere; photolyzes in the stratosphere.

· **Max light absorption (nm):** lambda max below 240 nm
*Comment:* Hubrich, C and Stuhl, F (1980)

· **Aq photol half-life:**                    High:        Infinite
                                              Low:

*Comment:* Does not directly photolyze in the troposphere.

## Photooxidation half-life:
· **Water:**                                  High:        No data
                                              Low:

*Comment:*

· **Air:**                                    High: $8.8 \times 10^6$ hours         (1000 years)
                                              Low: $3.5 \times 10^5$ hours          (40 years)

*Comment:* Tropospheric rate based upon measured reaction rates with O(1D) radicals (Davidson, JA et al. (1978); Pitts, JN Jr et al. (1974)). Does not react with ·OH or ozone in the troposphere.

## Reduction half-life:                        High:        No data
                                              Low:

*Comment:*

## Hydrolysis:
· **First-order hydr half-life:**             No data
*Comment:*

· **Acid rate const $(M(H+)-hr)^{-1}$:**
*Comment:*

· **Base rate const $(M(OH-)-hr)^{-1}$:**
*Comment:*

# Heptachlor

CAS Registry Number: 76-44-8

Structure:

## Half-lives:
**· Soil:**    High:    129.4 hours    (5.4 days)
              Low:    23.1 hours
*Comment:* Scientific judgement based upon hydrolysis half-lives (low $t_{1/2}$: Kollig, HP et al. (1987A); high $t_{1/2}$: Chapman, RA and Cole, CM (1982)).

**· Air:**    High:    9.8 hours
              Low:    59 minutes
*Comment:* Scientific judgement based upon estimated photooxidation half-life in air.

**· Surface Water:**    High:    129.4 hours    (5.4 days)
                        Low:    23.1 hours
*Comment:* Scientific judgement based upon hydrolysis half-lives (low $t_{1/2}$: Kollig, HP et al. (1987A); high $t_{1/2}$: Chapman, RA and Cole, CM (1982)).

**· Ground Water:**    High:    129.4 hours    (5.4 days)
                       Low:    23.1 hours
*Comment:* Scientific judgement based upon hydrolysis half-lives (low $t_{1/2}$: Kollig, HP et al. (1987A); high $t_{1/2}$: Chapman, RA and Cole, CM (1982)).

## Aqueous Biodegradation (unacclimated):
**· Aerobic half-life:**    High:    1567 hours    (65 days)
                           Low:    360 hours    (15 days)
*Comment:* Scientific judgement based upon unacclimated aerobic soil grab sample data (Castro, TF and Yoshida, T (1971)).

**· Anaerobic half-life:**    High:    6268 hours    (260 days)
                             Low:    1440 hours    (60 days)
*Comment:* Scientific judgement based upon unacclimated aerobic biodegradation half-life.

**· Removal/secondary treatment:**    High:    65%
                                      Low:    53%
*Comment:* Based upon % degraded under aerobic continuous flow conditions (Hannah, SA et al.

(1986)).

## Photolysis:
· **Atmos photol half-life:**   High:   No data
                                Low:
*Comment:*

· **Max light absorption (nm):**   Lambda max = 236, 309, 328
*Comment:* Reported absorption maxima (IARC (1979)).

· **Aq photol half-life:**   High:   No data
                             Low:
*Comment:*

## Photooxidation half-life:
· **Water:**   High:   No data
               Low:
*Comment:*

· **Air:**   High:   9.8 hours
             Low:    59 minutes
*Comment:* Scientific judgement based upon an estimated rate constant for the vapor phase reaction with hydroxyl radicals in air (Atkinson, R (1987A)).

## Reduction half-life:            High:   No data
                                   Low:
*Comment:*

## Hydrolysis:
· **First-order hydr half-life:** 23.1 hours
*Comment:* Scientific judgement based upon reported rate constant that is independent of pH $(2.97 \times 10^{-2} \text{ hr}^{-1})$ at pH 7 and 25 °C (Kollig, HP et al. (1987A)).

· **First-order hydr half-life:** 129.4 hours (5.4 days)
*Comment:* Scientific judgement based upon reported rate constant $(5.36 \times 10^{-3} \text{ hr}^{-1})$ at pH 4.5 and 25 °C (Chapman, RA and Cole, CM (1982)).

· **Acid rate const $(M(H+)-hr)^{-1}$:**
*Comment:*

· **Base rate const $(M(OH-)-hr)^{-1}$:**
*Comment:*

# Hexachlorocyclopentadiene

**CAS Registry Number:** 77-47-4

**Half-lives:**

· **Soil:**                                 High:     672 hours              (4 weeks)
                                            Low:      168 hours              (7 days)
*Comment:* Scientific judgement based upon aerobic aqueous biodegradation half-life.

· **Air:**                                  High:     8.9 hours
                                            Low:      1.0 hours
*Comment:* Scientific judgement based upon estimated photooxidation half-life in air.

· **Surface Water:**                        High:     173 hours              (7.2 days)
                                            Low:      1.0 minutes
*Comment:* Scientific judgement based upon photolysis (low $t_{1/2}$) and hydrolysis (high $t_{1/2}$) half-lives.

· **Ground Water:**                         High:     1344 hours             (8 weeks)
                                            Low:      173 hours              (7.2 days)
*Comment:* Scientific judgement based upon aerobic aqueous biodegradation (high $t_{1/2}$) and hydrolysis (low $t_{1/2}$) half-lives.

**Aqueous Biodegradation (unacclimated):**

· **Aerobic half-life:**                    High:     672 hours              (4 weeks)
                                            Low:      168 hours              (7 days)
*Comment:* Scientific judgement based upon aerobic aqueous screening test data (Tabak, HH et al. (1981), Freitag, D et al. (1982)).

· **Anaerobic half-life:**                  High:     2688 hours             (16 weeks)
                                            Low:      672 hours              (28 days)
*Comment:* Scientific judgement based upon aerobic aqueous biodegradation half-life.

· **Removal/secondary treatment:**          High:     No data
                                            Low:
*Comment:*

**Photolysis:**

· **Atmos photol half-life:**               High:     No data
                                            Low:
*Comment:*

· **Max light absorption (nm):** lambda max = 323

*Comment:* Absorbance in cyclohexane (Sadtler UV No. 1397).

· **Aq photol half-life:**                    High:    10.7 minutes
                                              Low:      1.0 minutes
*Comment:* Scientific judgement based upon photolysis studies in aqueous solutions (high $t_{1/2}$: $k_p =$ 3.9 $hr^{-1}$, Wolfe, NL et al. (1982); low $t_{1/2}$: Butz, RG et al. (1982)).

## Photooxidation half-life:
· **Water:**                                  High:    No data
                                              Low:
*Comment:*

· **Air:**                                    High:    8.9 hours
                                              Low:     1.0 hours
*Comment:* Scientific judgement based upon calculated rate constants for vapor phase reactions with hydroxyl radicals and ozone in air (Cupitt, LT (1980)).

## Reduction half-life:                       High:    No data
                                              Low:
*Comment:*

## Hydrolysis:
· **First-order hydr half-life:** 173 hours                    (7.2 days)
*Comment:* Based upon a first order rate constant ($4.0 \times 10^{-3}$ $s^{-1}$) extrapolated to 25 °C and pH independent between a pH of 2.9 and 9.8 (Wolfe, NL et al. (1982)).

· **Acid rate const $(M(H+)-hr)^{-1}$:**                       No rate data
*Comment:* ($t_{1/2}$ = 7.2 days) Based upon a first order rate constant extrapolated to 25 °C and independent of hydronium ion concn to a pH of 2.9 (Wolfe, NL et al. (1982)).

· **Base rate const $(M(OH-)-hr)^{-1}$:**                      No rate data
*Comment:* ($t_{1/2}$ = 7.2 days) Based upon a first order rate constant extrapolated to 25 °C and independent of hydroxyl ion concn to a pH of 9.8 (Wolfe, NL et al. (1982)).

169

# Dimethyl sulfate

**CAS Registry Number:** 77-78-1

**Half-lives:**

· **Soil:**
        High:     12 hours
        Low:      1.2 hours

*Comment:* Low $t_{1/2}$ based upon measured overall hydrolysis rate constant for pH 7 at 25 °C (Mabey W and Mill T (1978)). High $t_{1/2}$ based upon maximum time observed for complete hydrolysis in neutral, slightly basic, and acidic aqueous solutions (Lee, ML et al. (1980)).

· **Air:**
        High:     365 hours     (15.2 days)
        Low:      36.5 hours    (1.5 days)

*Comment:* Scientific judgement based upon estimated photooxidation half-life in air.

· **Surface Water:**
        High:     12 hours
        Low:      1.2 hours

*Comment:* Low $t_{1/2}$ based upon measured overall hydrolysis rate constant for pH 7 at 25 °C (Mabey W and Mill T (1978)). High $t_{1/2}$ based upon maximum time observed for complete hydrolysis in neutral, slightly basic, and acidic aqueous solutions (Lee, ML et al. (1980)).

· **Ground Water:**
        High:     12 hours
        Low:      1.2 hours

*Comment:* Low $t_{1/2}$ based upon measured overall hydrolysis rate constant for pH 7 at 25 °C (Mabey W and Mill T (1978)). High $t_{1/2}$ based upon maximum time observed for complete hydrolysis in neutral, slightly basic, and acidic aqueous solutions (Lee, ML et al. (1980)).

**Aqueous Biodegradation (unacclimated):**

· **Aerobic half-life:**
        High:     672 hours     (4 weeks)
        Low:      168 hours     (7 days)

*Comment:* Scientific judgement.

· **Anaerobic half-life:**
        High:     2688 hours   (16 weeks)
        Low:      672 hours     (28 days)

*Comment:* Scientific judgement.

· **Removal/secondary treatment:**
        High:     No data
        Low:

*Comment:*

**Photolysis:**

· **Atmos photol half-life:**
        High:
        Low:

*Comment:*

· **Max light absorption (nm):**     No data
*Comment:*

· **Aq photol half-life:**     High:
                                     Low:

*Comment:*

## Photooxidation half-life:
· **Water:**     High:     No data
                                     Low:

*Comment:*

· **Air:**     High:     365 hours         (15.2 days)
                Low:     36.5 hours       (1.5 days)
*Comment:* Based upon estimated rate constant for vapor phase reaction with hydroxyl radicals in air (Atkinson, R (1987A)).

## Reduction half-life:
                                          High:     No data
                                        Low:
*Comment:*

## Hydrolysis:
· **First-order hydr half-life:**     1.2 hours
*Comment:* Based upon overall rate constant ($K_h$ = 1.6X10-4 sec$^{-1}$) at pH 7 and 25 °C (Mabey W and Mill T (1978)).

· **Acid rate const (M(H+)-hr)$^{-1}$:**     No data
*Comment:*

· **Base rate const (M(OH-)-hr)$^{-1}$:**     No data
*Comment:*

# Tetraethyl lead

**CAS Registry Number:** 78-00-2

**Half-lives:**
- **Soil:**               High:    672 hours          (4 weeks)
                          Low:     168 hours          (7 days)
  *Comment:* Scientific judgement based upon unacclimated aqueous aerobic biodegradation half-life.

- **Air:**                High:    9.0 hours
                          Low:     2.3 hours
  *Comment:* Based upon photolysis half-life in air.

- **Surface Water:**      High:    9.0 hours
                          Low:     2.3 hours
  *Comment:* Based upon photolysis half-life in water.

- **Ground Water:**       High:    1344 hours         (8 weeks)
                          Low:     336 hours          (14 days)
  *Comment:* Scientific judgement based upon unacclimated aqueous aerobic biodegradation half-life.

**Aqueous Biodegradation (unacclimated):**
- **Aerobic half-life:**  High:    672 hours          (4 weeks)
                          Low:     168 hours          (7 days)
  *Comment:* Scientific judgement.

- **Anaerobic half-life:** High:   2688 hours         (16 weeks)
                          Low:     672 hours          (28 days)
  *Comment:* Scientific judgement based upon unacclimated aqueous aerobic biodegradation half-life.

- **Removal/secondary treatment:**  High:   No data
                          Low:
  *Comment:*

**Photolysis:**
- **Atmos photol half-life:** High:  9.0 hours
                          Low:     2.3 hours
  *Comment:* Based upon measured rate of aqueous photolysis under simulated sunlight (Harrison, RM and Laxen, DPH (1978)).

· **Max light absorption (nm):** No data
*Comment:*

· **Aq photol half-life:**           High:      9.0 hours
                                          Low:      2.3 hours
*Comment:* Based upon measured rate of aqueous photolysis under simulated sunlight (Harrison, RM and Laxen, DPH (1978)).

**Photooxidation half-life:**
· **Water:**                       High:      No data
                                          Low:
*Comment:*

· **Air:**                              High:      22.3 hours
                                          Low:      2.99 hours
*Comment:* Based upon measured rate constants for reaction with hydroxyl radical (Hewitt, CN and Harrison, RM (1986)) and ozone (Atkinson, R and Carter, WPL (1984)).

**Reduction half-life:**            High:      No data
                                          Low:
*Comment:*

**Hydrolysis:**
· **First-order hydr half-life:** 14.5 hours
*Comment:* ($t_{1/2}$ in dilute sea water) (Tiravanti, G and Boari, G (1979)).

· **Acid rate const $(M(H+)-hr)^{-1}$:** No data
*Comment:*

· **Base rate const $(M(OH-)-hr)^{-1}$:** No data
*Comment:*

# Isophorone

CAS Registry Number: 78-59-1

Structure:

Half-lives:
- **· Soil:**　　　　　　　　　　　　High: 672 hours 　　　(4 weeks)
　　　　　　　　　　　　　　　　　　Low: 168 hours 　　　(7 days)
Comment: Scientific judgement based upon estimated unacclimated aqueous aerobic biodegradation half-life.

- **· Air:**　　　　　　　　　　　　　High: 3.13 hours
　　　　　　　　　　　　　　　　　　Low: 0.411 hours
Comment: Scientific judgment based upon estimated photooxidation half-life in air.

- **· Surface Water:**　　　　　　　　High: 672 hours 　　　(4 weeks)
　　　　　　　　　　　　　　　　　　Low: 168 hours 　　　(7 days)
Comment: Scientific judgement based upon estimated unacclimated aqueous aerobic biodegradation half-life.

- **· Ground Water:**　　　　　　　　High: 1344 hours 　　　(8 weeks)
　　　　　　　　　　　　　　　　　　Low: 336 hours 　　　(14 days)
Comment: Scientific judgement based upon estimated unacclimated aqueous aerobic biodegradation half-life.

Aqueous Biodegradation (unacclimated):
- **· Aerobic half-life:**　　　　　　High: 672 hours 　　　(4 weeks)
　　　　　　　　　　　　　　　　　　Low: 168 hours 　　　(7 days)
Comment: Scientific judgement based upon aqueous aerobic biodegradation screening test data (Price, KS et al. (1974)).

- **· Anaerobic half-life:**　　　　　High: 2688 hours 　　　(16 weeks)
　　　　　　　　　　　　　　　　　　Low: 672 hours 　　　(28 days)
Comment: Scientific judgement based upon estimated unacclimated aqueous aerobic biodegradation half-life.

- **· Removal/secondary treatment:**　High:　　　98%
　　　　　　　　　　　　　　　　　　Low:
Comment: Removal percentage based upon data from a continuous activated sludge biological

treatment simulator (Hannah, SA et al. (1986)).

## Photolysis:
· **Atmos photol half-life:**          High:      No data

                                                  Low:

*Comment:*

· **Max light absorption (nm):** lambda max = 235 nm (methanol)
*Comment:* Absorption extends to approximately 290 nm (Sadtler UV No. 44).

· **Aq photol half-life:**            High:      No data

                                                  Low:

*Comment:*

## Photooxidation half-life:
· **Water:**            High:      No data

                                                  Low:

*Comment:*

· **Air:**            High:      3.13 hours

                                                  Low:      0.411 hours

*Comment:* Scientific judgement based upon estimated rate constants for reaction with hydroxyl radical (Atkinson, R (1987A)) and ozone (Atkinson, R and Carter, WPL (1984)) in air.

## Reduction half-life:
                                       High:      No data

                                       Low:

*Comment:*

## Hydrolysis:
· **First-order hydr half-life:**
*Comment:*

· **Acid rate const (M(H+)-hr)$^{-1}$:**      No hydrolyzable groups
*Comment:*

· **Base rate const (M(OH-)-hr)$^{-1}$:**
*Comment:*

# Isoprene

CAS Registry Number:  78-79-5

## Half-lives:
· **Soil:**                              High:        672 hours            (4 weeks)
                                         Low:         168 hours            (7 days)
*Comment:*  Scientific judgement based upon estimated unacclimated aqueous aerobic biodegradation half-life.

· **Air:**                               High:        5.14 hours
                                         Low:         0.514 hours
*Comment:*  Based upon photooxidation half-life in air.

· **Surface Water:**                     High:        672 hours            (4 weeks)
                                         Low:         168 hours            (7 days)
*Comment:*  Scientific judgement based upon estimated unacclimated aqueous aerobic biodegradation half-life.

· **Ground Water:**                      High:        1344 hours           (8 weeks)
                                         Low:         336 hours            (14 days)
*Comment:*  Scientific judgement based upon estimated unacclimated aqueous aerobic biodegradation half-life.

## Aqueous Biodegradation (unacclimated):
· **Aerobic half-life:**                 High:        672 hours            (4 weeks)
                                         Low:         168 hours            (7 days)
*Comment:*  Scientific judgement.

· **Anaerobic half-life:**               High:        2688 hours           (16 weeks)
                                         Low:         672 hours            (28 days)
*Comment:*  Scientific judgement based upon estimated unacclimated aqueous aerobic biodegradation half-life.

· **Removal/secondary treatment:**       High:        No data
                                         Low:
*Comment:*

## Photolysis:
· **Atmos photol half-life:**            High:
                                         Low:
*Comment:*

· **Max light absorption (nm):**　　　　No data
*Comment:*

· **Aq photol half-life:**　　　　High:
　　　　　　　　　　　　　　　　Low:

*Comment:*

**Photooxidation half-life:**
　· **Water:**　　　　　　　　High:　　　No data
　　　　　　　　　　　　　　Low:
*Comment:*

　· **Air:**　　　　　　　　　High:　　5.14 hours
　　　　　　　　　　　　　　Low:　　0.556 hours
*Comment:* Based upon measured rate constants for reaction with hydroxyl radical (Atkinson, R (1985)) and ozone (Atkinson, R and Carter, WPL (1984)) in air.

**Reduction half-life:**　　　　　　High:　　　No data
　　　　　　　　　　　　　　　　Low:
*Comment:*

**Hydrolysis:**
　· **First-order hydr half-life:**
*Comment:*

　· **Acid rate const (M(H+)-hr)$^{-1}$:**　　No hydrolyzable groups
*Comment:*

　· **Base rate const (M(OH-)-hr)$^{-1}$:**
*Comment:*

# Isobutyl alcohol

**CAS Registry Number:** 78-83-1

**Half-lives:**
- **Soil:**                                     High:      173 hours

                                                      Low:       43 hours

*Comment:* Scientific judgement based upon estimated unacclimated aqueous aerobic biodegradation half-life.

- **Air:**         High:      99.6 hours      (4.2 days)

    Low:     9.96 hours

*Comment:* Scientific judgement based upon estimated photooxidation half-life in air.

- **Surface Water:**     High:    173 hours

    Low:     43 hours

*Comment:* Scientific judgement based upon estimated unacclimated aqueous aerobic biodegradation half-life.

- **Ground Water:**   High:    346 hours    (14.4 days)

    Low:     86 hours     (3.6 days)

*Comment:* Scientific judgement based upon estimated unacclimated aqueous aerobic biodegradation half-life.

**Aqueous Biodegradation (unacclimated):**
- **Aerobic half-life:**   High:    173 hours

    Low:     43 hours

*Comment:* Based upon river die-away test data for one sample of water from one river (Hammerton, C (1955)).

- **Anaerobic half-life:**   High:    692 hours    (28.8 days)

    Low:     172 hours    (7.17 days)

*Comment:* Scientific judgement based upon estimated unacclimated aqueous aerobic biodegradation half-life.

- **Removal/secondary treatment:**   High:    99%

    Low:

*Comment:* Based upon biological oxygen demand results from a semicontinuous activated sludge biological treatment simulator (Hatfield, R (1957)).

**Photolysis:**
- **Atmos photol half-life:**   High:

    Low:

*Comment:*

· **Max light absorption (nm):**     No data
*Comment:*

· **Aq photol half-life:**     High:
                                   Low:
*Comment:*

**Photooxidation half-life:**
  · **Water:**     High: $1.9 \times 10^5$ hours     (22 years)
                         Low:    4813 hours      (201 days)
*Comment:* Based upon measured rate constant for reaction with hydroxyl radical in water (Anbar, M and Neta, P (1967)).

  · **Air:**     High:   99.6 hours     (4.2 days)
                        Low:   9.96 hours
*Comment:* Scientific judgement based upon estimated rate constant for reaction with hydroxyl radical in air (Atkinson, R (1987A)).

**Reduction half-life:**     High:     No data
                          Low:
*Comment:*

**Hydrolysis:**
  · **First-order hydr half-life:**
*Comment:*

  · **Acid rate const (M(H+)-hr)$^{-1}$:**     No hydrolyzable groups
*Comment:*

  · **Base rate const (M(OH-)-hr)$^{-1}$:**
*Comment:*

# Isobutyraldehyde

<u>CAS Registry Number:</u> 78-84-2

<u>Half-lives:</u>
  · **Soil:**                              High:     168 hours                    (7 days)
                                           Low:       24 hours                    (1 day)
  *Comment:* Scientific judgement based upon estimated aqueous aerobic biodegradation half-life.

  · **Air:**                               High:       24 hours
                                           Low:        2.4 hours
  *Comment:* Based upon measured photooxidation rate constant in air.

  · **Surface Water:**                     High:     168 hours                    (7 days)
                                           Low:       24 hours                    (1 day)
  *Comment:* Scientific judgement based upon estimated aqueous aerobic biodegradation half-life.

  · **Ground Water:**                      High:     336 hours                    (14 days)
                                           Low:       48 hours                    (2 days)
  *Comment:* Scientific judgement based upon estimated aqueous aerobic biodegradation half-life.

<u>Aqueous Biodegradation (unacclimated):</u>
  · **Aerobic half-life:**                 High:     168 hours                    (7 days)
                                           Low:       24 hours                    (1 day)
  *Comment:* Scientific judgement based upon aqueous screening studies (Heukelekian, H and Rand, MC (1955); Gerhold, RM and Malaney, GW (1966)).

  · **Anaerobic half-life:**               High:     672 hours                    (28 days)
                                           Low:       96 hours                    (4 days)
  *Comment:* Scientific judgement based upon estimated aqueous aerobic biodegradation half-life.

  · **Removal/secondary treatment:**       High:     No data
                                           Low:
  *Comment:*

<u>Photolysis:</u>
  · **Atmos photol half-life:**            High:     No data
                                           Low:
  *Comment:*

  · **Max light absorption (nm):**
  *Comment:*

**· Aq photol half-life:**          High:
                                         Low:

*Comment:*

## Photooxidation half-life:
**· Water:**          High:         No data
                                         Low:

*Comment:*

**· Air:**          High:         24 hours
                                         Low:         2.4 hours
*Comment:* Based upon measured reaction rate constants with ·OH (Atkinson, R (1985)).

## Reduction half-life:
                                         High:         No data
                                         Low:

*Comment:*

## Hydrolysis:
**· First-order hydr half-life:**          No hydrolyzable functions
*Comment:*

**· Acid rate const (M(H+)-hr)$^{-1}$:**
*Comment:*

**· Base rate const (M(OH-)-hr)$^{-1}$:**
*Comment:*

# 1,2-Dichloropropane

<u>CAS Registry Number:</u>  78-87-5

<u>Half-lives:</u>
   · **Soil:**                                 High:    30936 hours            (3.5 years)
                                         Low:     4008 hours              (167 days)
  *Comment:* Scientific judgement based upon estimated aqueous aerobic biodegradation half-life.

   · **Air:**                                   High:      646 hours           (26.9 days)
                                           Low:       65 hours             (2.7 days)
  *Comment:* Scientific judgement based upon estimated photooxidation half-life in air.

   · **Surface Water:**                      High:    30936 hours            (3.5 years)
                                         Low:     4008 hours              (167 days)
  *Comment:* Scientific judgement based upon estimated aqueous aerobic biodegradation half-life.

   · **Ground Water:**                      High:    61872 hours            (7.1 years)
                                         Low:     8016 hours              (334 days)
  *Comment:* Scientific judgement based upon estimated aqueous aerobic biodegradation half-life.

<u>Aqueous Biodegradation (unacclimated):</u>
   · **Aerobic half-life**:                    High:    30936 hours            (3.5 years)
                                         Low:     4008 hours              (167 days)
  *Comment:* Scientific judgement based upon acclimated aerobic soil grab sample data (Roberts, TR and Stoydin, G (1976)).

   · **Anaerobic half-life**:                 High: 123744 hours          (14.1 years)
                                         Low:   16032 hours            (668 days)
  *Comment:* Scientific judgement based upon aerobic biodegradation half-life.

   · **Removal/secondary treatment:**     High:
                                         Low:           0%
  *Comment:* Based upon % degraded under aerobic continuous flow conditions (Kincannon, DF et al. (1983)).

<u>Photolysis:</u>
   · **Atmos photol half-life:**            High:       No data
                                         Low:
  *Comment:*

   · **Max light absorption (nm):**     No data
  *Comment:*

· **Aq photol half-life:**       High:     No data
                                              Low:

*Comment:*

## Photooxidation half-life:
· **Water:**       High:     No data
                                             Low:

*Comment:*

· **Air:**       High:     646 hours           (26.9 days)
                          Low:      65 hours            (2.7 days)
*Comment:* Scientific judgement based upon an estimated rate constant for the vapor phase reaction with hydroxyl radicals in air (Atkinson, R (1987A)).

## Reduction half-life:
                                 High:     No data
                                 Low:

*Comment:*

## Hydrolysis:
· **First-order hydr half-life:**      15.8 years
*Comment:* Scientific judgement based upon reported rate constant ($k_N = 5.0 \times 10^{-6}$ hr$^{-1}$) at pH 7 to 9 and 25 °C (Ellington, JJ et al. (1987)).

· **Acid rate const (M(H+)-hr)$^{-1}$:**
*Comment:*

· **Base rate const (M(OH-)-hr)$^{-1}$:**     $4.3 \times 10^{-4}$ M$^{-1}$ hr$^{-1}$
*Comment:* Base reaction not expected to be important (Ellington, JJ et al. (1987)).

# sec-Butyl alcohol

**CAS Registry Number:** 78-92-2

**Half-lives:**
  **· Soil:**                                    High:     168 hours          (7 days)
                                                 Low:       24 hours          (1 day)
  *Comment:* Scientific judgement based upon estimated aerobic biodegradation half-life.

  **· Air:**                                     High:      72 hours
                                                 Low:        7.2 hours
  *Comment:* Based upon measured reaction rate constants with ·OH (Edney, EO and Corse, EW (1986)).

  **· Surface Water:**                           High:     168 hours          (7 days)
                                                 Low:       24 hours          (1 day)
  *Comment:* Scientific judgement based upon estimated aerobic biodegradation half-life.

  **· Ground Water:**                            High:     336 hours          (14 days)
                                                 Low:       48 hours          (2 days)
  *Comment:* Scientific judgement based upon estimated aerobic biodegradation half-life.

**Aqueous Biodegradation (unacclimated):**
  **· Aerobic half-life:**                       High:     168 hours          (7 days)
                                                 Low:       24 hours          (1 day)
  *Comment:* Scientific judgement based upon a river die-away study (Hammerton, C (1955)).

  **· Anaerobic half-life:**                     High:     672 hours          (28 days)
                                                 Low:       96 hours          (4 days)
  *Comment:* Scientific judgement based upon estimated aerobic biodegradation half-life.

  **· Removal/secondary treatment:**             High:      No data
                                                 Low:
  *Comment:*

**Photolysis:**
  **· Atmos photol half-life:**                  High: No photolyzable functions
                                                 Low:
  *Comment:*

  **· Max light absorption (nm):**
  *Comment:*

184

· **Aq photol half-life:**                    High:
                                              Low:

*Comment:*

## Photooxidation half-life:
    · **Water:**                              High: 2.0 E+6 hours              (23 years)
                                              Low:    3100 hours              (129 days)
*Comment:* Based upon measured reaction rate constants of sec-butyl alcohol with ·OH in
aqueous solution (Anbar, M and Neta, P (1967); Dorfman, LM and Adams, GE (1973)).

    · **Air:**                                High:        72 hours
                                              Low:         7.2 hours
*Comment:* Based upon measured reaction rate constants with ·OH (Edney, EO and Corse, EW
(1986)).

## Reduction half-life:                       High:        No data
                                              Low:

*Comment:*

## Hydrolysis:
    · **First-order hydr half-life:**          No hydrolyzable functions
*Comment:*

    · **Acid rate const (M(H+)-hr)$^{-1}$:**
*Comment:*

    · **Base rate const (M(OH-)-hr)$^{-1}$:**
*Comment:*

185

# Methyl ethyl ketone

<u>CAS Registry Number:</u> 78-93-3

<u>Half-lives:</u>
· **Soil:**                                   High:     168 hours                (7 days)
                                              Low:       24 hours                (1 day)
*Comment:* Scientific judgement based upon estimated unacclimated aqueous aerobic
biodegradation half-life.

· **Air:**                                    High:     642 hours                (26.7 days)
                                              Low:      64.2 hours               (2.7 days)
*Comment:* Based upon photooxidation half-life in air.

· **Surface Water:**                          High:     168 hours                (7 days)
                                              Low:       24 hours                (1 day)
*Comment:* Scientific judgement based upon unacclimated grab sample of aerobic fresh water (low
$t_{1/2}$: Dojlido, JR (1979)) and aerobic aqueous screening test data (high $t_{1/2}$: Takemoto, S et al.
(1981)).

· **Ground Water:**                           High:     336 hours                (14 days)
                                              Low:       48 hours                (2 days)
*Comment:* Scientific judgement based upon estimated unacclimated aqueous aerobic
biodegradation half-life.

<u>Aqueous Biodegradation (unacclimated):</u>
· **Aerobic half-life:**                      High:     168 hours                (7 days)
                                              Low:       24 hours                (1 day)
*Comment:* Scientific judgement based upon unacclimated grab sample of aerobic fresh water (low
$t_{1/2}$: Dojlido, JR (1979)) and aerobic aqueous screening test data (high $t_{1/2}$: Takemoto, S et al.
(1981)).

· **Anaerobic half-life:**                    High:     672 hours                (28 days)
                                              Low:       96 hours                (4 days)
*Comment:* Scientific judgement based upon estimated aqueous unacclimated aerobic
biodegradation half-life.

· **Removal/secondary treatment:**            High:     100%
                                              Low:       86%
*Comment:* Based upon % degraded in an 8 day period under acclimated aerobic semi-continuous
flow conditions (high %: Dojlido, JR (1979); low %: Dojlido, J et al. (1984)).

<u>Photolysis:</u>

186

· **Atmos photol half-life:**  High:
                                                     Low:

*Comment:*

· **Max light absorption (nm):**  No data
*Comment:*

· **Aq photol half-life:**  High:
                                           Low:

*Comment:*

## Photooxidation half-life:
· **Water:**  High: $7.1 \times 10^5$ hours          (81.4 years)
                    Low: $1.8 \times 10^4$ hours          (48.8 years)
*Comment:* Based upon measured rate data for hydroxyl radicals in aqueous solution (Anbar, M and Neta, P (1967)).

· **Air:**  High:      642 hours          (26.7 days)
                 Low:      64.2 hours          (2.7 days)
*Comment:* Based upon measured rate data for the vapor phase reaction with hydroxyl radicals in air (Atkinson, R (1985)).

## Reduction half-life:  High:      No data
                                          Low:
*Comment:*

## Hydrolysis:
· **First-order hydr half-life:**      >50 years (if at all)
*Comment:* Unreactive towards hydrolysis from pH 5 to 9 at 15 °C (Kollig, HP et al. (1987A)).

· **Acid rate const (M(H+)-hr)$^{-1}$:**      0.0
*Comment:* Unreactive towards hydrolysis at pH 5 and 15 °C (Kollig, HP et al. (1987A)).

· **Base rate const (M(OH-)-hr)$^{-1}$:**      0.0
*Comment:* Unreactive towards hydrolysis at pH 9 and 15 °C (Kollig, HP et al. (1987A)).

# 1,1,2-Trichloroethane

CAS Registry Number: 79-00-5

## Half-lives:
**· Soil:**                           High:    8760 hours              (1 year)

Low:     3263 hours              (4.5 months)

*Comment:* Scientific judgement based upon estimated hydrolysis half-life at pH 9 and 25°C (low $t_{1/2}$) and data from the estimated unacclimated aerobic aqueous biodegradation half-life (high $t_{1/2}$) and a soil column test in which no biodegradation was observed (Wilson et al. (1981)).

**· Air:**                            High:    1956 hours              (81.5 days)

Low:      196 hours              (8.2 days)

*Comment:* Scientific judgement based upon estimated photooxidation half-life in air.

**· Surface Water:**                  High:    8760 hours              (1 year)

Low:     3263 hours              (4.5 months)

*Comment:* Scientific judgement based upon estimated hydrolysis half-life at pH 9 and 25°C (low $t_{1/2}$) and estimated unacclimated aerobic aqueous biodegradation half-life (high $t_{1/2}$).

**· Ground Water:**                   High:    17520 hours             (2 years)

Low:     3263 hours              (4.5 months)

*Comment:* Scientific judgement based upon estimated hydrolysis half-life at pH 9 and 25°C (low $t_{1/2}$) and data from the estimated unacclimated aerobic aqueous biodegradation half-life (high $t_{1/2}$) and a ground water die-away study in which no biodegradation was observed (Wilson et al. (1984)).

## Aqueous Biodegradation (unacclimated):
**· Aerobic half-life:**              High:    8760 hours              (1 year)

Low:     4320 hours              (6 months)

*Comment:* Scientific judgement based upon the extremely slow or no biodegradation which was observed in screening tests and a river die-away test (Tabak, HH et al. (1981); Kawasaki, M (1980); Mudder, TI and Musterman, JL (1982)).

**· Anaerobic half-life:**            High:    35040                   (4 years)

Low:     17280                   (24 months)

*Comment:* Scientific judgement based upon estimated unacclimated aerobic aqueous biodegradation half-life.

**· Removal/secondary treatment:**    High:    No data

Low:

*Comment:*

## Photolysis:
**· Atmos photol half-life:**      High:

Low:

*Comment:*


**· Max light absorption (nm):**      No data
*Comment:*


**· Aq photol half-life:**      High:

Low:

*Comment:*


## Photooxidation half-life:
**· Water:**      High:      No data

Low:

*Comment:*


**· Air:**      High:      1956 hours            (81.5 hours)

Low:      196 hours            (8.2 days)

*Comment:*  Based upon measured rate constants for reaction with hydroxyl radicals in air (Atkinson, R (1985)).


## Reduction half-life:
                        High:      No data

Low:

*Comment:*


## Hydrolysis:
**· First-order hydr half-life:**      $3.26 \times 10^5$ hours (37 years) at pH 7
*Comment:*  Scientific judgement based upon base catalyzed hydrolysis rate constant at 25°C (Mabey, WR et al. (1983A)).


**· Acid rate const $(M(H+)\text{-hr})^{-1}$:**      No data
*Comment:*


**· Base rate const $(M(OH-)\text{-hr})^{-1}$:**      $5.9 \times 10^{-3}$ $M^{-1}$ $s^{-1}$
*Comment:*  ($t_{1/2}$ 3263 hours (4.5 months) at pH 9)  Scientific judgement based upon base catalyzed hydrolysis rate constant at 25°C (Mabey, WR et al. (1983A)).

# Trichloroethylene

CAS Registry Number: 79-01-6

## Half-lives:

· **Soil:**            High:    8640 hours          (1 year)

                                Low:     4320 hours         (6 months)

*Comment:* Scientific judgement based upon estimated aqueous aerobic biodegradation half-life.

· **Air:**             High:    272 hours            (11.3 days)

                                Low:     27 hours            (1.1 days)

*Comment:* Based upon photooxidation half-life in air.

· **Surface Water:**       High:    8640 hours          (1 year)

                                Low:     4320 hours         (6 months)

*Comment:* Scientific judgement based upon estimated aqueous aerobic biodegradation half-life.

· **Ground Water:**       High:   39672 hours         (4.5 years)

                                Low:     7704 hours      (10.7 months)

*Comment:* Scientific judgement based upon hydrolysis half-life (low $t_{1/2}$: Dilling, WL et al. (1975)) and anaerobic sediment grab sample data (high $t_{1/2}$: Barrio-Lage, G et al. (1986)).

## Aqueous Biodegradation (unacclimated):

· **Aerobic half-life:**     High:    8640 hours          (1 year)

                                Low:     4320 hours         (6 months)

*Comment:* Scientific judgement based upon acclimated aerobic soil screening test data (Tabak, HH et al. (1981)).

· **Anaerobic half-life:**    High:   39672 hours         (4.5 years)

                                Low:     2352 hours          (98 days)

*Comment:* Scientific judgement based upon anaerobic sediment grab sample data (Barrio-Lage, G et al. (1986)).

· **Removal/secondary treatment:**    High:        No data

                                          Low:

*Comment:*

## Photolysis:

· **Atmos photol half-life:**    High:        No data

                                          Low:

*Comment:*

· **Max light absorption (nm):**    No data

*Comment:*

**· Aq photol half-life:**        High:     No data
                                      Low:

*Comment:*

## Photooxidation half-life:
**· Water:**        High:     No data
                                      Low:

*Comment:*

**· Air:**        High:     272 hours     (11.3 days)
                        Low:      27 hours      (1.1 days)

*Comment:* Based upon measured rate data for the vapor phase reaction with hydroxyl radicals in air (Atkinson, R (1985)).

## Reduction half-life:
                                High:     No data
                                  Low:

*Comment:*

## Hydrolysis:
**· First-order hydr half-life:**     7704 hours     (10.7 months)

*Comment:* Based upon a first order rate constant ($k = 0.065$ mo$^{-1}$) at 25 °C (Dilling, WL et al. (1975)).

**· Acid rate const (M(H+)-hr)$^{-1}$:**
*Comment:*

**· Base rate const (M(OH-)-hr)$^{-1}$:**
*Comment:*

# Acrylic acid

**CAS Registry Number:** 79-10-7

**Half-lives:**
  · **Soil:**                          High:     168 hours          (7 days)
                                       Low:       24 hours          (1 day)
  *Comment:* Scientific judgement based upon estimated unacclimated aqueous aerobic
  biodegradation half-life.

  · **Air:**                           High:     23.8 hours
                                       Low:       2.5 hours
  *Comment:* Scientific judgement based upon estimated photooxidation half-life in air.

  · **Surface Water:**                 High:     168 hours          (7 days)
                                       Low:       24 hours          (1 day)
  *Comment:* Scientific judgement based upon estimated unacclimated aqueous aerobic
  biodegradation half-life.

  · **Ground Water:**                  High:    4320 hours          (6 months)
                                       Low:       48 hours          (2 days)
  *Comment:* Scientific judgement based upon estimated unacclimated aqueous aerobic and
  anaerobic biodegradation half-lives.

**Aqueous Biodegradation (unacclimated):**
  · **Aerobic half-life:**             High:     168 hours          (7 days)
                                       Low:       24 hours          (1 day)
  *Comment:* Scientific judgement based upon unacclimated aqueous screening test data (Sasaki, S
  (1978); Dore, M et al. (1975)).

  · **Anaerobic half-life:**           High:    4320 hours          (6 months)
                                       Low:      672 hours          (4 weeks)
  *Comment:* Scientific judgement based upon unacclimated anaerobic reactor test data (Chou, WL
  et al. (1979)).

  · **Removal/secondary treatment:**   High:     No data
                                       Low:
  *Comment:*

**Photolysis:**
  · **Atmos photol half-life:**        High:     No data
                                       Low:
  *Comment:*

· **Max light absorption (nm):**          lambda max = 250
*Comment:* No absorbance occurs above 320 nm in methanol (Sadtler UV No. 2994).

· **Aq photol half-life:**          High:          No data
                                    Low:
*Comment:*

**Photooxidation half-life:**
· **Water:**          High:          No data
                      Low:
*Comment:*

· **Air:**          High:          23.8 hours
                    Low:          2.5 hours
*Comment:* Scientific judgement based upon an estimated rate constant for the vapor phase reaction with hydroxyl radicals and ozone in air (Atkinson, R (1987A); Atkinson, R and Carter, WPL (1984)).

**Reduction half-life:**          High:          No data
                                  Low:
*Comment:*

**Hydrolysis:**
· **First-order hydr half-life:**          No hydrolyzable groups
*Comment:*

· **Acid rate const (M(H+)-hr)$^{-1}$:**
*Comment:*

· **Base rate const (M(OH-)-hr)$^{-1}$:**
*Comment:*

# Chloroacetic acid

**CAS Registry Number:** 79-11-8

**Half-lives:**
- **· Soil:**            High:    168 hours        (7 days)

                                       Low:     24 hours         (1 day)

*Comment:* Scientific judgement based upon estimated aqueous aerobic biodegradation half-life.

- **· Air:**            High:    2050 hours      (85.5 days)

                      Low:     205 hours       (8.6 days)

*Comment:* Based upon combined, estimated rates for photooxidation and photolysis in air.

- **· Surface Water:**    High:    168 hours        (7 days)

                      Low:     24 hours         (1 day)

*Comment:* Scientific judgement based upon estimated aqueous aerobic biodegradation half-life.

- **· Ground Water:**    High:    336 hours        (14 days)

                      Low:      48 hours        (2 days)

*Comment:* Scientific judgement based upon estimated aqueous aerobic biodegradation half-life.

**Aqueous Biodegradation (unacclimated):**
- **· Aerobic half-life:**    High:    168 hours        (7 days)

                      Low:     24 hours         (1 day)

*Comment:* Scientific judgement based upon river die-away tests using radio-labeled material (Boethling, RS and Alexander, M (1979A)).

- **· Anaerobic half-life:**    High:    672 hours        (28 days)

                      Low:      96 hours        (4 days)

*Comment:* Scientific judgement based upon estimated aqueous aerobic biodegradation half-life.

- **· Removal/secondary treatment:**    High:    No data

                                                Low:

*Comment:*

**Photolysis:**
- **· Atmos photol half-life:**    High:    19000 hours      (790 days)

                                   Low:     1900 hours       ( 79 days)

*Comment:* Based upon estimated aqueous photolysis half-lives.

- **· Max light absorption (nm):**    lambda max = 360 nm

*Comment:* Extinction coefficients at 300-360 nm are less than 0.2 L/M-cm (Draper, WM and Crosby, DG (1983A)).

194

· **Aq photol half-life:**          High:   19000 hours          (790 days)

                                    Low:    1900 hours           ( 79 days)

*Comment:* Low value based upon experimental photolysis data utilizing an artificial light source (Draper, WM and Crosby, DG (1983A)); high value assumes a diminished light intensity.

## Photooxidation half-life:

· **Water:**          High:     No data

                       Low:

*Comment:*

· **Air:**          High:     2300 hours          (96 days)

                     Low:      230 hours          (9.6 days)

*Comment:* Based upon estimated reaction rate constant with ·OH (Atkinson, R (1987A)).

## Reduction half-life:          High:     No data

                                  Low:

*Comment:*

## Hydrolysis:

· **First-order hydr half-life:**          23000 hours          (960 days)

*Comment:* Approximated half-life based upon losses in dark control tests during photolysis experiments (Draper, WM and Crosby, DG (1983A)).

· **Acid rate const $(M(H+)-hr)^{-1}$:**

*Comment:*

· **Base rate const $(M(OH-)-hr)^{-1}$:**

*Comment:*

# Peracetic acid

<u>CAS Registry Number:</u>  79-21-0

<u>Half-lives:</u>
  · **Soil:**                                    High:        48 hours
                                                 Low:          4 hour
*Comment:*  Scientific judgement based upon peracetic acid's expected high reactivity with organic matter in soil.

  · **Air:**                                     High:       138 hours              (5.75 days)
                                                 Low:        13.8 hours
*Comment:*  Based upon estimated photooxidation half-life in air.

  · **Surface Water:**                           High:       168 hours              (7 days)
                                                 Low:          4 hour
*Comment:*  High value based upon estimated high value for aqueous aerobic biodegradation. Low value based upon estimated low value for aqueous photooxidation.

  · **Ground Water:**                            High:       336 hours              (14 days)
                                                 Low:         48 hours
*Comment:*  High value based upon estimated aqueous aerobic biodegradation half-life. Low value is scientific judgement based upon potential reactivity with organic material in ground water.

<u>Aqueous Biodegradation (unacclimated):</u>
  · **Aerobic half-life:**                       High:       168 hours              (7 days)
                                                 Low:         24 hours              (1 day)
*Comment:*  Scientific judgement based upon rapid aerobic biodegradation of acetic acid.

  · **Anaerobic half-life:**                     High:       672 hours              (28 days)
                                                 Low:         96 hours              (4 days)
*Comment:*  Scientific judgement based upon estimated aqueous aerobic biodegradation half-lives.

  · **Removal/secondary treatment:**             High:       No data
                                                 Low:
*Comment:*

<u>Photolysis:</u>
  · **Atmos photol half-life:**                  High:       No data
                                                 Low:
*Comment:*

  · **Max light absorption (nm):**

*Comment:*

**· Aq photol half-life:**             High:
                                       Low:
*Comment:*

## Photooxidation half-life:
   **· Water:**                        High:      198 hours            (8.25 days)
                                       Low:         4 hours
*Comment:* Scientific judgement based upon reported reaction of the hydroperoxide class with $RO_2$· in natural water (Mill, T and Mabey, W (1985)).

   **· Air:**                          High:      138 hours            (5.75 days)
                                       Low:      13.8 hours
*Comment:* Based upon estimated reaction rate constant with ·OH (PCFAP-USEPA, 1987).

## Reduction half-life:                High:      No data
                                       Low:
*Comment:*

## Hydrolysis:
   **· First-order hydr half-life:**   No data
*Comment:*

   **· Acid rate const $(M(H+)-hr)^{-1}$:**
*Comment:*

   **· Base rate const $(M(OH-)-hr)^{-1}$:**
*Comment:*

# 1,1,2,2-Tetrachloroethane

**CAS Registry Number:** 79-34-5

## Half-lives:
· **Soil:**                    High:    1056 hours              (45 days)
                               Low:     10.7 hours
*Comment:* Scientific judgement based upon hydrolysis half-lives at pH 7 and 9 (Haag, WR and Mill, T (1988)).

· **Air:**                     High:    2131 hours              (88.8 days)
                               Low:     213 hours               (8.9 days)
*Comment:* Scientific judgement based upon estimated photooxidation half-life in air.

· **Surface Water:**           High:    1056 hours              (45 days)
                               Low:     10.7 hours
*Comment:* Scientific judgement based upon hydrolysis half-lives at pH 7 and 9 (Haag, WR and Mill, T (1988)).

· **Ground Water:**            High:    1056 hours              (45 days)
                               Low:     10.7 hours
*Comment:* Scientific judgement based upon hydrolysis half-lives at pH 7 and 9 (Haag, WR and Mill, T (1988)).

## Aqueous Biodegradation (unacclimated):
· **Aerobic half-life**:       High:    4320 hours              (6 months)
                               Low:     672 hours               (4 weeks)
*Comment:* Scientific judgement based upon acclimated river die-away rate data (Mudder, T (1981)).

· **Anaerobic half-life**:     High:    672 hours               (4 weeks)
                               Low:     168 hours               (7 days)
*Comment:* Scientific judgement based upon anaerobic sediment grab sample data (low $t_{1/2}$) (Jafvert, CT and Wolfe, NL (1987)), and anaerobic screening test data (high $t_{1/2}$) (Hallen, RT et al. (1986)).

· **Removal/secondary treatment:**     High:    No data
                                       Low:
*Comment:*

## Photolysis:
· **Atmos photol half-life:**          High:
                                       Low:

*Comment:*

· **Max light absorption (nm):**     No data
*Comment:*

· **Aq photol half-life:**     High:
                                       Low:

*Comment:*

## Photooxidation half-life:
  · **Water:**                 High:     No data
                                    Low:

*Comment:*

  · **Air:**                   High:     2131 hours         (88.8 days)
                                    Low:      213 hours          (8.9 days)
*Comment:* Scientific judgement based upon an estimated rate constant for vapor phase reaction with hydroxyl radicals in air (Atkinson, R (1987A)).

## Reduction half-life:
                                         High:     No data
                                       Low:
*Comment:*

## Hydrolysis:
  · **First-order hydr half-life:**     1056 hours       (45 days)
*Comment:* Based upon rate constant ($k_B = 1.8$ $M^{-1}$ $s^{-1}$) for base reaction at 25 °C and pH of 7 (Haag, WR and Mill, T (1988)).

  · **Acid rate const $(M(H+)\text{-}hr)^{-1}$:**     No data
*Comment:*

  · **Base rate const $(M(OH\text{-})\text{-}hr)^{-1}$:**     $k_B = 6480$
*Comment:* ($t_{1/2}= 10.7$ hours) Rate constant for base reaction at 25 °C and pH of 9 (Haag, WR and Mill, T (1988)).

# Dimethylcarbamyl chloride

**CAS Registry Number:** 79-44-7

**Structure:**

$$H_3C-N(CH_3)-C(=O)-Cl$$

**Half-lives:**
 · **Soil:**                          High:    195 seconds        (3.2 minutes)
                                      Low:     14 seconds
*Comment:* Scientific judgement based upon measured overall hydrolysis rate constant for pH 7 at 25 (high $t_{1/2}$) and 5 °C (low $t_{1/2}$) (Queen, A (1967)).

 · **Air:**                           High:    10.2 hours
                                      Low:     1.0 hours
*Comment:* Scientific judgement based upon estimated photooxidation half-life in air.

 · **Surface Water:**                 High:    195 seconds        (3.2 minutes)
                                      Low:     14 seconds
*Comment:* Scientific judgement based upon measured overall hydrolysis rate constant for pH 7 at 25 (high $t_{1/2}$) and 5 °C (low $t_{1/2}$) (Queen, A (1967)).

 · **Ground Water:**                  High:    195 seconds        (3.2 minutes)
                                      Low:     14 seconds
*Comment:* Scientific judgement based upon measured overall hydrolysis rate constant for pH 7 at 25 (high $t_{1/2}$) and 5 °C (low $t_{1/2}$) (Queen, A (1967)).

**Aqueous Biodegradation (unacclimated):**
 · **Aerobic half-life:**             High:    672 hours          (4 weeks)
                                      Low:     168 hours          (7 days)
*Comment:* Scientific judgement.

 · **Anaerobic half-life:**           High:    2688 hours         (16 weeks)
                                      Low:     672 hours          (28 days)
*Comment:* Scientific judgement based upon estimated unacclimated aqueous aerobic biodegradation half-life.

 · **Removal/secondary treatment:**   High:    No data
                                      Low:
*Comment:*

## Photolysis:
· **Atmos photol half-life:**               High:       No data
                                               Low:

*Comment:*

· **Max light absorption (nm):**      No data
*Comment:*

· **Aq photol half-life:**                 High:       No data
                                               Low:

*Comment:*

## Photooxidation half-life:
· **Water:**                            High:       No data
                                               Low:

*Comment:*

· **Air:**                                High:    10.2 hours
                                             Low:     1.0 hours
*Comment:* Scientific judgement based upon an estimated rate constant for vapor phase reaction with hydroxyl radicals in air (Atkinson, R (1987A)).

## Reduction half-life:
                                       High:       No data
                                       Low:

*Comment:*

## Hydrolysis:
· **First-order hydr half-life:**     14 seconds
*Comment:* Based upon overall rate constant ($k_h = 4.97 \times 10^{-2}$) at pH 7 and 25 °C (Queen, A (1967)).

· **First-order hydr half-life:**     195 seconds             (3.2 minutes)
*Comment:* Based upon overall rate constant ($k_h = 3.56 \times 10^{-3}$) at pH 7 and 5 °C (Queen, A (1967)).

· **Acid rate const $(M(H+)-hr)^{-1}$:**     No data
*Comment:*

· **Base rate const $(M(OH-)-hr)^{-1}$:**     No data
*Comment:*

# 2-Nitropropane

**CAS Registry Number:** 79-46-9

**Half-lives:**
 · **Soil:**                            High:    4320 hours              (6 months)
                                         Low:      672 hours              (4 weeks)
 *Comment:* Scientific judgement based upon estimated unacclimated aqueous aerobic
 biodegradation half-life.

 · **Air:**                             High:    48.7 hours              (2.03 days)
                                         Low:     4.87 hours
 *Comment:* Scientific judgement based upon estimated photooxidation half-life in air (Atkinson, R
 (1987A)).

 · **Surface Water:**                   High:    4320 hours              (6 months)
                                         Low:      672 hours              (4 weeks)
 *Comment:* Scientific judgement based upon estimated unacclimated aqueous aerobic
 biodegradation half-life.

 · **Ground Water:**                    High:    8640 hours              (12 months)
                                         Low:     1344 hours             (8 weeks)
 *Comment:* Scientific judgement based upon estimated unacclimated aqueous aerobic
 biodegradation half-life.

**Aqueous Biodegradation (unacclimated):**
 · **Aerobic half-life:**               High:    4320 hours              (6 months)
                                         Low:      672 hours              (4 weeks)
 *Comment:* Scientific judgement based upon limited data from an aqueous activated sludge
 screening test (Freitag, D et al. (1985)).

 · **Anaerobic half-life:**             High:    17280 hours             (24 months)
                                         Low:     2688 hours             (16 weeks)
 *Comment:* Scientific judgement based upon estimated unacclimated aqueous aerobic
 biodegradation half-life.

 · **Removal/secondary treatment:**     High:      No data
                                         Low:
 *Comment:*

**Photolysis:**
 · **Atmos photol half-life:**          High:      No data
                                         Low:

*Comment:*

· **Max light absorption (nm):**       lambda max = 278 (cyclohexane)
*Comment:* 2-Nitropropane absorbs light up to approximately 350 nm (Sadtler UV No. 1).

· **Aq photol half-life:**         High:     No data
                                     Low:
*Comment:*

## Photooxidation half-life:
· **Water:**                    High:     No data
                                     Low:
*Comment:*

· **Air:**                       High:     48.7 hours        (2.03 days)
                                     Low:      4.87 hours
*Comment:* Scientific judgement based upon estimated rate constant for reaction with hydroxyl radical in air (Atkinson, R 1987A)).

## Reduction half-life:              High:     No data
                                     Low:
*Comment:*

## Hydrolysis:
· **First-order hydr half-life:**
*Comment:*

· **Acid rate const $(M(H+)\text{-}hr)^{-1}$:**     No hydrolyzable groups
*Comment:*

· **Base rate const $(M(OH\text{-})\text{-}hr)^{-1}$:**
*Comment:*

# 4,4'-Isopropylidenediphenol

<u>CAS Registry Number:</u> 80-05-7

<u>Structure:</u>

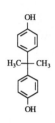

<u>Half-lives:</u>
· **Soil:**                                    High:    4320 hours              (6 months)
                                                 Low:      24 hours               (1 day)
*Comment:* Scientific judgement based upon estimated unacclimated aqueous aerobic
biodegradation half-life.

· **Air:**                                     High:      7.4 hours
                                                 Low:      0.74 hours              (44 minutes)
*Comment:* Scientific judgement based upon estimated photooxidation half-life in air.

· **Surface Water:**                    High:    3833 hours              (5.3 months)
                                                 Low:      24 hours               (1 day)
*Comment:* Scientific judgement based upon estimated photooxidation half-life in water (high $t_{1/2}$)
and estimated unacclimated aqueous aerobic biodegradation half-life (low $t_{1/2}$).

· **Ground Water:**                    High:    8640 hours              (12 months)
                                                 Low:      48 hours               (2 days)
*Comment:* Scientific judgement based upon estimated unacclimated aqueous aerobic
biodegradation half-life.

<u>Aqueous Biodegradation (unacclimated):</u>
· **Aerobic half-life**:                 High:    4320 hours              (6 months)
                                                 Low:      24 hours               (1 day)
*Comment:* Scientific judgement based upon screening test data which indicate either rapid
degradation (Dorn, PB et al. (1987) or resistance to or slow degradation (Sasaki, S (1978)).

· **Anaerobic half-life**:             High:    17280 hours            (24 months)
                                                 Low:      96 hours               (4 days)
*Comment:* Scientific judgement based upon estimated unacclimated aerobic aqueous
biodegradation half-life.

· **Removal/secondary treatment:**    High:          72%

Low:

*Comment:* Removal percentage based upon semi-continuous activated sludge treatment simulator data (Matsui, S et al. (1975)).

## Photolysis:

· **Atmos photol half-life:**          High:     No data
                                       Low:

*Comment:*

· **Max light absorption (nm):**     lambda max = 227, 278.5 nm (methanol)
*Comment:* Sadtler UV No. 325.

· **Aq photol half-life:**            High:     No data
                                      Low:

*Comment:*

## Photooxidation half-life:

Water:                               High:     3840 hours          (160 days)
                                     Low:      66 hours           (2.75 days)
*Comment:* Scientific judgement based upon reported reaction rate constants for ·OH and $RO_2$·
(low $t_{1/2}$) and $RO_2^-$ (high $t_{1/2}$) with the phenol class (Mill, T and Mabey, W (1985); Guesten, H et
al. (1981)).

· **Air:**                           High:     7.4 hours
                                     Low:      0.74 hours          (44 minutes)
*Comment:* Based upon estimated rate constant for reaction with hydroxyl radicals in air
(Atkinson, R (1987A)).

## Reduction half-life:                High:     No data
                                       Low:

*Comment:*

## Hydrolysis:

· **First-order hydr half-life:**     No hydrolyzable groups
*Comment:*

· **Acid rate const $(M(H+)-hr)^{-1}$:**
*Comment:*

· **Base rate const $(M(OH-)-hr)^{-1}$:**
*Comment:*

# Cumene hydroperoxide

**CAS Registry Number:**  80-15-9

**Structure:**

## Half-lives:
  · **Soil:**                             High:    672 hours       (4 weeks)

                                         Low:     168 hours       (7 days)

     *Comment:*  Scientific judgement based upon estimated aerobic aqueous biodegradation half-life.

  · **Air:**                              High:    130 hours       (5.4 days)

                                         Low:      13 hours

     *Comment:*  Scientific judgement based upon photooxidation half-life in air.

  · **Surface Water:**               High:    672 hours       (4 weeks)

                                         Low:     168 hours       (7 days)

     *Comment:*  Scientific judgement based upon estimated aerobic aqueous biodegradation half-life.

  · **Ground Water:**              High:   1344 hours      (8 weeks)

                                         Low:     336 hours      (14 days)

     *Comment:*  Scientific judgement based upon estimated aerobic aqueous biodegradation half-life.

## Aqueous Biodegradation (unacclimated):
  · **Aerobic half-life:**           High:    672 hours       (4 weeks)

                                         Low:     168 hours       (7 days)

     *Comment:*  Scientific judgement.

  · **Anaerobic half-life:**       High:  2688 hours      (16 weeks)

                                         Low:     672 hours      (28 days)

     *Comment:*  Scientific judgement based upon estimated aerobic aqueous biodegradation half-life.

  · **Removal/secondary treatment:**   High:

                                         Low:

     *Comment:*

## Photolysis:
  · **Atmos photol half-life:**     High:      No data

                                         Low:

*Comment:*

· **Max light absorption (nm):**     lambda max is approximately 242, 247, 252, 259, 264 (n-hexane)
*Comment:* (Norrish, RGW and Searby, MH (1956)).

· **Aq photol half-life:**     High:     No data
                              Low:
*Comment:*

**Photooxidation half-life:**
   · **Water:**     High:     No data
                    Low:
*Comment:*

· **Air:**     High:     130 hours          (5.4 days)
              Low:     13 hours
*Comment:* Scientific judgement based upon estimated rate constant for reaction with hydroxyl radical in air (Atkinson, R 1987A)).

**Reduction half-life:**     High:     No data
                            Low:
*Comment:*

**Hydrolysis:**
   · **First-order hydr half-life:**
*Comment:*

· **Acid rate const (M(H+)-hr)$^{-1}$:**     No data
*Comment:*

· **Base rate const (M(OH-)-hr)$^{-1}$:**
*Comment:*

# Methyl methacrylate

**CAS Registry Number:** 80-62-6

**Half-lives:**
  · **Soil:**                          High:    672 hours        (4 weeks)

High: 672 hours (4 weeks)
Low: 168 hours (7 days)

*Comment:* Scientific judgement based upon estimated aqueous unacclimated aerobic biodegradation half-life.

  · **Air:**

High: 9.7 hours
Low: 1.1 hours (66 minutes)

*Comment:* Scientific judgement based upon estimated photooxidation half-life in air.

  · **Surface Water:**

High: 672 hours (4 weeks)
Low: 168 hours (7 days)

*Comment:* Scientific judgement based upon estimated aqueous unacclimated aerobic biodegradation half-life.

  · **Ground Water:**

High: 1344 hours (8 weeks)
Low: 336 hours (2 weeks)

*Comment:* Scientific judgement based upon estimated aqueous unacclimated aerobic biodegradation half-life.

**Aqueous Biodegradation (unacclimated):**
  · **Aerobic half-life:**

High: 672 hours (4 weeks)
Low: 168 hours (7 days)

*Comment:* Scientific judgement based upon unacclimated screening test data (Pahren, HR and Bloodgood, DE (1961), Sasaki, S (1978)).

  · **Anaerobic half-life:**

High: 2688 hours (16 weeks)
Low: 672 hours (4 weeks)

*Comment:* Scientific judgement based upon estimated aqueous unacclimated aerobic biodegradation half-life.

  · **Removal/secondary treatment:**

High: No data
Low:

*Comment:*

**Photolysis:**
  · **Atmos photol half-life:**

High:
Low:

*Comment:*

**· Max light absorption (nm):**            No data
*Comment:*


**· Aq photol half-life:**            High:
                                      Low:

*Comment:*


**Photooxidation half-life:**
  **· Water:**                        High:    No data
                                      Low:

*Comment:*


  **· Air:**                          High:    9.7 hours
                                      Low:     1.1 hours            (66 minutes)
*Comment:* Scientific judgement based upon estimated rate constants for vapor phase reactions
with hydroxyl radicals and ozone in air (Atkinson, R (1987A)).


**Reduction half-life:**              High:    No data
                                      Low:

*Comment:*


**Hydrolysis:**
  **· First-order hydr half-life:**       4 years
*Comment:* Based upon base rate constant ($k_B$ = 200 $M^{-1}$ $hr^{-1}$) at 25 °C and pH of 7 (Ellington, JJ
et al. (1987)).


  **· Acid rate const $(M(H+)-hr)^{-1}$:**       No data
*Comment:*


  **· Base rate const $(M(OH-)-hr)^{-1}$:**       200 $M^{-1}$ $hr^{-1}$
*Comment:* ($t_{1/2}$ = 14.4 days at pH 9) Measured rate constant at 25 °C and pH 11 (Ellington, JJ et
al. (1987)).

# Saccharin

**CAS Registry Number:** 81-07-2

**Structure:**

**Half-lives:**
  · **Soil:**  High:  672 hours  (4 weeks)
  Low:  168 hours  (7 days)
  *Comment:* Scientific judgement based upon estimated aqueous aerobic biodegradation half-life.

  · **Air:**  High:  10 hours
  Low:  1 hour
  *Comment:* Scientific judgement based upon estimated photooxidation half-life in air for a saccharin analog containing a deoxygenated sulfur.

  · **Surface Water:**  High:  672 hours  (4 weeks)
  Low:  168 hours  (7 days)
  *Comment:* Scientific judgement based upon estimated aqueous aerobic biodegradation half-life.

  · **Ground Water:**  High:  1344 hours  (8 weeks)
  Low:  336 hours  (14 days)
  *Comment:* Scientific judgement based upon estimated aqueous aerobic biodegradation half-life.

**Aqueous Biodegradation (unacclimated):**
  · **Aerobic half-life:**  High:  672 hours  (4 weeks)
  Low:  168 hours  (7 days)
  *Comment:* Scientific judgement.

  · **Anaerobic half-life:**  High:  2688 hours  (16 weeks)
  Low:  672 hours  (28 days)
  *Comment:* Scientific judgement based upon aqueous aerobic biodegradation half-life.

  · **Removal/secondary treatment:**  High:  No data
  Low:
  *Comment:*

**Photolysis:**
  · **Atmos photol half-life:**  High:
  Low:

210

*Comment:*

**· Max light absorption (nm):**                     No data
*Comment:*

**· Aq photol half-life:**               High:
                                         Low:

*Comment:*

## Photooxidation half-life:
**· Water:**                             High:        No data
                                         Low:

*Comment:*

**· Air:**                               High:        10 hours
                                         Low:          1 hour
*Comment:* Scientific judgement based upon an estimated rate constant for the reaction of a saccharin analog containing a deoxygenated sulfur with hydroxyl radicals in air (Atkinson, R (1987A)).

## Reduction half-life:                  High:        No data
                                         Low:
*Comment:*

## Hydrolysis:
**· First-order hydr half-life:**        No data
*Comment:*

**· Acid rate const $(M(H+)-hr)^{-1}$:**
*Comment:*

**· Base rate const $(M(OH-)-hr)^{-1}$:**
*Comment:*

# Warfarin

**CAS Registry Number:** 81-81-2

**Structure:**

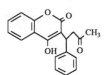

**Half-lives:**
  **· Soil:**                     High:     672 hours          (4 weeks)
                                  Low:      168 hours          (7 days)
  *Comment:*  Scientific judgement based upon estimated unacclimated aqueous aerobic
  biodegradation half-life.

  **· Air:**                      High:     1.87 hours
                                  Low:      0.254 hours
  *Comment:*  Scientific judgement based upon estimated photooxidation half-life in air.

  **· Surface Water:**            High:     672 hours          (4 weeks)
                                  Low:      168 hours          (7 days)
  *Comment:*  Scientific judgement based upon estimated unacclimated aqueous aerobic
  biodegradation half-life.

  **· Ground Water:**             High:     1344 hours         (8 weeks)
                                  Low:      336 hours          (14 days)
  *Comment:*  Scientific judgement based upon estimated aqueous aerobic biodegradation half-lives.

**Aqueous Biodegradation (unacclimated):**
  **· Aerobic half-life:**        High:     672 hours          (4 weeks)
                                  Low:      168 hours          (7 days)
  *Comment:*  Scientific judgement.

  **· Anaerobic half-life:**      High:     2688 hours         (16 weeks)
                                  Low:      672 hours          (28 days)
  *Comment:*  Scientific judgement based upon estimated aqueous aerobic biodegradation half-lives.

  **· Removal/secondary treatment:**   High:     No data
                                       Low:
  *Comment:*

**Photolysis:**

212

**· Atmos photol half-life:**        High:     No data
                                        Low:
*Comment:*

**· Max light absorption (nm):**      lambda max = 205, 282, 305 nm (methanol)
*Comment:* Absorption extends to 328 nm (Gore, RC et al. (1971)).

**· Aq photol half-life:**          High:     No data
                                        Low:
*Comment:*

## Photooxidation half-life:
**· Water:**                     High:     No data
                                        Low:
*Comment:*

**· Air:**                         High:    1.87 hours
                                      Low:    0.254 hours
*Comment:* Scientific judgement based upon estimated rate constant for reaction with hydroxyl radicals (Atkinson, R (1987A)) and ozone (Atkinson, R and Carter, WPL (1984)) in air.

## Reduction half-life:
                                      High:     No data
                                      Low:
*Comment:*

## Hydrolysis:
**· First-order hydr half-life:**      $1.41 \times 10^5$ hours (16.1 years)
*Comment:* ($t_{1/2}$ at pH 5-7 and 25°C) Based upon measured neutral and base catalyzed hydrolysis rate constants (Ellington, JJ et al. (1987A)).

**· Acid rate const (M(H+)-hr)$^{-1}$:**     No data
*Comment:*

**· Base rate const (M(OH-)-hr)$^{-1}$:**     0.026
*Comment:* ($t_{1/2}$ = $1.34 \times 10^5$ hours (15.2 years) at pH 9 and 25°C) Based upon measured neutral and base catalyzed hydrolysis rate constants (Ellington, JJ et al. (1987A)).

# 1-Amino-2-methylanthraquinone

<u>CAS Registry Number:</u>  82-28-0

<u>Structure:</u>

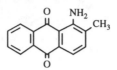

<u>Half-lives:</u>
  · **Soil:**                                          High:        672 hours                    (4 weeks)
                                                        Low:         168 hours                    (7 days)
  *Comment:*  Scientific judgement based upon estimated unacclimated aqueous aerobic
  biodegradation half-life.

  · **Air:**                                           High:         22 hours
                                                        Low:          2.2 hours
  *Comment:*  Scientific judgement based upon estimated photooxidation half-life in air.

  · **Surface Water:**                                 High:        672 hours                    (4 weeks)
                                                        Low:         62.4 hours                   (2.6 days)
  *Comment:*  Scientific judgement based upon estimated rate constants for reactions of
  representative aromatic amines with ·OH and $RO_2$· (low $t_{1/2}$) (Mill, T and Mabey, W (1985)) and
  estimated unacclimated aqueous aerobic biodegradation half-life (high $t_{1/2}$).

  · **Ground Water:**                                  High:       1344 hours                    (8 weeks)
                                                        Low:         336 hours                   (14 days)
  *Comment:*  Scientific judgement based upon estimated unacclimated aqueous aerobic
  biodegradation half-life.

<u>Aqueous Biodegradation (unacclimated):</u>
  · **Aerobic half-life:**                             High:        672 hours                    (4 weeks)
                                                        Low:         168 hours                    (7 days)
  *Comment:*  Scientific judgement based upon unacclimated aerobic aqueous screening test data for
  9,10-anthraquinone (Blok, J and Booy, M (1984); Zanella, EF et al. (1979)).

  · **Anaerobic half-life:**                           High:       2688 hours                    (16 weeks)
                                                        Low:         672 hours                   (28 days)
  *Comment:*  Scientific judgement based upon estimated unacclimated aqueous aerobic
  biodegradation half-life.

  · **Removal/secondary treatment:**        High:        No data

214

                                                  Low:
*Comment:*

## Photolysis:
· **Atmos photol half-life:**            High:        No data
                                         Low:
*Comment:*

· **Max light absorption (nm):**         No data
*Comment:*

· **Aq photol half-life:**               High:        No data
                                         Low:
*Comment:*

## Photooxidation half-life:
· **Water:**                             High:        3480 hours              (145 days)
                                         Low:         62.4 hours              (2.6 days)
*Comment:* Scientific judgement based upon estimated rate constants for reactions of representative aromatic amines with ·OH and $RO_2$· (Mill, T and Mabey, W (1985)).

· **Air:**                               High:        22 hours
                                         Low:         2.2 hours
*Comment:* Scientific judgement based upon estimated rate constant for reaction with hydroxyl radicals in air (Atkinson, R (1987A)).

## Reduction half-life:                  High:        No data
                                         Low:
*Comment:*

## Hydrolysis:
· **First-order hydr half-life:**        No hydrolyzable groups
*Comment:*

· **Acid rate const $(M(H+)-hr)^{-1}$:**
*Comment:*

· **Base rate const $(M(OH-)-hr)^{-1}$:**
*Comment:*

215

# Pentachloronitrobenzene

**CAS Registry Number:** 82-68-8

**Half-lives:**

· **Soil:**                              High:   16776 hours          (699 days)
                                          Low:    5112 hours           (213 days)
*Comment:* Scientific judgement based upon unacclimated aerobic soil grab sample data (Beck, J and Hansen, KE (1974)).

· **Air:**                               High:   87912 hours          (10 years)
                                          Low:    8791 hours           (366 days)
*Comment:* Scientific judgement based upon estimated photooxidation half-life in air.

· **Surface Water:**                     High:   16776 hours          (699 days)
                                          Low:    5112 hours           (213 days)
*Comment:* Scientific judgement based upon estimated aqueous aerobic biodegradation half-life.

· **Ground Water:**                      High:   16776 hours          (699 days)
                                          Low:    216 hours           (9 days)
*Comment:* Scientific judgement based upon estimated aqueous aerobic and anaerobic biodegradation half-lives.

**Aqueous Biodegradation (unacclimated):**

· **Aerobic half-life:**                 High:   16776 hours          (699 days)
                                          Low:    5112 hours           (213 days)
*Comment:* Scientific judgement based upon unacclimated aerobic soil grab sample data (Beck, J and Hansen, KE (1974)).

· **Anaerobic half-life:**               High:   816 hours            (34 days)
                                          Low:    216 hours           (9 days)
*Comment:* Scientific judgement based upon estimated aqueous aerobic biodegradation half-life.

· **Removal/secondary treatment:**       High:       No data
                                          Low:
*Comment:*

**Photolysis:**

· **Atmos photol half-life:**            High:       No data
                                          Low:
*Comment:*

· **Max light absorption (nm):**         lambda max = 301.0, 212.5

216

*Comment:* Reported absorbance maxima (Howard, PH et al. (1976)).

· **Aq photol half-life:**  High: Probably not important
                            Low:
*Comment:* Stable under aqueous photolysis conditions (Hamadmad, N (1967)).

## Photooxidation half-life:
· **Water:**  High: No data
              Low:

*Comment:*

· **Air:**  High:  87912 hours  (10 years)
            Low:  8791 hours  (366 days)
*Comment:* Scientific judgement based upon an estimated rate constant for the vapor phase reaction with hydroxyl radicals in air (Atkinson, R (1987A)).

## Reduction half-life:
              High: No data
              Low:

*Comment:*

## Hydrolysis:
· **First-order hydr half-life:**  13182  (2.8 years)
*Comment:* Scientific judgement based upon rate constant ($2.8 \times 10^{-5}$ hr$^{-1}$) measured at pH 7 and 69 °C extrapolated 25 °C (Ellington, JJ et al. (1988A)).

· **Acid rate const (M(H+)-hr)$^{-1}$:**
*Comment:*

· **Base rate const (M(OH-)-hr)$^{-1}$:**
*Comment:*

# Acenaphthene

**CAS Registry Number:** 83-32-9

**Structure:**

**Half-lives:**
· **Soil:**                               High:      2448 hours            (102 days)
                                          Low:       295 hours             (12.3 days)
*Comment:* Based upon aerobic soil column test data (Kincannon, DF and Lin, YS (1985)).

· **Air:**                                High:      8.79 hours
                                          Low:       0.879 hours
*Comment:* Scientific judgement based upon estimated photooxidation half-life in air.

· **Surface Water:**                      High:      300 hours             (12.5 days)
                                          Low:       3 hours
*Comment:* Scientific judgement based upon photolysis half-life in water.

· **Ground Water:**                       High:      4896 hours            (204 days)
                                          Low:       590 hours             (24.6 days)
*Comment:* Scientific judgement based upon estimated unacclimated aqueous aerobic biodegradation half-life.

**Aqueous Biodegradation (unacclimated):**
· **Aerobic half-life:**                  High:      2448 hours            (102 days)
                                          Low:       295 hours             (12.3 days)
*Comment:* Based upon aerobic soil column test data (Kincannon, DF and Lin, YS (1985)).

· **Anaerobic half-life:**                High:      9792 hours            (408 days)
                                          Low:       1180 hours            (49.2 days)
*Comment:* Scientific judgement based upon estimated unacclimated aqueous aerobic biodegradation half-life.

· **Removal/secondary treatment:**        High:      52%
                                          Low:
*Comment:* Removal percentage based upon data from a continuous activated sludge biological treatment simulator (Petrasek, AC et al. (1983)).

**Photolysis:**

· **Atmos photol half-life:**  High:  60 hours  (2.5 days)
Low:  3 hours
*Comment:* Scientific judgement based upon measured rate of photolysis in water irradiated with light >290 nm (Fukuda, K et al. (1988)).

· **Max light absorption (nm):**  lambda max = 320, 313.5, 306, 300, 288 nm (methanol).
*Comment:* Sadtler UV No. 272.

· **Aq photol half-life:**  High:  60 hours  (2.5 days)
Low:  3 hours
*Comment:* Scientific judgement based upon measured rate of photolysis in water irradiated with light >290 nm (Fukuda, K et al. (1988)).

## Photooxidation half-life:
· **Water:**  High:  No data
Low:

*Comment:*

· **Air:**  High:  8.79 hours
Low:  0.879 hours
*Comment:* Scientific judgement based upon estimated rate constant for reaction with hydroxyl radical in air (Atkinson, R (1987A)).

## Reduction half-life:  High:  No data
Low:

*Comment:*

## Hydrolysis:
· **First-order hydr half-life:**
*Comment:*

· **Acid rate const $(M(H+)-hr)^{-1}$:**  No hydrolyzable groups
*Comment:*

· **Base rate const $(M(OH-)-hr)^{-1}$:**
*Comment:*

219

# Diethyl phthalate

<u>CAS Registry Number:</u> 84-66-2

<u>Half-lives:</u>
  · **Soil:**                        High:      1344 hours             (8 weeks)
                                      Low:       72 hours               (3 days)
*Comment:* Scientific judgement based upon estimated unacclimated aqueous aerobic biodegradation half-life.

  · **Air:**                          High:      212 hours             (8.8 days)
                                      Low:       21 hours
*Comment:* Scientific judgement based upon estimated photooxidation half-life in air.

  · **Surface Water:**            High:      1344 hours             (8 weeks)
                                      Low:       72 hours               (3 days)
*Comment:* Scientific judgement based on aerobic aqueous grab sample data for fresh (low $t_{1/2}$) and marine (high $t_{1/2}$) water (Hattori, Y et al. (1975)).

  · **Ground Water:**           High:      2688 hours            (16 weeks)
                                      Low:       144 hours             (6 days)
*Comment:* Scientific judgement based upon estimated unacclimated aqueous aerobic biodegradation half-life.

## Aqueous Biodegradation (unacclimated):
  · **Aerobic half-life**:          High:      1344 hours             (8 weeks)
                                      Low:       72 hours               (3 days)
*Comment:* Scientific judgement based on aerobic aqueous grab sample data for fresh (low $t_{1/2}$) and marine (high $t_{1/2}$) water (Hattori, Y et al. (1975)).

  · **Anaerobic half-life**:      High:      5376 hours            (32 weeks)
                                      Low:       672 hours             (4 weeks)
*Comment:* Scientific judgement based upon estimated unacclimated aqueous aerobic biodegradation half-life (high $t_{1/2}$) and anaerobic screening test data (low $t_{1/2}$) (Horowitz, A et al. (1982)).

  · **Removal/secondary treatment:**    High:          98%
                                        Low:           20%
*Comment:* Based upon % degraded under aerobic continuous flow conditions (Bouwer, EJ et al. (1984); Patterson, JW and Kodukalas, PS (1981)).

<u>Photolysis:</u>
  · **Atmos photol half-life:**      High:          No data

Low:

*Comment:*

· **Max light absorption (nm):** lambda max = 274, 224.5
*Comment:* No absorbance occurs above 320 nm in methanol (Sadtler UV No. 150).

· **Aq photol half-life:** High: No data
Low:
*Comment:*

**Photooxidation half-life:**
· **Water:** High: $1.1 \times 10^6$ hours (12.2 years)
Low: $2.1 \times 10^4$ hours (2.4 years)
*Comment:* Scientific judgement based upon estimated rate data for alkylperoxyl radicals in aqueous solution (Wolfe, NL et al. (1980A)).

· **Air:** High: 212 hours (8.8 days)
Low: 21 hours
*Comment:* Scientific judgement based upon an estimated rate constant for vapor phase reaction with hydroxyl radicals in air (Atkinson, R (1987A)).

**Reduction half-life:** High: No data
Low:
*Comment:*

**Hydrolysis:**
· **rst-order hydr half-life:** 8.8 years
*Comment:* Based upon base rate constant ($k_B = 2.5 \times 10^{-2}$ $M^{-1}$ $s^{-1}$) at 30 °C and pH of 7 (Wolfe, NL et al. (1980B)).

· **Acid rate const $(M(H+)-hr)^{-1}$:** No data
*Comment:*

· **Base rate const $(M(OH-)-hr)^{-1}$:** 90
*Comment:* At pH 9, $t_{1/2}$ = 32 days based upon measured rate constant at 30 °C and pH from 10 to 12 (Wolfe, NL et al. (1980B)).

221

# Ethyl carbethoxymethyl phthalate

**CAS Registry Number:** 84-72-0

**Structure:**

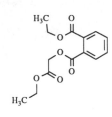

**Half-lives:**
 · **Soil:**                          High:    672 hours           (4 weeks)
                                      Low:     168 hours           (1 week)
 *Comment:* Scientific judgement based upon estimated aqueous aerobic biodegradation half-life.

 · **Air:**                           High:    212 hours           (8.8 days)
                                      Low:      21 hours
 *Comment:* Scientific judgement based upon estimated photooxidation half-life in air.

 · **Surface Water:**                 High:    672 hours           (4 weeks)
                                      Low:     168 hours           (1 week)
 *Comment:* Scientific judgement based upon estimated aqueous aerobic biodegradation half-life.

 · **Ground Water:**                  High:    8640 hours          (8 weeks)
                                      Low:     336 hours           (2 weeks)
 *Comment:* Scientific judgement based upon estimated aqueous aerobic biodegradation half-life.

**Aqueous Biodegradation (unacclimated):**
 · **Aerobic half-life:**             High:    672 hours           (4 weeks)
                                      Low:     168 hours           (1 week)
 *Comment:* Scientific judgement.

 · **Anaerobic half-life:**           High:    2688 hours          (16 weeks)
                                      Low:     672 hours           (4 weeks)
 *Comment:* Scientific judgement based upon estimated aqueous aerobic biodegradation half-life.

 · **Removal/secondary treatment:**   High:    No data
                                      Low:
 *Comment:*

**Photolysis:**
 · **Atmos photol half-life:**        High:

222

Low:

*Comment:*

· **Max light absorption (nm):**　　No data
*Comment:*

· **Aq photol half-life:**　　High:
　　　　　　　　　　　　Low:

*Comment:*

**Photooxidation half-life:**
　· **Water:**　　　　　　High:　　No data
　　　　　　　　　　　　Low:

*Comment:*

　· **Air:**　　　　　　High:　　212 hours　　　　(8.8 days)
　　　　　　　　　　　Low:　　21 hours
*Comment:*  Scientific judgement based upon estimated rate data for hydroxyl radicals in air (Atkinson, R (1987A)).

**Reduction half-life:**　　　High:　　No data
　　　　　　　　　　　　　Low:
*Comment:*

**Hydrolysis:**
　· **First-order hydr half-life:**　　No data
　*Comment:*

　· **Acid rate const (M(H+)-hr)$^{-1}$:**
　*Comment:*

　· **Base rate const (M(OH-)-hr)$^{-1}$:**
　*Comment:*

# Di-n-butyl phthalate

<u>CAS Registry Number:</u>  84-74-2

<u>Half-lives:</u>
  · **Soil:**                                                High:    552 hours         (23 days)

High: 552 hours (23 days)
Low: 48 hours (2 days)

*Comment:* Scientific judgement based upon unacclimated aerobic soil grab sample data (low $t_{1/2}$: Shanker, R et al. (1985), high $t_{1/2}$: Inman, JC et al. (1984)).

  · **Air:**

High: 74 hours (3.1 days)
Low: 7.4 hours

*Comment:* Scientific judgement based upon estimated photooxidation half-life in air.

  · **Surface Water:**

High: 336 hours (14 days)
Low: 24 hours (1 day)

*Comment:* Scientific judgement based upon unacclimated aerobic river die-away test (low $t_{1/2}$: Schouten, MJ, et al. (1979)) and freshwater/sediment grab sample data (high $t_{1/2}$: Johnson, BT et al. (1984)).

  · **Ground Water:**

High: 552 hours (23 days)
Low: 48 hours (2 days)

*Comment:* Scientific judgement based upon estimated unacclimated aqueous aerobic and anaerobic biodegradation half-lives.

<u>Aqueous Biodegradation (unacclimated):</u>
  · **Aerobic half-life:**

High: 552 hours (23 days)
Low: 24 hours (1 day)

*Comment:* Scientific judgement based upon unacclimated aerobic river die-away test (low $t_{1/2}$: Schouten, MJ et al. (1979)) and soil grab sample data (high $t_{1/2}$: Inman, JC et al. (1984)).

  · **Anaerobic half-life:**

High: 552 hours (23 days)
Low: 48 hours (2 days)

*Comment:* Scientific judgement based upon unacclimated anaerobic grab sample data for soil and sediment (low $t_{1/2}$: Johnson, BT and Lulves, W (1975), high $t_{1/2}$: Shanker, R et al. (1985)).

  · **Removal/secondary treatment:**

High: 95%
Low: 35%

*Comment:* Based upon % degraded under aerobic semi-continuous flow conditions (O'Grady, DP et al. (1985); Hutchins, SR et al. (1983)).

<u>Photolysis:</u>
  · **Atmos photol half-life:**              High:  Not expected to be important

Low:
*Comment:* Scientific judgement based upon aqueous photolysis data.

· **Max light absorption (nm):**           lambda max = 274, 224.5
*Comment:* No absorbance occurs above 310 nm in methanol (Sadtler UV No. 529).

· **Aq photol half-life:**           High:  Not expected to be important
                                     Low:      3466 hours                (144 days)
*Comment:* Scientific judgement based upon estimated maximum aqueous photolysis rate data
$(2.0X10^{-4}$ $hr^{-1})$ (Wolfe, NL et al. (1980A)).

## Photooxidation half-life:
· **Water:**           High: $1.1X10^6$ hours          (12.2 years)
                       Low: $2.1X10^4$ hours          (2.4 years)
*Comment:* Scientific judgement based upon estimated rate data for alkylperoxyl radicals in
aqueous solution (Wolfe, NL et al. (1980A)).

· **Air:**           High:      74 hours          (2.5 days)
                     Low:      7.4 hours
*Comment:* Scientific judgement based upon estimated rate data for hydroxyl radicals in air
(Atkinson, R (1987A)).

## Reduction half-life:           High:      No data
                                  Low:
*Comment:*

## Hydrolysis:
· **First-order hydr half-life:**           10 years
*Comment:* Based upon overall rate constant $(1.0X10^{-2}$ $mol^{-1}$ $s^{-1})$ at pH 7 and 30 °C (Wolfe,
NL et al. (1980B)).

· **Acid rate const $(M(H+)-hr)^{-1}$:**
*Comment:*

· **Base rate const $(M(OH-)-hr)^{-1}$:**
*Comment:*

# Phenanthrene

CAS Registry Number: 85-01-8

Structure:

Half-lives:
· **Soil:**                                    High:    4800 hours            (200 days)
                                               Low:      384 hours            (16 days)
*Comment:* Based upon aerobic soil die-away test data (Coover, MP and Sims, RC (1987); Sims, RC (1990)).

· **Air:**                                     High:    20.1 hours
                                               Low:     2.01 hours
*Comment:* Based upon photooxidation half-life in air.

· **Surface Water:**                           High:     25 hours
                                               Low:       3 hours
*Comment:* Based upon aqueous photolysis half-life.

· **Ground Water:**                            High:    9600 hours            (1.10 years)
                                               Low:      768 hours            (32 days)
*Comment:* Scientific judgement based upon estimated unacclimated aqueous aerobic biodegradation half-life.

Aqueous Biodegradation (unacclimated):
· **Aerobic half-life:**                       High:    4800 hours            (200 days)
                                               Low:      384 hours            (16 days)
*Comment:* Based upon aerobic soil die-away test data (Coover, MP and Sims, RC (1987); Sims, RC (1990)).

· **Anaerobic half-life:**                     High:    19200 hours           (2.19 years)
                                               Low:     1536 hours            (64 hours)
*Comment:* Scientific judgement based upon estimated unacclimated aqueous aerobic biodegradation half-life.

· **Removal/secondary treatment:**             High:         37%
                                               Low:
*Comment:* Removal percentage based upon data from a continuous activated sludge biological treatment simulator (Petrasek, AC et al. (1983)).

## Photolysis:
· **Atmos photol half-life:**           High:      25 hours

                                                Low:       3 hours

*Comment:* Based upon measured aqueous photolysis quantum yields and calculated for midday summer sunlight at 40°N latitude (low $t_{1/2}$) (Zepp, RG and Schlotzhauer, PF (1979)) and adjusted for approximate winter sunlight intensity (high $t_{1/2}$) (Lyman, WJ et al. (1982)).

· **Max light absorption (nm):**      lambda max = 210, 219, 242, 251, 273.5, 281, 292.5,
308.5,                                      314, 322.5, 329.5, 337, 345 nm   (methanol/ethanol
mixture)

*Comment:* IARC (1983A).

· **Aq photol half-life:**              High:      25 hours

                                                Low:       3 hours

*Comment:* Based upon measured aqueous photolysis quantum yields and calculated for midday summer sunlight at 40°N latitude (low $t_{1/2}$) (Zepp, RG and Schlotzhauer, PF (1979)) and adjusted for approximate winter sunlight intensity (Lyman, WJ et al. (1982)).

## Photooxidation half-life:
· **Water:**                           High:      No data

                                                Low:

*Comment:*

· **Air:**                               High:      20.1 hours

                                                Low:      2.01 hours

*Comment:* Scientific judgement based upon measured rate constant for reaction with hydroxyl radical in air (Atkinson, R (1987A)).

## Reduction half-life:
                                                High:      No data

                                                Low:

*Comment:*

## Hydrolysis:
· **First-order hydr half-life:**

*Comment:*

· **Acid rate const (M(H+)-hr)$^{-1}$:**      No hydrolyzable groups

*Comment:*

· **Base rate const (M(OH-)-hr)$^{-1}$:**

*Comment:*

# Phthalic anhydride

CAS Registry Number: 85-44-9

· **Soil:**                          High:    27 minutes
                                     Low:     32 seconds
*Comment:* Low $t_{1/2}$ based upon measured first order hydrolysis rate constant for pH 5.2 at 28 °C (Hawkins, MD (1975)). High $t_{1/2}$ based upon measured first order hydrolysis rate constant for pH 7 at 25 °C (Bunton, CA et al. (1963)).

· **Air:**                           High:    4847 hours          (202 days)
                                     Low:     485 hours           (20 days)
*Comment:* Scientific judgement based upon estimated photooxidation half-life in air.

· **Surface Water:**                 High:    27 minutes
                                     Low:     32 seconds
*Comment:* Low $t_{1/2}$ based upon measured first order hydrolysis rate constant for pH 5.2 at 28 °C (Hawkins, MD (1975)). High $t_{1/2}$ based upon measured first order hydrolysis rate constant for pH 7 at 25 °C (Bunton, CA et al. (1963)).

· **Ground Water:**                  High:    27 minutes
                                     Low:     32 seconds
*Comment:* Low $t_{1/2}$ based upon measured first order hydrolysis rate constant for pH 5.2 at 28 °C (Hawkins, MD (1975)). High $t_{1/2}$ based upon measured first order hydrolysis rate constant for pH 7 at 25 °C (Bunton, CA et al. (1963)).

## Aqueous Biodegradation (unacclimated):
· **Aerobic half-life:**             High:    168 hours           (7 days)
                                     Low:     24 hours            (1 day)
*Comment:* Scientific judgement based upon limited aqueous screening test data (given the fast hydrolysis rate this may be for the hydrolysis product) (Sasaki, S (1978); Matsui, S et al. (1975)).

· **Anaerobic half-life:**           High:    672 hours           (28 days)
                                     Low:     96 hours            (4 days)
*Comment:* Scientific judgement based upon estimated unacclimated aqueous aerobic biodegradation half-life.

· **Removal/secondary treatment:**   High:    No data
                                     Low:
*Comment:*

## Photolysis:
· **Atmos photol half-life:**        High:    No data

*Comment:*

· **Max light absorption (nm):**        lambda max = 274, 224
*Comment:* No absorbance occurs above 310 nm in methanol (Sadtler UV No. 18).

· **Aq photol half-life:**        High:      No data
                                         Low:
*Comment:*

**Photooxidation half-life:**
· **Water:**        High:      No data
                                         Low:
*Comment:*

· **Air:**        High:      4847 hours        (202 days)
                                       Low:      485 hours         (20 days)
*Comment:* Scientific judgement based upon an estimated rate constant for vapor phase reaction with hydroxyl radicals in air (Atkinson, R (1987A)).

**Reduction half-life:**        High:      No data
                                         Low:
*Comment:*

**Hydrolysis:**
· **First-order hydr half-life:**        27 minutes
*Comment:* Scientific judgement based upon first order rate data measured at pH of 7 and 25 °C ($k = 4.29 \times 10^{-4}$ s$^{-1}$) (Bunton, CA et al. (1963)).

· **Acid rate const (M(H+)-hr)$^{-1}$:**        $k = 4.29 \times 10^{-4}$ s$^{-1}$
*Comment:* Scientific judgement based upon first order rate data ($t_{1/2} = 32$ seconds) measured at pH of 5.2 and 28 °C (Hawkins, MD (1975)).

· **Base rate const (M(OH-)-hr)$^{-1}$:**        No data
*Comment:*

# Butyl benzyl phthalate

**CAS Registry Number:** 85-68-7

**Half-lives:**

 **· Soil:**       High:  168 hours    (7 days)
             Low:   24 hours    (1 day)
 *Comment:* Scientific judgement based upon estimated unacclimated aqueous aerobic biodegradation half-life.

 **· Air:**        High:  60 hours    (2.5 days)
             Low:   6 hours
 *Comment:* Scientific judgement based upon estimated photooxidation half-life in air.

 **· Surface Water:**    High:  168 hours    (7 days)
             Low:   24 hours    (1 day)
 *Comment:* Scientific judgement based upon unacclimated aerobic river die-away test (Saeger, VW and Tucker, ES (1976)).

 **· Ground Water:**    High:  4320 hours   (6 months)
             Low:   48 hours    (2 days)
 *Comment:* Scientific judgement based upon estimated unacclimated aqueous aerobic and anaerobic biodegradation half-lives.

**Aqueous Biodegradation (unacclimated):**

 **· Aerobic half-life:**   High:  168 hours    (7 days)
             Low:   24 hours    (1 day)
 *Comment:* Scientific judgement based upon unacclimated aerobic river die-away test (Saeger, VW and Tucker, ES (1976)).

 **· Anaerobic half-life:**  High:  4320 hours   (6 months)
             Low:   672 hours    (4 weeks)
 *Comment:* Scientific judgement based upon unacclimated anaerobic screening test data (Shelton, DR et al. (1984), Horowitz, A et al. (1982)).

 **· Removal/secondary treatment:** High:  No data
             Low:
 *Comment:*

**Photolysis:**

 **· Atmos photol half-life:**  High: Not expected to be important
             Low:
 *Comment:* Scientific judgement based upon aqueous photolysis data.

· **Max light absorption (nm):**          No data
*Comment:*

· **Aq photol half-life:**          High:  Not expected to be important
                                    Low:  876000 hours          (100 years)
*Comment:* Stable in sunlit aqueous solution (Gledhill, WE et al. (1980)).

## Photooxidation half-life:
· **Water:**          High:          No data
                      Low:

*Comment:*

· **Air:**          High:          60 hours          (2.5 days)
                    Low:          6 hours
*Comment:* Scientific judgement based upon estimated rate data for hydroxyl radicals in air (Atkinson, R (1987A)).

## Reduction half-life:          High:          No data
                                 Low:

*Comment:*

## Hydrolysis:
· **First-order hydr half-life:**          Not expected to be important
*Comment:* Stable in aqueous solution (Gledhill, WE et al. (1980)).

· **Acid rate const (M(H+)-hr)$^{-1}$:**
*Comment:*

· **Base rate const (M(OH-)-hr)$^{-1}$:**
*Comment:*

# Butylglycolyl butyl phthalate

<u>CAS Registry Number:</u>  85-70-1

<u>Structure:</u>

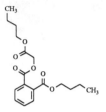

<u>Half-lives:</u>
  · **Soil:**                      High:    168 hours         (7 days)
                                   Low:     24 hours          (1 day)
  *Comment:*  Scientific judgement based upon estimated unacclimated aqueous aerobic biodegradation half-life.

  · **Air:**                       High:    73.7 hours        (3.1 days)
                                   Low:     7.4 hours
  *Comment:*  Scientific judgement based upon estimated photooxidation half-life in air.

  · **Surface Water:**             High:    168 hours         (7 days)
                                   Low:     24 hours          (1 day)
  *Comment:*  Scientific judgement based upon unacclimated river die-away test data (Saeger, VW and Tucker ES (1973)).

  · **Ground Water:**              High:    336 hours         (14 days)
                                   Low:     48 hours          (2 days)
  *Comment:*  Scientific judgement based upon estimated unacclimated aqueous aerobic biodegradation half-life.

<u>Aqueous Biodegradation (unacclimated):</u>
  · **Aerobic half-life:**         High:    168 hours         (7 days)
                                   Low:     24 hours          (1 day)
  *Comment:*  Scientific judgement based upon unacclimated river die-away test data (Saeger, VW and Tucker ES (1973)).

  · **Anaerobic half-life:**       High:    672 hours         (28 days)
                                   Low:     96 hours          (4 days)
  *Comment:*  Scientific judgement based upon unacclimated aerobic biodegradation half-life.

  · **Removal/secondary treatment:**   High:        99%
                                       Low:

232

*Comment:* Based upon % removal under aerobic semi-continuous flow conditions (Saeger, VW and Tucker ES (1973)).

## Photolysis:
  · **Atmos photol half-life:**          High:     No data
                                         Low:
  *Comment:*

  · **Max light absorption (nm):**      No data
  *Comment:*

  · **Aq photol half-life:**             High:     No data
                                         Low:
  *Comment:*

## Photooxidation half-life:
  · **Water:**                       High:     No data
                                         Low:
  *Comment:*

  · **Air:**                          High:   73.7 hours        (3.1 days)
                                         Low:    7.4 hours
  *Comment:* Scientific judgement based upon an estimated rate constant for the vapor phase reaction with hydroxyl radicals in air (Atkinson, R (1987A)).

## Reduction half-life:
                                          High:     No data
                                         Low:
  *Comment:*

## Hydrolysis:
  · **First-order hydr half-life:**    No data
  *Comment:*

  · **Acid rate const (M(H+)-hr)$^{-1}$:**
  *Comment:*

  · **Base rate const (M(OH-)-hr)$^{-1}$:**
  *Comment:*

# N-Nitrosodiphenylamine

CAS Registry Number: 86-30-6

Half-lives:
- Soil:                          High:    816 hours              (34 days)
                                 Low:     240 hours              (10 days)

  Comment: Scientific judgement based upon estimated unacclimated aqueous aerobic biodegradation half-life.

- Air:                           High:    7.0 hours
                                 Low:     0.70 hours

  Comment: Scientific judgement based upon estimated photooxidation half-life in air.

- Surface Water:                 High:    816 hours              (34 days)
                                 Low:     240 hours              (10 days)

  Comment: Scientific judgement based upon estimated unacclimated aqueous aerobic biodegradation half-life.

- Ground Water:                  High:    1632 hours             (68 days)
                                 Low:     480 hours              (20 days)

  Comment: Scientific judgement based upon estimated unacclimated aqueous aerobic biodegradation half-life.

Aqueous Biodegradation (unacclimated):
- Aerobic half-life:             High:    816 hours              (34 days)
                                 Low:     240 hours              (10 days)

  Comment: Scientific judgement based upon data from one soil die-away test; a range was bracketed around the reported half-life of 22 days (Mallik, MAB and Tesfai, K (1981)).

- Anaerobic half-life:           High:    3264 hours             (136 days)
                                 Low:     960 hours              (40 days)

  Comment: Scientific judgement based upon estimated unacclimated aqueous aerobic biodegradation half-life.

- Removal/secondary treatment:   High:    >84%
                                 Low:

  Comment: Removal percentage based upon data from an activated sludge biological treatment simulator (Patterson, JW and Kodukala, PS (1981)) .

Photolysis:
- Atmos photol half-life:        High:    No data
                                 Low:

234

*Comment:*

· **Max light absorption (nm):**          lambda max = 262 nm (methanol)
*Comment:* Sadtler UV No. 153.

  · **Aq photol half-life:**          High:          No data
                                      Low:

*Comment:*

**Photooxidation half-life:**
  · **Water:**                       High:          No data
                                     Low:

*Comment:*

  · **Air:**                         High:          7.0 hours
                                     Low:           0.70 hours
*Comment:* Scientific judgement based upon estimated rate constant for reaction with hydroxyl radical in air (Atkinson, R (1987A)).

**Reduction half-life:**             High:          No data
                                     Low:

*Comment:*

**Hydrolysis:**
  · **First-order hydr half-life:**          No data
*Comment:*

  · **Acid rate const (M(H+)-hr)$^{-1}$:**
*Comment:*

  · **Base rate const (M(OH-)-hr)$^{-1}$:**
*Comment:*

# Fluorene

**CAS Registry Number:** 86-73-7

**Structure:**

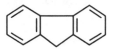

**Half-lives:**
  · **Soil:**                                      High:    1440 hours        (60 days)

                                                  Low:     768 hours        (32 days)

*Comment:* Based upon aerobic soil die-away test data (Coover, MP and Sims, RC (1987)).

  · **Air:**        High:    68.1 hours        (2.8 days)

        Low:    6.81 hours

*Comment:* Scientific judgement based upon estimated photooxidation half-life in air.

  · **Surface Water:**    High:    1440 hours        (60 days)

        Low:     768 hours        (32 days)

*Comment:* Based upon aerobic soil die-away test data (Sims, RC (1990)).

  · **Ground Water:**    High:    2880 hours        (120 days)

        Low:    1536 hours        (64 days)

*Comment:* Scientific judgement based upon estimated unacclimated aqueous aerobic biodegradation half-life.

**Aqueous Biodegradation (unacclimated):**
  · **Aerobic half-life:**    High:    1440 hours        (60 days)

        Low:     768 hours        (32 days)

*Comment:* Based upon aerobic soil die-away test data (Coover, MP and Sims, RC (1987)).

  · **Anaerobic half-life:**    High:    5760 hours        (240 days)

        Low:    3072 hours        (128 days)

*Comment:* Scientific judgement based upon estimated unacclimated aqueous aerobic biodegradation half-life.

  · **Removal/secondary treatment:**    High:    44%

        Low:

*Comment:* (Removal percentage based upon data from a continuous activated sludge biological treatment simulator (Petrasek, AC et al. (1983)).

## Photolysis:
· **Atmos photol half-life:**  High:　　No data
　　　　　　　　　　　　　　　Low:

*Comment:*

· **Max light absorption (nm):**  lambda max = 208, 261, 289, 301 nm (ethanol).
*Comment:* IARC (1983A).

· **Aq photol half-life:**  High:　　No data
　　　　　　　　　　　　　Low:

*Comment:*

## Photooxidation half-life:
· **Water:**  High:　　No data
　　　　　　　Low:

*Comment:*

· **Air:**  High:　　68.1 hours　　　　(2.8 days)
　　　　　Low:　　6.81 hours

*Comment:* Scientific judgement based upon estimated rate constant for reaction with hydroxyl radical in air (Atkinson, R (1987A)).

## Reduction half-life:
　　　　　　　　　　　　　High:　　No data
　　　　　　　　　　　　　Low:

*Comment:*

## Hydrolysis:
· **First-order hydr half-life:**
*Comment:*

· **Acid rate const $(M(H+)-hr)^{-1}$:**  No hydrolyzable groups
*Comment:*

· **Base rate const $(M(OH-)-hr)^{-1}$:**
*Comment:*

# 2,6-Xylidine

<u>CAS Registry Number:</u>  87-62-7

<u>Structure:</u>

<u>Half-lives:</u>
  · **Soil:**                              High:     7584 hours              (316 days)
                                           Low:        72 hours             (3 days)
  *Comment:* High value based upon radio-labeled transformation to $CO_2$ in a soil degradation study
  (Bollag, JM et al. (1978)). Low value based upon measured soil persistence in a soil which may
  have been acclimated to chemicals from the coke industry (Medvedev, VA and Davidov, VD
  (1981B)).

  · **Air:**                               High:      3.3 hours
                                           Low:      0.33 hours
  *Comment:*  Based upon estimated photooxidation half-lives in air.

  · **Surface Water:**                     High:     3480 hours              (145 days)
                                           Low:       62.4 hours            (2.6 days)
  *Comment:*  Based upon estimated photooxidation half-lives in water.

  · **Ground Water:**                      High:     8640 hours              (12 months)
                                           Low:      1344 hours             (8 weeks)
  *Comment:*  Scientific judgement based upon estimated aqueous aerobic biodegradation half-lives.

<u>Aqueous Biodegradation (unacclimated):</u>
  · **Aerobic half-life**:                 High:     4320 hours              (6 months)
                                           Low:       672 hours             (4 weeks)
  *Comment:* Scientific judgement based upon a biological screening study (Baird R et al. (1977))
  and a soil degradation study (Bollag JM et al. (1978)).

  · **Anaerobic half-life**:               High:    17280 hours             (24 months)
                                           Low:      2688 hours             (16 weeks)
  *Comment:*  Scientific judgement based upon estimated aqueous biodegradation half-lives.

  · **Removal/secondary treatment:**       High:        No data
                                           Low:
  *Comment:*

## Photolysis:
· **Atmos photol half-life:**          High:      No data

                                                Low:

*Comment:*

· **Max light absorption (nm):**

*Comment:*

· **Aq photol half-life:**             High:

                                                Low:

*Comment:*

## Photooxidation half-life:
· **Water:**                        High:     3480 hours          (145 days)

                                              Low:     62.4 hours          (2.6 days)

*Comment:* Scientific judgement based upon reaction rate constants of the aromatic amine class with $RO_2 \cdot$ and $\cdot OH$ (Mill, T and Mabey W (1985); Guesten H et al. (1981)).

· **Air:**                            High:     3.3 hours

                                              Low:     0.33 hours

*Comment:* Based upon estimated reaction rate constant with $\cdot OH$ in air (Atkinson, R (1987A)).

## Reduction half-life:
                                          High:      No data

                                          Low:

*Comment:*

## Hydrolysis:
· **First-order hydr half-life:**        No hydrolyzable groups

*Comment:*

· **Acid rate const (M(H+)-hr)$^{-1}$:**

*Comment:*

· **Base rate const (M(OH-)-hr)$^{-1}$:**

*Comment:*

# Hexachlorobutadiene

**CAS Registry Number:** 87-68-3

**Half-lives:**

  · **Soil:**                             High:      4320 hours           (6 months)

                                         Low:       672 hours            (4 weeks)

*Comment:* Scientific judgement based upon estimated aqueous aerobic biodegradation half-life.

  · **Air:**                              High:    28650 hours        (3.3 years)

                                         Low:     2865 hours      (119.4 days)

*Comment:* Scientific judgement based upon estimated photooxidation half-life in air.

  · **Surface Water:**                High:      4320 hours           (6 months)

                                       Low:       672 hours            (4 weeks)

*Comment:* Scientific judgement based upon estimated aqueous aerobic biodegradation half-life.

  · **Ground Water:**               High:      8640 hours         (12 months)

                                       Low:     1344 hours         (8 weeks)

*Comment:* Scientific judgement based upon estimated aqueous aerobic biodegradation half-life.

**Aqueous Biodegradation (unacclimated):**

  · **Aerobic half-life:**           High:      4320 hours           (6 months)

                                       Low:       672 hours            (4 weeks)

*Comment:* Scientific judgement based upon monitoring data (Zoeteman, BCJ et al. (1980)) and acclimated aqueous screening test data (Tabak, HH et al. (1981)).

  · **Anaerobic half-life:**         High:    17280 hours        (24 months)

                                       Low:     2688 hours       (16 weeks)

*Comment:* Scientific judgement based upon estimated aqueous aerobic biodegradation half-life.

  · **Removal/secondary treatment:**    High:

                                       Low:           4%

*Comment:* Based upon % degraded under aerobic continuous flow conditions (Schroeder, HF (1987)).

**Photolysis:**

  · **Atmos photol half-life:**        High:      No data

                                       Low:

*Comment:*

  · **Max light absorption (nm):**      lambda max = 249, 220

*Comment:* Very little absorbance above 290 nm and did not absorb UV light at wavelengths greater than 305 nm in methanol (Sadtler UV No. 956).

· **Aq photol half-life:**          High:     No data
                                         Low:
*Comment:*

## Photooxidation half-life:
· **Water:**                  High:     No data
                                         Low:
*Comment:*

· **Air:**        High:    28650 hours        (3.3 years)
                         Low:     2865 hours       (119.4 days)
*Comment:* Scientific judgement based upon an estimated rate constant for vapor phase reaction with hydroxyl radicals in air (Atkinson, R (1987A)).

## Reduction half-life:
                                  High:     No data
                                  Low:
*Comment:*

## Hydrolysis:
· **First-order hydr half-life:**     No hydrolyzable groups
*Comment:* Rate constant at neutral pH is zero (Kollig, HP et al. (1987A)).

· **Acid rate const $(M(H+)-hr)^{-1}$:**    0.0
*Comment:* Based upon measured rate data at acid pH (Kollig, HP et al. (1987A)).

· **Base rate const $(M(OH-)-hr)^{-1}$:**    0.0
*Comment:* Based upon measured rate data at basic pH (Kollig, HP et al. (1987A)).

# Pentachlorophenol

CAS Registry Number: 87-86-5

Half-lives:
- **Soil:**          High:    4272 hours       (178 days)
                                     Low:     552 hours        (23 days)

*Comment:* Scientific judgement based upon estimated unacclimated aqueous aerobic biodegradation half-life.

- **Air:**          High:    1392 hours       (58 days)
                                       Low:    139.2 hours      (5.8 days)

*Comment:* Scientific judgement based upon estimated photooxidation half-life in air.

- **Surface Water:**          High:    110 hours       (4.6 days)
                                       Low:     1 hour

*Comment:* Scientific judgement based upon aqueous photolysis half-life.

- **Ground Water:**          High:   36480 hours      (4.2 years)
                                       Low:    1104 hours      (46 days)

*Comment:* Scientific judgement based upon estimated unacclimated aqueous aerobic sediment grab sample data (low $t_{1/2}$: Delaune, RD et al. (1983)) and unacclimated anaerobic grab sample data for ground water (high $t_{1/2}$: Baker, MD and Mayfield, CI (1980)).

## Aqueous Biodegradation (unacclimated):
- **Aerobic half-life:**          High:    4272 hours       (178 days)
                                       Low:     552 hours        (23 days)

*Comment:* Scientific judgement based upon unacclimated and acclimated aerobic sediment grab sample data (low $t_{1/2}$: Delaune, RD et al. (1983), high $t_{1/2}$: Baker, MD et al. (1980)).

- **Anaerobic half-life:**          High:   36480 hours      (4.2 years)
                                       Low:    1008 hours      (42 days)

*Comment:* Scientific judgement based upon unacclimated anaerobic grab sample data for soil and ground water (low $t_{1/2}$: Ide, A et al. (1972) high $t_{1/2}$: Baker, MD and Mayfield, CI (1980)).

- **Removal/secondary treatment:**     High:       84%
                                       Low:        0%

*Comment:* Based upon % degraded under aerobic continuous flow conditions (low %: Petrasek, AC et al. (1983), high %: Brink, RH Jr (1976)).

## Photolysis:
- **Atmos photol half-life:**          High:     No data
                                       Low:

*Comment:*

· **Max light absorption (nm):**        lambda max = 280.0, 273.0, 224.0
*Comment:* No absorbance of UV light at wavelengths greater than 310 in methanol (Sadtler UV No. 112).

· **Aq photol half-life:**        High:        110 hours        (4.6 days)
                                                Low:        1 hour
*Comment:* Low $t_{1/2}$ based upon photolysis rate constant for transformation summer sunlight conditions at 25 °C (Hwang H et al. (1986)). High $t_{1/2}$ derived from predicted photolysis rate constant (based on laboratory data) for an environmental pond at noon under fall sunlight conditions at 40° N (Sugiura, K et al. (1984)).

**Photooxidation half-life:**
· **Water:**        High:        3480 hours        (145 days)
                        Low:        66.0 hours        (2.75 days)
*Comment:* Scientific judgement based upon reported reaction rate constants for ·OH and $RO_2$· with the phenol class (Mill, T and Mabey, W (1985); Guesten, H et al. (1981)).

· **Air:**        High:        1392 hours        (58 days)
                    Low:        139.2 hours        (5.8 days)
*Comment:* Scientific judgement based upon an estimated rate constant for the vapor phase reaction with hydroxyl radicals in air (Atkinson, R (1987A)).

**Reduction half-life:**        High:        No data
                                        Low:
*Comment:*

**Hydrolysis:**
· **First-order hydr half-life:**        No hydrolyzable groups
*Comment:* Rate constant at neutral pH is zero (Kollig, HP et al. (1987A)).

· **Acid rate const $(M(H+)-hr)^{-1}$:**        0.0
*Comment:* Based upon measured rate data at acid pH (Kollig, HP et al. (1987A)).

· **Base rate const $(M(OH-)-hr)^{-1}$:**        0.0
*Comment:* Based upon measured rate data at basic pH (Kollig, HP et al. (1987A)).

243

# 2,4,6-Trichlorophenol

<u>CAS Registry Number:</u>  88-06-2

<u>Half-lives:</u>
· **Soil:**                          High:    1680 hours          (70 days)
                                     Low:      168 hours          (7 days)
*Comment:* Scientific judgement based upon estimated unacclimated aqueous aerobic
biodegradation half-life including soil grab sample data (Haider, K et al. (1974)).

· **Air:**                           High:    1234 hours          (51.4 days)
                                     Low:    123.4 hours          (5.1 days)
*Comment:* Scientific judgement based upon estimated photooxidation half-life in air.

· **Surface Water:**                 High:      96 hours          (4 days)
                                     Low:       2 hours
*Comment:* Scientific judgement based upon aqueous photolysis half-life.

· **Ground Water:**                  High:   43690 hours          (5 years)
                                     Low:      336 hours          (14 days)
*Comment:* Scientific judgement based upon estimated unacclimated aqueous aerobic
biodegradation half-life (low $t_{1/2}$) and aqueous anaerobic half-life (high $t_{1/2}$).

<u>Aqueous Biodegradation (unacclimated):</u>
· **Aerobic half-life:**             High:    1680 hours          (70 days)
                                     Low:      168 hours          (7 days)
*Comment:* Scientific judgement based upon unacclimated aerobic river die-away test and soil
grab sample data (low $t_{1/2}$: Blades-Fillmore, LA et al. (1982), Haider, K et al. (1974)).

· **Anaerobic half-life:**           High:   43690 hours          (5 years)
                                     Low:    4050 hours           (169 days)
*Comment:* Scientific judgement based upon unacclimated anaerobic grab sample data for soil
(Baker, MD and Mayfield, CI (1980)).

· **Removal/secondary treatment:**   High:        60%
                                     Low:
*Comment:* Based upon % degraded under aerobic continuous flow conditions (Salkinoja-Salonen,
MS et al. (1983)).

<u>Photolysis:</u>
· **Atmos photol half-life:**        High:    No data
                                     Low:
*Comment:*

· **Max light absorption (nm):**     lambda max = 296.0, 288.5
*Comment:* In methanol (Sadtler UV No. 11905).

· **Aq photol half-life:**     High:     96 hours          (4 days)
                                              Low:      2 hours
*Comment:* Both high $t_{1/2}$ and low $t_{1/2}$ derived from predicted photolysis rate constants (based on laboratory data) for an environmental pond at noon under fall sunlight conditions and at 40° N (Sugiura, K et al. (1984)).

## Photooxidation half-life:
· **Water:**     High:     2027 hours          (84.5 days)
                            Low:      20.3 hours
*Comment:* Scientific judgement based upon measured rate data for reaction with singlet oxygen in aqueous solution (Scully, FE Jr and Hoigne, J (1987)).

· **Air:**     High:     1234 hours          (51.4 days)
                         Low:      123.4 hours          (5.1 days)
*Comment:* Scientific judgement based upon an estimated rate constant for the vapor phase reaction with hydroxyl radicals in air (Atkinson, R (1987A)).

## Reduction half-life:     High:     No data
                                           Low:
*Comment:*

## Hydrolysis:
· **First-order hydr half-life:**     $>8 \times 10^6$ years
*Comment:* Rate constant at neutral pH is zero (Kollig, HP et al. (1987A)).

· **Acid rate const $(M(H+)-hr)^{-1}$:**     0.0
*Comment:* Based upon measured rate data at acid pH (Kollig, HP et al. (1987A)).

· **Base rate const $(M(OH-)-hr)^{-1}$:**     0.0
*Comment:* Based upon measured rate data at basic pH (Kollig, HP et al. (1987A)).

# 2-Nitrophenol

<u>CAS Registry Number:</u> 88-75-5

<u>Half-lives:</u>
- **· Soil:**  High:  672 hours  (4 weeks)
  
  Low:  168 hours  (1 week)

*Comment:* Scientific judgement based upon estimated aqueous aerobic biodegradation half-life and anaerobic soil grab sample data (Sudhakar-Barik, R and Sethunathan, N (1978A)).

- **· Air:**  High:  71 hours
  
  Low:  7 hours

*Comment:* Based upon photooxidation half-life in air.

- **· Surface Water:**  High:  672 hours  (4 weeks)
  
  Low:  168 hours  (1 week)

*Comment:* Scientific judgement based upon aerobic river die-away test data (Ludzack, FJ et al. (1958)).

- **· Ground Water:**  High:  672 hours  (4 weeks)
  
  Low:  336 hours  (2 weeks)

*Comment:* Scientific judgement based upon estimated aqueous aerobic and anaerobic biodegradation half-lives.

<u>Aqueous Biodegradation (unacclimated):</u>
- **· Aerobic half-life:**  High:  672 hours  (4 weeks)
  
  Low:  168 hours  (1 week)

*Comment:* Scientific judgement based upon unacclimated aerobic screening test data (Sasaki, S (1978), Gerike, P and Fischer, WK (1979)).

- **· Anaerobic half-life:**  High:  672 hours  (4 weeks)
  
  Low:  168 hours  (1 week)

*Comment:* Scientific judgement based upon anaerobic soil grab sample data (Sudhakar-Barik, R and Sethunathan, N (1978A)).

- **· Removal/secondary treatment:**  High:
  
  Low:  39%

*Comment:* Based upon % degraded under aerobic continuous flow conditions (Gerike, P and Fischer, WK (1979)).

<u>Photolysis:</u>
- **· Atmos photol half-life:**  High:  No data
  
  Low:

*Comment:*

· **Max light absorption (nm):**       lambda max = 408, 276, 230
*Comment:* In methanol-potassium hydroxide (Sadtler UV No. 15826).

· **Aq photol half-life:**       High:    No data
                                    Low:
*Comment:*

## Photooxidation half-life:
· **Water:**       High:    No data
                    Low:
*Comment:*

· **Air:**       High:    71 hours
                 Low:     7 hours
*Comment:* Based upon measured rate data for the vapor phase reaction with hydroxyl radicals in air (Atkinson, R (1985)).

## Reduction half-life:
                                High:    No data
                                Low:
*Comment:*

## Hydrolysis:
· **First-order hydr half-life:**     No data
*Comment:*

· **Acid rate const (M(H+)-hr)$^{-1}$:**
*Comment:*

· **Base rate const (M(OH-)-hr)$^{-1}$:**
*Comment:*

# Dinoseb

**CAS Registry Number:** 88-85-7

**Structure:**

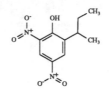

## Half-lives:

**· Soil:**
|  | High: | 2952 hours | (123 days) |
|---|---|---|---|
|  | Low: | 1032 hours | (43 days) |

*Comment:* Scientific judgement based upon aerobic soil mineralization data for one soil (Doyle, RC et al. (1978)).

**· Air:**
|  | High: | 122 hours | (5.1 days) |
|---|---|---|---|
|  | Low: | 12.2 hours |  |

*Comment:* Scientific judgement based upon estimated photooxidation half-life in air.

**· Surface Water:**
|  | High: | 2952 hours | (123 days) |
|---|---|---|---|
|  | Low: | 1032 hours | (43 days) |

*Comment:* Scientific judgement based upon aerobic soil mineralization data for one soil (Doyle, RC et al. (1978)).

**· Ground Water:**
|  | High: | 5904 hours | (246 days) |
|---|---|---|---|
|  | Low: | 96 hours | (4 days) |

*Comment:* Scientific judgement based upon unacclimated aqueous aerobic (high $t_{1/2}$) and anaerobic (low $t_{1/2}$) biodegradation half-lives.

## Aqueous Biodegradation (unacclimated):

**· Aerobic half-life:**
|  | High: | 2952 hours | (123 days) |
|---|---|---|---|
|  | Low: | 1032 hours | (43 days) |

*Comment:* Scientific judgement based upon aerobic soil mineralization data for one soil (Doyle, RC et al. (1978)).

**· Anaerobic half-life:**
|  | High: | 360 hours | (15 days) |
|---|---|---|---|
|  | Low: | 96 hours | (4 days) |

*Comment:* Scientific judgement based upon anaerobic soil die-away test data for isopropalin (Gingerich, LL and Zimdahl, RL (1976)).

**· Removal/secondary treatment:**
|  | High: | No data |
|---|---|---|
|  | Low: |  |

*Comment:*

## Photolysis:
**· Atmos photol half-life:**        High:      No data

                                                Low:

*Comment:*

**· Max light absorption (nm):**       lambda max = 375 (water)
*Comment:* Schneider, M and Smith, GW (1978)).

**· Aq photol half-life:**            High:      No data

                                                Low:

*Comment:*

## Photooxidation half-life:
**· Water:**                          High:      No data

                                                Low:

*Comment:*

**· Air:**                              High:      122 hours                (5.1 days)

                                             Low:        12.2 hours

*Comment:* Scientific judgement based upon estimated rate constant for reaction with hydroxyl radical in air (Atkinson, R (1987A)).

## Reduction half-life:
                                                 High:      No data

                                                 Low:

*Comment:*

## Hydrolysis:
**· First-order hydr half-life:**
*Comment:*

**· Acid rate const $(M(H+)-hr)^{-1}$:**      No hydrolyzable groups
*Comment:*

**· Base rate const $(M(OH-)-hr)^{-1}$:**
*Comment:*

# Picric acid

<u>CAS Registry Number:</u>  88-89-1

<u>Structure:</u>

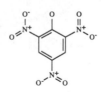

<u>Half-lives:</u>
  · **Soil:**                           High:     4320 hours        (6 months)
                                         Low:       672 hours        (4 weeks)
  *Comment:* Scientific judgement based upon estimated aqueous aerobic biodegradation half-lives.

  · **Air:**                            High:     4320 hours        (6 months)
                                         Low:       672 hours        (4 weeks)
  *Comment:*  Based upon estimated photooxidation half-lives in air.

  · **Surface Water:**              High:     4320 hours        (6 months)
                                         Low:       672 hours        (4 weeks)
  *Comment:*  Scientific judgement based upon estimated aqueous aerobic biodegradation half-lives.

  · **Ground Water:**               High:     8640 hours        (12 months)
                                         Low:        48 hours        ( 2 days)
  *Comment:*  High value is scientific judgement based upon estimated high value for aerobic
  aqueous biodegradation. Low value based upon estimated low value for anaerobic biodegradation.

<u>Aqueous Biodegradation (unacclimated):</u>
  · **Aerobic half-life**:            High:     4320 hours        (6 months)
                                         Low:       672 hours        (4 weeks)
  *Comment:* Scientific judgement based upon aerobic biodegradation screening tests (Pitter, P
  (1976); Chambers, CW et al. (1963)).

  · **Anaerobic half-life**:         High:      300 hours        (12.5 days)
                                         Low:        48 hours        ( 2 days)
  *Comment:*  Scientific judgement based upon anaerobic natural water die-away test data for the
  dinitrotoluenes (Spanggord, RJ et al. (1980)). Reduction of the nitro group observed in these
  studies may not be due to biodegradation.

  · **Removal/secondary treatment:**     High:      No data
                                              Low:
  *Comment:*

## Photolysis:

· **Atmos photol half-life:**       High:       No data
                                    Low:

*Comment:*

· **Max light absorption (nm):**       lambda max = 360 nm
*Comment:*  Absorption maxima in alcohol (HSDB, 1988).

· **Aq photol half-life:**       High:       No data
                                 Low:

*Comment:*

## Photooxidation half-life:

· **Water:**       High:       No data
                   Low:

*Comment:*

· **Air:**       High:       4320 hours       (6 months)
                 Low:        677 hours        (4 weeks)

*Comment:*  Scientific judgement based upon estimated reaction rates with ·OH and nitrate radicals.

## Reduction half-life:       High:       No data
                              Low:

*Comment:*

## Hydrolysis:

· **First-order hydr half-life:**       No hydrolyzable groups
*Comment:*

· **Acid rate const $(M(H+)-hr)^{-1}$:**
*Comment:*

· **Base rate const $(M(OH-)-hr)^{-1}$:**
*Comment:*

# o-Anisidine

CAS Registry Number:  90-04-0

Half-lives:
  · Soil:                              High:    4320 hours          (6 months)
                                       Low:      672 hours          (4 weeks)
  *Comment:* Scientific judgement based upon estimated aqueous aerobic biodegradation half-lives.

  · Air:                               High:     5.3 hours
                                       Low:     0.53 hours
  *Comment:*  Based upon estimated photooxidation half-lives in air.

  · Surface Water:                     High:    4320 hours          (6 months)
                                       Low:     62.4 hours          (2.6 days)
  *Comment:*  High value based upon scientific judgement of estimated high aqueous biodegradation
  half-life. Low value based upon estimated low value for aqueous photooxidation.

  · Ground Water:                      High:    8640 hours          (12 months)
                                       Low:     1344 hours          (8 weeks)
  *Comment:*  Scientific judgement based upon estimated aqueous aerobic biodegradation half-lives.

Aqueous Biodegradation (unacclimated):
  · Aerobic half-life:                 High:    4320 hours          (6 months)
                                       Low:      672 hours          (4 weeks)
  *Comment:* Scientific judgement based upon aerobic aqueous screening studies (Kondo, M et al.
  (1988A); Kool, HJ (1984)).

  · Anaerobic half-life:               High:    17280 hours         (24 months)
                                       Low:     2688 hours          (8 weeks)
  *Comment:*  Scientific judgement based upon estimated aerobic aqueous biodegradation half-lives.

  · Removal/secondary treatment:       High:    No data
                                       Low:
  *Comment:*

Photolysis:
  · Atmos photol half-life:            High:    No data
                                       Low:
  *Comment:*

  · Max light absorption (nm):
  *Comment:*

**· Aq photol half-life:**

                                           High:

                                           Low:

*Comment:*

## Photooxidation half-life:
**· Water:**

                                           High:     3480 hours          (145 days)

                                           Low:      62.4 hours         (2.6 days)

*Comment:* Scientific judgement based upon reaction rate constants of the aromatic amine class with $RO_2\cdot$ and $\cdot OH$ (Mill, T and Mabey, W (1985); Guesten, H et al. (1981)).

**· Air:**

                                           High:     5.3 hours

                                           Low:     0.53 hours

*Comment:*  Based upon estimated reaction rate constant with $\cdot OH$ (Atkinson, R 1987A)).

## Reduction half-life:

                                           High:     No data

                                           Low:

*Comment:*

## Hydrolysis:
**· First-order hydr half-life:**             No hydrolyzable groups
*Comment:*

**· Acid rate const $(M(H+)-hr)^{-1}$:**
*Comment:*

**· Base rate const $(M(OH-)-hr)^{-1}$:**
*Comment:*

# 2-Phenylphenol

CAS Registry Number: 90-43-7

Half-lives:
  · Soil:                                High:      168 hours              (7 days)
                                         Low:        24 hours              (1 day)
  Comment: Scientific judgement based upon estimated aqueous aerobic biodegradation half-lives.

  · Air:                                 High:       22 hours
                                         Low:        0.1 hours
  Comment: Based upon estimated photooxidation half-lives in air.

  · Surface Water:                       High:      168 hours              (7 days)
                                         Low:        24 hours              (1 day)
  Comment: Scientific judgement based upon estimated aqueous aerobic biodegradation half-life.

  · Ground Water:                        High:      336 hours              (14 days)
                                         Low:        48 hours              (2 days)
  Comment: Scientific judgement based upon estimated aqueous aerobic biodegradation half-lives.

Aqueous Biodegradation (unacclimated):
  · Aerobic half-life:                   High:      168 hours              (7 days)
                                         Low:        24 hours              (1 day)
  Comment: Based upon a river die-away study in which a 50% degradation was observed over a one week period (Gonsior, SJ et al. (1984)).

  · Anaerobic half-life:                 High:      672 hours              (28 days)
                                         Low:        96 hours              (4 days)
  Comment: Scientific judgement based upon estimated aqueous aerobic biodegradation half-lives.

  · Removal/secondary treatment:         High:        98%
                                         Low:         50%
  Comment: Semi-continuous activated sludge simulator (Gonsior, SJ et al. (1984)); low percentage refers to non-acclimated sludge; high percentage refers to acclimated sludge.

Photolysis:
  · Atmos photol half-life:              High:      No data
                                         Low:
  Comment:

  · Max light absorption (nm):           lambda max = 284 nm

*Comment:* Approximate upper absorption maxima in hexane solution; absorption extends to approximately 310 nm (Gore, RC et al. (1971)).

**· Aq photol half-life:**　　　　　　High:　　No data
　　　　　　　　　　　　　　　　　　Low:

*Comment:*

## Photooxidation half-life:

Water:　　　　　　　　　　　　High:　　3840 hours　　　　　(160 days)
　　　　　　　　　　　　　　　　Low:　　66 hours　　　　　　(2.75 days)
*Comment:* Scientific judgement based upon reported reaction rate constants for ·OH and RO$_2$· (low t$_{1/2}$) and RO$_2$· (high t$_{1/2}$) with the phenol class (Mill, T and Mabey, W (1985); Guesten, H et al. (1981)).

**· Air:**　　　　　　　　　　　　High:　　22 hours
　　　　　　　　　　　　　　　　Low:　　0.1 hours
*Comment:* High value based upon estimated reaction rate constant with ·OH (Atkinson, R (1987A)); in addition to ·OH, low value assumes additional reaction with nitrate radicals in night-time air based upon scientific judgement of measured rates for the phenol and cresol classes.

## Reduction half-life:　　　　　　High:　　No data
　　　　　　　　　　　　　　　　Low:

*Comment:*

## Hydrolysis:
**· First-order hydr half-life:**　　No hydrolyzable groups
*Comment:*

**· Acid rate const (M(H+)-hr)$^{-1}$:**
*Comment:*

**· Base rate const (M(OH-)-hr)$^{-1}$:**
*Comment:*

# Michler's ketone

<u>CAS Registry Number:</u>  90-94-8

<u>Structure:</u>

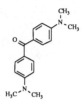

<u>Half-lives:</u>
  **· Soil:**                                   High:        672 hours           (4 weeks)
                                                Low:         168 hours           (7 days)
  *Comment:*  Scientific judgement based upon estimated aqueous aerobic biodegradation half-lives.

  **· Air:**                                    High:          2 hours
                                                Low:         0.2 hours
  *Comment:*  Based upon estimated photooxidation half-lives in air.

  **· Surface Water:**                          High:        672 hours           (4 weeks)
                                                Low:        31.2 hours
  *Comment:*  High value is scientific judgement based upon the high estimated value for aqueous biodegradation. The low value is based upon the low value estimated for photooxidation half-life in water.

  **· Ground Water:**                           High:       1344 hours           (12 months)
                                                Low:         336 hours           (14 days)
  *Comment:*  Scientific judgement based upon estimated aqueous aerobic biodegradation half-lives.

<u>Aqueous Biodegradation (unacclimated):</u>
  **· Aerobic half-life:**                      High:        672 hours           (4 weeks)
                                                Low:         168 hours           (7 days)
  *Comment:*  Scientific judgement based upon aerobic screening tests utilizing activated sludge (Malaney, GW et al. (1967); Lutin, PA et al. (1965)).

  **· Anaerobic half-life:**                    High:       2688 hours           (16 weeks)
                                                Low:         672 hours           (28 days)
  *Comment:*  Scientific judgement based upon estimated aqueous aerobic biodegradation half-life.

  **· Removal/secondary treatment:**            High:        No data
                                                Low:
  *Comment:*

## Photolysis:
· **Atmos photol half-life:**  High:  No data
    Low:

*Comment:*

· **Max light absorption (nm):**
*Comment:*

· **Aq photol half-life:**  High:
    Low:

*Comment:*

## Photooxidation half-life:
· **Water:**  High:  1740 hours  (72.5 days)
    Low:  31.2 hours

*Comment:* Scientific judgement based upon reaction rates of ·OH and $RO_2$· with the aromatic amine class (Mill, T and Mabey, W (1985); Guesten, H et al. (1981)). It is assumed that Michler's ketone reacts twice as fast as aniline.

· **Air:**  High:  2.0 hours
    Low:  0.2 hours

*Comment:* Based upon estimated reaction rate constant with ·OH (Atkinson, R (1987A)).

## Reduction half-life:  High:  No data
    Low:

*Comment:*

## Hydrolysis:
· **First-order hydr half-life:**  No hydrolyzable functions
*Comment:*

· **Acid rate const $(M(H+)-hr)^{-1}$:**
*Comment:*

· **Base rate const $(M(OH-)-hr)^{-1}$:**
*Comment:*

257

# Toluene-2,6-diisocyanate

<u>CAS Registry Number:</u> 91-08-7

<u>Half-lives:</u>
  · **Soil:**                          High:        24 hours
                                        Low:         12 hours
  *Comment:* Scientific judgement based upon estimated hydrolysis half-life in water.

  · **Air:**                           High:        3.21 hours
                                        Low:         0.321 hours
  *Comment:* Scientific judgement based upon estimated photooxidation half-life in air.

  · **Surface Water:**                 High:        24 hours
                                        Low:         12 hours
  *Comment:* Scientific judgement based upon estimated hydrolysis half-life in water.

  · **Ground Water:**                  High:        24 hours
                                        Low:         12 hours
  *Comment:* Scientific judgement based upon estimated hydrolysis half-life in water.

<u>Aqueous Biodegradation (unacclimated):</u>
  · **Aerobic half-life:**             High:        672 hours        (4 weeks)
                                        Low:         168 hours        (7 days)
  *Comment:* Scientific judgement.

  · **Anaerobic half-life:**           High:        2688 hours       (16 weeks)
                                        Low:         672 hours        (28 days)
  *Comment:* Scientific judgement based upon estimated unacclimated aerobic aqueous
  biodegradation half-life.

  · **Removal/secondary treatment:**   High:        No data
                                        Low:
  *Comment:*

<u>Photolysis:</u>
  · **Atmos photol half-life:**        High:
                                        Low:
  *Comment:*

  · **Max light absorption (nm):**     No data
  *Comment:*

258

**· Aq photol half-life:**              High:
                                         Low:

*Comment:*

## Photooxidation half-life:
   **· Water:**                          High:       No data
                                         Low:

   *Comment:*

   **· Air:**                            High:       3.21 hours
                                         Low:        0.321 hours

   *Comment:* Scientific judgement based upon estimated rate constant for reaction with hydroxyl radical in air (Atkinson, R (1987A)).

## Reduction half-life:                  High:       No data
                                         Low:

   *Comment:*

## Hydrolysis:
   **· First-order hydr half-life:**     12 hours

   *Comment:* Scientific judgement based upon observed disappearance of toluene diisocyanate (a mixture of toluene-2,6- and 2,4-diisocyanates) in a model river (Duff, PB (1983)).

   **· Acid rate const $(M(H+)-hr)^{-1}$:**     No data
   *Comment:*

   **· Base rate const $(M(OH-)-hr)^{-1}$:**     No data
   *Comment:*

# Naphthalene

<u>CAS Registry Number:</u>  91-20-3

<u>Half-lives:</u>
**· Soil:**                          High:      1152 hours            (48 days)
                                 Low:        398 hours             (16.6 days)
*Comment:* Based upon soil-die away test data (Kincannon, DF and Lin, YS (1985)).

**· Air:**                           High:      29.6 hours
                                 Low:        2.96 hours
*Comment:* Based upon photooxidation half-life in air.

**· Surface Water:**                 High:      480 hours             (20 days)
                                 Low:        12 hours
*Comment:* Scientific judgement based upon estimated unacclimated aqueous aerobic
biodegradation half-life.

**· Ground Water:**                  High:      6192 hours            (258 days)
                                 Low:        24 hours
*Comment:* Scientific judgement based upon estimated unacclimated aerobic (low $t_{1/2}$) and
anaerobic (high $t_{1/2}$) biodegradation half-lives.

<u>Aqueous Biodegradation (unacclimated):</u>
**· Aerobic half-life:**             High:      480 hours             (20 days)
                                 Low:        12 hours
*Comment:* Based upon die-away test data for an oil-polluted creek (low $t_{1/2}$) (Walker, JD and
Colwell, RR (1976)) and for an estuarine river (Lee, RF and Ryan, C (1976)).

**· Anaerobic half-life:**           High:      6192 hours            (258 days)
                                 Low:        600 hours             (25 days)
*Comment:* Based upon anaerobic estuarine sediment die-away test data at pH 8 (low $t_{1/2}$) and pH
5 (high $t_{1/2}$) (Hambrick, GA et al. (1980)).

**· Removal/secondary treatment:**   High:      98.6%
                                 Low:        77%
*Comment:* Removal percentages based upon data from continuous activated sludge biological
treatment simulators (Kincannon, DF et al. (1983B); Petrasek, AC et al. (1983)).

<u>Photolysis:</u>
**· Atmos photol half-life:**        High:      13200 hours           (550 days)
                                 Low:        1704 hours            (71 days)
*Comment:* Calculated from quantum yield for photolysis in water irradiated at 313 nm and for

sunlight photolysis at latitude 40°N at midday in the summer in near-surface water (low $t_{1/2}$) and in a 5 meter deep inland water body (high $t_{1/2}$) (Zepp, RG and Schlotzhauer, PF (1979)).

· **Max light absorption (nm):**      lambda max = 220.5, 258, 265.5, 275, 283, 285.5, 297,
301,                                         303.5, 310.5 (methanol)

*Comment:* Sadtler UV No. 265.

· **Aq photol half-life:**      High:    13200 hours           (550 days)
                                         Low:     1704 hours            (71 days)

*Comment:* Calculated from quantum yield for photolysis in water irradiated at 313 nm and for sunlight photolysis at latitude 40°N at midday in the summer in near-surface water (low $t_{1/2}$) and in a 5 meter deep inland water body (high $t_{1/2}$) (Zepp, RG and Schlotzhauer, PF (1979)).

**Photooxidation half-life:**
· **Water:**      High:      No data
                             Low:

*Comment:*

· **Air:**      High:      29.6 hours
                       Low:      2.96 hours

*Comment:* Based upon measured rate constants for reaction with hydroxyl radical in air (Atkinson, R (1985)).

**Reduction half-life:**      High:      No data
                                       Low:

*Comment:*

**Hydrolysis:**
· **First-order hydr half-life:**
*Comment:*

· **Acid rate const (M(H+)-hr)$^{-1}$:**      No hydrolyzable groups
*Comment:*

· **Base rate const (M(OH-)-hr)$^{-1}$:**
*Comment:*

# Quinoline

<u>CAS Registry Number:</u>  91-22-5

<u>Half-lives:</u>
   · **Soil:**
                                           High:     240 hours        (10 days)
                                           Low:      72 hours         (3 days)
   *Comment:*  Scientific judgement based upon estimated aqueous aerobic biodegradation half-life.

   · **Air:**
                                           High:     99 hours        (4.1 days)
                                           Low:      10 hours
   *Comment:*  Scientific judgement based upon estimated photooxidation half-life in air.

   · **Surface Water:**
                                         High:     240 hours        (10 days)
                                           Low:      72 hours         (3 days)
   *Comment:*  Scientific judgement based upon an acclimated fresh water grab sample data (Rogers, JE et al. (1984)).

   · **Ground Water:**
                                        High:     480 hours        (20 days)
                                           Low:     144 hours       (6 days)
   *Comment:*  Scientific judgement based upon estimated aqueous aerobic biodegradation half-life.

<u>Aqueous Biodegradation (unacclimated):</u>
   · **Aerobic half-life:**
                                       High:     240 hours        (10 days)
                                         Low:      72 hours         (3 days)
   *Comment:*  Scientific judgement based upon an acclimated fresh water grab sample data (Rogers, JE et al. (1984)).

   · **Anaerobic half-life**:
                                       High:     960 hours        (40 days)
                                         Low:     288 hours       (12 days)
   *Comment:*  Scientific judgement based upon estimated aqueous aerobic biodegradation half-life.

   · **Removal/secondary treatment:**
                                       High:     No data
                                       Low:
   *Comment:*

<u>Photolysis:</u>
   · **Atmos photol half-life:**
                                     High:     No data
                                     Low:
   *Comment:*

   · **Max light absorption (nm):**     lambda max = 340
   *Comment:*  UV absorbance in water (Mill, T et al. (1981)).

**· Aq photol half-life:**         High:    3851 hours         (160.4 days)
                                   Low:     535 hours          (22.3 days)
*Comment:* Based upon aqueous photolysis rate constants at pH 6.9 for summer ($k_p = 3.6 \times 10^{-7}$ s$^{-1}$; low $t_{1/2}$) and winter ($k_p = 5.0 \times 10^{-8}$ s$^{-1}$; high $t_{1/2}$) sunlight at 40 °N (Mill, T et al. (1981)).

## Photooxidation half-life:
**· Water:**                       High:    No data
                                   Low:
*Comment:*

**· Air:**                         High:    99 hours           (4.1 days)
                                   Low:     10 hours
*Comment:* Scientific judgement based upon an estimated rate constant for vapor phase reaction with hydroxyl radicals in air (Atkinson, R (1987A)).

## Reduction half-life:              High:    No data
                                   Low:
*Comment:*

## Hydrolysis:
**· First-order hydr half-life:**    No hydrolyzable groups
*Comment:*

**· Acid rate const (M(H+)-hr)$^{-1}$:**
*Comment:*

**· Base rate const (M(OH-)-hr)$^{-1}$:**
*Comment:*

# beta-Naphthylamine

<u>CAS Registry Number:</u>  91-59-8

<u>Half-lives:</u>

· **Soil:**                              High:    4320 hours        (6 months)

Low:     672 hours         (4 weeks)

*Comment:* Scientific judgement based upon unacclimated aerobic soil grab sample data (Medvedev, VA and Davidov, VD (1981B)) and unacclimated aerobic screening test data (Fochtman, EG and Eisenberg, W (1979)).

· **Air:**                               High:    2.9 hours

Low:     0.3 hours

*Comment:* Scientific judgement based upon estimated photooxidation half-life in air.

· **Surface Water:**                     High:    3480 hours        (145 days)

Low:      62 hours         (2.6 days)

*Comment:* Scientific judgement based upon photooxidation rate constants with ·OH and $RO_2$· for the aromatic amine class (Mill, T and Mabey, W (1985); Guesten, H et al. (1981)).

· **Ground Water:**                      High:    8640 hours        (12 months)

Low:     1344 hours        (8 weeks)

*Comment:* Scientific judgement based upon estimated unacclimated aqueous aerobic biodegradation half-life.

<u>Aqueous Biodegradation (unacclimated):</u>

· **Aerobic half-life:**                 High:    4320 hours        (6 months)

Low:     672 hours         (4 weeks)

*Comment:* Scientific judgement based upon unacclimated aerobic soil grab sample data (Medvedev, VA and Davidov, VD (1981B)) and unacclimated aerobic screening test data (Fochtman, EG and Eisenberg, W (1979)).

· **Anaerobic half-life:**               High:    17280 hours       (24 months)

Low:     2880 hours        (16 weeks)

*Comment:* Scientific judgement based upon estimated unacclimated aqueous aerobic biodegradation half-life.

· **Removal/secondary treatment:**       High:    100%

Low:     87%

*Comment:* Based upon % degraded in a 24 hr period under continuous reactor conditions after 1 week (low % ) and 6 weeks (high %) of operation (Fochtman, EG and Eisenberg, W (1979)).

<u>Photolysis:</u>

· **Atmos photol half-life:**          High:
                                       Low:
*Comment:*

· **Max light absorption (nm):**     No data
*Comment:*

· **Aq photol half-life:**           High:
                                       Low:
*Comment:*

## Photooxidation half-life:
· **Water:**                   High:     3480 hours         (145 days)
                                       Low:      62 hours          (2.6 days)
*Comment:* Scientific judgement based upon photooxidation rate constants with ·OH and $RO_2$·
for the aromatic amine class (Mill, T and Mabey, W (1985); Guesten, H et al. (1981)).

· **Air:**                      High:     2.9 hours
                                       Low:      0.3 hours
*Comment:* Scientific judgement based upon an estimated rate constant for hydroxyl radicals in air
(Atkinson, R (1987A)).

## Reduction half-life:
                                         High:      No data
                                       Low:
*Comment:*

## Hydrolysis:
· **First-order hydr half-life:**     No hydrolyzable groups
*Comment:*

· **Acid rate const (M(H+)-hr)$^{-1}$:**
*Comment:*

· **Base rate const (M(OH-)-hr)$^{-1}$:**
*Comment:*

# 3,3'-Dichlorobenzidine

<u>CAS Registry Number:</u> 91-94-1

<u>Half-lives:</u>

· **Soil:**           High:    4320 hours        (6 months)
              Low:     672 hours          (4 weeks)
*Comment:* Scientific judgement based upon estimated unacclimated aqueous aerobic biodegradation half-life.

· **Air:**            High:    0.075 hours        (4.5 minutes)
              Low:     0.025 hours        (1.5 minutes)
*Comment:* Based upon measured half-life for direct photolysis in distilled water in midday summer sunlight (low $t_{1/2}$) (Banerjee, S et al. (1978A); Sikka, HC et al. (1978)); high $t_{1/2}$: scientific judgement based upon the direct photolysis half-life and approximate winter sunlight intensity (Banerjee, S et al. (1978A); Sikka, HC et al. (1978); Lyman, WJ et al. (1982)).

· **Surface Water:**     High:    0.075 hours        (4.5 minutes)
              Low:     0.025 hours        (1.5 minutes)
*Comment:* Based upon measured half-life for direct photolysis in distilled water in midday summer sunlight (low $t_{1/2}$) (Banerjee, S et al. (1978A); Sikka, HC et al. (1978)); high $t_{1/2}$: scientific judgement based upon the direct photolysis half-life and approximate winter sunlight intensity (Banerjee, S et al. (1978A); Sikka, HC et al. (1978); Lyman, WJ et al. (1982)).

· **Ground Water:**      High:    8640 hours        (12 months)
              Low:     1344 hours        (8 weeks)
*Comment:* Scientific judgement based upon estimated unacclimated aqueous aerobic biodegradation half-life.

<u>Aqueous Biodegradation (unacclimated):</u>

· **Aerobic half-life:**     High:    4320 hours        (6 months)
              Low:     672 hours          (4 weeks)
*Comment:* Scientific judgement based upon data from a lake die-away study (Appleton, H et al. (1978)) and a soil die-away test (Boyd, SA et al. (1984)).

· **Anaerobic half-life:**   High:    17280 hours       (24 months)
              Low:     2688 hours        (16 weeks)
*Comment:* Scientific judgement based upon estimated unacclimated aqueous aerobic biodegradation half-life.

· **Removal/secondary treatment:**   High:      No data
              Low:
*Comment:*

## Photolysis:
**· Atmos photol half-life:**       High:    0.075 hours       (4.5 minutes)

                                         Low:    0.025 hours       (1.5 minutes)

*Comment:* Based upon measured half-life for direct photolysis in distilled water in midday summer sunlight (low $t_{1/2}$) (Banerjee, S et al. (1978A); Sikka, HC et al. (1978)); high $t_{1/2}$: scientific judgement based upon the direct photolysis half-life and approximate winter sunlight intensity (Banerjee, S et al. (1978A); Sikka, HC et al. (1978); Lyman, WJ et al. (1982)).

**· Max light absorption (nm):**       lambda max = 290 (aqueous ethanol)

*Comment:* Significant absorption to 340 nm (Callahan, MA et al. (1979A)).

**· Aq photol half-life:**       High:    0.075 hours       (4.5 minutes)

                                 Low:    0.025 hours       (1.5 minutes)

*Comment:* Based upon measured half-life for direct photolysis in distilled water in midday summer sunlight (low $t_{1/2}$) (Banerjee, S et al. (1978A); Sikka, HC et al. (1978)); high $t_{1/2}$: scientific judgement based upon the direct photolysis half-life and approximate winter sunlight intensity (Banerjee, S et al. (1978A); Sikka, HC et al. (1978); Lyman, WJ et al. (1982)).

## Photooxidation half-life:
**· Water:**       High:    1740 hours       (72.5 days)

                         Low:    31.2 hours       (1.3 days)

*Comment:* Scientific judgement based upon estimated rate constants for reactions of representative aromatic amines with $\cdot OH$ and $RO_2 \cdot$ (Mill, T and Mabey, W (1985)). It is assumed that 3,3'-dichlorobenzidine reacts twice as fast as aniline.

**· Air:**       High:    9.05 hours

              Low:    0.905 hours

*Comment:* Scientific judgement based upon estimated rate constant for reaction with hydroxyl radicals in air (Atkinson, R (1987A)).

## Reduction half-life:       High:    No data

                               Low:

*Comment:*

## Hydrolysis:
**· First-order hydr half-life:**

*Comment:*

**· Acid rate const $(M(H+)\text{-hr})^{-1}$:**       No hydrolyzable groups

*Comment:*

**· Base rate const $(M(OH\text{-})\text{-hr})^{-1}$:**

*Comment:*

# Biphenyl

92-52-4

## Half-lives:
· **Soil:**                                    High:    168 hours          (7 days)

Low:      36 hours          (1.5 days)

*Comment:* Scientific judgement based upon unacclimated aqueous aerobic biodegradation half-life.

· **Air:**                                     High:    110 hours          (4.6 days)

Low:      7.8 hours

*Comment:* Based upon photooxidation half-life in air.

· **Surface Water:**                           High:    168 hours          (7 days)

Low:      36 hours          (1.5 days)

*Comment:* Scientific judgement based upon unacclimated aqueous aerobic biodegradation half-life.

· **Ground Water:**                            High:    336 hours          (14 days)

Low:      72 hours          (3 days)

*Comment:* Scientific judgement based upon unacclimated aqueous aerobic biodegradation half-life.

## Aqueous Biodegradation (unacclimated):
· **Aerobic half-life:**                       High:    168 hours          (7 days)

Low:      36 hours          (1.5 days)

*Comment:* Scientific judgement based upon river die-away test data (low $t_{1/2}$) (Bailey, RE et al. (1983)) and activated sludge screening test data (high $t_{1/2}$) (Sasaki, S (1978); Gaffney, PE (1976)).

· **Anaerobic half-life:**                     High:    672 hours          (28 days)

Low:      144 hours         (6 days)

*Comment:* Scientific judgement based upon estimated unacclimated aqueous aerobic biodegradation half-life.

· **Removal/secondary treatment:**             High:       74%

Low:

*Comment:* Activated sludge treatment simulator (Garrison, AW (1969)).

## Photolysis:
· **Atmos photol half-life:**                  High:    No data

Low:

*Comment:*

268

· **Max light absorption (nm):**          lambda max = 247 nm (methanol); 248, 264 nm
(chloroform).
*Comment:*  In methanol, Sadtler UV No. 255; in chloroform, Gore, RC et al. (1971).

· **Aq photol half-life:**          High:          No data
                                    Low:
*Comment:*

**Photooxidation half-life:**
· **Water:**                        High:          No data
                                    Low:
*Comment:*

· **Air:**                          High:          110 hours          (4.6 days)
                                    Low:           7.8 hours
*Comment:*  Based upon measured rates of reaction with hydroxyl radicals in air (Atkinson, R
(1987)).

**Reduction half-life:**            High:          No data
                                    Low:
*Comment:*

**Hydrolysis:**
· **First-order hydr half-life:**          No hydrolyzable groups
*Comment:*

· **Acid rate const (M(H+)-hr)$^{-1}$:**
*Comment:*

· **Base rate const (M(OH-)-hr)$^{-1}$:**
*Comment:*

# 4-Aminobiphenyl

<u>CAS Registry Number:</u>  92-67-1

<u>Half-lives:</u>
- **· Soil:**                               High:      168 hours                  (7 days)
                                            Low:        24 hours                   (1 day)
*Comment:* Scientific judgement based upon estimated unacclimated aerobic aqueous biodegradation half-life.

- **· Air:**                                High:        6 hours
                                            Low:        0.60 hours              (36 minutes)
*Comment:* Scientific judgement based upon estimated photooxidation half-life in air.

- **· Surface Water:**                      High:      168 hours                  (7 days)
                                            Low:        24 hours                   (1 day)
*Comment:* Scientific judgement based upon estimated unacclimated aqueous aerobic biodegradation half-life.

- **· Ground Water:**                       High:      336 hours                 (14 days)
                                            Low:        48 hours                   (2 days)
*Comment:* Scientific judgement based upon estimated unacclimated aerobic aqueous biodegradation half-life.

<u>Aqueous Biodegradation (unacclimated):</u>
- **· Aerobic half-life:**                  High:      168 hours                  (7 days)
                                            Low:        24 hours                   (1 day)
*Comment:* Scientific judgement based upon aerobic aqueous screening test data for 4-aminobiphenyl (Tabak, HH et al. (1981A)) and surface water die-away test data for aniline (Subba-Rao, RV et al. (1982); Kondo, M (1978)) in which rapid biodegradation of the chemical was observed.

- **· Anaerobic half-life:**                High:      672 hours                 (28 days)
                                            Low:        96 hours                   (4 days)
*Comment:* Scientific judgement based upon estimated unacclimated aerobic aqueous biodegradation half-life.

- **· Removal/secondary treatment:**        High:      No data
                                            Low:
*Comment:*

<u>Photolysis:</u>
- **· Atmos photol half-life:**             High:      No data

*Comment:*

· **Max light absorption (nm):**     No data
*Comment:*

· **Aq photol half-life:**     High:     No data
                               Low:

*Comment:*

## Photooxidation half-life:
   · **Water:**     High:     3480 hours          (145 days)
                     Low:     62.4 hours          (2.6 days)
*Comment:* Scientific judgement based upon estimated rate constants for reactions of representative aromatic amines with ·OH and $RO_2$· (Mill, T and Mabey, W (1985)).

   · **Air:**     High:     6 hours
                     Low:     0.6 hours          (36 minutes)
*Comment:* Scientific judgement based upon estimated rate for reaction with hydroxyl radicals in air (Atkinson, R (1987A)).

## Reduction half-life:     High:     No data
                            Low:

*Comment:*

## Hydrolysis:
   · **First-order hydr half-life:**     No hydrolyzable groups
*Comment:*

   · **Acid rate const $(M(H+)-hr)^{-1}$:**
*Comment:*

   · **Base rate const $(M(OH-)-hr)^{-1}$:**
*Comment:*

# Benzidine

<u>CAS Registry Number:</u>  92-87-5

<u>Structure:</u>

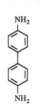

<u>Half-lives:</u>
| · **Soil:** | High: | 192 hours | (8 days) |
| | Low: | 48 hours | (2 days) |

*Comment:*  Based upon aerobic soil die-away test data (Lu, PY et al. (1977)).

| · **Air:** | High: | 3.12 hours | |
| | Low: | 0.312 hours | |

*Comment:*  Scientific judgement based upon estimated photooxidation half-life in air.

| · **Surface Water:** | High: | 192 hours | (8 days) |
| | Low: | 31.2 hours | (1.3 days) |

*Comment:*  Scientific judgement based upon estimated photooxidation half-life in water (low $t_{1/2}$) and estimated unacclimated aqueous aerobic biodegradation half-life (high $t_{1/2}$).

| · **Ground Water:** | High: | 384 hours | (16 days) |
| | Low: | 96 hours | (4 days) |

*Comment:*  Scientific judgement based upon estimated unacclimated aqueous aerobic biodegradation half-life.

<u>Aqueous Biodegradation (unacclimated):</u>
| · **Aerobic half-life:** | High: | 192 hours | (8 days) |
| | Low: | 48 hours | (2 days) |

*Comment:*  Based upon aerobic soil die-away test data (Lu, PY et al. (1977)).

| · **Anaerobic half-life:** | High: | 768 hours | (32 days) |
| | Low: | 192 hours | (8 days) |

*Comment:*  Scientific judgement based upon estimated unacclimated aqueous aerobic biodegradation half-life.

| · **Removal/secondary treatment:** | High: | 91% | |
| | Low: | 65% | |

*Comment:*  Removal percentages based upon data from continuous biological treatment simulators utilizing activated sludge (low percentage) and domestic sewage (high percentage) inoculums (low:high: Tabak, HH and Barth, EF (1978)).

## Photolysis:
· **Atmos photol half-life:**         High:      No data
                                      Low:
*Comment:*

· **Max light absorption (nm):**      lambda max = 287 (aqueous ethanol)
*Comment:* Significant absorption up to 340 nm (Callahan, MA et al. (1979A)).

· **Aq photol half-life:**            High:      No data
                                      Low:
*Comment:*

## Photooxidation half-life:
· **Water:**                          High:    1740 hours          (72.5 days)
                                      Low:      31.2 hours          (1.3 days)
*Comment:* Scientific judgement based upon estimated rate constants for reactions of representative aromatic amines with ·OH and $RO_2$· (Mill, T and Mabey, W (1985)). It is assumed that benzidine reacts twice as fast as aniline.

· **Air:**                            High:      3.12 hours
                                      Low:       0.312 hours
*Comment:* Scientific judgement based upon estimated rate constant for reaction with hydroxyl radical in air (Atkinson, R (1987A)).

## Reduction half-life:                High:      No data
                                      Low:
*Comment:*

## Hydrolysis:
· **First-order hydr half-life:**
*Comment:*

· **Acid rate const $(M(H+)-hr)^{-1}$:**     No hydrolyzable groups
*Comment:*

· **Base rate const $(M(OH-)-hr)^{-1}$:**
*Comment:*

273

# 4-Nitrobiphenyl

CAS Registry Number: 92-93-3

Half-lives:

· Soil:
|  | | |
|---|---|---|
| High: | 672 hours | (4 weeks) |
| Low: | 24 hours | (1 day) |

Comment: Scientific judgement based upon estimated unacclimated aqueous aerobic biodegradation half-life.

· Air:
|  | | |
|---|---|---|
| High: | 271 hours | (11.3 days) |
| Low: | 27.1 hours | (1.1 days) |

Comment: Scientific judgement based upon estimated photooxidation half-life in air.

· Surface Water:
|  | | |
|---|---|---|
| High: | 672 hours | (4 weeks) |
| Low: | 24 hours | (1 day) |

Comment: Scientific judgement based upon estimated unacclimated aqueous aerobic biodegradation half-life.

· Ground Water:
|  | | |
|---|---|---|
| High: | 1344 hours | (8 weeks) |
| Low: | 48 hours | (2 days) |

Comment: Scientific judgement based upon estimated unacclimated aqueous anaerobic (low $t_{1/2}$) and aerobic (high $t_{1/2}$) biodegradation half-lives.

Aqueous Biodegradation (unacclimated):

· Aerobic half-life:
|  | | |
|---|---|---|
| High: | 672 hours | (4 weeks) |
| Low: | 24 hours | (1 day) |

Comment: Scientific judgement based upon aqueous aerobic screening test data for 4-nitrobiphenyl which indicate rapid biodegradation (Tabak, HH et al. (1981A)) and aqueous aerobic screening test data for nitrobenzene which indicate either rapid biodegradation (Pitter, P (1976); Hallas, LE and Alexander, M (1983)) or no or slow degradation (Kawasaki, M (1980); Urano, K and Kato, Z (1986)).

· Anaerobic half-life:
|  | | |
|---|---|---|
| High: | 240 hours | (10 days) |
| Low: | 48 hours | (2 days) |

Comment: Scientific judgement based upon anaerobic natural water die-away test data for 2,4-dinitrotoluene (Spanggord, RJ et al. (1980)). Reduction of the nitro group observed in these studies may not be due to biodegradation.

· Removal/secondary treatment:
|  | |
|---|---|
| High: | No data |
| Low: | |

Comment:

## Photolysis:
· **Atmos photol half-life:**              High:      No data
                                           Low:

*Comment:*

· **Max light absorption (nm):**           No data
*Comment:*

· **Aq photol half-life:**                 High:      No data
                                           Low:

*Comment:*

## Photooxidation half-life:
· **Water:**                               High:      No data
                                           Low:

*Comment:*

· **Air:**                                 High:      271 hours          (11.3 days)
                                           Low:       27.1 hours         (1.1 days)
*Comment:*  Scientific judgement based upon estimated rate of reaction with hydroxyl radicals in air (Atkinson, R (1987A)).

## Reduction half-life:                    High:      No data
                                           Low:

*Comment:*

## Hydrolysis:
· **First-order hydr half-life:**          No hydrolyzable groups
*Comment:*

· **Acid rate const (M(H+)-hr)$^{-1}$:**
*Comment:*

· **Base rate const (M(OH-)-hr)$^{-1}$:**
*Comment:*

# Mecoprop

**CAS Registry Number:** 93-65-2

**Structure:**

**Half-lives:**

**· Soil:**   High:   240 hours   (10 days)
             Low:    168 hours   (7 days)
*Comment:* Scientific judgement based upon aerobic soil grab sample data (Kirkland, K and Fryer, JD (1972); Smith, AE and Hayden, BJ (1981)).

**· Air:**   High:   37.8 hours   (1.6 days)
             Low:    3.8 hours
*Comment:* Scientific judgement based upon estimated photooxidation half-life in air.

**· Surface Water:**   High:   240 hours   (10 days)
                       Low:    168 hours   (7 days)
*Comment:* Scientific judgement based upon estimated aqueous aerobic biodegradation half-life.

**· Ground Water:**   High:   4320 hours   (6 months)
                      Low:    336 hours    (14 days)
*Comment:* Scientific judgement based upon estimated aqueous aerobic and anaerobic biodegradation half-lives.

**Aqueous Biodegradation (unacclimated):**

**· Aerobic half-life:**   High:   240 hours   (10 days)
                           Low:    168 hours   (7 days)
*Comment:* Scientific judgement based upon aerobic soil grab sample data (Kirkland, K and Fryer, JD (1972); Smith, AE and Hayden, BJ (1981)).

**· Anaerobic half-life:**   High:   4320 hours   (6 months)
                             Low:    672 hours    (4 weeks)
*Comment:* Scientific judgement based upon anaerobic digestor sludge data (Battersby, NS and Wilson, V (1989)).

**· Removal/secondary treatment:**   High:   No data
                                     Low:
*Comment:*

<u>Photolysis:</u>
   · **Atmos photol half-life:**        High:
                                        Low:

  *Comment:*

   · **Max light absorption (nm):**     No data
  *Comment:*

   · **Aq photol half-life:**           High:
                                         Low:

  *Comment:*

<u>Photooxidation half-life:</u>
   · **Water:**                    High:      No data
                                         Low:

  *Comment:*

   · **Air:**                      High:     37.8 hours           (1.6 days)
                                         Low:      3.8 hours
  *Comment:* Based upon an estimated rate constant for reaction with hydroxyl radicals in air (Atkinson, R (1987A)).

<u>Reduction half-life:</u>               High:     No data
                                         Low:

  *Comment:*

<u>Hydrolysis:</u>
   · **First-order hydr half-life:**    No data
  *Comment:*

   · **Acid rate const $(M(H+)-hr)^{-1}$:**
  *Comment:*

   · **Base rate const $(M(OH-)-hr)^{-1}$:**
  *Comment:*

# 2,4,5-Trichlorophenoxyacetic acid

**CAS Registry Number:** 93-76-5

**Half-lives:**

| · Soil: | High: | 480 hours | (20 days) |
|---|---|---|---|
| | Low: | 240 hours | (10 days) |

*Comment:* Scientific judgement based upon unacclimated soil grab sample data (low $t_{1/2}$: Smith, AE (1978); high $t_{1/2}$: Smith, AE (1979A)).

| · Air: | High: | 122 hours | (5.1 days) |
|---|---|---|---|
| | Low: | 12.2 hours | |

*Comment:* Scientific judgement based upon estimated photooxidation half-life in air.

| · Surface Water: | High: | 480 hours | (20 days) |
|---|---|---|---|
| | Low: | 240 hours | (10 days) |

*Comment:* Scientific judgement based upon estimated unacclimated aqueous aerobic biodegradation half-life.

| · Ground Water: | High: | 4320 hours | (6 months) |
|---|---|---|---|
| | Low: | 480 hours | (20 days) |

*Comment:* Scientific judgement based upon estimated aqueous aerobic and anaerobic biodegradation half-lives.

**Aqueous Biodegradation (unacclimated):**

| · Aerobic half-life: | High: | 480 hours | (20 days) |
|---|---|---|---|
| | Low: | 240 hours | (10 days) |

*Comment:* Scientific judgement based upon unacclimated soil grab sample data (low $t_{1/2}$: Smith, AE (1978); high $t_{1/2}$: Smith, AE (1979A)).

| · Anaerobic half-life: | High: | 4320 hours | (6 months) |
|---|---|---|---|
| | Low: | 672 hours | (4 weeks) |

*Comment:* Scientific judgement based upon anaerobic digestor sludge data (Battersby, NS and Wilson, V (1989)).

| · Removal/secondary treatment: | High: | No data |
|---|---|---|
| | Low: | |

*Comment:*

**Photolysis:**

| · Atmos photol half-life: | High: | No data |
|---|---|---|
| | Low: | |

*Comment:*

· **Max light absorption (nm):**    No data
*Comment:*

· **Aq photol half-life:**    High: 1080 hours   (45 days)
               Low:  360 hours    (15 days)
*Comment:* Scientific judgement based upon reported photolysis half-life for summer sunlight at 40 °N (Skurlatov, YI et al. (1983)).

**Photooxidation half-life:**
 · **Water:**        High:  No data
             Low:

*Comment:*

 · **Air:**         High:  122 hours   (5.1 days)
             Low:  12.2 hours
*Comment:* Scientific judgement based upon an estimated rate constant for the vapor phase reaction with hydroxyl radicals in air (Atkinson, R (1987A)).

**Reduction half-life:**    High:  No data
             Low:
 *Comment:*

**Hydrolysis:**
 · **First-order hydr half-life:**   Not expected to be important
 *Comment:* Stable in distilled water for 40 days (Chau, ASY and Thomson, K (1978)).

 · **Acid rate const (M(H+)-hr)$^{-1}$:**
 *Comment:*

 · **Base rate const (M(OH-)-hr)$^{-1}$:**
 *Comment:*

# Benzoyl peroxide

**CAS Registry Number:** 94-36-0

**Half-lives:**

  · **Soil:**                      High:     48 hours
                                      Low:      4 hours

*Comment:* Scientific judgement based upon benzoyl peroxide's expected high reactivity with organic matter in soil.

  · **Air:**                       High:     510 hours        (21.3 days)
                                      Low:      51.0 hours      (2.1 days)

*Comment:* Based upon estimated photooxidation half-life in air.

  · **Surface Water:**         High:     168 hours        (7 days)
                                      Low:      24 hours        (1 day)

*Comment:* Scientific judgement based upon measured half-lives for benzoic acid from aerobic natural water die-away tests (Banerjee, S et al. (1984); Subba-Rao, RV et al. (1982)).

  · **Ground Water:**         High:     336 hours       (14 days)
                                      Low:      48 hours        (2 days)

*Comment:* Scientific judgement based upon measured half-lives for benzoic acid from aerobic natural water die-away tests (Banerjee, S et al. (1984); Subba-Rao, RV et al. (1982)).

**Aqueous Biodegradation (unacclimated):**

  · **Aerobic half-life:**       High:     168 hours        (7 days)
                                      Low:      24 hours        (1 day)

*Comment:* Scientific judgement based upon measured half-lives for benzoic acid from aerobic natural water die-away tests (Banerjee, S et al. (1984); Subba-Rao, RV et al. (1982)).

  · **Anaerobic half-life:**     High:     672 hours       (28 days)
                                      Low:      96 hours        (4 days)

*Comment:* Scientific judgement based upon measured half-lives for benzoic acid from aerobic natural water die-away tests (Banerjee, S et al. (1984); Subba-Rao, RV et al. (1982)).

  · **Removal/secondary treatment:**    High:     No data
                                        Low:

*Comment:*

**Photolysis:**

  · **Atmos photol half-life:**    High:     No data
                                        Low:

*Comment:*

**· Max light absorption (nm):**          lambda max = 235, 275 nm (1,4-dioxane)
*Comment:* Sadtler UV No. 127.

**· Aq photol half-life:**          High:          No data
                                    Low:

*Comment:*

## Photooxidation half-life:
**· Water:**                        High:          No data
                                    Low:

*Comment:*

**· Air:**                          High:          510 hours          (21.3 days)
                                    Low:          51.0 hours          (2.1 days)

*Comment:* Based upon estimated rate constant for reaction with hydroxyl radicals in air (Atkinson, R (1987A)).

## Reduction half-life:              High:          No data
                                    Low:

*Comment:*

## Hydrolysis:
**· First-order hydr half-life:**          No data
*Comment:*

**· Acid rate const $(M(H+)-hr)^{-1}$:**
*Comment:*

**· Base rate const $(M(OH-)-hr)^{-1}$:**
*Comment:*

# Dihydrosafrole

<u>CAS Registry Number:</u> 94-58-6

<u>Structure:</u>

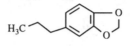

<u>Half-lives:</u>
- **Soil:**               High:    672 hours          (4 weeks)
                          Low:     168 hours          (7 days)
  *Comment:* Scientific judgement based upon estimated aqueous aerobic biodegradation half-life.

- **Air:**                High:    7.7 hours
                          Low:     0.8 hour           (46 minutes)
  *Comment:* Scientific judgement based upon estimated photooxidation half-life in air.

- **Surface Water:**      High:    672 hours          (4 weeks)
                          Low:     168 hours          (7 days)
  *Comment:* Scientific judgement based upon estimated aqueous aerobic biodegradation half-life.

- **Ground Water:**       High:    1344 hours         (8 weeks)
                          Low:     336 hours          (14 days)
  *Comment:* Scientific judgement based upon estimated aqueous aerobic biodegradation half-life.

<u>Aqueous Biodegradation (unacclimated):</u>
- **Aerobic half-life:**  High:    672 hours          (4 weeks)
                          Low:     168 hours          (7 days)
  *Comment:* Scientific judgement.

- **Anaerobic half-life:** High:   2688 hours         (16 weeks)
                          Low:     672 hours          (28 days)
  *Comment:* Scientific judgement based upon estimated aqueous aerobic biodegradation half-life.

- **Removal/secondary treatment:**  High:    No data
                                    Low:
  *Comment:*

<u>Photolysis:</u>
- **Atmos photol half-life:**       High:

282

|                                  | Low:             |            |              |
| -------------------------------- | ---------------- | ---------- | ------------ |

*Comment:*

**· Max light absorption (nm):**     No data
*Comment:*

**· Aq photol half-life:**     High:
                               Low:

*Comment:*

## Photooxidation half-life:
**· Water:**     High:     No data
                 Low:

*Comment:*

**· Air:**     High:     7.7 hours
               Low:     0.8 hours     (46 minutes)
*Comment:* Scientific judgement based upon an estimated rate constant for vapor phase reaction with hydroxyl radicals in air (Atkinson, R (1987A)).

## Reduction half-life:     High:     No data
                            Low:

*Comment:*

## Hydrolysis:
**· First-order hydr half-life:**     Not expected to be important
*Comment:* Scientific judgement based upon zero hydrolysis of safrole after 26 days for pHs of 3, 7 and 11 at 85 °C (Ellington, JJ et al. (1987)).

**· Acid rate const $(M(H+)-hr)^{-1}$:**     Not expected to be important
*Comment:* Scientific judgement based upon zero hydrolysis of safrole was observed after 26 days at 85 °C for a pH of 3 (Ellington, JJ et al. (1987)).

**· Base rate const $(M(OH-)-hr)^{-1}$:**     Not expected to be important
*Comment:* Scientific judgement based upon zero hydrolysis of safrole was observed after 26 days at 85 °C for a pH of 11 (Ellington, JJ et al. (1987)).

283

# Safrole

**CAS Registry Number:** 94-59-7

**Structure:**

**Half-lives:**
  **· Soil:**                                High:    672 hours        (4 weeks)
                                          Low:     168 hours        (7 days)
*Comment:* Scientific judgement based upon estimated aqueous aerobic biodegradation half-life.

  **· Air:**      High: 6 hours
             Low: 0.6 hour (36 minutes)
*Comment:* Scientific judgement based upon estimated photooxidation half-life in air.

  **· Surface Water:**  High: 672 hours (4 weeks)
             Low: 168 hours (7 days)
*Comment:* Scientific judgement based upon estimated aqueous aerobic biodegradation half-life.

  **· Ground Water:**  High: 1344 hours (8 weeks)
             Low: 336 hours (14 days)
*Comment:* Scientific judgement based upon estimated aqueous aerobic biodegradation half-life.

**Aqueous Biodegradation (unacclimated):**
  **· Aerobic half-life:**  High: 672 hours (4 weeks)
             Low: 168 hours (7 days)
*Comment:* Scientific judgement.

  **· Anaerobic half-life:**  High: 2688 hours (16 weeks)
             Low: 672 hours (28 days)
*Comment:* Scientific judgement based upon estimated aqueous aerobic biodegradation half-life.

  **· Removal/secondary treatment:**  High: No data
             Low:
*Comment:*

**Photolysis:**
  **· Atmos photol half-life:**  High:
             Low:

284

*Comment:*

· **Max light absorption (nm):**     No data
*Comment:*

· **Aq photol half-life:**     High:
                                   Low:

*Comment:*

**Photooxidation half-life:**
· **Water:**     High:     No data
                  Low:

*Comment:*

· **Air:**     High:     6 hours
          Low:     0.6 hours     (36 minutes)
*Comment:* Scientific judgement based upon an estimated rate constant for vapor phase reaction with hydroxyl radicals in air (Atkinson, R (1987A)).

**Reduction half-life:**     High:     No data
                  Low:
*Comment:*

**Hydrolysis:**
· **First-order hydr half-life:**     Not expected to be important
*Comment:* Scientific judgement based upon zero hydrolysis after 26 days for pHs of 3, 7 and 11 at 85 °C (Ellington, JJ et al. (1987)).

· **Acid rate const (M(H+)-hr)$^{-1}$:**     Not expected to be important
*Comment:* Scientific judgement based upon zero hydrolysis was observed after 26 days at 85 °C for a pH of 3 (Ellington, JJ et al. (1987)).

· **Base rate const (M(OH-)-hr)$^{-1}$:**     Not expected to be important
*Comment:* Scientific judgement based upon zero hydrolysis was observed after 26 days at 85 °C for a pH of 11 (Ellington, JJ et al. (1987)).

# 2-Methyl-4-chlorophenoxyacetic acid

<u>CAS Registry Number:</u> 94-74-6

<u>Half-lives:</u>
  · **Soil:**                            High:      168 hours              (7 days)
                                         Low:        96 hours              (4 days)
*Comment:* Scientific judgement based upon unacclimated soil grab sample data (Kirkland, K and Fryer, JD (1972); Smith, AE and Hayden, BJ (1981)).

  · **Air:**                             High:      52.6 hours             (2.2 days)
                                         Low:        5.3 hours
*Comment:* Scientific judgement based upon estimated photooxidation half-life in air.

  · **Surface Water:**                   High:      168 hours              (7 days))
                                         Low:        96 hours              (4 days)
*Comment:* Scientific judgement based upon estimated unacclimated aqueous aerobic biodegradation half-life.

  · **Ground Water:**                    High:      4320 hours             (6 months)
                                         Low:        192 hours             (8 days)
*Comment:* Scientific judgement based upon estimated aqueous aerobic and anaerobic biodegradation half-lives.

<u>Aqueous Biodegradation (unacclimated):</u>
  · **Aerobic half-life:**               High:      168 hours              (7 days)
                                         Low:        96 hours              (4 days)
*Comment:* Scientific judgement based upon aerobic soil grab sample data (Kirkland, K and Fryer, JD (1972); Smith, AE and Hayden, BJ (1981)).

  · **Anaerobic half-life:**             High:      4320 hours             (6 months)
                                         Low:        672 hours             (4 weeks)
*Comment:* Scientific judgement based upon anaerobic digestor sludge data (Battersby, NS and Wilson, V (1989)).

  · **Removal/secondary treatment:**     High:      No data
                                         Low:
*Comment:*

<u>Photolysis:</u>
  · **Atmos photol half-life:**          High:      No data
                                         Low:
*Comment:*

· **Max light absorption (nm):**　　　　lambda max = 278, 227
*Comment:*  In methanol (Sadtler UV No. 3794).

· **Aq photol half-life:**　　　　　　High:　　No data
　　　　　　　　　　　　　　　　　　Low:
*Comment:*

**Photooxidation half-life:**
　· **Water:**　　　　　　　　　　High:　　No data
　　　　　　　　　　　　　　　　　Low:
*Comment:*

　· **Air:**　　　　　　　　　　　High:　52.6 hours　　　　(2.2 days)
　　　　　　　　　　　　　　　　　Low:　　5.3 hours
*Comment:*  Scientific judgement based upon an estimated rate constant for the vapor phase reaction with hydroxyl radicals in air (Atkinson, R (1987A)).

**Reduction half-life:**　　　　　　High:　　No data
　　　　　　　　　　　　　　　　　Low:
*Comment:*

**Hydrolysis:**
　· **First-order hydr half-life:**　　　Not expected to be important
*Comment:*  Stable in distilled water for 40 days (Chau, ASY and Thomson, K (1978)).

　· **Acid rate const (M(H+)-hr)$^{-1}$:**
*Comment:*

　· **Base rate const (M(OH-)-hr)$^{-1}$:**
*Comment:*

# 2,4-Dichlorophenoxyacetic acid

<u>CAS Registry Number:</u> 94-75-7

<u>Half-lives:</u>

   · **Soil:**
        High:    1200 hours      (50 days)
        Low:     240 hours       (10 days)

*Comment:* Scientific judgement based upon estimated unacclimated aqueous aerobic biodegradation half-life.

   · **Air:**
        High:    18 hours
        Low:     1.8 hours

*Comment:* Scientific judgement based upon estimated photooxidation half-life in air.

   · **Surface Water:**
        High:    96 hours       (4 days)
        Low:     48 hours       (2 days)

*Comment:* Scientific judgement based upon reported photolysis half-lives for aqueous solution irradiated at UV wavelength of 356 nm (Baur, JR and Bovey, RW (1974)).

   · **Ground Water:**
        High:    4320 hours      (6 months)
        Low:     480 hours      (20 days)

*Comment:* Scientific judgement based upon estimated unacclimated aqueous aerobic and anaerobic (high $t_{1/2}$) biodegradation half-lives.

<u>Aqueous Biodegradation (unacclimated):</u>

   · **Aerobic half-life:**
        High:    1200 hours      (50 days)
        Low:     240 hours       (10 days)

*Comment:* Scientific judgement based upon unacclimated aerobic river die away test (Nesbitt, HJ and Watson, JR (1980)).

   · **Anaerobic half-life:**
        High:    4320 hours      (6 months)
        Low:     672 hours       (4 weeks)

*Comment:* Reported half-lives based upon unacclimated aqueous screening test data (Liu D, et al. (1981A)).

   · **Removal/secondary treatment:**
        High:    No data
        Low:

*Comment:*

<u>Photolysis:</u>

   · **Atmos photol half-life:**
        High:    No data
        Low:

*Comment:*

**· Max light absorption (nm):**  lambda max = 283, 291
*Comment:* Reported absorption maxima (Mookerjee, SK (1985)).

**· Aq photol half-life:**  High:  96 hours  (4 days)
  Low:  48 hours  (2 days)
*Comment:* Reported photolysis half-lives for aqueous solution irradiated at UV wavelength of 356 nm (Baur, JR and Bovey, RW (1974)).

## Photooxidation half-life:
**· Water:**  High:  No data
  Low:
*Comment:*

**· Air:**  High:  18 hours
  Low:  1.8 hours
*Comment:* Scientific judgement based upon an estimated rate constant for the vapor phase reaction with hydroxyl radicals in air (Atkinson, R (1987A)).

## Reduction half-life:
  High:  No data
  Low:
*Comment:*

## Hydrolysis:
**· First-order hydr half-life:**  No hydrolyzable groups
*Comment:* Rate constant at neutral pH is zero (Kollig, HP et al. (1987)).

**· Acid rate const (M(H+)-hr)$^{-1}$:**
*Comment:*

**· Base rate const (M(OH-)-hr)$^{-1}$:**
*Comment:*

# 2,4-DB

**CAS Registry Number:** 94-82-6

**Structure:**

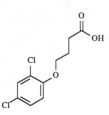

**Half-lives:**

· **Soil:**

|  | High: | 168 hours | (7 days) |
|---|---|---|---|
|  | Low: | 24 hours | (1 day) |

*Comment:* Scientific judgement based upon unacclimated soil grab sample data (Smith, AE (1978)).

· **Air:**

|  | High: | 60 hours | (2.5 days) |
|---|---|---|---|
|  | Low: | 6 hours | |

*Comment:* Scientific judgement based upon estimated photooxidation half-life in air.

· **Surface Water:**

|  | High: | 168 hours | (7 days) |
|---|---|---|---|
|  | Low: | 24 hours | (1 day) |

*Comment:* Scientific judgement based upon estimated unacclimated aqueous aerobic biodegradation half-life.

· **Ground Water:**

|  | High: | 336 hours | (14 days) |
|---|---|---|---|
|  | Low: | 48 hours | (2 days) |

*Comment:* Scientific judgement based upon estimated unacclimated aqueous aerobic biodegradation half-life.

**Aqueous Biodegradation (unacclimated):**

· **Aerobic half-life:**

|  | High: | 168 hours | (7 days) |
|---|---|---|---|
|  | Low: | 24 hours | (1 day) |

*Comment:* Scientific judgement based upon unacclimated soil grab sample data (Smith, AE (1978)).

· **Anaerobic half-life:**

|  | High: | 672 hours | (28 days) |
|---|---|---|---|
|  | Low: | 96 hours | (4 days) |

*Comment:* Scientific judgement based upon unacclimated aqueous aerobic biodegradation half-life.

· **Removal/secondary treatment:**        High:        No data

Low:
*Comment:*

## Photolysis:
· **Atmos photol half-life:**　　　　　High:
　　　　　　　　　　　　　　　　　　　　　Low:
*Comment:*

· **Max light absorption (nm):**　　　No data
*Comment:*

· **Aq photol half-life:**　　　　　　High:
　　　　　　　　　　　　　　　　　　　　　Low:

*Comment:*

## Photooxidation half-life:
· **Water:**　　　　　　　　　　　　　High:　　　No data
　　　　　　　　　　　　　　　　　　　　Low:
*Comment:*

· **Air:**　　　　　　　　　　　　　　High:　　　60 hours　　　　　(2.5 days)
　　　　　　　　　　　　　　　　　　　　Low:　　　6 hours
*Comment:* Scientific judgement based upon an estimated rate constant for the vapor phase reaction with hydroxyl radicals in air (Atkinson, R (1987A)).

## Reduction half-life:　　　　　　　　High:　　　No data
　　　　　　　　　　　　　　　　　　　　Low:
*Comment:*

## Hydrolysis:
· **First-order hydr half-life:**　　　Not expected to be important
*Comment:* Stable in distilled water for 40 days (Chau, ASY and Thomson, K (1978)).

· **Acid rate const $(M(H+)-hr)^{-1}$:**
*Comment:*

· **Base rate const $(M(OH-)-hr)^{-1}$:**
*Comment:*

# o-Xylene

CAS Registry Number: 95-47-6

## Half-lives:
- **· Soil:**  High: 672 hours  (4 weeks)
  Low: 168 hours  (1 week)
  *Comment:* Scientific judgement based upon estimated aqueous aerobic biodegradation half-life.

- **· Air:**  High: 44 hours  (1.8 days)
  Low: 4.4 hours
  *Comment:* Based upon photooxidation half-life in air.

- **· Surface Water:**  High: 672 hours  (4 weeks)
  Low: 168 hours  (1 week)
  *Comment:* Scientific judgement based upon estimated aqueous aerobic biodegradation half-life.

- **· Ground Water:**  High: 8640 hours  (12 months)
  Low: 336 hours  (2 weeks)
  *Comment:* Scientific judgement based upon estimated aqueous aerobic and anaerobic biodegradation half-lives.

## Aqueous Biodegradation (unacclimated):
- **· Aerobic half-life:**  High: 672 hours  (4 weeks)
  Low: 168 hours  (1 week)
  *Comment:* Scientific judgement based upon soil column study simulating an aerobic river/ground water infiltration system (high $t_{1/2}$: Kuhn, EP et al. (1985)) and aqueous screening test data (Bridie, AL et al. (1979)).

- **· Anaerobic half-life:**  High: 8640 hours  (12 months)
  Low: 4320 hours  (6 months)
  *Comment:* Scientific judgement based upon acclimated grab sample data for anaerobic soil from a ground water aquifer receiving landfill leachate (Wilson, BH et al. (1986)) and a soil column study simulating an anaerobic river/ground water infiltration system (Kuhn, EP et al. (1985)).

- **· Removal/secondary treatment:**  High: No data
  Low:
  *Comment:*

## Photolysis:
- **· Atmos photol half-life:**  High: No data
  Low:
  *Comment:*

**· Max light absorption (nm):**  lambda max = 269.5, 262
*Comment:* No absorbance occurs above 310 nm in methanol (Sadtler UV No. 7).

**· Aq photol half-life:**  High:  No data
Low:
*Comment:*

**Photooxidation half-life:**
**· Water:**  High: $2.7 \times 10^8$ hours  (31397 years)
Low: $3.9 \times 10^5$ hours  (43 years)
*Comment:* Scientific judgement based upon estimated rate data for alkylperoxyl radicals in aqueous solution (Hendry, DG et al. (1974)).

**· Air:**  High:  44 hours  (1.8 days)
Low:  4.4 hours
*Comment:* Based upon measured rate data for vapor phase reaction with hydroxyl radicals in air (Atkinson, R (1985)).

**Reduction half-life:**  High:  No data
Low:
*Comment:*

**Hydrolysis:**
**· First-order hydr half-life:**  No hydrolyzable groups
*Comment:*

**· Acid rate const $(M(H+)-hr)^{-1}$:**
*Comment:*

**· Base rate const $(M(OH-)-hr)^{-1}$:**
*Comment:*

# o-Cresol

<u>CAS Registry Number:</u>  95-48-7

<u>Half-lives:</u>
· **Soil:**                                 High:     168 hours          (7 days)
                                            Low:      24 hours           (1 day)
*Comment:*  Scientific judgement based upon estimated unacclimated aqueous aerobic
biodegradation half-life.

· **Air:**                                  High:     16 hours
                                            Low:      1.6 hours
*Comment:*  Based upon photooxidation half-life in air.

· **Surface Water:**                        High:     168 hours          (7 days)
                                            Low:      24 hours           (1 day)
*Comment:*  Scientific judgement based upon estimated unacclimated aqueous aerobic
biodegradation half-life.

· **Ground Water:**                         High:     336 hours          (14 days)
                                            Low:      48 hours           (2 days)
*Comment:*  Scientific judgement based upon estimated unacclimated aqueous aerobic
biodegradation half-life (low $t_{1/2}$) and aqueous anaerobic half-life (high $t_{1/2}$).

<u>Aqueous Biodegradation (unacclimated):</u>
· **Aerobic half-life:**                    High:     168 hours          (7 days)
                                            Low:      24 hours           (1 day)
*Comment:*  Scientific judgement based upon unacclimated aerobic screening test data (Takemoto,
S et al. (1981), Urushigawa, Y et al. (1983)).

· **Anaerobic half-life:**                  High:     672 hours          (28 days)
                                            Low:      96 hours           (4 days)
*Comment:*  Scientific judgement based upon unacclimated aqueous aerobic biodegradation half-
life.

· **Removal/secondary treatment:**          High:     No data
                                            Low:
*Comment:*

<u>Photolysis:</u>
· **Atmos photol half-life:**               High:     No data
                                            Low:
*Comment:*

294

· **Max light absorption (nm):**         lambda max = 282.0, 272.5, 238.0, 214.0
*Comment:* Very little absorbance of UV light at wavelengths greater than 290 nm and none above 295 nm in methanol (Sadtler UV No. 259).

· **Aq photol half-life:**          High:        No data
                                    Low:

*Comment:*

## Photooxidation half-life:
· **Water:**                        High:     3480 hours          (145 days)
                                    Low:       66.0 hours          (2.75 days)
*Comment:* Scientific judgement based upon reported reaction rate constants for ·OH and $RO_2$· with the phenol class (Mill, T and Mabey, W (1985); Guesten, H et al. (1981)).

· **Air:**                          High:        16 hours
                                    Low:        1.6 hours
*Comment:* Based upon measured rate data for the vapor phase reaction with hydroxyl radicals in air (Atkinson, R (1985)).

## Reduction half-life:           High:        No data
                                    Low:

*Comment:*

## Hydrolysis:
· **First-order hydr half-life:**           No hydrolyzable groups
*Comment:* Rate constant at neutral pH is zero (Kollig, HP et al. (1987A)).

· **Acid rate const $(M(H+)-hr)^{-1}$:**        0.0
*Comment:* Based upon measured rate data at acid pH (Kollig, HP et al. (1987A)).

· **Base rate const $(M(OH-)-hr)^{-1}$:**        0.0
*Comment:* Based upon measured rate data at basic pH (Kollig, HP et al. (1987A)).

# 1,2-Dichlorobenzene

<u>CAS Registry Number:</u>  95-50-1

<u>Half-lives:</u>
  · **Soil:**                                    High:    4320 hours              (6 months)
                                                 Low:      672 hours              (4 weeks)
  *Comment:*  Scientific judgement based upon unacclimated aerobic screening test data (Canton, JH et al. (1985)) and aerobic soil grab sample data (Haider, K et al. (1981)).

  · **Air:**                                     High:    1528 hours              (63.7 days)
                                                 Low:     152.8 hours             (6.4 days)
  *Comment:*  Based upon photooxidation half-life in air.

  · **Surface Water:**                           High:    4320 hours              (6 months)
                                                 Low:      672 hours              (4 weeks)
  *Comment:*  Scientific judgement based upon estimated unacclimated aqueous aerobic biodegradation half-life.

  · **Ground Water:**                            High:    8640 hours              (12 months)
                                                 Low:     1344 hours              (8 weeks)
  *Comment:*  Scientific judgement based upon estimated unacclimated aqueous aerobic biodegradation half-life.

<u>Aqueous Biodegradation (unacclimated):</u>
  · **Aerobic half-life:**                       High:    4320 hours              (6 months)
                                                 Low:      672 hours              (4 weeks)
  *Comment:*  Scientific judgement based upon unacclimated aerobic screening test data (Canton, JH et al. (1985)) and aerobic soil grab sample data (Haider, K et al. (1981)).

  · **Anaerobic half-life:**                     High:   17280 hours              (24 months)
                                                 Low:     2880 hours              (16 weeks)
  *Comment:*  Scientific judgement based upon estimated unacclimated aqueous aerobic biodegradation half-life.

  · **Removal/secondary treatment:**             High:          76%
                                                 Low:
  *Comment:*  Based upon % degraded in an 8 day period under aerobic continuous flow conditions (Stover, EL, Kinncannon, DF (1983)).

<u>Photolysis:</u>
  · **Atmos photol half-life:**                  High:        No data
                                                 Low:

296

*Comment:*

**· Max light absorption (nm):**         lambda max = 269, 219.5
*Comment:* No absorbance of UV light at wavelengths greater than 310 nm in iso-octane (Sadtler UV No. 303).

**· Aq photol half-life:**          High:     No data
                                    Low:
*Comment:*

## Photooxidation half-life:
**· Water:**                        High:     No data
                                    Low:
*Comment:*

**· Air:**                          High:     1528 hours        (63.7 days)
                                    Low:      152.8 hours       (6.4 days)
*Comment:* Based upon measured rate data for the vapor phase reaction with hydroxyl radicals in air (Atkinson, R et al. (1985B)).

## Reduction half-life:           High:     No data
                                  Low:
*Comment:*

## Hydrolysis:
**· First-order hydr half-life:**       >879 years
*Comment:* Scientific judgement based upon rate constant (<0.9 $M^{-1}$ $hr^{-1}$) extrapolated to pH 7 at 25 °C from 1% disappearance after 16 days at 85 °C and pH 9.7 (Ellington, JJ et al. (1988)).

**· Acid rate const (M(H+)-hr)$^{-1}$:**
*Comment:*

**· Base rate const (M(OH-)-hr)$^{-1}$:**       <0.9
*Comment:* Scientific judgement based upon 1% disappearance after 16 days at 85 °C and pH 9.7 (Ellington, JJ et al. (1988)).

# o-Toluidine

**CAS Registry Number:** 95-53-4

## Half-lives:

· **Soil:**                    High:    168 hours            (7 days)

                               Low:     24 hours             (1 day)

*Comment:* Scientific judgement based upon estimated unacclimated aqueous aerobic biodegradation half-life.

· **Air:**                     High:    3.94 hours

                               Low:     0.394 hours          (23.6 minutes)

*Comment:* Scientific judgement based upon estimated photooxidation half-life in air.

· **Surface Water:**           High:    68 hours             (7 days)

                               Low:     24 hours             (1 day)

*Comment:* Scientific judgement based upon estimated unacclimated aqueous aerobic biodegradation half-life.

· **Ground Water:**            High:    336 hours            (14 days)

                               Low:     48 hours             (2 days)

*Comment:* Scientific judgement based upon estimated unacclimated aqueous aerobic biodegradation half-life.

## Aqueous Biodegradation (unacclimated):

· **Aerobic half-life:**       High:    168 hours            (7 days)

                               Low:     24 hours             (1 day)

*Comment:* Scientific judgement based upon aqueous aerobic screening test data (Baird, R et al. (1977); Sasaki, S (1978)).

· **Anaerobic half-life:**     High:    672 hours            (28 days)

                               Low:     96 hours             (4 days)

*Comment:* Scientific judgement based upon estimated unacclimated aqueous aerobic biodegradation half-life.

· **Removal/secondary treatment:**   High:    92%

                               Low:

*Comment:* Removal percentage based upon data from a continuous activated sludge biological treatment simulator (Matsui, S et al. (1975)).

## Photolysis:

· **Atmos photol half-life:**  High:    No data

                               Low:

*Comment:*

· **Max light absorption (nm):**     lambda max = 234, 284 nm (ethanol)
*Comment:* Absorption up to approximately 400 nm (Meallier,P (1969)).

· **Aq photol half-life:**     High:     No data
                               Low:
*Comment:*

## Photooxidation half-life:
· **Water:**     High:     3480 hours          (145 days)
                 Low:      62.4 hours          (2.6 days)
*Comment:* Scientific judgement based upon estimated rate constants for reactions of representative aromatic amines with ·OH and $RO_2$· (Mill, T and Mabey, W (1985)).

· **Air:**     High:     3.94 hours
               Low:      0.394 hours     (23.6 minutes)
*Comment:* Scientific judgement based upon estimated rate constant for reaction with hydroxyl radical in air (Atkinson, R (1987A)).

## Reduction half-life:     High:     No data
                           Low:
*Comment:*

## Hydrolysis:
· **First-order hydr half-life:**
*Comment:*

· **Acid rate const $(M(H+)\text{-}hr)^{-1}$:**     No hydrolyzable groups
*Comment:*

· **Base rate const $(M(OH\text{-})\text{-}hr)^{-1}$:**
*Comment:*

# 1,2,4-Trimethylbenzene

<u>CAS Registry Number:</u> 95-63-6

<u>Half-lives:</u>
 · **Soil:**                          High:     672 hours          (4 weeks)
                                      Low:      168 hours          (7 days)
 *Comment:* Scientific judgement based upon estimated aqueous aerobic biodegradation half-lives.


 · **Air:**                           High:      16 hours
                                      Low:       1.6 hours
 *Comment:* Based upon measured photooxidation rate constant with ·OH (Atkinson, R (1985)).


 · **Surface Water:**                 High:     672 hours          (4 weeks)
                                      Low:      168 hours          (7 days)
 *Comment:* Scientific judgement based upon estimated aqueous aerobic biodegradation half-lives.


 · **Ground Water:**                  High:    1344 hours          (8 weeks)
                                      Low:      336 hours          (14 days)
 *Comment:* Scientific judgement based upon estimated aqueous aerobic biodegradation half-lives.

<u>Aqueous Biodegradation (unacclimated):</u>
 · **Aerobic half-life:**             High:     672 hours          (4 weeks)
                                      Low:      168 hours          (7 days)
 *Comment:* Intermediate degradation rate selected based upon scientific judgement of conflicting aqueous screening studies (Kitano, M (1978); Van der Linden, AC (1978); Tester, DJ and Harker, RJ (1981); Trzilova, B and Horska, E (1988); Marion, CV and Malaney, GW (1964)).


 · **Anaerobic half-life:**           High:    2688 hours          (16 weeks)
                                      Low:      672 hours          (28 days)
 *Comment:* Scientific judgement based upon estimated aqueous aerobic biodegradation half-lives.


 · **Removal/secondary treatment:**   High:     No data
                                      Low:
 *Comment:*


<u>Photolysis:</u>
 · **Atmos photol half-life:**        High: No photolyzable functions
                                      Low:
 *Comment:*


 · **Max light absorption (nm):**
 *Comment:*

· **Aq photol half-life:**              High:
                                        Low:

*Comment:*

## Photooxidation half-life:
· **Water:**                            High:   43000 hours              (4.9 years)
                                        Low:    1056 hours              (44 days)
*Comment:* Based upon measured aqueous photooxidation rate constant with hydroxyl radicals
(Guesten H et al. (1981)).

· **Air:**                              High:   16 hours
                                        Low:    1.6 hours
*Comment:* Based upon measured vapor-phase photooxidation rate constant with ·OH (Atkinson,
R (1985)).

## Reduction half-life:                 High:   No data
                                        Low:
*Comment:*

## Hydrolysis:
· **First-order hydr half-life:**       No hydrolyzable functions
*Comment:*

· **Acid rate const $(M(H+)-hr)^{-1}$:**
*Comment:*

· **Base rate const $(M(OH-)-hr)^{-1}$:**
*Comment:*

# 2,4-Diaminotoluene

<u>CAS Registry Number:</u>  95-80-7

<u>Half-lives:</u>
  · **Soil:**                    High:    4320 hours        (6 months)
                                     Low:     672 hours         (4 weeks)
*Comment:* Scientific judgement based upon estimated unacclimated aqueous aerobic biodegradation half-life.

  · **Air:**                    High:    2.7 hours
                                       Low:     0.27 hours
*Comment:* Based upon estimated photooxidation half-life in air.

  · **Surface Water:**           High:    1740 hours        (72 days)
                                     Low:      31 hours       (1.3 days)
*Comment:* Based upon estimated photooxidation half-life in water.

  · **Ground Water:**           High:    8640 hours      (12 months)
                                     Low:    1344 hours        (8 weeks)
*Comment:* Scientific judgement based upon estimated unacclimated aqueous aerobic biodegradation half-life.

<u>Aqueous Biodegradation (unacclimated):</u>
  · **Aerobic half-life:**       High:    4320 hours        (6 months)
                                     Low:     672 hours         (4 weeks)
*Comment:* Scientific judgement based upon unacclimated aerobic aqueous screening test data which confirmed resistance to biodegradation (Sasaki, S (1978)).

  · **Anaerobic half-life:**     High:  17,280 hours     (24 months)
                                     Low:    2688 hours        (8 weeks)
*Comment:* Scientific judgement based upon estimated unacclimated aerobic aqueous biodegradation half-life.

  · **Removal/secondary treatment:**   High:         34%
                                     Low:
*Comment:* Based upon activated sludge degradation results using a fill and draw method (Matsui, S et al. (1975)).

<u>Photolysis:</u>
  · **Atmos photol half-life:**     High:
                                     Low:
*Comment:*

**· Max light absorption (nm):**
*Comment:*

**· Aq photol half-life:**                              High:
                                                       Low:

*Comment:*

**Photooxidation half-life:**
  **· Water:**                                         High:    1740 hours              (72 days)
                                                       Low:      31 hours               (1.3 days)
*Comment:* Scientific judgement based upon estimated half-life for reaction of aromatic amines
with hydroxyl radicals in water (Mill, T (1989); Guesten, H et al. (1981)). It is assumed that 2,4-
diaminotoluene reacts twice as fast as aniline.

  **· Air:**                                           High:    2.7 hours
                                                       Low:     0.27 hours
*Comment:* Based upon estimated rate constant for reaction with hydroxyl radicals in air
(Atkinson, R (1987A)).

**Reduction half-life:**                               High:  Not expected to be significant
                                                       Low:

*Comment:*

**Hydrolysis:**
  **· First-order hydr half-life:**
*Comment:*

  **· Acid rate const $(M(H+)-hr)^{-1}$:**              No hydrolyzable groups
*Comment:*

  **· Base rate const $(M(OH-)-hr)^{-1}$:**
*Comment:*

# 1,2,4,5-Tetrachlorobenzene

**CAS Registry Number:** 95-94-3

**Half-lives:**

  · **Soil:**                            High:    4320 hours         (6 months)
                                       Low:     672 hours          (4 weeks)

*Comment:* Scientific judgement based upon estimated unacclimated aqueous aerobic biodegradation half-life.

  · **Air:**                              High:    7631 hours         (318 days)
                                         Low:    763.1 hours      (31.8 days)

*Comment:* Based upon photooxidation half-life in air.

  · **Surface Water:**                 High:    4320 hours         (6 months)
                                         Low:     672 hours          (4 weeks)

*Comment:* Scientific judgement based upon estimated unacclimated aqueous aerobic biodegradation half-life.

  · **Ground Water:**                 High:    8640 hours        (12 months)
                                         Low:    1344 hours        (8 weeks)

*Comment:* Scientific judgement based upon estimated unacclimated aqueous aerobic biodegradation half-life.

**Aqueous Biodegradation (unacclimated):**

  · **Aerobic half-life:**             High:    4320 hours         (6 months)
                                         Low:     672 hours          (4 weeks)

*Comment:* Scientific judgement based upon unacclimated aerobic screening test data (Kitano, M (1978)).

  · **Anaerobic half-life:**           High:   17280 hours      (24 months)
                                        Low:    2880 hours       (16 weeks)

*Comment:* Scientific judgement based upon estimated unacclimated aqueous aerobic biodegradation half-life.

  · **Removal/secondary treatment:**     High:    No data
                                         Low:

*Comment:*

**Photolysis:**

  · **Atmos photol half-life:**       High:    No data
                                         Low:

*Comment:*

· **Max light absorption (nm):**       No data
*Comment:*

· **Aq photol half-life:**        High:      No data
                                  Low:
*Comment:*

**Photooxidation half-life:**
· **Water:**                      High:      No data
                                  Low:
*Comment:*

· **Air:**                        High:      7631 hours          (318 days)
                                  Low:       763.1 hours         (31.8 days)
*Comment:* Based upon measured rate data for the vapor phase reaction with hydroxyl radicals in air (Atkinson, R et al.(1985B)).

**Reduction half-life:**           High:      No data
                                  Low:
*Comment:*

**Hydrolysis:**
· **First-order hydr half-life:**      >879 years
*Comment:* Scientific judgement based upon rate constant (<0.9 $M^{-1}$ $hr^{-1}$) extrapolated to pH 7 at 25 °C from 1% disappearance after 16 days at 85 °C and pH 9.7 (Ellington, JJ et al. (1988)).

· **Acid rate const $(M(H+)-hr)^{-1}$:**
*Comment:*

· **Base rate const $(M(OH-)-hr)^{-1}$:**      <0.9
*Comment:* Scientific judgement based upon 1% disappearance after 16 days at 85 °C and pH 9.7 (Ellington, JJ et al. (1988)).

# 2,4,5-Trichlorophenol

**CAS Registry Number:** 95-95-4

**Half-lives:**

- **Soil:**  High:  16560 hours  (690 days)
              Low:  552 hours  (23 days)

*Comment:* Scientific judgement based upon estimated unacclimated aqueous aerobic biodegradation half-life.

- **Air:**  High:  301 hours  (12.6 days)
             Low:  30.1 hours  (1.3 days)

*Comment:* Scientific judgement based upon photooxidation half-life in air.

- **Surface Water:**  High:  336 hours  (14 days)
                      Low:  0.5 hour

*Comment:* Scientific judgement based upon aqueous photolysis half-life.

- **Ground Water:**  High:  43690 hours  (5 years)
                     Low:  1104 hours  (46 days)

*Comment:* Scientific judgement based upon estimated unacclimated aqueous aerobic biodegradation half-life (low $t_{1/2}$) and aqueous anaerobic half-life (high $t_{1/2}$).

**Aqueous Biodegradation (unacclimated):**

- **Aerobic half-life:**  High:  16560 hours  (690 days)
                          Low:  552 hours  (23 days)

*Comment:* Scientific judgement based upon unacclimated aerobic river die-away test data (Lee, RF and Ryan, C (1979)).

- **Anaerobic half-life:**  High:  43690 hours  (5 years)
                            Low:  3028 hours  (126 days)

*Comment:* Scientific judgement based upon unacclimated anaerobic grab sample data for soil and ground water (low $t_{1/2}$: Gibson, SA and Suflita, JM (1986) high $t_{1/2}$: Baker, MD and Mayfield, CI (1980)).

- **Removal/secondary treatment:**  High:  No data
                                    Low:

*Comment:*

**Photolysis:**

- **Atmos photol half-life:**  High:  No data
                               Low:

*Comment:*

**· Max light absorption (nm):**   lambda max = 299.0, 292.5
*Comment:* In methanol (Sadtler UV No. 3859).

**· Aq photol half-life:**   High:   336 hours      (14 days)
                             Low:    0.5 hour
*Comment:* High $t_{1/2}$ and low $t_{1/2}$ based upon photolysis rate constants for transformation and mineralization under summer and winter sunlight conditions at 25 and 18° C, respectively (Hwang H et al. (1986)).

## Photooxidation half-life:
**· Water:**   High:   3480 hours      (145 days)
               Low:    66 hours        (2.75 days)
*Comment:* Scientific judgement based upon reported reaction rate constants for ·OH and $RO_2$· with the phenol class (Mill, T and Mabey, W (1985); Guesten, H et al. (1981)).

**· Air:**   High:   301 hours       (12.6 days)
             Low:    30.1 hours      (1.3 days)
*Comment:* Scientific judgement based upon an estimated rate constant for the vapor phase reaction with hydroxyl radicals in air (Atkinson, R (1987A)).

## Reduction half-life:
   High:   No data
   Low:
*Comment:*

## Hydrolysis:
**· First-order hydr half-life:**   $>8 \times 10^6$ years
*Comment:* Rate constant at neutral pH is zero (Kollig, HP et al. (1987A)).

**· Acid rate const $(M(H+)-hr)^{-1}$:**   0.0
*Comment:* Based upon measured rate data at acid pH (Kollig, HP et al. (1987A)).

**· Base rate const $(M(OH-)-hr)^{-1}$:**   0.0
*Comment:* Based upon measured rate data at basic pH (Kollig, HP et al. (1987A)).

# Styrene oxide

CAS Registry Number: 96-09-3

**Half-lives:**
  **· Soil:**                    High:    27.5 hours
                                 Low: 0.00385 hours          (2.31 minutes)
  *Comment:* Based upon measured aqueous hydrolysis half-lives.

  **· Air:**                     High:    123 hours          (5.1 days)
                                 Low:    12.3 hours
  *Comment:* Based upon estimated photooxidation half-lives in air.

  **· Surface Water:**           High:    27.5 hours
                                 Low: 0.00385 hours          (2.31 minutes)
  *Comment:* Based upon measured aqueous hydrolysis half-lives.

  **· Ground Water:**            High:    27.5 hours
                                 Low: 0.00385 hours          (2.31 minutes)
  *Comment:* Based upon measured aqueous hydrolysis half-lives.

**Aqueous Biodegradation (unacclimated):**
  **· Aerobic half-life:**       High:    168 hours          (7 days)
                                 Low:     24 hours           (1 day)
  *Comment:* Scientific judgement based upon biological screening test data (Schmidt-Bleek, F et al. (1982)).

  **· Anaerobic half-life:**     High:    672 hours          (28 days)
                                 Low:     96 hours           (4 days)
  *Comment:* Scientific judgement based upon estimated aqueous aerobic biodegradation half-lives.

  **· Removal/secondary treatment:**   High:    No data
                                        Low:
  *Comment:*

**Photolysis:**
  **· Atmos photol half-life:**   High:    No data
                                  Low:
  *Comment:*

  **· Max light absorption (nm):**
  *Comment:*

**· Aq photol half-life:**　　　　　　　　High:
　　　　　　　　　　　　　　　　　　　　Low:
*Comment:*

## Photooxidation half-life:
　**· Water:**　　　　　　　　　　　　High:　　No data
　　　　　　　　　　　　　　　　　　　Low:
　*Comment:*

　**· Air:**　　　　　　　　　　　　　High:　　123 hours　　　　　(5.1 days)
　　　　　　　　　　　　　　　　　　　Low:　　12.3 hours
　*Comment:* Based upon estimated photooxidation rate constant with ·OH (Atkinson, R (1987A)).

## Reduction half-life:
　　　　　　　　　　　　　　　　　　　High:　　No data
　　　　　　　　　　　　　　　　　　　Low:
　*Comment:*

## Hydrolysis:
　**· First-order hydr half-life:**　　　　0.00385 to 27.5 hours
　*Comment:* Based upon the measured first-order rate constants at 25°C, the hydrolysis half-lives at pH 5, pH 7, and pH 9 are 2.31 minutes (0.00385 hours), 21.4 hours, and 27.5 hours, respectively (Haag, WR and Mill, T (1988)).

　**· Acid rate const $(M(H+)-hr)^{-1}$:**
　*Comment:*

　**· Base rate const $(M(OH-)-hr)^{-1}$:**
　*Comment:*

# 1,2-Dibromo-3-chloropropane

CAS Registry Number:  96-12-8

Half-lives:
    · Soil:                        High:    4320 hours         (6 months)

|  | High: | 4320 hours | (6 months) |
|---|---|---|---|
| Soil: | Low: | 672 hours | (4 weeks) |

Comment:  Scientific judgement based upon unacclimated aerobic screening test data (Castro, CE and Belser, NO (1968)) and aerobic soil field data (Liu, CCK et al. (1987)).

| · Air: | High: | 1459 hours | (60.8 days) |
|---|---|---|---|
|  | Low: | 146 hours | (6.1 days) |

Comment:  Based upon photooxidation half-life in air.

| · Surface Water: | High: | 4320 hours | (6 months) |
|---|---|---|---|
|  | Low: | 672 hours | (4 weeks) |

Comment:  Scientific judgement based upon estimated unacclimated aqueous aerobic biodegradation half-life.

| · Ground Water: | High: | 8640 hours | (12 months) |
|---|---|---|---|
|  | Low: | 1344 hours | (8 weeks) |

Comment:  Scientific judgement based upon estimated unacclimated aqueous aerobic biodegradation half-life.

Aqueous Biodegradation (unacclimated):

| · Aerobic half-life: | High: | 4320 hours | (6 months) |
|---|---|---|---|
|  | Low: | 672 hours | (4 weeks) |

Comment:  Scientific judgement based upon unacclimated aerobic screening test data (Castro, CE and Belser, NO (1968)) and aerobic soil field data (Liu, CCK et al. (1987)).

| · Anaerobic half-life: | High: | 17280 hours | (24 months) |
|---|---|---|---|
|  | Low: | 2880 hours | (16 weeks) |

Comment:  Scientific judgement based upon estimated unacclimated aqueous aerobic biodegradation half-life.

| · Removal/secondary treatment: | High: | No data |
|---|---|---|
|  | Low: | |

Comment:

Photolysis:

| · Atmos photol half-life: | High: |
|---|---|
|  | Low: |

Comment:

**· Max light absorption (nm):**     No data
*Comment:*

**· Aq photol half-life:**     High:
                               Low:
*Comment:*

**Photooxidation half-life:**
  **· Water:**     High:     No data
                            Low:
*Comment:*

  **· Air:**     High:     1459 hours     (60.8 days)
                          Low:     146 hours     (6.1 days)
*Comment:* Based upon measured rate data for the vapor phase reaction with hydroxyl radicals in air (Atkinson, R (1985)).

**Reduction half-life:**     High:     No data
                            Low:
*Comment:*

**Hydrolysis:**
  **· First-order hydr half-life:**     38.4 years
*Comment:* Scientific judgement based upon reported rate constant (20.6 $M^{-1}$ $hr^{-1}$) at pH 7 and 25 °C (Ellington, JJ et al. (1987A)).

  **· Acid rate const $(M(H+)-hr)^{-1}$:**
*Comment:*

  **· Base rate const $(M(OH-)-hr)^{-1}$:**     20.6
*Comment:* Half-life of 140 days at 25 °C and pH 9.0 (Ellington, JJ et al. (1987A)).

# 1,2,3-Trichloropropane

<u>CAS Registry Number:</u>  96-18-4

<u>Half-lives:</u>
  · **Soil:**                                High:    8640 hours            (1 year)
                                       Low:     4320 hours          (6 months)
*Comment:* Scientific judgement based upon estimated aqueous aerobic biodegradation half-life.

  · **Air:**                                 High:    613 hours          (25.5 days)
                                       Low:      61 hours          (2.6 days)
*Comment:* Scientific judgement based upon estimated photooxidation half-life in air.

  · **Surface Water:**                  High:    8640 hours            (1 year)
                                       Low:     4320 hours          (6 months)
*Comment:* Scientific judgement based upon estimated aqueous aerobic biodegradation half-life.

  · **Ground Water:**                 High:   17280 hours         (2 years)
                                       Low:     8640 hours        (12 months)
*Comment:* Scientific judgement based upon estimated aqueous aerobic biodegradation half-life.

<u>Aqueous Biodegradation (unacclimated):</u>
  · **Aerobic half-life:**              High:    8640 hours            (1 year)
                                       Low:     4320 hours          (6 months)
*Comment:* Scientific judgement based upon acclimated aerobic soil grab sample data for 1,2-dichloropropane (Roberts, TR and Stoydin, G (1976)).

  · **Anaerobic half-life:**           High:   34560 hours         (4 years)
                                     Low:   17280 hours        (24 months)
*Comment:* Scientific judgement based upon aerobic biodegradation half-life.

  · **Removal/secondary treatment:**   High:     No data
                                       Low:
*Comment:*

<u>Photolysis:</u>
  · **Atmos photol half-life:**        High:     No data
                                       Low:
*Comment:*

  · **Max light absorption (nm):**   No data
  *Comment:*

**· Aq photol half-life:**    High:    No data
                             Low:

*Comment:*

## Photooxidation half-life:
**· Water:**    High:    No data
               Low:

*Comment:*


**· Air:**    High:    613 hours    (25.5 days)
             Low:    61 hours    (2.6 days)

*Comment:* Scientific judgement based upon an estimated rate constant for the vapor phase reaction with hydroxyl radicals in air (Atkinson, R (1987A)).

## Reduction half-life:    High:    No data
                          Low:

*Comment:*

## Hydrolysis:
**· First-order hydr half-life:**    44 years

*Comment:* Scientific judgement based upon reported rate constant ($k_N = 1.8 \times 10^{-6}$ hr$^{-1}$) at pH 7 to 9 and 25 °C (Ellington, JJ et al. (1987A)).

**· Acid rate const (M(H+)-hr)$^{-1}$:**
*Comment:*

**· Base rate const (M(OH-)-hr)$^{-1}$:**    $9.9 \times 10^{-4}$ M$^{-1}$ hr$^{-1}$

*Comment:* Base reaction not expected to be important (Ellington, JJ et al. (1987A)).

# Methyl acrylate

CAS Registry Number: 96-33-3

Half-lives:
  · Soil:                           High:    168 hours            (7 days)
                                    Low:      24 hours            (1 day)
  Comment: Scientific judgement based upon aqueous aerobic biodegradation half-life.

  · Air:                            High:    27.0 hours
                                    Low:      2.7 hours
  Comment: Based upon estimated photooxidation half-lives in air.

  · Surface Water:                  High:    168 hours            (7 days)
                                    Low:      24 hours            (1 day)
  Comment: Scientific judgement based upon aqueous aerobic biodegradation half-life.

  · Ground Water:                   High:    336 hours            (14 days)
                                    Low:      48 hours            (2 days)
  Comment: High value is scientific judgement based upon aqueous aerobic biodegradation half-
  life. Low value is based upon measured hydrolysis data.

Aqueous Biodegradation (unacclimated):
  · Aerobic half-life:              High:    168 hours            (7 days)
                                    Low:      24 hours            (1 day)
  Comment: Scientific judgement based upon a biological screening test (Sasaki, S (1978)).

  · Anaerobic half-life:            High:    672 hours            (28 days)
                                    Low:      96 hours            (4 days)
  Comment: Scientific judgement based upon aqueous aerobic biodegradation half-lives.

  · Removal/secondary treatment:    High:    No data
                                    Low:
  Comment:

Photolysis:
  · Atmos photol half-life:         High:    No data
                                    Low:
  Comment:

  · Max light absorption (nm):
  Comment:

· **Aq photol half-life:**  High:
  Low:

*Comment:*

**Photooxidation half-life:**
  · **Water:**  High:  No data
  Low:

*Comment:*

  · **Air:**  High:  27.0 hours
  Low:  2.70 hours

*Comment:* Based upon estimated photooxidation rate constant with ·OH (Atkinson, R (1987A)).

**Reduction half-life:**  High:  No data
  Low:

*Comment:*

**Hydrolysis:**
  · **First-order hydr half-life:**  24,700 hours  (2.8 years)

*Comment:* Half-life at 25°C and pH 7 based upon measured base rate constant.

  · **Acid rate const (M(H+)-hr)$^{-1}$:**  0.00000012

*Comment:* Measured rate constant for ethyl acrylate at 25°C (Mabey, W and Mill, T (1978)). At 25°C and pH 5, the half-life would be 280 years; this half-life results primarily from the base rate constant; the acid rate constant is too slow to be important.

  · **Base rate const (M(OH-)-hr)$^{-1}$:**  0.0779

*Comment:* Measured rate constant for methyl acrylate at 25°C (Roy, RS (1972)). At pH 9 and 25°C, the half-life would be 240 hours (10 days).

# Ethylenethiourea

**CAS Registry Number:** 96-45-7

**Structure:**

**Half-lives:**

- **Soil:**
  High: 672 hours (4 weeks)
  Low: 168 hours (7 days)
  *Comment:* Scientific judgement based upon field test data in soil (Rhodes, RC (1977)).

- **Air:**
  High: 4.7 hours
  Low: 0.5 hour (30 minutes)
  *Comment:* Scientific judgement based upon estimated photooxidation half-life in air.

- **Surface Water:**
  High: 672 hours (4 weeks)
  Low: 168 hours (7 days)
  *Comment:* Scientific judgement based upon estimated aqueous aerobic biodegradation half-life.

- **Ground Water:**
  High: 1344 hours (8 weeks)
  Low: 336 hours (14 days)
  *Comment:* Scientific judgement based upon estimated aqueous aerobic biodegradation half-life.

**Aqueous Biodegradation (unacclimated):**

- **Aerobic half-life:**
  High: 672 hours (4 weeks)
  Low: 168 hours (7 days)
  *Comment:* Scientific judgement based upon field test data in soil (Rhodes, RC (1977)).

- **Anaerobic half-life:**
  High: 2688 hours (16 weeks)
  Low: 672 hours (28 days)
  *Comment:* Scientific judgement based upon estimated aqueous aerobic biodegradation half-life.

- **Removal/secondary treatment:**
  High: No data
  Low:
  *Comment:*

**Photolysis:**

- **Atmos photol half-life:**
  High: Not expected to be important
  Low:
  *Comment:*

- **Max light absorption (nm):**
  lambda max = 239

*Comment:* Very little absorbance occurs above 290 nm in methanol (Sadtler UV No. 4571).

· **Aq photol half-life:**              High:  Not expected to be important
                                        Low:
*Comment:* Stable under aqueous photolysis conditions (Ross, RD and Crosby DG (1973), Ross, RD and Crosby DG (1985)).

**Photooxidation half-life:**
· **Water:**                           High:  No available rate data, may be important
                                        Low:
*Comment:* Rates and oxidants were not identified (Ross, RD and Crosby DG (1973), Ross, RD and Crosby DG (1985)).

· **Air:**                             High:    4.7 hours
                                        Low:     0.5 hours           (30 minutes)
*Comment:* Scientific judgement based upon an estimated rate constant for vapor phase reaction with hydroxyl radicals in air (Atkinson, R (1987A)).

**Reduction half-life:**               High:    No data
                                        Low:
*Comment:*

**Hydrolysis:**
· **First-order hydr half-life:**       Not expected to be important
*Comment:* Scientific judgement based upon zero hydrolysis was observed after 3 months for pHs of 5, 7, and 9 at 90 °C (Cruickshank, PA and Jarrow, HC (1973)) and zero hydrolysis was observed after 90 days for pHs of 3, 7 and 11 at 90 °C (Ellington, JJ et al. (1987)).

· **Acid rate const $(M(H+)-hr)^{-1}$:**       Not expected to be important
*Comment:* Scientific judgement based upon zero hydrolysis was observed after 3 months at 90 °C for pHs of 3 (Ellington, JJ et al. (1987)) and 5 (Cruickshank, PA and Jarrow, HC (1973)).

· **Base rate const $(M(OH-)-hr)^{-1}$:**       Not expected to be important
*Comment:* Scientific judgement based upon zero hydrolysis was observed after 3 months at 90 °C for pHs of 9 (Cruickshank, PA and Jarrow, HC (1973)) and 11 (Ellington, JJ et al. (1987)).

# C.I. Solvent Yellow 3

**CAS Registry Number:** 97-56-3

**Structure:**

**Half-lives:**
· **Soil:**　　　　　　　　　　　　High:　　672 hours　　　　　(4 weeks)
　　　　　　　　　　　　　　　　Low:　　168 hours　　　　　(7 days)
*Comment:* Scientific judgement based upon estimated aqueous aerobic biodegradation half-lives.

· **Air:**　　　　　　　　　　　　High:　　8.2 hours
　　　　　　　　　　　　　　　　Low:　　0.82 hours
*Comment:* Based upon estimated photooxidation half-lives in air.

· **Surface Water:**　　　　　　　High:　　672 hours　　　　　(4 weeks)
　　　　　　　　　　　　　　　　Low:　　62.4 hours　　　　　(2.6 days)
*Comment:* The high value is scientific judgement based upon estimated aqueous aerobic biodegradation half-life. The low value is based upon aqueous photooxidation half-life.

· **Ground Water:**　　　　　　　High:　　1344 hours　　　　(8 weeks)
　　　　　　　　　　　　　　　　Low:　　336 hours　　　　　(14 days)
*Comment:* Scientific judgement based upon estimated aqueous aerobic biodegradation half-lives.

**Aqueous Biodegradation (unacclimated):**
· **Aerobic half-life**:　　　　　High:　　672 hours　　　　　(4 weeks)
　　　　　　　　　　　　　　　　Low:　　168 hours　　　　　(7 days)
*Comment:* Scientific judgement based upon an estimated biodegradation classification of "biodegrades fast with adaptation" for 4-aminoazobenzene from the BIODEG file.

· **Anaerobic half-life**:　　　　High:　　2688 hours　　　　(16 weeks)
　　　　　　　　　　　　　　　　Low:　　672 hours　　　　　(28 days)
*Comment:* Scientific judgement based upon estimated aqueous aerobic biodegradation half-lives.

· **Removal/secondary treatment:**　High:　　No data
　　　　　　　　　　　　　　　　Low:
*Comment:*

**Photolysis:**

318

**· Atmos photol half-life:**  High:  No data

Low:

*Comment:*

**· Max light absorption (nm):**

*Comment:*

**· Aq photol half-life:**  High:

Low:

*Comment:*

**Photooxidation half-life:**

**· Water:**  High:  3480 hours  (145 days)

Low:  62.4 hours  (2.6 days)

*Comment:*  Scientific judgement based upon photooxidation rate constants with ·OH and $RO_2$·
for the aromatic amine class (Mill, T and Mabey, W (1985); Guesten, H et al. (1981)).

**· Air:**  High:  8.20 hours

Low:  0.82 hours

*Comment:*  Based upon estimated photooxidation rate constant with ·OH (Atkinson, R (1987A)).

**Reduction half-life:**  High:  No data

Low:

*Comment:*

**Hydrolysis:**

**· First-order hydr half-life:**  No hydrolyzable functions

*Comment:*

**· Acid rate const $(M(H+)-hr)^{-1}$:**

*Comment:*

**· Base rate const $(M(OH-)-hr)^{-1}$:**

*Comment:*

319

# Benzotrichloride

<u>CAS Registry Number:</u> 98-07-7

<u>Half-lives:</u>
- **Soil:** 

|  | High: | 3 minutes |
|---|---|---|
|  | Low: | 11 seconds |

*Comment:* Low $t_{1/2}$ based upon measured overall hydrolysis rate constant for pH 7 at 25 °C (Mabey W and Mill T (1978)). High $t_{1/2}$ based upon measured overall hydrolysis rate constant for pH 7 at 5 °C (Laughton and PM, Robertson, RE (1959)).

- **Air:**

|  | High: | 1737 hours | (72.4 days) |
|---|---|---|---|
|  | Low: | 173.7 hours | (7.24 days) |

*Comment:* Scientific judgement based upon estimated photooxidation half-life in air.

- **Surface Water:**

|  | High: | 3 minutes |
|---|---|---|
|  | Low: | 11 seconds |

*Comment:* Low $t_{1/2}$ based upon measured overall hydrolysis rate constant for pH 7 at 25 °C (Mabey W and Mill T (1978)). High $t_{1/2}$ based upon measured overall hydrolysis rate constant for pH 7 at 5 °C (Laughton, PM and Robertson, RE (1959)).

- **Ground Water:**

|  | High: | 3 minutes |
|---|---|---|
|  | Low: | 11 seconds |

*Comment:* Low $t_{1/2}$ based upon measured overall hydrolysis rate constant for pH 7 at 25 °C (Mabey W and Mill T (1978)). High $t_{1/2}$ based upon measured overall hydrolysis rate constant for pH 7 at 5 °C (Laughton, PM and Robertson, RE (1959)).

<u>Aqueous Biodegradation (unacclimated):</u>
- **Aerobic half-life:**

|  | High: | 168 hours | (7 days) |
|---|---|---|---|
|  | Low: | 24 hours | (1 day) |

*Comment:* Scientific judgement based upon limited aqueous screening test data (given the fast hydrolysis rate this may be for the hydrolysis product) (Steinhauser, KG et al. (1986)).

- **Anaerobic half-life:**

|  | High: | 672 hours | (4 weeks) |
|---|---|---|---|
|  | Low: | 96 hours | (4 days) |

*Comment:* Scientific judgement based upon estimated unacclimated aqueous aerobic biodegradation half-life.

- **Removal/secondary treatment:**

|  | High: | No data |
|---|---|---|
|  | Low: |  |

*Comment:*

<u>Photolysis:</u>

· **Atmos photol half-life:**        High:     No data
                                               Low:

*Comment:*

· **Max light absorption (nm):**      lambda max = 274, 267, 260.5, 225
*Comment:* Very little absorption occurs above 290 nm in methanol (Sadtler UV No. 380).

· **Aq photol half-life:**          High:     No data
                                               Low:

*Comment:*

## Photooxidation half-life:
· **Water:**                    High:     No data
                                             Low:

*Comment:*

· **Air:**                      High:    1737 hours        (72.4 days)
                                               Low:     173.7 hours     (7.24 days)
*Comment:* Scientific judgement based upon an estimated rate constant for vapor phase reaction with hydroxyl radicals in air (Atkinson, R (1987A)).

## Reduction half-life:
                                          High:     No data
                                          Low:

*Comment:*

## Hydrolysis:
· **First-order hydr half-life:**      11 seconds
*Comment:* Based upon overall rate constant ($k_h = 6.3 \times 10^{-2}$) at pH 7 and 25 °C (Mabey W and Mill T (1978)).

· **First-order hydr half-life:**      3.0 minutes
*Comment:* Based upon overall rate constant ($k_h = 3.87 \times 10^{-3}$) at pH 7 and 5 °C (Laughton, PM and Robertson, RE (1959)).

· **Acid rate const (M(H+)-hr)$^{-1}$:**     No data
*Comment:*

· **Base rate const (M(OH-)-hr)$^{-1}$:**     No data
*Comment:*

# Cumene

<u>CAS Registry Number:</u>  98-82-8

<u>Structure:</u>

<u>Half-lives:</u>
  · **Soil:**                         High:      192 hours           (8 days)
                                       Low:        48 hours           (2 days)
*Comment:*  Based upon data from a soil column study in which aerobic ground water was continuously percolated through quartz sand (Kappeler, T and Wuhrmann, K (1978)).

  · **Air:**                          High:      97.2 hours          (4.05 days)
                                       Low:       9.72 hours
*Comment:*  Scientific judgement based upon estimated photooxidation half-life in air.

  · **Surface Water:**                High:      192 hours           (8 days)
                                       Low:        48 hours           (2 days)
*Comment:*  Based upon data from a soil column study in which aerobic ground water was continuously percolated through quartz sand (Kappeler, T and Wuhrmann, K (1978)).

  · **Ground Water:**                 High:      384 hours           (16 days)
                                       Low:        96 hours           (4 days)
*Comment:*  Based upon data from a soil column study in which aerobic ground water was continuously percolated through quartz sand (Kappeler, T and Wuhrmann, K (1978)).

<u>Aqueous Biodegradation (unacclimated):</u>
  · **Aerobic half-life:**            High:      192 hours           (8 days)
                                       Low:        48 hours           (2 days)
*Comment:*  Based upon data from a soil column study in which aerobic ground water was continuously percolated through quartz sand (Kappeler, T and Wuhrmann, K (1978)).

  · **Anaerobic half-life:**          High:      768 hours           (32 days)
                                       Low:       192 hours           (8 days)
*Comment:*  Scientific judgement based upon unacclimated aqueous aerobic biodegradation half-life.

  · **Removal/secondary treatment:**  High:         100%
                                       Low:
*Comment:*  Removal percentage based upon data from a continuous activated sludge biological

322

treatment simulator (Kappeler, T and Wuhrmann, K (1978)).

**Photolysis:**
  · **Atmos photol half-life:**          High:     No data
                                              Low:
*Comment:*

  · **Max light absorption (nm):**     lambda max = 248, 252, 258, 260, 264, 267 nm
(cyclohexane).
*Comment:* Absorption extends to approximately 292 nm (Sadtler UV No. 95).

  · **Aq photol half-life:**            High:     No data
                                              Low:
*Comment:*

**Photooxidation half-life:**
  · **Water:**                     High: $1.3 \times 10^5$ hours       (14.6 years)
                                          Low:    3208 hours         (134 days)
*Comment:* Based upon measured rate constant for reaction with hydroxyl radical in water (Mill, T et al. (1978)).

  · **Air:**                        High:    97.2 hours        (4.05 days)
                                          Low:    9.72 hours
*Comment:* Scientific judgement based upon estimated rate constant for reaction with hydroxyl radical in air (Atkinson, R (1987A)).

**Reduction half-life:**             High:     No data
                                            Low:
*Comment:*

**Hydrolysis:**
  · **First-order hydr half-life:**
*Comment:*

  · **Acid rate const $(M(H+)\text{-}hr)^{-1}$:**     No data
*Comment:*

  · **Base rate const $(M(OH\text{-})\text{-}hr)^{-1}$:**
*Comment:*

# (Dichloromethyl)benzene

<u>CAS Registry Number:</u>  98-87-3

<u>Half-lives:</u>
- **· Soil:**                    High:     0.6 hours          (36 minutes)
                              Low:      0.1 hours          (6 minutes)

*Comment:* Scientific judgement based upon estimated hydrolysis half-lives at 5 and 25 °C (Mabey W and Mill T (1978)).

- **· Air:**                     High:     270 hours          (11.3 days)
                              Low:      27 hours            (1.1 days)

*Comment:* Scientific judgement based upon estimated photooxidation half-life in air.

- **· Surface Water:**           High:     0.6 hours          (36 minutes)
                              Low:      0.1 hours          (6 minutes)

*Comment:* Scientific judgement based upon estimated hydrolysis half-lives at 5 and 25 °C (Mabey W and Mill T (1978)).

- **· Ground Water:**            High:     0.6 hours          (36 minutes)
                              Low:      0.1 hours          (6 minutes)

*Comment:* Scientific judgement based upon estimated hydrolysis half-lives at 5 and 25 °C (Mabey W and Mill T (1978)).

<u>Aqueous Biodegradation (unacclimated):</u>
- **· Aerobic half-life:**       High:     672 hours          (4 weeks)
                              Low:      168 hours          (7 days)

*Comment:* Scientific judgement.

- **· Anaerobic half-life:**     High:     2688 hours         (16 weeks)
                              Low:      672 hours          (4 weeks)

*Comment:* Scientific judgement based upon estimated aqueous aerobic biodegradation half-life.

- **· Removal/secondary treatment:**   High:     No data
                                    Low:

*Comment:*

<u>Photolysis:</u>
- **· Atmos photol half-life:**  High:
                              Low:

*Comment:*

- **· Max light absorption (nm):**   No data

*Comment:*

**· Aq photol half-life:**  High:
  Low:

*Comment:*

## Photooxidation half-life:
**· Water:**  High:  No data
  Low:

*Comment:*

**· Air:**  High:  270 hours  (11.3 days)
  Low:  27 hours  (1.1 days)
*Comment:* Scientific judgement based upon estimated rate data for hydroxyl radicals in air (Atkinson, R (1987A)).

## Reduction half-life:  High:  No data
  Low:

*Comment:*

## Hydrolysis:
**· First-order hydr half-life:**  0.1 hours  (6 minutes)
*Comment:* Based upon overall rate constant ($k_h$ = 1.56X10$^{-3}$) at pH 7 and 25 °C (Mabey W and Mill T (1978)).

**· First-order hydr half-life:**  0.6 hours  (36 minutes)
*Comment:* Based upon overall rate constant ($k_h$ = 3.2X10$^{-4}$) at pH 7 and 5 °C (Mabey W and Mill T (1978)).

**· Acid rate const (M(H+)-hr)$^{-1}$:**
*Comment:*

**· Base rate const (M(OH-)-hr)$^{-1}$:**
*Comment:*

# Benzoyl chloride

CAS Registry Number: 98-88-4

## Half-lives:
· **Soil:**               High:    150 seconds          (2.5 minutes)
                          Low:      17 seconds
*Comment:* Scientific judgement based upon hydrolysis half-lives (low $t_{1/2}$: Mabey, W and Mill, T (1978); High $t_{1/2}$: Butler, AR and Gold, V (1962A)).

· **Air:**                High:    1024 hours           (42.7 days)
                          Low:      102 hours           (4.3 days)
*Comment:* Scientific judgement based upon estimated photooxidation half-life in air.

· **Surface Water:**      High:    150 seconds          (2.5 minutes)
                          Low:      17 seconds
*Comment:* Scientific judgement based upon hydrolysis half-lives (low $t_{1/2}$: Mabey and W, Mill, T (1978); High $t_{1/2}$: Butler, AR and Gold, V (1962A)).

· **Ground Water:**       High:    150 seconds          (2.5 minutes)
                          Low:      17 seconds
*Comment:* Scientific judgement based upon hydrolysis half-lives (low $t_{1/2}$: Mabey, W and Mill, T (1978); High $t_{1/2}$: Butler, AR and Gold, V (1962A)).

## Aqueous Biodegradation (unacclimated):
· **Aerobic half-life:**          High:    672 hours    (4 weeks)
                                  Low:     168 hours    (7 days)
*Comment:* Scientific judgement.

· **Anaerobic half-life:**        High:    2688 hours   (16 weeks)
                                  Low:     672 hours    (28 days)
*Comment:* Scientific judgement based upon aerobic aqueous biodegradation half-life.

· **Removal/secondary treatment:**   High:    No data
                                     Low:
*Comment:*

## Photolysis:
· **Atmos photol half-life:**     High:    No data
                                  Low:
*Comment:*

· **Max light absorption (nm):**     lambda max = 293.0, 282.9, 249.9, 246.6, 242.0

*Comment:* UV absorbance measured in methyl cyclohexane (Shashidar, MA (1971)).

· **Aq photol half-life:**           High:      No data
                                     Low:

*Comment:*

## Photooxidation half-life:
· **Water:**                       High:      No data
                                     Low:

*Comment:*

· **Air:**                         High:      1024 hours        (42.7 days)
                                     Low:       102 hours         (4.3 days)
*Comment:* Scientific judgement based upon an estimated rate constant for vapor phase reaction with hydroxyl radicals in air (Atkinson, R (1987A)).

## Reduction half-life:
                                     High:      No data
                                     Low:

*Comment:*

## Hydrolysis:
· **First-order hydr half-life:**       17 seconds
*Comment:* Scientific judgement based upon measured rate data at pH of 7 ($k_N = 4.2 \times 10^{-2}$ s$^{-1}$) extrapolated to 25 °C (Mabey, W and Mill, T (1978)).

· **First-order hydr half-life:**       150 seconds
*Comment:* Scientific judgement based upon measured rate data at pH of 7 and 25 °C ($k_N = 4.62 \times 10^{-3}$ s$^{-1}$) for a 25.6 M water-dioxan solution (Butler, AR and Gold, V (1962A)).

· **Acid rate const (M(H+)-hr)$^{-1}$:**       No data
*Comment:*

· **Base rate const (M(OH-)-hr)$^{-1}$:**       No data
*Comment:*

327

# Nitrobenzene

**CAS Registry Number:** 98-95-3

**Half-lives:**

  · **Soil:**                     High:    4728 hours        (6.6 months)
                                  Low:     322 hours        (13.4 days)

*Comment:* Based upon aerobic soil column biodegradation study data (Kincannon, DF and Lin, YS (1985)).

  · **Air:**                      High:    5.44 hours
                                    Low:     0.544 hours

*Comment:* Based upon photooxidation half-life in air.

  · **Surface Water:**         High:    4728 hours        (6.6 months)
                                  Low:     322 hours        (13.4 days)

*Comment:* Scientific judgement based upon estimated unacclimated aqueous aerobic biodegradation half-life.

  · **Ground Water:**         High:    9456 hours        (13 months)
                                  Low:      48 hours        (2 days)

*Comment:* Scientific judgement based upon estimated unacclimated aqueous anaerobic biodegradation half-life for 2,4-dinitrotoluene (low $t_{1/2}$) and estimated unacclimated aqueous aerobic biodegradation half-life (high $t_{1/2}$).

**Aqueous Biodegradation (unacclimated):**

  · **Aerobic half-life:**       High:    4728 hours        (6.6 months)
                                  Low:     322 hours        (13.4 days)

*Comment:* Based upon aerobic soil column biodegradation study data (Kincannon, DF and Lin, YS (1985)).

  · **Anaerobic half-life:**     High:    300 hours         (13 days)
                                    Low:      48 hours        (2 days)

*Comment:* Scientific judgement based upon anaerobic natural water die-away test data for 2,4-dinitrotoluene (Spanggord, RJ et al. (1980)). Reduction of the nitro group observed in these studies may not be due to biodegradation.

  · **Removal/secondary treatment:**    High:    97.8%
                                    Low:     76%

*Comment:* Removal percentages based upon data from continuous activated sludge biological treatment simulators (Kincannon, DF et al. (1983); Stover, EL and Kincannon, DF (1983)).

**Photolysis:**

· **Atmos photol half-life:**    High:    4800 hours          (200 days)
                                 Low:     1608 hours          (67 days)
*Comment:* Scientific judgement based upon measured photolysis rate constant in distilled water under midday sun at 40°N latitude (Simmons, MS and Zepp, RG (1986)) adjusted for sunlight intensity in summer and winter (Lyman, WJ et al. (1982)).

· **Max light absorption (nm):**    lambda max = 259 nm (light petroleum solvent)
*Comment:* Absorption extends to approximately 395 nm (Mabey, WR et al. (1981)).

· **Aq photol half-life:**    High:    4800 hours          (200 days)
                              Low:     1608 hours          (67 days)
*Comment:* Scientific judgement based upon measured photolysis rate constant in distilled water under midday sun at 40°N latitude (Simmons, MS and Zepp, RG (1986)) adjusted for sunlight intensity in summer and winter (Lyman, WJ et al. (1982)).

## Photooxidation half-life:
· **Water:**    High: $1.9 \times 10^5$ hours          (22 years)
                Low:     3009 hours          (125 days)
*Comment:* Based upon measured rate constant for reaction with hydroxyl radical in water (Dorfman, LM and Adams, GE (1973); Anbar, M and Neta, P (1967)).

· **Air:**    High:    5.44 hours
              Low:     0.544 hours
*Comment:* Based upon measured rate constant for reaction with hydroxyl radical in air (Atkinson, R et al. (1987A)).

## Reduction half-life:    High:    No data
                          Low:
*Comment:*

## Hydrolysis:
· **First-order hydr half-life:**
*Comment:*

· **Acid rate const (M(H+)-hr)$^{-1}$:**    No hydrolyzable groups
*Comment:*

· **Base rate const (M(OH-)-hr)$^{-1}$:**
*Comment:*

# 5-Nitro-o-toluidine

<u>CAS Registry Number:</u>  99-55-8

<u>Structure:</u>

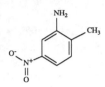

<u>Half-lives:</u>
   · **Soil:**                                     High:      672 hours          (4 weeks)
                                                   Low:       24 hours           (1 day)
   *Comment:*  Scientific judgement based upon estimated unacclimated aqueous aerobic
   biodegradation half-life.

   · **Air:**                                      High:      19.8 hours
                                                   Low:       1.98 hours
   *Comment:*  Based upon estimated photooxidation half-life in air.

   · **Surface Water:**                            High:      672 hours          (4 weeks)
                                                   Low:       24 hours           (1 day)
   *Comment:*  Scientific judgement based upon estimated unacclimated aqueous aerobic
   biodegradation half-life.

   · **Ground Water:**                             High:      1344 hours         (8 weeks)
                                                   Low:       48 hours           (2 days)
   *Comment:*  Scientific judgement based upon estimated unacclimated aqueous anaerobic (low $t_{1/2}$)
   and aerobic (high $t_{1/2}$) biodegradation half-lives.

<u>Aqueous Biodegradation (unacclimated):</u>
   · **Aerobic half-life:**                        High:      672 hours          (4 weeks)
                                                   Low:       24 hours           (1 day)
   *Comment:*  Scientific judgement.

   · **Anaerobic half-life:**                      High:      240 hours          (10 days)
                                                   Low:       48 hours           (2 days)
   *Comment:*  Scientific judgement based upon anaerobic natural water die-away test data for 2,4-
   dinitrotoluene (Spanggord, RJ et al. (1980)).  Reduction of the nitro group observed in these
   studies may not be due to biodegradation.

   · **Removal/secondary treatment:**              High:      No data
                                                   Low:

*Comment:*

## Photolysis:
· **Atmos photol half-life:**        High:
                                                 Low:

*Comment:*

· **Max light absorption (nm):**      No data
*Comment:*

· **Aq photol half-life:**           High:
                                                 Low:

*Comment:*

## Photooxidation half-life:
· **Water:**                       High:     3480 hours            (145 days)
                                         Low:      62.4 hours            (2.6 days)
*Comment:* Scientific judgement based upon estimated rate constants for reactions of representative aromatic amines with ·OH and $RO_2$· (Mill, T and Mabey, W (1985)).

· **Air:**                             High:     19.8 hours
                                         Low:      1.98 hours
*Comment:* Scientific judgement based upon estimated rate constant for reaction with hydroxyl radical in air (Atkinson, R (1987A)).

## Reduction half-life:
                                               High:     No data
                                               Low:
*Comment:*

## Hydrolysis:
· **First-order hydr half-life:**
*Comment:*

· **Acid rate const $(M(H+)-hr)^{-1}$:**      No hydrolyzable groups
*Comment:*

· **Base rate const $(M(OH-)-hr)^{-1}$:**
*Comment:*

# 5-Nitro-o-anisidine

<u>**CAS Registry Number:**</u>  99-59-2

<u>**Structure:**</u>

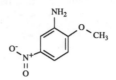

<u>**Half-lives:**</u>
- **Soil:**               High:      672 hours          (4 weeks)
                          Low:       24 hours           (1 day)
  *Comment:* Scientific judgement based upon estimated unacclimated aqueous aerobic biodegradation half-life.

- **Air:**                High:      21.7 hours
                          Low:       2.17 hours
  *Comment:* Based upon estimated photooxidation half-life in air.

- **Surface Water:**      High:      672 hours          (4 weeks)
                          Low:       24 hours           (1 day)
  *Comment:* Scientific judgement based upon estimated unacclimated aqueous aerobic biodegradation half-life.

- **Ground Water:**       High:      1344 hours         (8 weeks)
                          Low:       48 hours           (2 days)
  *Comment:* Scientific judgement based upon estimated unacclimated aqueous anaerobic (low $t_{1/2}$) and aerobic (high $t_{1/2}$) biodegradation half-lives.

<u>**Aqueous Biodegradation (unacclimated):**</u>
- **Aerobic half-life:**  High:      672 hours          (4 weeks)
                          Low:       24 hours           (1 day)
  *Comment:* Scientific judgement based upon aqueous aerobic screening test data for 5-nitro-o-anisidine (Howard, PH et al. (1976)) and nitrobenzene (Pitter, P (1976); Hallas, LE and Alexander, M (1983); (Kawasaki, M (1980); Urano, K and Kato, Z (1986)), and based upon surface water die-away test data for aniline (Subba-Rao, RV et al. (1982); Kondo, M (1978)).

- **Anaerobic half-life:** High:     240 hours          (10 days)
                          Low:       48 hours           (2 days)
  *Comment:* Scientific judgement based upon anaerobic natural water die-away test data for 2,4-dinitrotoluene (Spanggord, RJ et al. (1980)).  Reduction of the nitro group observed in these studies may not be due to biodegradation.

332

**· Removal/secondary treatment:**  High:  No data

Low:

*Comment:*

## Photolysis:

**· Atmos photol half-life:**  High:  No data

Low:

*Comment:*

**· Max light absorption (nm):**  lambda max = 374, 303, 257, 227 nm
*Comment:* Howard, PH et al. (1976)

**· Aq photol half-life:**  High:  No data

Low:

*Comment:*

## Photooxidation half-life:

**· Water:**  High:  3480 hours  (145 days)

Low:  62.4 hours  (2.6 days)

*Comment:* Scientific judgement based upon estimated rate constants for reactions of representative aromatic amines with ·OH and $RO_2$· (Mill, T and Mabey, W (1985)).

**· Air:**  High:  21.7 hours

Low:  2.17 hours

*Comment:* Based upon estimated rate constant for reaction with hydroxyl radicals in air (Atkinson, R (1987A)).

## Reduction half-life:  High:  No data

Low:

*Comment:*

## Hydrolysis:

**· First-order hydr half-life:**  Not expected to be significant.
*Comment:*

**· Acid rate const $(M(H+)-hr)^{-1}$:**
*Comment:*

**· Base rate const $(M(OH-)-hr)^{-1}$:**
*Comment:*

# 1,3-Dinitrobenzene

<u>CAS Registry Number:</u> 99-65-0

<u>Half-lives:</u>
· **Soil:**                           High:     4320 hours            (6 months)
                                       Low:       672 hours            (4 weeks)
*Comment:* Scientific judgement based upon estimated unacclimated aqueous aerobic biodegradation half-life.

· **Air:**                            High:      720 hours            (30 days)
                                       Low:       554 hours            (23 days)
*Comment:* Based upon measured rate of photolysis in distilled water under sunlight (low $t_{1/2}$: Simmons, MS and Zepp, RG (1986); high $t_{1/2}$: Spanggord, RJ et al. (1980)).

· **Surface Water:**                  High:      720 hours            (30 days)
                                       Low:       554 hours            (23 days)
*Comment:* Based upon measured rate of photolysis in distilled water under sunlight (low $t_{1/2}$: Simmons, MS and Zepp, RG (1986); high $t_{1/2}$: Spanggord, RJ et al. (1980)).

· **Ground Water:**                   High:     8640 hours            (12 months)
                                       Low:        48 hours            (2 days)
*Comment:* Scientific judgement based upon estimated unacclimated aqueous aerobic (high $t_{1/2}$) and anaerobic (low $t_{1/2}$) biodegradation half-lives.

<u>Aqueous Biodegradation (unacclimated):</u>
· **Aerobic half-life:**              High:     4320 hours            (6 months)
                                       Low:       672 hours            (4 weeks)
*Comment:* Scientific judgement based upon aerobic natural water die-away test data (Mitchell, WR and Dennis, WH Jr (1982)).

· **Anaerobic half-life:**            High:      300 hours            (13 days)
                                       Low:        48 hours            (2 days)
*Comment:* Scientific judgement based upon anaerobic natural water die-away test data for 2,4-dinitrotoluene (Spanggord, RJ et al. (1980)). Reduction of the nitro group observed in these studies may not be due to biodegradation.

· **Removal/secondary treatment:**    High:      No data
                                       Low:
*Comment:*

<u>Photolysis:</u>
· **Atmos photol half-life:**         High:      720 hours            (30 days)

334

Low:     554 hours        (23 days)

*Comment:* Based upon measured rate of photolysis in distilled water under sunlight (low $t_{1/2}$: Simmons, MS and Zepp, RG (1986); high $t_{1/2}$: Spanggord, RJ et al. (1980)).

· **Max light absorption (nm):**     No data
*Comment:*

· **Aq photol half-life:**     High:    720 hours     (30 days)
                                Low:    554 hours     (23 days)

*Comment:* Based upon measured rate of photolysis in distilled water under sunlight (low $t_{1/2}$: Simmons, MS and Zepp, RG (1986); high $t_{1/2}$: Spanggord, RJ et al. (1980)).

## Photooxidation half-life:
· **Water:**     High:    No data
               Low:
*Comment:*

· **Air:**     High: $2.1 \times 10^4$ hours     (2.37 years)
          Low:    2077 hours     (86.5 days)

*Comment:* Scientific judgement based upon estimated rate constant for reaction with hydroxyl radical in air (Atkinson, R (1987A)).

## Reduction half-life:
                            High:    No data
                            Low:
*Comment:*

## Hydrolysis:
· **First-order hydr half-life:**
*Comment:*

· **Acid rate const (M(H+)-hr)$^{-1}$:**     No hydrolyzable groups
*Comment:*

· **Base rate const (M(OH-)-hr)$^{-1}$:**
*Comment:*

# 4-Nitrophenol

<u>CAS Registry Number:</u>  100-02-7

<u>Half-lives:</u>

 · **Soil:**                           High:      29 hours              (1.2 days)
                                       Low:       17 hours
*Comment:*  Based upon aerobic soil-die away test data (Loekke, H (1985)).


 · **Air:**                            High:      145 hours             (6 days)
                                       Low:       3.1 hours
*Comment:*  Scientific judgement based upon estimated photooxidation half-life in air (high $t_{1/2}$) and measured photolysis half-life in water (low $t_{1/2}$).


 · **Surface Water:**                  High:      168 hours             (7 days)
                                       Low:       18.2 hours
*Comment:*  Scientific judgement based upon estimated unacclimated aqueous aerobic biodegradation half-life.


 · **Ground Water:**                   High:      235 hours             (9.8 days)
                                       Low:       36.4 hours            (1.5 days)
*Comment:*  Based upon pond die-away test data (low $t_{1/2}$) (Paris, DF et al. (1983)) and anaerobic die-away data in two different flooded soils (high $t_{1/2}$) (Sudhakar-Barik, R and Sethunathan, N (1978A)).


<u>Aqueous Biodegradation (unacclimated):</u>

 · **Aerobic half-life:**              High:      168 hours             (7 days)
                                       Low:       18.2 hours
*Comment:*  Based upon pond die-away test data (low $t_{1/2}$: Paris, DF et al. (1983); high $t_{1/2}$: Bourquin, AW (1984)).


 · **Anaerobic half-life:**            High:      235 hours             (9.8 days)
                                       Low:       163 hours             (6.8 days)
*Comment:*  Based upon anaerobic die-away data in two different flooded soils (Sudhakar-Barik, R and Sethunathan, N (1978A)).


 · **Removal/secondary treatment:**    High:      100%
                                       Low:       90%
*Comment:*  Removal percentages based upon data from continuous or semicontinuous activated sludge biological treatment simulators (Means, JL and Anderson, SJ (1981); Gericke, P and Fischer, WK; (1979); Wilderer, P (1981)).


<u>Photolysis:</u>

336

**· Atmos photol half-life:**          High:     329 hours          (13.7 days)
                                        Low:      3.1 hours

*Comment:* Based upon measured rate of sunlight photolysis at pH 9 (high $t_{1/2}$) (Hustert, K et al. (1981)) and pH 4 (low $t_{1/2}$) (Lemaire, J et al. (1985)).

**· Max light absorption (nm):**          lambda max = 227, 310 nm

*Comment:* Absorption extends to approximately 400 nm (Callahan, MA et al. (1979A) and Sadtler UV No. 1684)).

**· Aq photol half-life:**          High:     329 hours          (13.7 days)
                                    Low:      3.1 hours

*Comment:* Based upon measured rate of sunlight photolysis at pH 9 (high $t_{1/2}$) (Hustert, K et al. (1981)) and pH 4 (low $t_{1/2}$) (Lemaire, J et al. (1985)).

## Photooxidation half-life:

**· Water:**          High: $4.9 \times 10^4$ hours          (5.6 years)
                      Low:      642 hours          (21 days)

*Comment:* Based upon measured rate constants for reactions in water with hydroxyl radical (Dorfman, LM and Adams, GE (1973)) and singlet oxygen (Scully, FE Jr and Hoigne, J (1987)).

**· Air:**          High:     145 hours          (6.0 days)
                    Low:      14.5 hours

*Comment:* Scientific judgement based upon estimated rate constant for reaction with hydroxyl radical in air (Atkinson, R (1987A)).

## Reduction half-life:          High:          No data
                                 Low:

*Comment:*

## Hydrolysis:

**· First-order hydr half-life:**
*Comment:*

**· Acid rate const $(M(H+)-hr)^{-1}$:**          No hydrolyzable groups
*Comment:*

**· Base rate const $(M(OH-)-hr)^{-1}$:**
*Comment:*

# Terephthalic acid

**CAS Registry Number:** 100-21-0

**Half-lives:**
   · **Soil:**                                  High:     168 hours           (7 days)
                                                   Low:      24 hours            (1 day)

*Comment:* Scientific judgement based upon estimated unacclimated aqueous aerobic biodegradation half-life.

   · **Air:**                                     High:     2326 hours       (97 days)
                                                   Low:     232.6 hours    (9.7 days)

*Comment:* Based upon estimated photooxidation half-life in air.

   · **Surface Water:**                      High:     168 hours           (7 days)
                                                     Low:      24 hours            (1 day)

*Comment:* Scientific judgement based upon estimated unacclimated aqueous aerobic biodegradation half-life.

   · **Ground Water:**                      High:     336 hours         (14 days)
                                                     Low:      48 hours           (2 days)

*Comment:* Scientific judgement based upon estimated unacclimated aqueous aerobic biodegradation half-life.

**Aqueous Biodegradation (unacclimated):**
   · **Aerobic half-life:**                  High:     168 hours           (7 days)
                                                     Low:      24 hours            (1 day)

*Comment:* Scientific judgement based upon aqueous screening test data (Gerike, P and Fischer, WK (1979); Sasaki, S (1978)).

   · **Anaerobic half-life:**              High:     672 hours         (28 days)
                                                     Low:      96 hours           (4 days)

*Comment:* Scientific judgement based upon estimated unacclimated aqueous aerobic biodegradation half-life.

   · **Removal/secondary treatment:**     High:         93%
                                                     Low:

*Comment:* Removal percentages based upon continuous activated sludge treatment simulator data (Gerike, P and Fischer, WK (1979)).

**Photolysis:**
   · **Atmos photol half-life:**           High:     No data
                                                    Low:

*Comment:*

**· Max light absorption (nm):**       lambda max = 285.5, 240.5 nm (1,4- dioxane)
*Comment:*  Sadtler UV No. 2382.

**· Aq photol half-life:**          High:      No data
                                     Low:
*Comment:*

**Photooxidation half-life:**
**· Water:**                   High:      No data
                                     Low:
*Comment:*

**· Air:**                      High:   2326 hours        (97 days)
                                Low:    232.6 hours     (9.7 hours)
*Comment:*  Based upon estimated rate constant for reaction with hydroxyl radicals in air
(Atkinson, R (1987A)).

**Reduction half-life:**           High:      No data
                                     Low:
*Comment:*

**Hydrolysis:**
**· First-order hydr half-life:**      No hydrolyzable groups
*Comment:*

**· Acid rate const (M(H+)-hr)$^{-1}$:**
*Comment:*

**· Base rate const (M(OH-)-hr)$^{-1}$:**
*Comment:*

# Ethylbenzene

CAS Registry Number: 100-41-4

Half-lives:
 · Soil:                                    High:     240 hours          (10 days)
                                            Low:       72 hours          (3 days)
 Comment: Scientific judgement based upon unacclimated aqueous aerobic biodegradation half-life.

 · Air:                                     High:     85.6 hours         (3.57 days)
                                            Low:      8.56 hours
 Comment: Based upon photooxidation half-life in air.

 · Surface Water:                           High:     240 hours          (10 days)
                                            Low:       72 hours          (3 days)
 Comment: Scientific judgement based upon unacclimated aqueous aerobic biodegradation half-life.

 · Ground Water:                            High:     5472 hours         (228 days)
                                            Low:      144 hours          (6 days)
 Comment: Scientific judgement based upon sea water die-away test data (high $t_{1/2}$) (Van der Linden, AC (1978)) and upon unacclimated aqueous aerobic biodegradation half-life.

Aqueous Biodegradation (unacclimated):
 · Aerobic half-life:                       High:     240 hours          (10 days)
                                            Low:       72 hours          (3 days)
 Comment: Scientific judgement based upon sea water die-away test data (Van der Linden, AC (1978)).

 · Anaerobic half-life:                     High:     5472 hours         (228 days)
                                            Low:      4224 hours         (176 days)
 Comment: Scientific judgement based upon anaerobic ground water die-away test data (Wilson, BH et al. (1986)).

 · Removal/secondary treatment:             High:     95%
                                            Low:      72%
 Comment: Removal percentages based upon data from continuous activated sludge biological treatment simulators (Hannah, SA et al. (1986)).

Photolysis:
 · Atmos photol half-life:                  High:     No data
                                            Low:

*Comment:*

· **Max light absorption (nm):**          lambda max = 269.5, 260.5, 254.5, 208 nm (methanol).
*Comment:* Absorption extends to approximately 295 nm (Sadtler UV No. 97).

· **Aq photol half-life:**          High:          No data
                                          Low:
*Comment:*

**Photooxidation half-life:**
· **Water:**          High:          No data
                             Low:
*Comment:*

· **Air:**          High:          85.6 hours          (3.57 days)
                       Low:          8.56 hours
*Comment:* Scientific judgement based upon measured rate for reaction with hydroxyl radical in air (Atkinson, R (1985)).

**Reduction half-life:**          High:          No data
                                           Low:
*Comment:*

**Hydrolysis:**
· **First-order hydr half-life:**
*Comment:*

· **Acid rate const (M(H+)-hr)$^{-1}$:**          No hydrolyzable groups
*Comment:*

· **Base rate const (M(OH-)-hr)$^{-1}$:**
*Comment:*

341

# Styrene

<u>CAS Registry Number:</u>  100-42-5

<u>Half-lives:</u>
| | | | |
|---|---|---|---|
| · **Soil:** | High: | 672 hours | (4 weeks) |
| | Low: | 336 hours | (2 weeks) |

*Comment:* Scientific judgement based upon unacclimated grab samples of aerobic soil (low $t_{1/2}$) (Sielicki, M et al. (1978) and acclimated aqueous screening test data (Bridie, et al. (1979)).

| | | | |
|---|---|---|---|
| · **Air:** | High: | 7.3 hours | |
| | Low: | 0.9 hours | (51 minutes) |

*Comment:* Based upon photooxidation half-life in air.

| | | | |
|---|---|---|---|
| · **Surface Water:** | High: | 672 hours | (4 weeks) |
| | Low: | 336 hours | (2 weeks) |

*Comment:* Scientific judgement based upon estimated unacclimated aqueous aerobic biodegradation half-life.

| | | | |
|---|---|---|---|
| · **Ground Water:** | High: | 5040 hours | (30 weeks) |
| | Low: | 672 hours | (4 weeks) |

*Comment:* Scientific judgement based upon estimated unacclimated aqueous aerobic biodegradation half-life and a subsurface sample (high $t_{1/2}$: Wilson, JT et al. (1983)).

<u>Aqueous Biodegradation (unacclimated):</u>
| | | | |
|---|---|---|---|
| · **Aerobic half-life:** | High: | 672 hours | (4 weeks) |
| | Low: | 336 hours | (2 weeks) |

*Comment:* Scientific judgement based upon unacclimated grab samples of aerobic soil (low $t_{1/2}$) (Sielicki, M et al. (1978) and acclimated aqueous screening test data (Bridie, et al. (1979)).

| | | | |
|---|---|---|---|
| · **Anaerobic half-life:** | High: | 2688 hours | (16 weeks) |
| | Low: | 1344 hours | (8 weeks) |

*Comment:* Scientific judgement based upon estimated unacclimated aqueous aerobic biodegradation half-life.

| | | | |
|---|---|---|---|
| · **Removal/secondary treatment:** | High: | 99% | |
| | Low: | 8% | |

*Comment:* Based upon % degraded under aerobic (high %) and anaerobic (low % ) continuous flow conditions (Bouwer, EJ and McCarty, PL (1983)).

<u>Photolysis:</u>
| | | |
|---|---|---|
| · **Atmos photol half-life:** | High: | |
| | Low: | |

*Comment:*

· **Max light absorption (nm):**     No data
*Comment:*

· **Aq photol half-life:**     High:
     Low:

*Comment:*

## Photooxidation half-life:
· **Water:**     High:     No data
     Low:

*Comment:*

· **Air:**     High:     7.3 hours
     Low:     0.9 hours     (51 minutes)
*Comment:* Based upon measured rate data for hydroxyl radicals (Atkinson, R (1985)) and ozone in air (Atkinson, R and Carter, WPL (1984)).

## Reduction half-life:
     High:     No data
     Low:
*Comment:*

## Hydrolysis:
· **First-order hydr half-life:**     No hydrolyzable groups
*Comment:*

· **Acid rate const (M(H+)-hr)$^{-1}$:**
*Comment:*

· **Base rate const (M(OH-)-hr)$^{-1}$:**
*Comment:*

# Benzyl chloride

CAS Registry Number:  100-44-7

Half-lives:
· **Soil:**                                         High:       290 hours                    (12.1 days)
                                                    Low:         15 hours
*Comment:* Scientific judgement based upon estimated hydrolysis half-lives (low $t_{1/2}$: Mabey W and Mill T (1978); high $t_{1/2}$: Ohnishi, R and Tanabe, K (1971)).

· **Air:**                                          High:       218 hours                    (9.1 days)
                                                    Low:         22 hours
*Comment:* Based upon photooxidation half-life in air.

· **Surface Water:**                                High:       290 hours                    (12.1 days)
                                                    Low:         15 hours
*Comment:* Scientific judgement based upon estimated hydrolysis half-lives (low $t_{1/2}$: Mabey W and Mill T (1978); high $t_{1/2}$: Ohnishi, R and Tanabe, K (1971)).

· **Ground Water:**                                 High:       290 hours                    (12.1 days)
                                                    Low:         15 hours
*Comment:* Scientific judgement based upon estimated hydrolysis half-lives (low $t_{1/2}$: Mabey W and Mill T (1978); high $t_{1/2}$: Ohnishi, R and Tanabe, K (1971)).

Aqueous Biodegradation (unacclimated):
· **Aerobic half-life**:                            High:       672 hours                    (4 weeks)
                                                    Low:        168 hours                    (7 days)
*Comment:* Scientific judgement based upon unacclimated aqueous screening test data (Sasaki, S (1978)).

· **Anaerobic half-life**:                          High:      2688 hours                    (16 weeks)
                                                    Low:        672 hours                    (4 weeks)
*Comment:* Scientific judgement based upon estimated aqueous aerobic biodegradation half-life.

· **Removal/secondary treatment:**                  High:       No data
                                                    Low:
*Comment:*

Photolysis:
· **Atmos photol half-life:**                       High:
                                                    Low:
*Comment:*

· **Max light absorption (nm):**          No data
*Comment:*

· **Aq photol half-life:**          High:
                                    Low:
*Comment:*

**Photooxidation half-life:**
   · **Water:**                     High:     No data
                                    Low:
*Comment:*

   · **Air:**                       High:     218 hours          (9.1 days)
                                    Low:      22 hours
*Comment:* Based upon measured rate constant for vapor phase reaction with hydroxyl radicals in air (Edney, EO et al. (1986)).

**Reduction half-life:**             High:     No data
                                    Low:
*Comment:*

**Hydrolysis:**
   · **First-order hydr half-life:**          15 hours
*Comment:* Based upon overall rate constant ($k_h$ = 1.28X10$^{-5}$) at pH 7 and 25 °C (Mabey W and Mill T (1978)).

   · **First-order hydr half-life:**          290 hours          (12.1 days)
*Comment:* Based upon overall rate constant ($k_h$ = 3.98X10$^{-5}$) at pH 7 and 5 °C (Ohnishi, R and Tanabe, K (1971)).

   · **Acid rate const (M(H+)-hr)$^{-1}$:**          $k_h$ = 1.28X10$^{-5}$
*Comment:* Based upon overall rate constant that is independent of pH (Mabey W and Mill T (1978)).

   · **Base rate const (M(OH-)-hr)$^{-1}$:**          $k_h$ = 1.28X10$^{-5}$
*Comment:* Based upon overall rate constant that is independent of pH (Mabey W and Mill, T (1978)).

# N-Nitrosopiperidine

<u>CAS Registry Number:</u>  100-75-4

<u>Structure:</u>

$$N\!\!\nearrow^{\!\!O}$$

<u>Half-lives:</u>
  · **Soil:**                                       High:     4320 hours              (6 months)
                                                    Low:       672 hours              (4 weeks)
*Comment:* Based upon aerobic soil die-away data for N-nitrosodimethylamine and N-nitrosodipropylamine (high $t_{1/2}$: Tate, RL III and Alexander, M (1975); low $t_{1/2}$: Oliver, JE et al. (1979)).

  · **Air:**                                        High:     24.9 hours
                                                    Low:       2.49 hours
*Comment:* Scientific judgement based upon estimated photooxidation half-life in air.

  · **Surface Water:**                              High:      225 hours              (9.4 days)
                                                    Low:       2.25 hours
*Comment:* Scientific judgement based upon measured photolysis in water irradiated with artificial light at >290 nm (Challis, BC and Li, BFL (1982)).

  · **Ground Water:**                               High:     8640 hours              (12 months)
                                                    Low:      1344 hours              (8 weeks)
*Comment:* Based upon aerobic soil die-away data for N-nitrosodimethylamine and N-nitrosodipropylamine (high $t_{1/2}$: Tate, RL III and Alexander, M (1975); low $t_{1/2}$: Oliver, JE et al. (1979)).

<u>Aqueous Biodegradation (unacclimated):</u>
  · **Aerobic half-life**:                          High:     4320 hours              (6 months)
                                                    Low:       672 hours              (4 weeks)
*Comment:* Based upon aerobic soil die-away data for N-nitrosodimethylamine and N-nitrosodipropylamine (high $t_{1/2}$: Tate, RL III and Alexander, M (1975); low $t_{1/2}$: Oliver, JE et al. (1979)).

  · **Anaerobic half-life**:                        High:    17280 hours              (24 months)
                                                    Low:      2688 hours              (16 weeks)
*Comment:* Scientific judgement based upon estimated unacclimated aqueous aerobic biodegradation half-lives.

346

**· Removal/secondary treatment:**     High:     No data
                                            Low:

*Comment:*

## Photolysis:

**· Atmos photol half-life:**     High:     225 hours         (9.4 days)
                                            Low:     2.25 hours

*Comment:* Scientific judgement based upon measured photolysis in water irradiated with artificial light at >290 nm (Challis, BC and Li, BFL (1982)).

**· Max light absorption (nm):**     lambda max = 235, 337 nm (water)
*Comment:* IARC (1978).

**· Aq photol half-life:**     High:     225 hours         (9.4 days)
                                            Low:     2.25 hours

*Comment:* Scientific judgement based upon measured photolysis in water irradiated with artificial light at >290 nm (Challis, BC and Li, BFL (1982)).

## Photooxidation half-life:

**· Water:**     High:     No data
                                          Low:

*Comment:*

**· Air:**     High:     24.9 hours
                                          Low:     2.49 hours

*Comment:* Scientific judgement based upon estimated rate constant for reaction with hydroxyl radical in air (Atkinson, R (1987A)).

## Reduction half-life:

                                          High:     No data
                                          Low:

*Comment:*

## Hydrolysis:

**· First-order hydr half-life:**     Not expected to be significant
*Comment:* IARC (1978).

**· Acid rate const (M(H+)-hr)$^{-1}$:**     No data
*Comment:*

**· Base rate const (M(OH-)-hr)$^{-1}$:**     No data
*Comment:*

# 4,4'-Methylenebis(2-chloroaniline)

CAS Registry Number: 101-14-4

## Half-lives:
  · **Soil:**                            High:    4320 hours         (6 months)
                                      Low:     672 hours          (4 weeks)
*Comment:* Scientific judgement based upon unacclimated aqueous aerobic biodegradation half-life.

  · **Air:**                              High:    2.90 hours
                                        Low:    0.290 hours
*Comment:* Scientific judgement based upon estimated photooxidation half-life in air.

  · **Surface Water:**             High:    1740 hours         (72 days)
                                    Low:      31 hours         (1.3 days)
*Comment:* Scientific judgement based upon estimated photooxidation half-life in water.

  · **Ground Water:**             High:    8640 hours        (12 months)
                                    Low:    1344 hours        (8 weeks)
*Comment:* Scientific judgement based upon unacclimated aqueous aerobic biodegradation half-life.

## Aqueous Biodegradation (unacclimated):
  · **Aerobic half-life:**         High:    4320 hours         (6 months)
                                    Low:     672 hours          (4 weeks)
*Comment:* Scientific judgement based upon unacclimated aqueous aerobic biodegradation screening test data (Fochtman, EG and Eisenberg, W (1979)).

  · **Anaerobic half-life:**       High:   17280 hours      (24 months)
                                    Low:    2688 hours
*Comment:* Scientific judgement based upon unacclimated aqueous aerobic biodegradation half-life.

  · **Removal/secondary treatment:**   High:       100%
                                    Low:       77%
*Comment:* Removal percentages based upon data from a continuous activated sludge biological treatment simulator (Fochtman, EG and Eisenberg, W (1979)).

## Photolysis:
  · **Atmos photol half-life:**       High:    No data
                                    Low:
*Comment:*

**· Max light absorption (nm):**      lambda max = 247, 299 nm (methanol)
*Comment:* Sadtler UV No. 2914.

**· Aq photol half-life:**        High:      No data

                                    Low:

*Comment:*

## Photooxidation half-life:

**· Water:**             High:    1740 hours        (72 days)

                                 Low:     31 hours        (1.3 days)

*Comment:* Scientific judgement based upon photooxidation rate constants with ·OH and $RO_2$· for the aromatic amine class (Mill, T and Mabey, W (1985); Guesten, H et al. (1981)).

**· Air:**                High:    2.90 hours

                                 Low:    0.290 hours

*Comment:* Scientific judgement based upon estimated rate constant for reaction with hydroxyl radical in air (Atkinson, R (1987A)).

## Reduction half-life:

                                  High:      No data

                                  Low:

*Comment:*

## Hydrolysis:

**· First-order hydr half-life:**     No data
*Comment:*

**· Acid rate const $(M(H+)-hr)^{-1}$:**
*Comment:*

**· Base rate const $(M(OH-)-hr)^{-1}$:**
*Comment:*

# Bis-(4-dimethylaminophenyl)methane

<u>CAS Registry Number:</u> 101-61-1

<u>Half-lives:</u>
&middot; **Soil:**                     High:    4320 hours         (6 months)
                            Low:      672  hours         (4 weeks)
*Comment:* Scientific judgement based upon estimated unacclimated aqueous aerobic biodegradation half-life.

&middot; **Air:**                      High:    2.0 hours
                            Low:     0.20 hours
*Comment:* Based upon estimated photooxidation half-life in air.

&middot; **Surface Water:**            High:    2626 hours         (109 days)
                            Low:     26.3 hours         (1.1 days)
*Comment:* Scientific judgement based upon measured rate constant for reaction of dimethylaniline with singlet oxygen in water (Haag, WR and Hoigne, J (1985A)).

&middot; **Ground Water:**             High:    8640 hours         (12 months)
                            Low:     1344 hours         (8 weeks)
*Comment:* Scientific judgement based upon estimated unacclimated aqueous aerobic biodegradation half-life.

<u>Aqueous Biodegradation (unacclimated):</u>
&middot; **Aerobic half-life**:        High:    4320 hours         (6 months)
                            Low:      672 hours         (4 weeks)
*Comment:* Scientific judgement based upon aqueous screening test data for dimethylaniline (Kitano, M (1978); Heukelekian, H and Rand, MC (1955)).

&middot; **Anaerobic half-life**:      High:    17280 hours        (24 months)
                            Low:     2688 hours         (16 weeks)
*Comment:* Scientific judgement based upon estimated unacclimated aqueous aerobic biodegradation half-life.

&middot; **Removal/secondary treatment:**    High:     No data
                                    Low:
*Comment:*

<u>Photolysis:</u>
&middot; **Atmos photol half-life:**   High:     No data
                            Low:
*Comment:*

350

**· Max light absorption (nm):**
*Comment:*

**· Aq photol half-life:**                High:
                                          Low:
*Comment:*

## Photooxidation half-life:
**· Water:**                              High:     2626 hours              (109 days)
                                          Low:      26.3 hours              (1.1 days)
*Comment:* Scientific judgement based upon measured rate constant for reaction of dimethylaniline with singlet oxygen in water (Haag, WR and Hoigne, J (1985A)).

**· Air:**                                High:     2.0 hours
                                          Low:      0.20 hours
*Comment:* Based upon estimated rate constant for reaction with hydroxyl radicals in air (Atkinson, R (1987A)).

## Reduction half-life:                   High:     No data
                                          Low:
*Comment:*

## Hydrolysis:
**· First-order hydr half-life:**         No hydrolyzable groups
*Comment:*

**· Acid rate const $(M(H+)-hr)^{-1}$:**
*Comment:*

**· Base rate const $(M(OH-)-hr)^{-1}$:**
*Comment:*

# Methylenebis(phenylisocyanate)

**CAS Registry Number:** 101-68-8

**Half-lives:**
- **Soil:**                         High:        24 hours
                                    Low:          6 hours
*Comment:* Scientific judgement based upon disappearance rate in water for methylenebis(phenylisocyanate) (Brochhagen, FK and Grieveson, BM (1984)) and toluene diisocyanate (Duff, PB (1983)).

- **Air:**                          High:        5.8 hours
                                    Low:         0.58 hours
*Comment:* Based upon estimated photooxidation half-life in air.

- **Surface Water:**                High:        24 hours
                                    Low:          6 hours
*Comment:* Scientific judgement based upon disappearance rate in water for methylenebis(phenylisocyanate) (Brochhagen, FK and Grieveson, BM (1984)) and toluene diisocyanate (Duff, PB (1983)).

- **Ground Water:**                 High:        24 hours
                                    Low:          6 hours
*Comment:* Scientific judgement based upon disappearance rate in water for methylenebis(phenylisocyanate) (Brochhagen, FK and Grieveson, BM (1984)) and toluene diisocyanate (Duff, PB (1983)).

**Aqueous Biodegradation (unacclimated):**
- **Aerobic half-life:**            High:        672 hours         (4 weeks)
                                    Low:         168 hours         (7 days)
*Comment:* Scientific judgement.

- **Anaerobic half-life:**          High:        2688 hours        16 weeks)
                                    Low:         672 hours         (28 days)
*Comment:* Scientific judgement based upon estimated unacclimated aqueous aerobic biodegradation half-life.

- **Removal/secondary treatment:**  High:        No data
                                    Low:
*Comment:*

**Photolysis:**
- **Atmos photol half-life:**       High:        No data

Low:

*Comment:*

· **Max light absorption (nm):**          lambda max = 240 nm (methanol)
*Comment:* Sadtler UV No. 24524.

· **Aq photol half-life:**          High:          No data
                                    Low:
*Comment:*

**Photooxidation half-life:**
· **Water:**          High:          No data
                      Low:
*Comment:*

· **Air:**          High:          5.8 hours
                    Low:          0.58 hours
*Comment:*  Based upon estimated rate constant for reaction with hydroxyl radicals in air (Atkinson, R (1987A)).

**Reduction half-life:**          High:          No data
                                  Low:
*Comment:*

**Hydrolysis:**
· **First-order hydr half-life:**          High:          12 hours
                                           Low:          6 hours
*Comment:*  Scientific judgement based upon disappearance rate in water for methylenebis(phenylisocyanate) (Brochhagen, FK and Grieveson, BM (1984)) and toluene diisocyanate (Duff, PB (1983)).

· **Acid rate const (M(H+)-hr)$^{-1}$:**          No data
*Comment:*

· **Base rate const (M(OH-)-hr)$^{-1}$:**          No data
*Comment:*

# 4,4'-Methylenedianiline

**CAS Registry Number:** 101-77-9

## Half-lives:
· **Soil:**  High:  168 hours  (7 days)
  Low:  24 hours  (1 day)
*Comment:* Scientific judgement based upon estimated unacclimated aqueous aerobic biodegradation half-life.

· **Air:**  High:  2.7 hours
  Low:  0.27 hours
*Comment:* Based upon estimated photooxidation half-life in air.

· **Surface Water:**  High:  168 hours  (7 days)
  Low:  24 hours  (1 day)
*Comment:* Scientific judgement based upon estimated unacclimated aqueous aerobic biodegradation half-life.

· **Ground Water:**  High:  336 hours  (14 days)
  Low:  48 hours  (2 days)
*Comment:* Scientific judgement based upon estimated unacclimated aqueous aerobic biodegradation half-life.

## Aqueous Biodegradation (unacclimated):
· **Aerobic half-life:**  High:  168 hours  (7 days)
  Low:  24 hours  (1 day)
*Comment:* Scientific judgement based upon aqueous screening test data for toluidines (Sasaki, S (1978); Pitter, P (1976)).

· **Anaerobic half-life:**  High:  672 hours  (28 days)
  Low:  96 hours  (4 days)
*Comment:* Scientific judgement based upon estimated unacclimated aqueous aerobic biodegradation half-life.

· **Removal/secondary treatment:**  High:  No data
  Low:
*Comment:*

## Photolysis:
· **Atmos photol half-life:**  High:  No data
  Low:
*Comment:*

· **Max light absorption (nm):**
*Comment:*

· **Aq photol half-life:**                               High:
                                                         Low:

*Comment:*

**Photooxidation half-life:**
  · **Water:**                                            High:     3480 hours              (145 days)
                                                          Low:      62.4 hours              (2.6 days)
*Comment:* Scientific judgement based upon estimated rate constants for reactions of representative aromatic amines with ·OH and $RO_2$· (Mill, T and Mabey, W (1985)).

  · **Air:**                                              High:     2.7 hours
                                                          Low:      0.27 hours
*Comment:* Based upon estimated rate constant for reaction with hydroxyl radicals in air (Atkinson, R (1987A)).

**Reduction half-life:**                                  High:     No data
                                                          Low:

*Comment:*

**Hydrolysis:**
  · **First-order hydr half-life:**                       No hydrolyzable groups
*Comment:*

  · **Acid rate const $(M(H+)-hr)^{-1}$:**
*Comment:*

  · **Base rate const $(M(OH-)-hr)^{-1}$:**
*Comment:*

# 4,4'-Diaminodiphenyl ether

<u>CAS Registry Number:</u>  101-80-4

<u>Half-lives:</u>

· **Soil:**          High:    4320 hours          (6 months)

Low:      672 hours          (4 weeks)

*Comment:* Scientific judgement based upon estimated unacclimated aqueous aerobic biodegradation half-life.

· **Air:**          High:      3 hours

Low:      0.3 hours

*Comment:* Based upon estimated photooxidation half-life in air.

· **Surface Water:**          High:    3480 hours          (145 days)

Low:      62.4 hours          (2.6 days)

*Comment:* Scientific judgement based upon estimated rate constants for reactions of representative aromatic amines with ·OH and RO2· (Mill, T and Mabey, W (1985)).

· **Ground Water:**          High:    8640 hours          (12 months)

Low:    1344 hours          (8 weeks)

*Comment:* Scientific judgement based upon estimated unacclimated aqueous aerobic biodegradation half-life.

<u>Aqueous Biodegradation (unacclimated):</u>

· **Aerobic half-life:**          High:    4320 hours          (6 months)

Low:      672 hours          (4 weeks)

*Comment:* Scientific judgement based upon aqueous aerobic screening test data for benzidine (Howard, PH and Saxena, J (1976)) and grab sample test data for diphenyl ether (Ludzack, FJ and Ettinger, MB (1963)).

· **Anaerobic half-life:**          High:    17280 hours          (24 months)

Low:    2688 hours          (16 weeks)

*Comment:* Scientific judgement based upon estimated unacclimated aqueous aerobic biodegradation half-life.

· **Removal/secondary treatment:**          High:    No data

Low:

*Comment:*

<u>Photolysis:</u>

· **Atmos photol half-life:**          High:    No data

Low:

356

*Comment:*

**· Max light absorption (nm):**
*Comment:*

**· Aq photol half-life:**          High:
                                       Low:

*Comment:*

## Photooxidation half-life:
**· Water:**                High:     3480 hours        (145 days)
                                  Low:      62.4 hours        (2.6 days)
*Comment:* Scientific judgement based upon estimated rate constants for reactions of representative aromatic amines with ·OH and $RO_2$· (Mill, T and Mabey, W (1985)). It is assumed that 4,4'-diaminodiphenyl ether reacts twice as fast as aniline.

**· Air:**                  High:     3 hours
                                  Low:      0.3 hours
*Comment:* Based upon estimated rate constant for reaction with hydroxyl radicals in air (Atkinson, R (1987A)).

## Reduction half-life:            High:     No data
                                  Low:
*Comment:*

## Hydrolysis:
**· First-order hydr half-life:**      Not expected to be significant
*Comment:*

**· Acid rate const $(M(H+)-hr)^{-1}$:**
*Comment:*

**· Base rate const $(M(OH-)-hr)^{-1}$:**
*Comment:*

# Bis(2-ethylhexyl)adipate

**CAS Registry Number:** 103-23-1

**Half-lives:**

· **Soil:**         High:    672 hours         (4 weeks)
              Low:     19.2 hours
*Comment:* High value is scientific judgement based upon estimated aqueous aerobic biodegradation half-life. Low value is based upon the estimated low value for aqueous hydrolysis.

· **Air:**          High:    26 hours
              Low:     2.6 hours
*Comment:* Based upon estimated photooxidation half-lives in air.

· **Surface Water:**  High:    672 hours         (4 weeks)
              Low:     19.2 hours
*Comment:* High value is scientific judgement based upon estimated aqueous aerobic biodegradation half-life. Low value is based upon the estimated low value for aqueous hydrolysis.

· **Ground Water:**   High:    1344 hours        (8 weeks)
              Low:     19.2 hours
*Comment:* High value is scientific judgement based upon estimated aqueous aerobic biodegradation half-life. Low value is based upon the estimated low value for aqueous hydrolysis.

**Aqueous Biodegradation (unacclimated):**

· **Aerobic half-life:**   High:    672 hours         (4 weeks)
              Low:     168 hours          (7 days)
*Comment:* Scientific judgement based upon biological screening test data (Saeger, VW and Zucker, ES (1976)).

· **Anaerobic half-life:**  High:    2688 hours        (16 weeks)
              Low:     672 hours          (28 days)
*Comment:* Scientific judgement based upon estimated aqueous aerobic biodegradation half-lives.

· **Removal/secondary treatment:**   High:    No data
              Low:
*Comment:*

**Photolysis:**

· **Atmos photol half-life:**   High:    No data
              Low:
*Comment:*

358

· **Max light absorption (nm):**
*Comment:*

· **Aq photol half-life:**              High:
                                        Low:

*Comment:*

**Photooxidation half-life:**
· **Water:**                           High:      No data
                                        Low:

*Comment:*

· **Air:**                             High:      26 hours
                                        Low:       2.6 hours
*Comment:* Based upon estimated photooxidation rate constant with ·OH (Atkinson, R (1987A)).

**Reduction half-life:**               High:      No data
                                        Low:

*Comment:*

**Hydrolysis:**
· **First-order hydr half-life:**                1920 hours              (80 days)
*Comment:* Scientific judgement at 25°C and pH 7 based upon estimated base rate constant.

· **Acid rate const (M(H+)-hr)$^{-1}$:**     No data
*Comment:*

· **Base rate const (M(OH-)-hr)$^{-1}$:**    1.0
*Comment:* Scientific judgement based upon measured hydrolysis rates for diethyl adipate
(Venkoba Rao, G and Venkatasubramanian, N (1972)). At 25°C and pH 9, the half-life would be
19.2 hours.

# p-Anisidine

<u>CAS Registry Number:</u>  104-94-9

<u>Structure:</u>

<u>Half-lives:</u>
  · **Soil:**                                          High:    672 hours                (4 weeks)
                                                       Low:     168 hours                (7 days)
  *Comment:*  Scientific judgement based upon estimated aqueous aerobic biodegradation half-lives.

  · **Air:**                                           High:    5.3 hours
                                                       Low:     0.53 hours
  *Comment:*  Based upon estimated photooxidation half-lives in air.

  · **Surface Water:**                                 High:    672 hours                (4 weeks)
                                                       Low:     62.4 hours               (2.6 days)
  *Comment:*  The high value is scientific judgement based upon estimated aqueous aerobic biodegradation half-life. The low value is based upon aqueous photooxidation half-life.

  · **Ground Water:**                                  High:    1344 hours               (8 weeks)
                                                       Low:     336 hours                (14 days)
  *Comment:*  Scientific judgement based upon estimated aqueous aerobic biodegradation half-lives.

<u>Aqueous Biodegradation (unacclimated):</u>
  · **Aerobic half-life:**                             High:    672 hours                (4 weeks)
                                                       Low:     168 hours                (7 days)
  *Comment:*  Scientific judgement based upon aerobic biological screening studies (Sasaki, S (1978); Kitano, M (1978); Alexander, M and Lustigman, BK (1966); Brown, D and Laboureur, P (1983A); Kondo, M et al. (1988A)).

  · **Anaerobic half-life:**                           High:    2688 hours               (16 weeks)
                                                       Low:     672 hours                (28 days)
  *Comment:*  Scientific judgement based upon estimated aqueous aerobic biodegradation half-lives.

  · **Removal/secondary treatment:**                   High:    No data
                                                       Low:
  *Comment:*

## Photolysis:
  · **Atmos photol half-life:**    High:  No data
                  Low:

 *Comment:*

  · **Max light absorption (nm):**
 *Comment:*

  · **Aq photol half-life:**     High:
                  Low:

 *Comment:*

## Photooxidation half-life:
  · **Water:**         High:  3480 hours    (145 days)
                  Low:  62.4 hours    (2.6 days)
 *Comment:* Scientific judgement based upon photooxidation rate constants with ·OH and $RO_2$·
for the aromatic amine class (Mill, T and Mabey, W (1985); Guesten, H et al. (1981)).

  · **Air:**          High:  5.3 hours
                  Low:  0.53 hours
 *Comment:* Based upon estimated photooxidation rate constant with ·OH (Atkinson, R (1987A)).

## Reduction half-life:
                High:  No data
                Low:

 *Comment:*

## Hydrolysis:
  · **First-order hydr half-life:**   No hydrolyzable groups
 *Comment:*

  · **Acid rate const $(M(H+)-hr)^{-1}$:**
 *Comment:*

  · **Base rate const $(M(OH-)-hr)^{-1}$:**
 *Comment:*

# 2,4-Dimethylphenol

<u>CAS Registry Number:</u>  105-67-9

<u>Half-lives:</u>
- **Soil:**

| | High: | 168 hours | (7 days) |
| --- | --- | --- | --- |
| | Low: | 24 hours | (1 day) |

*Comment:* Scientific judgement based upon estimated aqueous aerobic biodegradation half-life.

- **Air:**

| | High: | 11.9 hours |
| --- | --- | --- |
| | Low: | 1.19 hours |

*Comment:* Scientific judgement based upon estimated photooxidation half-life in air.

- **Surface Water:**

| | High: | 168 hours | (7 days) |
| --- | --- | --- | --- |
| | Low: | 24 hours | (1 day) |

*Comment:* Scientific judgement based upon estimated aqueous aerobic biodegradation half-life.

- **Ground Water:**

| | High: | 336 hours | (14 days) |
| --- | --- | --- | --- |
| | Low: | 48 hours | (2 days) |

*Comment:* Scientific judgement based upon estimated aqueous aerobic biodegradation half-life.

<u>Aqueous Biodegradation (unacclimated):</u>
- **Aerobic half-life:**

| | High: | 168 hours | (7 days) |
| --- | --- | --- | --- |
| | Low: | 24 hours | (1 day) |

*Comment:* Scientific judgement based upon aqueous aerobic screening test data (Petrasek, AC et al. (1983); Chambers, CW et al. (1963)).

- **Anaerobic half-life:**

| | High: | 672 hours | (28 days) |
| --- | --- | --- | --- |
| | Low: | 96 hours | (4 days) |

*Comment:* Scientific judgement based upon estimated aqueous aerobic biodegradation half-life.

- **Removal/secondary treatment:**

| | High: | 98% |
| --- | --- | --- |
| | Low: | |

*Comment:* Removal percentage based upon data from a continuous activated sludge biological treatment simulator (Petrasek, AC et al. (1983)).

<u>Photolysis:</u>
- **Atmos photol half-life:**

| | High: | No data |
| --- | --- | --- |
| | Low: | |

*Comment:*

- **Max light absorption (nm):**       lambda max = 292 nm (methanol)

*Comment:* Absorption continues to >350 nm (Sadtler UV No. 178).

**· Aq photol half-life:**　　　　　　　　High:　　　No data
　　　　　　　　　　　　　　　　　　　　　Low:
*Comment:*

**Photooxidation half-life:**
　**· Water:**　　　　　　　　　　　　　High:　　　3840 hours　　　　　(160 days)
　　　　　　　　　　　　　　　　　　　　Low:　　　77 hours　　　　　　(3.2 days)
*Comment:* Scientific judgement based upon reported reaction rate constants for $RO_2\cdot$ with the phenol class (Mill, T and Mabey, W (1985)).

　**· Air:**　　　　　　　　　　　　　　High:　　　11.9 hours
　　　　　　　　　　　　　　　　　　　　Low:　　　1.19 hours
*Comment:* Scientific judgement based upon estimated rate constant for reaction with hydroxyl radical in air (Atkinson, R (1987A)).

**Reduction half-life:**　　　　　　　　High:　　　No data
　　　　　　　　　　　　　　　　　　　　Low:
*Comment:*

**Hydrolysis:**
　**· First-order hydr half-life:**
　*Comment:*

　**· Acid rate const $(M(H+)\text{-}hr)^{-1}$:**　　　No hydrolyzable groups
　*Comment:*

　**· Base rate const $(M(OH\text{-})\text{-}hr)^{-1}$:**
　*Comment:*

# p-Xylene

**CAS Registry Number:** 106-42-3

**Structure:**

## Half-lives:
**· Soil:** High: 672 hours (4 weeks)

Low: 168 hours (1 week)

*Comment:* Scientific judgement based upon estimated aqueous aerobic biodegradation half-life.

**· Air:** High: 42 hours (1.7 days)

Low: 4.2 hours

*Comment:* Based upon photooxidation half-life in air.

**· Surface Water:** High: 672 hours (4 weeks)

Low: 168 hours (1 week)

*Comment:* Scientific judgement based upon estimated aqueous aerobic biodegradation half-life.

**· Ground Water:** High: 8640 hours (8 weeks)

Low: 336 hours (2 weeks)

*Comment:* Scientific judgement based upon estimated aqueous aerobic and anaerobic biodegradation half-lives.

## Aqueous Biodegradation (unacclimated):
**· Aerobic half-life:** High: 672 hours (4 weeks)

Low: 168 hours (1 week)

*Comment:* Scientific judgement based upon soil column study simulating an aerobic river/groundwater infiltration system (high $t_{1/2}$: Kuhn, EP et al. (1985)) and aqueous screening test data (Bridie, AL et al. (1979)).

**· Anaerobic half-life:** High: 2688 hours (16 weeks)

Low: 672 hours (4 weeks)

*Comment:* Scientific judgement based upon unacclimated aqueous aerobic biodegradation half-life.

**· Removal/secondary treatment:** High: No data

Low:

*Comment:*

<u>Photolysis:</u>
  · **Atmos photol half-life:**           High:     No data
                                          Low:
  *Comment:*


  · **Max light absorption (nm):**        lambda max = 274.5, 211.5
  *Comment:* No absorbance occurs above 290 nm in cyclohexane (Sadtler UV No. 609).


  · **Aq photol half-life:**              High:     No data
                                          Low:
  *Comment:*


<u>Photooxidation half-life:</u>
  · **Water:**                            High: $1.4 \times 10^8$ hours          (15699 years)
                                          Low: $2.8 \times 10^6$ hours           (314 years)
  *Comment:* Scientific judgement based upon estimated rate data for alkylperoxyl radicals in aqueous solution (Hendry, DG et al. (1974)).


  · **Air:**                              High:     42 hours           (1.7 days)
                                          Low:      4.2 hours
  *Comment:* Based upon measured rate data for vapor phase reaction with hydroxyl radicals in air (Atkinson, R (1985)).


<u>Reduction half-life:</u>                High:     No data
                                          Low:
  *Comment:*


<u>Hydrolysis:</u>
  · **First-order hydr half-life:**       No hydrolyzable groups
  *Comment:*


  · **Acid rate const (M(H+)-hr)$^{-1}$:**
  *Comment:*


  · **Base rate const (M(OH-)-hr)$^{-1}$:**
  *Comment:*


365

# p-Cresol

**CAS Registry Number:** 106-44-5

**Structure:**

OH

CH₃

**Half-lives:**
- **Soil:** High: 16 hours
  Low: 1 hour

*Comment:* Scientific judgement based upon estimated unacclimated aqueous aerobic biodegradation half-life.

- **Air:** High: 15 hours
  Low: 1.5 hours

*Comment:* Based upon photooxidation half-life in air.

- **Surface Water:** High: 16 hours
  Low: 1 hour

*Comment:* Scientific judgement based upon unacclimated marine and freshwater grab sample data (low $t_{1/2}$: Vanveld, PA and Spain, JC (1983), high $t_{1/2}$: Rogers, JE et al. (1984)).

- **Ground Water:** High: 672 hours (28 days)
  Low: 2 hours

*Comment:* Scientific judgement based upon estimated unacclimated aqueous aerobic biodegradation half-life (low $t_{1/2}$) and aqueous anaerobic half-life (high $t_{1/2}$).

**Aqueous Biodegradation (unacclimated):**
- **Aerobic half-life:** High: 16 hours
  Low: 1 hour

*Comment:* Scientific judgement based upon unacclimated marine and freshwater grab sample data (low $t_{1/2}$: Vanveld, PA and Spain, JC (1983), high $t_{1/2}$: Rogers, JE et al. (1984)).

- **Anaerobic half-life:** High: 672 hours (28 days)
  Low: 240 hours (10 days)

*Comment:* Scientific judgement based upon anaerobic screening test data (low $t_{1/2}$: Boyd, SA et al. (1983), high $t_{1/2}$: Horowitz, A et al. (1982)).

- **Removal/secondary treatment:** High: 99.4%
  Low:

366

*Comment:* Based upon % degraded under aerobic continuous flow conditions (Chudoba, J et al. (1968)).

## Photolysis:
· **Atmos photol half-life:**          High:     No data
                                       Low:
*Comment:*


· **Max light absorption (nm):**       lambda max = 279.0
*Comment:* In methanol (Sadtler UV No. 15).


· **Aq photol half-life:**             High:     283 hours
                                       Low:
*Comment:* Based upon photolysis rate data for April sunlight (Smith, JH et al. (1978)).

## Photooxidation half-life:
· **Water:**                           High:   11325 hours        (1.3 years)
                                       Low:      144 hours        (6 days)
*Comment:* Scientific judgement based upon measured rate data for reactions with singlet oxygen and hydroxyl radicals in aqueous solution (Scully, FE Jr and Hoigne, J (1987), Anbar, M and Neta, P (1967)).


· **Air:**                             High:     15 hours
                                       Low:      1.5 hours
*Comment:* Based upon measured rate data for the vapor phase reaction with hydroxyl radicals in air (Atkinson, R (1985)).

## Reduction half-life:                High:     No data
                                       Low:
*Comment:*

## Hydrolysis:
· **First-order hydr half-life:**      No hydrolyzable groups
*Comment:* Rate constant at neutral pH is zero (Kollig, HP et al. (1987A)).


· **Acid rate const $(M(H+)-hr)^{-1}$:**     0.0
*Comment:* Based upon measured rate data at acid pH (Kollig, HP et al. (1987A)).


· **Base rate const $(M(OH-)-hr)^{-1}$:**     0.0
*Comment:* Based upon measured rate data at basic pH (Kollig, HP et al. (1987A)).

# p-Dichlorobenzene

<u>CAS Registry Number:</u>  106-46-7

<u>Half-lives:</u>
   **· Soil:**                             High:    4320 hours       (6 months)
                                       Low:     672 hours        (4 weeks)
*Comment:* Scientific judgement based upon unacclimated aerobic screening test data (Canton, JH et al. (1985)) and aerobic soil grab sample data (Haider, K et al. (1981)).

   **· Air:**                             High:    2006 hours       (83.6 days)
                                       Low:    200.6 hours     (8.4 days)
*Comment:* Based upon photooxidation half-life in air.

   **· Surface Water:**               High:    4320 hours       (6 months)
                                       Low:     672 hours       (4 weeks)
*Comment:* Scientific judgement based upon estimated unacclimated aqueous aerobic biodegradation half-life.

   **· Ground Water:**               High:    8640 hours       (12 months)
                                       Low:    1344 hours      (8 weeks)
*Comment:* Scientific judgement based upon estimated unacclimated aqueous aerobic biodegradation half-life.

<u>Aqueous Biodegradation (unacclimated):</u>
   **· Aerobic half-life**:            High:    4320 hours       (6 months)
                                       Low:     672 hours       (4 weeks)
*Comment:* Scientific judgement based upon unacclimated aerobic screening test data (Canton, JH et al. (1985)) and aerobic soil grab sample data (Haider, K et al. (1981)).

   **· Anaerobic half-life**:        High:   17280 hours     (24 months)
                                     Low:    2688 hours      (16 weeks)
*Comment:* Scientific judgement based upon estimated unacclimated aqueous aerobic biodegradation half-life.

   **· Removal/secondary treatment:**   High:     No data
                                       Low:
*Comment:*

<u>Photolysis:</u>
   **· Atmos photol half-life:**      High:     No data
                                       Low:
*Comment:*

· **Max light absorption (nm):**    lambda max = 280.0, 271.5, 264.0, 223.5
*Comment:* No absorbance of UV light at wavelengths greater than 300 nm in methanol (Sadtler UV No. 55).

· **Aq photol half-life:**    High:    No data
                              Low:
*Comment:*

**Photooxidation half-life:**
· **Water:**    High:    No data
               Low:
*Comment:*

· **Air:**    High:    2006 hours    (83.6 days)
             Low:    200.6 hours    (8.4 days)
*Comment:* Based upon measured rate data for the vapor phase reaction with hydroxyl radicals in air (Atkinson, R et al.(1985B)).

**Reduction half-life:**    High:    No data
                           Low:
*Comment:*

**Hydrolysis:**
· **First-order hydr half-life:**    >879 years
*Comment:* Scientific judgement based upon rate constant ($<0.9$ $M^{-1}$ $hr^{-1}$) extrapolated to pH 7 at 25 °C from 1% disappearance after 16 days at 85 °C and pH 9.7 (Ellington, JJ et al. (1988)).

· **Acid rate const $(M(H+)-hr)^{-1}$:**    Not expected to be important
*Comment:* Scientific judgement based upon rate constant ($<0.9$ $M^{-1}$ $hr^{-1}$) extrapolated to pH 7 at 25 °C from 1% disappearance after 16 days at 85 °C and pH 9.7 (Ellington, JJ et al. (1988)).

· **Base rate const $(M(OH-)-hr)^{-1}$:**    $<0.9$
*Comment:* Scientific judgement based upon 1% disappearance after 16 days at 85 °C and pH 9.7 (Ellington, JJ et al. (1988)).

369

# p-Phenylenediamine

CAS Registry Number: 106-50-3

Structure:

## Half-lives:
   · **Soil:**                              High:     672 hours        (4 weeks)
                                            Low:      168 hours        (7 days)
   *Comment:* Scientific judgement based upon estimated aqueous aerobic biodegradation half-lives.

   · **Air:**                               High:     2.80 hours
                                            Low:      0.28 hours
   *Comment:* Based upon estimated photooxidation half-lives in air.

   · **Surface Water:**                     High:     672 hours        (4 weeks)
                                            Low:      31 hours         (1.3 days)
   *Comment:* The high value is scientific judgement based upon estimated aqueous aerobic biodegradation half-life. The low value is based upon aqueous photooxidation half-life.

   · **Ground Water:**                      High:     1344 hours       (8 weeks)
                                            Low:      336 hours        (14 days)
   *Comment:* Scientific judgement based upon estimated aqueous aerobic biodegradation half-lives.

## Aqueous Biodegradation (unacclimated):
   · **Aerobic half-life:**                 High:     672 hours        (4 weeks)
                                            Low:      168 hours        (7 days)
   *Comment:* Scientific judgement based upon aerobic biological screening studies (Malaney, GW (1960); Marion, CV and Malaney, GW (1964); Pitter, P (1976)).

   · **Anaerobic half-life:**               High:     2688 hours       (16 weeks)
                                            Low:      672 hours        (28 days)
   *Comment:* Scientific judgement based upon estimated aqueous aerobic biodegradation half-lives.

   · **Removal/secondary treatment:**       High:     No data
                                            Low:
   *Comment:*

## Photolysis:

**· Atmos photol half-life:**           High:       No data
                                        Low:
*Comment:*


**· Max light absorption (nm):**        lambda max = 308 nm
*Comment:* Absorption maxima in methanol (Sadtler 1187 UV); absorption extends above 370 nm.


**· Aq photol half-life:**              High:
                                        Low:
*Comment:*


**Photooxidation half-life:**
  **· Water:**                          High:   1740 hours          (72 days)
                                        Low:      31 hours          (1.3 days)
*Comment:* Scientific judgement based upon photooxidation rate constants with ·OH and $RO_2$·
for the aromatic amine class (Mill, T and Mabey, W (1985); Guesten, H et al. (1981)). It is
assumed that p-Phenylenediamine reacts twice as fast as aniline.


  **· Air:**                            High:   2.80 hours
                                        Low:    0.28 hours
*Comment:* Based upon estimated photooxidation rate constant with ·OH (Atkinson, R (1987A)).


**Reduction half-life:**                High:       No data
                                        Low:
  *Comment:*


**Hydrolysis:**
  **· First-order hydr half-life:**     No hydrolyzable groups
  *Comment:*


  **· Acid rate const (M(H+)-hr)$^{-1}$:**
  *Comment:*


  **· Base rate const (M(OH-)-hr)$^{-1}$:**
  *Comment:*

# 1,4-Benzoquinone

CAS Registry Number:  106-51-4

Half-lives:
   · Soil:                             High:    120 hours        (5 days)
                                    Low:     1 hour

*Comment:*  Scientific judgement based upon aerobic soil die-away test data (Medvedev, VA and Davidov, VD (1981A)).

   · Air:                           High:    6.6 hours
                                 Low:    0.66 hours

*Comment:*  Scientific judgement based upon estimated photooxidation half-life in air.

   · Surface Water:             High:    120 hours        (5 days)
                                 Low:     1 hour

*Comment:*  Scientific judgement based upon unacclimated aerobic aqueous biodegradation half-life.

   · Ground Water:             High:    240 hours      (10 days)
                                 Low:     2 hours

*Comment:*  Scientific judgement based upon unacclimated aerobic aqueous biodegradation half-life.

Aqueous Biodegradation (unacclimated):
   · Aerobic half-life:          High:    120 hours        (5 days)
                                   Low:     1 hour

*Comment:*  Scientific judgement based upon aerobic soil die-away test data (Medvedev, VA and Davidov, VD (1981A)).

   · Anaerobic half-life:       High:    480 hours      (20 days)
                                   Low:     4 hours        (4 days)

*Comment:*  Scientific judgement based upon unacclimated aerobic aqueous biodegradation half-life.

   · Removal/secondary treatment:   High:    No data
                                   Low:

*Comment:*

Photolysis:
   · Atmos photol half-life:     High:
                                   Low:

*Comment:*

· **Max light absorption (nm):**       No data
*Comment:*

· **Aq photol half-life:**       High:
                                  Low:

*Comment:*

**Photooxidation half-life:**
· **Water:**       High:       No data
                                  Low:

*Comment:*

· **Air:**       High:       6.6 hours
                                  Low:       0.66 hours
*Comment:* Scientific judgement based upon estimated rate constants for reaction with hydroxyl radical (Atkinson, R (1987A) and ozone (Atkinson, R and Carter, WPL (1984)).

**Reduction half-life:**       High:       No data
                                  Low:
*Comment:*

**Hydrolysis:**
· **First-order hydr half-life:**
*Comment:*

· **Acid rate const $(M(H+)\text{-hr})^{-1}$:**       No data
*Comment:*

· **Base rate const $(M(OH-)\text{-hr})^{-1}$:**
*Comment:*

# 1,2-Butylene oxide

**CAS Registry Number:** 106-88-7

**Half-lives:**
 · **Soil:**                          High:     310 hours         (12.9 days)
                                      Low:      168 hours         (7.0 days)
 *Comment:* Based upon estimated aqueous hydrolysis half-lives.

 · **Air:**                           High:     305 hours         (12.7 days)
                                      Low:      30.5 hours        (1.27 days)
 *Comment:* Based upon estimated photooxidation half-lives in air.

 · **Surface Water:**                 High:     310 hours         (12.9 days)
                                      Low:      168 hours         (7.0 days)
 *Comment:* Based upon estimated aqueous hydrolysis half-lives.

 · **Ground Water:**                  High:     310 hours         (12.9 days)
                                      Low:      168 hours         (7.0 days)
 *Comment:* Based upon estimated aqueous hydrolysis half-lives.

**Aqueous Biodegradation (unacclimated):**
 · **Aerobic half-life:**             High:     672 hours         (4 weeks)
                                      Low:      168 hours         (7 days)
 *Comment:* Scientific judgement based upon limited biodegradation data for propylene oxide.

 · **Anaerobic half-life:**           High:     2688 hours        (16 weeks)
                                      Low:      672 hours         (28 days)
 *Comment:* Scientific judgement based upon estimated aqueous aerobic biodegradation half-lives.

 · **Removal/secondary treatment:**   High:     No data
                                      Low:
 *Comment:*

**Photolysis:**
 · **Atmos photol half-life:**        High:     No data
                                      Low:
 *Comment:*

 · **Max light absorption (nm):**
 *Comment:*

 · **Aq photol half-life:**           High:

Low:

*Comment:*

## Photooxidation half-life:
**· Water:**  High: 480,000 hours  (55 years)
Low:  12,000 hours  (1.4 years)
*Comment:* Based upon measured photooxidation rate constant with ·OH in water (Guesten, H et al. (1981)).

**· Air:**  High:  305 hours  (12.7 days)
Low:  30.5 hours  (1.27 days)
*Comment:* Based upon measured photooxidation rate constant with ·OH in air (Atkinson, R (1985)).

## Reduction half-life:
High:  No data
Low:
*Comment:*

## Hydrolysis:
**· First-order hydr half-life:**  310 hours  (12.9 days)
*Comment:* This hydrolysis half-life at 25°C and pH 7 is scientific judgement based upon measured hydrolysis rates for propylene oxide (Meylan, W et al. (1986)).

**· Acid rate const (M(H+)-hr)$^{-1}$:**  187.2
*Comment:* Rate for propylene oxide at 25°C (Meylan, W et al. (1986)). At 25°C and pH 5, the half-life would be 7.0 days; this half-life includes the effects of neutral, acid, and base hydrolysis.

**· Base rate const (M(OH-)-hr)$^{-1}$:**  0.313
*Comment:* Rate for propylene oxide at 25°C (Meylan, W et al. (1986)). At 25°C and pH 9, the half-life would be 12.9 days; this half-life includes the effects of neutral, acid, and base hydrolysis.

# Epichlorohydrin

<u>CAS Registry Number:</u>  106-89-8

<u>Half-lives:</u>
  · **Soil:**                    High:    672 hours         (4 weeks)
                                 Low:     168 hours         (7 days)
  *Comment:* Scientific judgement based upon estimated unacclimated aqueous aerobic biodegradation half-life.

  · **Air:**                     High:    1458 hours        (60.8 days)
                                 Low:     146 hours         (6.1 days)
  *Comment:* Based upon measured rate constant for reaction with hydroxyl radical in air (Atkinson, R (1985)).

  · **Surface Water:**           High:    672 hours         (4 weeks)
                                 Low:     168 hours         (7 days)
  *Comment:* Scientific judgement based upon estimated unacclimated aqueous aerobic biodegradation half-life.

  · **Ground Water:**            High:    1344 hours        (8 weeks)
                                 Low:     336 hours         (14 days)
  *Comment:* Scientific judgement based upon estimated unacclimated aqueous aerobic biodegradation half-life.

<u>Aqueous Biodegradation (unacclimated):</u>
  · **Aerobic half-life**:       High:    672 hours         (4 weeks)
                                 Low:     168 hours         (7 days)
  *Comment:* Scientific judgement based upon aqueous aerobic biodegradation screening test data (Bridie, AL et al. (1979); Sasaki, S (1978)).

  · **Anaerobic half-life**:     High:    2688 hours        (16 weeks)
                                 Low:     672 hours         (28 days)
  *Comment:* Scientific judgement based upon estimated unacclimated aqueous aerobic biodegradation half-life.

  · **Removal/secondary treatment:**   High:        89%
                                 Low:
  *Comment:* Removal percentage based upon data from a fill and draw activated sludge biological treatment simulator (Matsui, S et al. (1975)).

<u>Photolysis:</u>
  · **Atmos photol half-life:**        High:

376

Low:

*Comment:*

**· Max light absorption (nm):**     No data
*Comment:*

**· Aq photol half-life:**     High:
                               Low:

*Comment:*

**Photooxidation half-life:**
  **· Water:**     High:     No data
                   Low:

*Comment:*

  **· Air:**     High:     1458 hours     (60.8 days)
                 Low:      146 hours      (6.1 days)
*Comment:* Based upon measured rate constant for reaction with hydroxyl radical in air (Atkinson, R (1985)).

**Reduction half-life:**     High:     No data
                             Low:

*Comment:*

**Hydrolysis:**
  **· First-order hydr half-life:**          197 hours          (8.2 days)
*Comment:* ($t_{1/2}$ at pH 5-9 and 20°C) (Mabey, W and Mill, T (1978)).

  **· Acid rate const (M(H+)-hr)$^{-1}$:**     1.46
*Comment:* (Mabey, W and Mill, T (1978)).

  **· Base rate const (M(OH-)-hr)$^{-1}$:**     No data
*Comment:*

# Ethylene dibromide

<u>CAS Registry Number:</u>  106-93-4

<u>Half-lives:</u>
  · **Soil:**                                  High:    4320 hours          (6 months)
                                               Low:      672 hours          (4 weeks)
  *Comment:* Scientific judgement based upon unacclimated aqueous aerobic biodegradation half-life.

  · **Air:**                                   High:    2567 hours          (107 days)
                                               Low:      257 hours          (10.7 days)
  *Comment:* Scientific judgement based upon estimated photooxidation half-life in air.

  · **Surface Water:**                         High:    4320 hours          (6 months)
                                               Low:      672 hours          (4 weeks)
  *Comment:* Scientific judgement based upon unacclimated aqueous aerobic biodegradation half-life.

  · **Ground Water:**                          High:    2880 hours          (120 days)
                                               Low:      470 hours          (19.6 days)
  *Comment:* Based upon data from an anaerobic ground water ecosystem study (low $t_{1/2}$) (Wilson, BH et al. (1986)) and data from an aerobic ground water ecosystem study (Swindoll, CM et al. (1987)).

<u>Aqueous Biodegradation (unacclimated):</u>
  · **Aerobic half-life:**                     High:    4320 hours          (6 months)
                                               Low:      672 hours          (4 weeks)
  *Comment:* Scientific judgement based upon unacclimated aqueous aerobic biodegradation screening test data (Bouwer, EJ and McCarty, PL (1983)).

  · **Anaerobic half-life:**                   High:     360 hours          (15 days)
                                               Low:       48 hours          (2 days)
  *Comment:* Based upon anaerobic stream and pond water sediment die-away test data (Jafvert, CT and Wolfe, NL (1987)).

  · **Removal/secondary treatment:**           High:    No data
                                               Low:
  *Comment:*

<u>Photolysis:</u>
  · **Atmos photol half-life:**                High:
                                               Low:

*Comment:*

**· Max light absorption (nm):**     No data
*Comment:*

**· Aq photol half-life:**     High:
                                 Low:

*Comment:*

**Photooxidation half-life:**
  **· Water:**     High:     No data
                               Low:

*Comment:*

  **· Air:**     High:     2567 hours     (107 days)
                Low:     257 hours     (10.7 days)
*Comment:* Scientific judgement based upon estimated rate constant for reaction with hydroxyl radical in air (Atkinson, R (1987A)).

**Reduction half-life:**     High:     No data
                               Low:
*Comment:*

**Hydrolysis:**
  **· First-order hydr half-life:**     19272 hours (2.2 years)
*Comment:* ($t_{1/2}$ at pH 7.5 and 25°C) Based upon measured neutral hydrolysis rate constant (Weintraub, RA et al. (1986)).

**· Acid rate const $(M(H+)-hr)^{-1}$:**     No data
*Comment:*

**· Base rate const $(M(OH-)-hr)^{-1}$:**     No data
*Comment:*

# 1,3-Butadiene

<u>CAS Registry Number:</u>  106-99-0

<u>Half-lives:</u>

· **Soil:**  High:  672 hours  (4 weeks)
Low:  168 hours  (7 days)
*Comment:* Scientific judgement based upon estimated aqueous aerobic biodegradation half-lives.

· **Air:**  High:  7.8 hours  (0.32 days)
Low:  0.76 hours  (0.03 days)
*Comment:* Based upon measured photooxidation rate constants in air.

· **Surface Water:**  High:  672 hours  (4 weeks)
Low:  168 hours  (7 days)
*Comment:* Scientific judgement based upon estimated aqueous aerobic biodegradation half-lives.

· **Ground Water:**  High:  1344 hours  (8 weeks)
Low:  336 hours  (14 days)
*Comment:* Scientific judgement based upon estimated aqueous aerobic biodegradation half-lives.

<u>Aqueous Biodegradation (unacclimated):</u>

· **Aerobic half-life:**  High:  672 hours  (4 weeks)
Low:  168 hours  (7 days)
*Comment:* Scientific judgement.

· **Anaerobic half-life:**  High:  2688 hours  (16 weeks)
Low:  672 hours  (28 days)
*Comment:* Scientific judgement based upon estimated aqueous aerobic biodegradation half-lives.

· **Removal/secondary treatment:**  High:  No data
Low:
*Comment:*

<u>Photolysis:</u>

· **Atmos photol half-life:**  High:  No data
Low:
*Comment:*

· **Max light absorption (nm):**
*Comment:*

· **Aq photol half-life:**  High:

Low:

*Comment:*

## Photooxidation half-life:
**· Water:**          High:   48000 hours          (2000 days)
                      Low:    1200 hours           (50 days)
*Comment:* Based upon measured photooxidation rate constant with hydroxyl radicals in water (Guesten, H et al. (1981)).

**· Air:**            High:   7.8 hours            (0.32 days)
                      Low:    0.76 hours           (0.03 days)
*Comment:* Based upon measured photooxidation rate constants in air with ·OH, ozone, and nitrate radicals (Atkinson, R (1985); Atkinson, R and Carter, WPL (1984); Atkinson, R et al. (1984A)).

## Reduction half-life:          High:          No data
                                 Low:

*Comment:*

## Hydrolysis:
**· First-order hydr half-life:**          No hydrolyzable groups
*Comment:*

**· Acid rate const $(M(H+)-hr)^{-1}$:**
*Comment:*

**· Base rate const $(M(OH-)-hr)^{-1}$:**
*Comment:*

# Acrolein

CAS Registry Number: 107-02-8

## Half-lives:
· Soil:                                    High:      672 hours         (4 weeks)
                                           Low:       168 hours         (7 days)
*Comment:* Scientific judgement based upon estimated aqueous aerobic biodegradation half-life.

· Air:                                     High:      33.7 hours
                                           Low:        3.4 hours
*Comment:* Based upon photooxidation half-life in air.

· Surface Water:                           High:      672 hours         (4 weeks)
                                           Low:       168 hours         (7 days)
*Comment:* Scientific judgement based upon estimated aqueous aerobic biodegradation half-life.

· Ground Water:                            High:      1344 hours        (8 weeks)
                                           Low:        336 hours        (14 days)
*Comment:* Scientific judgement based upon estimated aqueous aerobic biodegradation half-life.

## Aqueous Biodegradation (unacclimated):
· Aerobic half-life:                       High:      672 hours         (4 weeks)
                                           Low:       168 hours         (7 days)
*Comment:* Scientific judgement based upon acclimated aqueous screening test data (Tabak, HH et al. (1981)).

· Anaerobic half-life:                     High:      2880 hours        (4 months)
                                           Low:        672 hours        (4 weeks)
*Comment:* Scientific judgement based upon aqueous aerobic biodegradation half-life.

· Removal/secondary treatment:             High:      99.9%
                                           Low:       45.0%
*Comment:* Removal percentages based upon continuous-flow activated sludge treatment simulator data (Stover, EL and Kincannon, DF (1983)).

## Photolysis:
· Atmos photol half-life:                  High:  Not expected to be important
                                           Low:
*Comment:* Acrolein was stable to photolysis at 30 °C and 313 nm in the presence and absence of oxygen (Osborne, AD et al. (1962)).

· Max light absorption (nm):               No data

*Comment:*

**· Aq photol half-life:**              High:      No data
                                        Low:
*Comment:*

**Photooxidation half-life:**
  Water:                                High:      No data
                                        Low:
*Comment:*

**· Air:**                              High:      33.7 hours
                                        Low:       3.4 hours
*Comment:* Based upon measured rate constant for reaction with hydroxyl radicals in air (Atkinson, R (1985)).

**Reduction half-life:**                High:      No data
                                        Low:
*Comment:*

**Hydrolysis:**
  **· First-order hydr half-life:**     No data
  *Comment:*

  **· Acid rate const (M(H+)-hr)$^{-1}$:**
  *Comment:*

  **· Base rate const (M(OH-)-hr)$^{-1}$:**
  *Comment:*

# 3-Chloropropene

CAS Registry Number: 107-05-1

**Half-lives:**
· **Soil:**                              High:      335 hours              (14 days)
                                        Low:       166 hours              (6.9 days)
*Comment:* (t$_{1/2}$ at pH 5-9) Based upon measured neutral hydrolysis rate constants in water at 20°C (low t$_{1/2}$) and 25°C (high t$_{1/2}$) (Robertson, RE and Scott, JMW (1961)).

· **Air:**                               High:      28.8 hours
                                        Low:       3.03 hours
*Comment:* Scientific judgement based upon estimated photooxidation half-life in air.

· **Surface Water:**                     High:      335 hours              (14 days)
                                        Low:       166 hours              (6.9 days)
*Comment:* (t$_{1/2}$ at pH 5-9) Based upon measured neutral hydrolysis rate constants at 20°C (low t$_{1/2}$) and 25°C (high t$_{1/2}$) (Robertson, RE and Scott, JMW (1961)).

· **Ground Water:**                      High:      335 hours              (14 days)
                                        Low:       166 hours              (6.9 days)
*Comment:* (t$_{1/2}$ at pH 5-9) Based upon measured neutral hydrolysis rate constants at 20°C (low t$_{1/2}$) and 25°C (high t$_{1/2}$) (Robertson, RE and Scott, JMW (1961)).

**Aqueous Biodegradation (unacclimated):**
· **Aerobic half-life:**                 High:      672 hours              (4 weeks)
                                        Low:       168 hours              (7 days)
*Comment:* Scientific judgement based upon unacclimated aqueous aerobic screening test data (Bridie, AL et al. (1979); Ilisescu, A (1971)).

· **Anaerobic half-life:**               High:      2688 hours             (16 weeks)
                                        Low:       672 hours              (28 days)
*Comment:* Scientific judgement based upon estimated unacclimated aqueous aerobic biodegradation half-life.

· **Removal/secondary treatment:**       High:      No data
                                        Low:
*Comment:*

**Photolysis:**
· **Atmos photol half-life:**            High:
                                        Low:
*Comment:*

**· Max light absorption (nm):**     No data
*Comment:*


**· Aq photol half-life:**          High:
                                    Low:

*Comment:*


**Photooxidation half-life:**
  **· Water:**             High:     No data
                                    Low:
*Comment:*


  **· Air:**              High:     28.8 hours
                                    Low:      3.03 hours
*Comment:* Scientific judgement based upon measured rate constant for reaction with hydroxyl radicals (Edney, EO et al. (1986)) and estimated rate constant for reaction with ozone (Atkinson, R and Carter, WPL (1984)).


**Reduction half-life:**            High:     No data
                                    Low:
*Comment:*


**Hydrolysis:**
  **· First-order hydr half-life:**   High:   335 hours        (14 days)
                                                 Low:    166 hours        (6.9 days)
*Comment:* ($t_{1/2}$ at pH 5-9) Based upon measured neutral hydrolysis rate constants at 20°C (low $t_{1/2}$) and 25°C (high $t_{1/2}$) (Robertson, RE and Scott, JMW (1961)).


**· Acid rate const $(M(H+)-hr)^{-1}$:**     No data
*Comment:*


**· Base rate const $(M(OH-)-hr)^{-1}$:**    No data
*Comment:*

# 1,2-Dichloroethane

**CAS Registry Number:** 107-06-2

## Half-lives:

**· Soil:**

|  | High: | 4320 hours | (6 months) |
|---|---|---|---|
|  | Low: | 2400 hours | (100 days) |

*Comment:* Scientific judgement based upon estimated aqueous aerobic biodegradation half-life.

**· Air:**

|  | High: | 2917 hours | (122 days) |
|---|---|---|---|
|  | Low: | 292 hours | (12.2 days) |

*Comment:* Based upon photooxidation half-life in air.

**· Surface Water:**

|  | High: | 4320 hours | (6 months) |
|---|---|---|---|
|  | Low: | 2400 hours | (100 days) |

*Comment:* Scientific judgement based upon unacclimated grab sample of aerobic soil from ground water aquifers (low $t_{1/2}$: Wilson, JT et al. (1983A)) and acclimated river die-away rate data (high $t_{1/2}$: Mudder, T (1981)).

**· Ground Water:**

|  | High: | 8640 hours | (12 months) |
|---|---|---|---|
|  | Low: | 2400 hours | (100 days) |

*Comment:* Scientific judgement based upon unacclimated grab sample of aerobic soil from ground water aquifers (low $t_{1/2}$: Wilson, JT et al. (1983A)) and estimated aqueous aerobic biodegradation half-life.

## Aqueous Biodegradation (unacclimated):

**· Aerobic half-life:**

|  | High: | 4320 hours | (6 months) |
|---|---|---|---|
|  | Low: | 2400 hours | (100 days) |

*Comment:* Scientific judgement based upon unacclimated grab sample of aerobic soil from ground water aquifers (low $t_{1/2}$: Wilson, JT et al. (1983A)) and acclimated river die-away rate data (high $t_{1/2}$: Mudder, T (1981)).

**· Anaerobic half-life:**

|  | High: | 17280 hours | (24 months) |
|---|---|---|---|
|  | Low: | 9600 hours | (400 days) |

*Comment:*

**· Removal/secondary treatment:**

|  | High: | No data |
|---|---|---|
|  | Low: | |

*Comment:*

## Photolysis:

**· Atmos photol half-life:**

|  | High: |
|---|---|
|  | Low: |

*Comment:*

**· Max light absorption (nm):**  No data
*Comment:*

**· Aq photol half-life:**  High:
Low:

*Comment:*

**Photooxidation half-life:**
　**· Water:**  High:  No data
Low:

*Comment:*

　**· Air:**  High:  2917 hours  (122 days)
Low:  292 hours  (12.2 days)
*Comment:* Based upon measured rate data for the vapor phase reaction with hydroxyl radicals in air (Atkinson, R (1985)).

**Reduction half-life:**  High:  No data
Low:

*Comment:*

**Hydrolysis:**
　**· First-order hydr half-life:**  1.1 year
*Comment:* Scientific judgement based upon rate constant at neutral pH ($0.63$ yr$^{-1}$) (Kollig, HP et al. (1987)).

**· Acid rate const (M(H+)-hr)$^{-1}$:**  0.0
*Comment:* Based upon measured rate data at acid pH (Kollig, HP et al. (1987)).

**· Base rate const (M(OH-)-hr)$^{-1}$:**  0.0
*Comment:* Based upon measured rate data at basic pH (Kollig, HP et al. (1987)).

387

# Acrylonitrile

CAS Registry Number: 107-13-1

**Half-lives:**

· Soil:                                    High:     552 hours              (23 days)
                                           Low:       30 hours              (1.25 days)
*Comment:* Scientific judgement based upon estimated aqueous aerobic biodegradation half-life.

· **Air:**                                 High:     189 hours              (8.25 days)
                                           Low:      13.4 hours             (0.56 days)
*Comment:* Based upon photooxidation half-life in air.

· **Surface Water:**                       High:     552 hours              (23 days)
                                           Low:       30 hours              (1.25 days)
*Comment:* Scientific judgement based upon estimated aqueous aerobic biodegradation half-life.

· **Ground Water:**                        High:    1104 hours              (46 days)
                                           Low:       60 hours              (2.5 days)
*Comment:* Scientific judgement based upon estimated aqueous aerobic biodegradation half-life.

**Aqueous Biodegradation (unacclimated):**

· **Aerobic half-life:**                   High:     552 hours              (23 days)
                                           Low:       30 hours              (1.25 days)
*Comment:* Scientific judgement based upon river die-away test data (Going, J et al. (1979); Ludzack, FJ et al. (1958)).

· **Anaerobic half-life:**                 High:    2208 hours              (92 days)
                                           Low:     120 hours              (5 days)
*Comment:* Scientific judgement based upon estimated aqueous aerobic biodegradation half-life.

· **Removal/secondary treatment:**         High:      99.9%
                                           Low:       75%
*Comment:* Removal percentages based upon continuous activated sludge treatment simulator data (Kincannon, DF et al. (1983)).

**Photolysis:**

· **Atmos photol half-life:**              High:     No data
                                           Low:
*Comment:*

· **Max light absorption (nm):**           lambda max = 203 nm (ethanol)
*Comment:* Grasselli, J and Ritchey, W (1975).

388

**· Aq photol half-life:**  High:  No data
Low:

*Comment:*

## Photooxidation half-life:
**· Water:**  High:  No data
Low:

*Comment:*

**· Air:**  High:  189 hours  (8.25 days)
Low:  13.4 hours  (0.56 days)
*Comment:* Based upon measured rate constant for reaction with hydroxyl radical in air (Atkinson, R (1985)).

## Reduction half-life:
High:  No data
Low:

*Comment:*

## Hydrolysis:
**· First-order hydr half-life:**  $1.06 \times 10^7$ hours  (1210 years)
*Comment:* ($t_{1/2}$ at pH 7.0) Based upon measured acid and base catalyzed hydrolysis rate constants (Ellington, JJ et al. (1987A)).

**· Acid rate const $(M(H+)\text{-hr})^{-1}$:**  $4.2 \times 10^{-2}$ $M^{-1}$ $hr^{-1}$
*Comment:* ($t_{1/2} = 1.65 \times 10^6$ hours (188 years) at pH 5.0) Based upon measured acid and base catalyzed hydrolysis rate constants (Ellington, JJ et al. (1987A)).

**· Base rate const $(M(OH-)\text{-hr})^{-1}$:**  $6.1 \times 10^{-1}$ $M^{-1}$ $hr^{-1}$
*Comment:* ($t_{1/2} = 1.14 \times 10^5$ hours (13 years) at pH 9.0) Based upon measured acid and base catalyzed hydrolysis rate constants (Ellington, JJ et al. (1987A)).

# Allyl alcohol

CAS Registry Number:  107-18-6

## Half-lives:
· Soil:                                    High:       168 hours              (7 days)
                                           Low:         24 hours              (1 day)
*Comment:*  Scientific judgement based upon estimated unacclimated aqueous aerobic biodegradation half-life.

· Air:                                     High:       22.0 hours
                                           Low:        2.20 hours
*Comment:*  Scientific judgement based upon estimated photooxidation half-life in air.

· Surface Water:                           High:       168 hours              (7 days)
                                           Low:         24 hours              (1 day)
*Comment:*  Scientific judgement based upon estimated unacclimated aqueous aerobic biodegradation half-life.

· Ground Water:                            High:       336 hours             (14 days)
                                           Low:         48 hours              (2 days)
*Comment:*  Scientific judgement based upon estimated unacclimated aqueous aerobic biodegradation half-life.

## Aqueous Biodegradation (unacclimated):
· Aerobic half-life:                       High:       168 hours              (7 days)
                                           Low:         24 hours              (1 day)
*Comment:*  Scientific judgement based upon unacclimated aqueous aerobic biodegradation screening test data (Sasaki, S (1978)).

· Anaerobic half-life:                     High:       672 hours             (28 days)
                                           Low:         96 hours              (4 days)
*Comment:*  Scientific judgement based upon estimated unacclimated aqueous aerobic biodegradation half-life.

· Removal/secondary treatment:             High:            73%
                                           Low:
*Comment:*  Based upon biological oxygen demand results from an activated sludge dispersed seed aeration treatment simulator (Mills, EJ Jr and Stack, VT Jr (1954)).

## Photolysis:
· Atmos photol half-life:                  High:
                                           Low:

*Comment:*

**· Max light absorption (nm):**          No data
*Comment:*

**· Aq photol half-life:**               High:
                                         Low:

*Comment:*

## Photooxidation half-life:
**· Water:**                             High: $3.2 \times 10^5$ hours          (37 years)
                                         Low:     8020 hours            (334 days)
*Comment:* Based upon measured rate constant for reaction with hydroxyl radical in water (Anbar, M and Neta, P (1967)).

**· Air:**                               High:     22.0 hours
                                         Low:      2.20 hours
*Comment:* Scientific judgement based upon estimated rate constant for reaction with hydroxyl radical in air (Atkinson, R (1987A)).

## Reduction half-life:
                                         High:       No data
                                         Low:
*Comment:*

## Hydrolysis:
**· First-order hydr half-life:**
*Comment:*

**· Acid rate const $(M(H+)-hr)^{-1}$:**     No hydrolyzable groups
*Comment:*

**· Base rate const $(M(OH-)-hr)^{-1}$:**
*Comment:*

# Ethylene glycol

CAS Registry Number: 107-21-1

## Half-lives:
**· Soil:** High: 288 hours (12 days)

Low: 48 hours ( 2 days)

*Comment:* Scientific judgement based upon aqueous aerobic biodegradation half-lives.

**· Air:** High: 83 hours (3.5 days)

Low: 8.3 hours (0.35 days)

*Comment:* Based upon measured photooxidation rate constant in air.

**· Surface Water:** High: 288 hours (12 days)

Low: 48 hours (2 days)

*Comment:* Scientific judgement based upon aqueous aerobic biodegradation half-lives.

**· Ground Water:** High: 576 hours (24 days)

Low: 96 hours (4 days)

*Comment:* Scientific judgement based upon aqueous aerobic biodegradation half-lives.

## Aqueous Biodegradation (unacclimated):
**· Aerobic half-life:** High: 288 hours (12 days)

Low: 48 hours (2 days)

*Comment:* Based upon grab sample, river die-away studies (Evans, WH and David, EJ (1974)).

**· Anaerobic half-life:** High: 1152 hours (48 days)

Low: 192 hours ( 8 days)

*Comment:* Scientific judgement based upon aqueous aerobic biodegradation half-lives.

**· Removal/secondary treatment:** High: 100%

Low: 88%

*Comment:* Based upon semi-continuous, biological treatment simulator data (Hatfield, R (1957); Matsui, S et al. (1975); Means, JL and Anderson, SJ (1981)).

## Photolysis:
**· Atmos photol half-life:** High: No data

Low:

*Comment:*

**· Max light absorption (nm):**
*Comment:*

· **Aq photol half-life:** High:
Low:

*Comment:*

## Photooxidation half-life:
· **Water:** High: 566,000 hours         (64.6 years)
Low:    6,400 hours         (267 days)

*Comment:* Based upon measured photooxidation rate constants with ·OH in water (Anbar, M and Neta, P (1967); Dorfman, LM and Adams, GE (1973)).

· **Air:** High:     83 hours         (3.5 days)
Low:     8.3 hours        (0.35 days)

*Comment:* Based upon measured photooxidation rate constant with ·OH in air (Atkinson, R (1985)).

## Reduction half-life:
                               High:     No data
Low:

*Comment:*

## Hydrolysis:
· **First-order hydr half-life:**    No hydrolyzable groups
*Comment:*

· **Acid rate const (M(H+)-hr)$^{-1}$:**
*Comment:*

· **Base rate const (M(OH-)-hr)$^{-1}$:**
*Comment:*

# Chloromethyl methyl ether

<u>CAS Registry Number:</u>  107-30-2

<u>Half-lives:</u>
· **Soil:**                                 High:     0.033 hours           (1.98 minutes)
                                            Low:    0.0108 hours           (39 seconds)
*Comment:*  High $t_{1/2}$: scientific judgement based upon measured neutral hydrolysis rate constants (Ellington, JJ et al. (1987)); low $t_{1/2}$: scientific judgement based upon measured neutral rate constant for bis(chloromethyl) ether (Tou, JC et al. (1974)).

· **Air:**                                  High:     227 hours             (9.5 days)
                                            Low:    22.7 hours
*Comment:*  Scientific judgement based upon estimated photooxidation half-life in air.

· **Surface Water:**                        High:     0.033 hours           (1.98 minutes)
                                            Low:    0.0108 hours           (39 seconds)
*Comment:*  High $t_{1/2}$: scientific judgement based upon measured neutral hydrolysis rate constants (Ellington, JJ et al. (1987)); low $t_{1/2}$: scientific judgement based upon measured neutral rate constant for bis(chloromethyl) ether (Tou, JC et al. (1974)).

· **Ground Water:**                         High:     0.033 hours           (1.98 minutes)
                                            Low:    0.0108 hours           (39 seconds)
*Comment:*  High $t_{1/2}$: scientific judgement based upon measured neutral hydrolysis rate constants (Ellington, JJ et al. (1987)); low $t_{1/2}$: scientific judgement based upon measured neutral rate constant for bis(chloromethyl) ether (Tou, JC et al. (1974)).

<u>Aqueous Biodegradation (unacclimated):</u>
· **Aerobic half-life:**                    High:     672 hours             (4 weeks)
                                            Low:     168 hours             (7 days)
*Comment:*  Scientific judgement.

· **Anaerobic half-life:**                  High:    2688 hours             (16 weeks)
                                            Low:     672 hours             (28 days)
*Comment:*  Scientific judgement based upon estimated unacclimated aerobic aqueous biodegradation half-life.

· **Removal/secondary treatment:**          High:     No data
                                            Low:
*Comment:*

<u>Photolysis:</u>
· **Atmos photol half-life:**               High:

|                                   | Low:                                  |
|:----------------------------------|:--------------------------------------|
| *Comment:*                        |                                       |

· **Max light absorption (nm):**      No data
*Comment:*

· **Aq photol half-life:**            High:
                                      Low:

*Comment:*

## Photooxidation half-life:
  · **Water:**                        High:      No data
                                      Low:

*Comment:*

  · **Air:**                          High:      227 hours          (9.5 days)
                                      Low:       22.7 hours

*Comment:* Scientific judgement based upon estimated rate constant for reaction with hydroxyl radical (Atkinson, R (1987A)).

## Reduction half-life:                High:      No data
                                      Low:

*Comment:*

## Hydrolysis:
  · **First-order hydr half-life:**   High:      0.033 hours        (1.98 min)
                                      Low:       0.0108 hours       (39 sec)

*Comment:* High $t_{1/2}$: scientific judgement based upon measured neutral hydrolysis rate constants (Ellington, JJ et al. (1987)); low $t_{1/2}$: scientific judgement based upon measured neutral rate constant for bis(chloromethyl) ether (Tou, JC et al. (1974)).

  · **Acid rate const (M(H+)-hr)$^{-1}$:**      No data
*Comment:*

  · **Base rate const (M(OH-)-hr)$^{-1}$:**     No data
*Comment:*

# Propylene glycol, monomethyl ether

**CAS Registry Number:** 107-98-2

**Half-lives:**

· **Soil:**                        High:    672 hours              (4 weeks)
                                  Low:     168 hours              (7 days)
*Comment:* Scientific judgement based upon estimated unacclimated aqueous aerobic biodegradation half-life.

· **Air:**                         High:    40.8 hours             (1.7 days)
                                  Low:     4.08 hours
*Comment:* Scientific judgement based upon estimated photooxidation half-life in air.

· **Surface Water:**               High:    672 hours              (4 weeks)
                                  Low:     168 hours              (7 days)
*Comment:* Scientific judgement based upon estimated unacclimated aqueous aerobic biodegradation half-life.

· **Ground Water:**                High:    1344 hours             (8 weeks)
                                  Low:     336 hours              (14 days)
*Comment:* Scientific judgement based upon estimated unacclimated aqueous aerobic biodegradation half-life.

**Aqueous Biodegradation (unacclimated):**

· **Aerobic half-life:**           High:    672 hours              (4 weeks)
                                  Low:     168 hours              (7 days)
*Comment:* Scientific judgement based upon unacclimated aqueous aerobic biodegradation screening test data (Dow Chemical Company (1981)).

· **Anaerobic half-life:**         High:    2688 hours             (16 weeks)
                                  Low:     672 hours              (28 days)
*Comment:* Scientific judgement based upon estimated unacclimated aqueous aerobic biodegradation half-life.

· **Removal/secondary treatment:**  High:    No data
                                  Low:
*Comment:*

**Photolysis:**

· **Atmos photol half-life:**      High:
                                  Low:
*Comment:*

· **Max light absorption (nm):**      No data
*Comment:*

· **Aq photol half-life:**      High:
                                    Low:
*Comment:*

**Photooxidation half-life:**
· **Water:**      High:      No data
                              Low:
*Comment:*

· **Air:**      High:    40.8 hours        (1.7 days)
                              Low:    4.08 hours
*Comment:* Scientific judgement based upon estimated rate constant for reaction with hydroxyl radical in air (Atkinson, R (1987A)).

**Reduction half-life:**      High:      No data
                              Low:
*Comment:*

**Hydrolysis:**
· **First-order hydr half-life:**
*Comment:*

· **Acid rate const (M(H+)-hr)$^{-1}$:**      No hydrolyzable groups
*Comment:*

· **Base rate const (M(OH-)-hr)$^{-1}$:**
*Comment:*

# Methyl isobutyl ketone

<u>CAS Registry Number:</u> 108-10-1

<u>Half-lives:</u>
  · **Soil:**                             High:     168 hours         (7 days)
                                        Low:      24 hours          (1 day)
*Comment:* Scientific judgement based upon estimated unacclimated aqueous aerobic biodegradation half-life.

  · **Air:**                              High:     45.5 hours      (1.9 days)
                                        Low:      4.6 hours
*Comment:* Based upon photooxidation half-life in air.

  · **Surface Water:**               High:     168 hours         (7 days)
                                          Low:      24 hours          (1 day)
*Comment:* Scientific judgement based upon estimated unacclimated aqueous aerobic biodegradation half-life.

  · **Ground Water:**             High:     336 hours        (14 days)
                                          Low:      48 hours         (2 days)
*Comment:* Scientific judgement based upon estimated unacclimated aqueous aerobic biodegradation half-life.

<u>Aqueous Biodegradation (unacclimated):</u>
  · **Aerobic half-life:**          High:     168 hours         (7 days)
                                          Low:      24 hours          (1 day)
*Comment:* Scientific judgement based upon unacclimated aerobic aqueous screening test data (Bridie, AL et al. (1979); Takemoto, S et al. (1981)).

  · **Anaerobic half-life:**        High:     672 hours        (28 days)
                                          Low:      96 hours          (4 days)
*Comment:* Scientific judgement based upon estimated aqueous unacclimated aerobic biodegradation half-life.

  · **Removal/secondary treatment:**   High:
                                          Low:      22%
*Comment:* Based upon % degraded in 10 day period under acclimated aerobic semi-continuous flow conditions (Mills, EJ Jr and Stack, VT Jr (1954)).

<u>Photolysis:</u>
  · **Atmos photol half-life:**      High:     No data
                                          Low:

*Comment:*

· **Max light absorption (nm):**  lambda max = 283, 232
*Comment:* No absorbance above 340 nm in cyclohexane (Sadtler UV No. 21)

· **Aq photol half-life:**  High: No data
Low:
*Comment:*

## Photooxidation half-life:
· **Water:**  High: No data
Low:
*Comment:*

· **Air:**  High: 45.5 hours  (1.9 days)
Low: 4.6 hours
*Comment:* Based upon measured rate data for the vapor phase reaction with hydroxyl radicals in air (Atkinson, R (1985)).

## Reduction half-life:  High: No data
Low:
*Comment:*

## Hydrolysis:
· **First-order hydr half-life:**  No hydrolyzable groups
*Comment:*

· **Acid rate const (M(H+)-hr)$^{-1}$:**
*Comment:*

· **Base rate const (M(OH-)-hr)$^{-1}$:**
*Comment:*

# m-Xylene

CAS Registry Number: 108-38-3

Structure:

Half-lives:
  · Soil:                           High:    672 hours          (4 weeks)
                                    Low:     168 hours          (1 week)
  Comment: Scientific judgement based upon estimated aqueous aerobic biodegradation half-life.

  · Air:                            High:    26 hours           (1.1 days)
                                    Low:     2.6 hours
  Comment: Based upon photooxidation half-life in air.

  · Surface Water:                  High:    672 hours          (4 weeks)
                                    Low:     168 hours          (1 week)
  Comment: Scientific judgement based upon estimated aqueous aerobic biodegradation half-life.

  · Ground Water:                   High:    8640 hours         (8 weeks)
                                    Low:     336 hours          (2 weeks)
  Comment: Scientific judgement based upon estimated aqueous aerobic and anaerobic
  biodegradation half-lives.

Aqueous Biodegradation (unacclimated):
  · Aerobic half-life:              High:    672 hours          (4 weeks)
                                    Low:     168 hours          (1 week)
  Comment: Scientific judgement based upon soil column study simulating an aerobic
  river/groundwater infiltration system (high $t_{1/2}$: Kuhn, EP et al. (1985)) and aqueous screening test
  data (Bridie, AL et al. (1979)).

  · Anaerobic half-life:            High:    12688 hours        (16 weeks)
                                    Low:     672 hours          (4 weeks)
  Comment: Scientific judgement based upon unacclimated aqueous aerobic biodegradation half-
  life.

  · Removal/secondary treatment:    High:    No data
                                    Low:
  Comment:

## Photolysis:

**· Atmos photol half-life:**        High:    No data

                                                Low:

*Comment:*

**· Max light absorption (nm):**       lambda max = 265, 268, 277

*Comment:* No absorbance occurs above 290 nm in cyclohexane (Sadtler UV No. 317).

**· Aq photol half-life:**           High:    No data

                                                Low:

*Comment:*

## Photooxidation half-life:

**· Water:**                      High: $2.4 \times 10^8$ hours     (27473 years)

                                                Low: $4.8 \times 10^6$ hours     (550 years)

*Comment:* Scientific judgement based upon estimated rate data for alkylperoxyl radicals in aqueous solution (Hendry, DG et al. (1974)).

**· Air:**                          High:    26 hours     (1.1 days)

                                                Low:    2.6 hours

*Comment:* Based upon measured rate data for vapor phase reaction with hydroxyl radicals in air (Atkinson, R (1985)).

## Reduction half-life:

                                          High:    No data

                                          Low:

*Comment:*

## Hydrolysis:

**· First-order hydr half-life:**      No hydrolyzable groups

*Comment:*

**· Acid rate const $(M(H+)\text{-}hr)^{-1}$:**

*Comment:*

**· Base rate const $(M(OH\text{-})\text{-}hr)^{-1}$:**

*Comment:*

# m-Cresol

**CAS Registry Number:** 108-39-4

**Structure:**

**Half-lives:**

· **Soil:**              High:    696 hours       (29 days)
                         Low:      48 hours        (2 days)

*Comment:* Scientific judgement based upon estimated unacclimated aqueous aerobic biodegradation half-life.

· **Air:**               High:    11.3 hours
                         Low:      1.1 hours

*Comment:* Based upon photooxidation half-life in air.

· **Surface Water:**      High:    696 hours       (29 days)
                          Low:      48 hours        (2 days)

*Comment:* Scientific judgement based upon estimated unacclimated aqueous aerobic biodegradation half-life.

· **Ground Water:**       High:    1176 hours      (49 days)
                          Low:      96 hours        (4 days)

*Comment:* Scientific judgement based upon estimated unacclimated aqueous aerobic biodegradation half-life (low $t_{1/2}$) and aqueous anaerobic half-life (high $t_{1/2}$).

**Aqueous Biodegradation (unacclimated):**

· **Aerobic half-life:**     High:    696 hours       (29 days)
                             Low:      48 hours        (2 days)

*Comment:* Scientific judgement based upon unacclimated marine water grab sample data (Pfaender, FK and Bartholomew, GW (1982)).

· **Anaerobic half-life:**   High:    1176 hours      (49 days)
                             Low:      360 hours       (15 days)

*Comment:* Scientific judgement based upon anaerobic screening test data (low $t_{1/2}$: Horowitz, A et al. (1982), high $t_{1/2}$: Shelton, DR and Tiedje, JM (1981)).

· **Removal/secondary treatment:**   High:    No data
                                     Low:

*Comment:*

## Photolysis:
  · **Atmos photol half-life:**        High:     No data
                                        Low:
*Comment:*

  · **Max light absorption (nm):**     lambda max = 273.0
*Comment:* Very little absorbance of UV light at wavelengths greater than 290 nm in methanol (Sadtler UV No. 622).

  · **Aq photol half-life:**          High:     No data
                                        Low:
*Comment:*

## Photooxidation half-life:
  · **Water:**                     High:    3480 hours           (145 days)
                                       Low:     66.0 hours          (2.75 days)
*Comment:* Scientific judgement based upon reported reaction rate constants for $\cdot OH$ and $RO_2 \cdot$ with the phenol class (Mill, T and Mabey, W (1985); Guesten, H et al. (1981)).

  · **Air:**                       High:    11.3 hours
                                       Low:     1.1 hours
*Comment:* Based upon measured rate data for the vapor phase reaction with hydroxyl radicals in air (Atkinson, R (1985)).

## Reduction half-life:              High:     No data
                                        Low:
*Comment:*

## Hydrolysis:
  · **First-order hydr half-life:**     No hydrolyzable groups
*Comment:* Rate constant at neutral pH is zero (Kollig, HP et al. (1987A)).

  · **Acid rate const $(M(H+)-hr)^{-1}$:**     0.0
*Comment:* Based upon measured rate data at acid pH (Kollig, HP et al. (1987A)).

  · **Base rate const $(M(OH-)-hr)^{-1}$:**     0.0
*Comment:* Based upon measured rate data at basic pH (Kollig, HP et al. (1987A)).

# 1,3-Benzenediamine

CAS Registry Number: 108-45-2

## Half-lives:
· **Soil:**                                  High:     672 hours          (4 weeks)

Low:      168 hours          (7 days)

*Comment:* Scientific judgement based upon unacclimated aqueous aerobic biodegradation half-life.

· **Air:**                       High:     2.78 hours

Low:      0.278 hours

*Comment:* Scientific judgement based upon estimated photooxidation half-life in air.

· **Surface Water:**          High:     672 hours          (4 weeks)

Low:      31 hours           (1.3 days)

*Comment:* Scientific judgement based upon unacclimated aqueous aerobic biodegradation half-life (high $t_{1/2}$) and estimated photooxidation half-life in water (low $t_{1/2}$).

· **Ground Water:**          High:     1344 hours      (8 weeks)

Low:      336 hours        (14 days)

*Comment:* Scientific judgement based upon unacclimated aqueous aerobic biodegradation half-life.

## Aqueous Biodegradation (unacclimated):
· **Aerobic half-life:**        High:     672 hours          (4 weeks)

Low:      168 hours          (7 days)

*Comment:* Scientific judgement based upon unacclimated aqueous aerobic biodegradation screening test data (Pitter, P (1976)).

· **Anaerobic half-life:**     High:     2688 hours      (16 weeks)

Low:      672 hours        (28 days)

*Comment:* Scientific judgement based upon unacclimated aqueous aerobic biodegradation half-life.

· **Removal/secondary treatment:**     High:     No data

Low:

*Comment:*

## Photolysis:
· **Atmos photol half-life:**     High:

Low:

*Comment:*

· **Max light absorption (nm):**          No data
*Comment:*

· **Aq photol half-life:**          High:
                                    Low:

*Comment:*

**Photooxidation half-life:**
    · **Water:**          High:    1740 hours          (72 days)
                          Low:      31 hours          (1.3 days)
*Comment:*  Scientific judgement based upon photooxidation rate constants with ·OH and RO$_2$·
for the aromatic amine class (Mill, T and Mabey, W (1985); Guesten, H et al. (1981)).

    · **Air:**          High:    2.78 hours
                        Low:     0.278 hours
*Comment:*  Scientific judgement based upon estimated rate constant for reaction with hydroxyl
radical in air (Atkinson, R (1987A)).

**Reduction half-life:**          High:    No data
                                  Low:

*Comment:*

**Hydrolysis:**
    · **First-order hydr half-life:**          No data
*Comment:*

    · **Acid rate const (M(H+)-hr)$^{-1}$:**
*Comment:*

    · **Base rate const (M(OH-)-hr)$^{-1}$:**
*Comment:*

# Bis(2-chloroisopropyl) ether

<u>CAS Registry Number:</u>  108-60-1

<u>Half-lives:</u>
 · **Soil:**                          High:    4320 hours            (6 months)
                                      Low:      432 hours             (18 days)
  *Comment:*  Scientific judgement based upon estimated unacclimated aerobic aqueous biodegradation half-life.

 · **Air:**                           High:    46.1 hours
                                      Low:     4.61 hours
  *Comment:*  Scientific judgement based upon estimated photooxidation half-life in air.

 · **Surface Water:**                 High:    4320 hours            (6 months)
                                      Low:      432 hours             (18 days)
  *Comment:*  Scientific judgement based upon estimated unacclimated aerobic aqueous biodegradation half-life.

 · **Ground Water:**                  High:    8640 hours            (12 months)
                                      Low:      864 hours             (36 days)
  *Comment:*  Scientific judgement based upon estimated unacclimated aerobic aqueous biodegradation half-life.

<u>Aqueous Biodegradation (unacclimated):</u>
 · **Aerobic half-life:**             High:    4320 hours            (6 months)
                                      Low:      432 hours             (18 days)
  *Comment:*  Scientific judgement based upon river die-away test data (high $t_{1/2}$) (Kleopfer, RD and Fairless, BJ (1972)) and aerobic soil column study data (low $t_{1/2}$) (Kincannon, DF and Lin, YS (1985)).

 · **Anaerobic half-life:**           High:    17280 hours           (24 months)
                                      Low:     1728 hours            (72 days)
  *Comment:*  Scientific judgement based upon estimated unacclimated aerobic aqueous biodegradation half-life.

 · **Removal/secondary treatment:**   High:      No data
                                      Low:
  *Comment:*

<u>Photolysis:</u>
 · **Atmos photol half-life:**        High:
                                      Low:

*Comment:*

**· Max light absorption (nm):**          No data
*Comment:*

**· Aq photol half-life:**               High:
                                         Low:

*Comment:*

**Photooxidation half-life:**
   **· Water:**                          High:      No data
                                         Low:

*Comment:*

   **· Air:**                            High:      46.1 hours
                                         Low:       4.61 hours

*Comment:*  Scientific judgement based upon estimated rate constant for reaction with hydroxyl radical in air (Atkinson, R (1987A)).

**Reduction half-life:**                 High:      No data
                                         Low:

*Comment:*

**Hydrolysis:**
   **· First-order hydr half-life:**
*Comment:*

   **· Acid rate const (M(H+)-hr)$^{-1}$:**          No data
*Comment:*

   **· Base rate const (M(OH-)-hr)$^{-1}$:**
*Comment:*

# Melamine

CAS Registry Number: 108-78-1

Structure:

NH$_2$

N

N

N

H$_2$N

N

NH$_2$

Half-lives:
   · Soil:                                       High:    4320 hours         (6 months)

                                             Low:     672 hours          (4 weeks)

*Comment:* Scientific judgement based upon estimated unacclimated aqueous aerobic biodegradation half-life.

   · Air:                                         High:    10.5 hours

                                           Low:     1.05 hours

*Comment:* Scientific judgement based upon estimated rate constant for reaction with hydroxyl radicals in air (Atkinson, R (1987A)).

                                       High:    3480 hours         (145 days)

                                         Low:     62.4 hours        (2.6 days)

*Comment:* Scientific judgement based upon estimated rate constants for reactions of representative aromatic amines with ·OH and RO$_2$· (Mill, T and Mabey, W (1985)).

   · Ground Water:                        High:    8640 hours         (12 months)

                                         Low:     1344 hours        (8 weeks)

*Comment:* Scientific judgement based upon estimated unacclimated aqueous aerobic biodegradation half-life.

Aqueous Biodegradation (unacclimated):
   · Aerobic half-life:                    High:    4320 hours         (6 months)

                                           Low:     672 hours          (4 weeks)

*Comment:* Scientific judgement based upon unacclimated aerobic aqueous screening test data which confirmed resistance to biodegradation (Sasaki, S (1978); Heukelekian, H and Rand, MC (1955)).

   · Anaerobic half-life:                 High:  17280 hours      (24 months)

                                         Low:   2688 hours      (16 weeks)

*Comment:* Scientific judgement based upon estimated unacclimated aqueous aerobic biodegradation half-life.

   · Removal/secondary treatment:     High:     No data

Low:

*Comment:*

## Photolysis:
· **Atmos photol half-life:**        High:     No data
                                      Low:

*Comment:*

· **Max light absorption (nm):**      absorption shoulder at 235 nm (0.01% aq. NaOH)
*Comment:* Sadtler UV No. 1499.

· **Aq photol half-life:**            High:     No data
                                        Low:

*Comment:*

## Photooxidation half-life:
· **Water:**                    High:    3480 hours        (145 days)
                                      Low:     62.4 hours        (2.6 days)
*Comment:* Scientific judgement based upon estimated rate constants for reactions of representative aromatic amines with ·OH and $RO_2$· (Mill, T and Mabey, W (1985)). It is assumed that melamine reacts three times as fast as aniline.

· **Air:**                         High:    10.5 hours
                                      Low:     1.05 hours
*Comment:* Scientific judgement based upon estimated rate constant for reaction with hydroxyl radicals in air (Atkinson, R (1987A)).

## Reduction half-life:
                                      High:     No data
                                      Low:

*Comment:*

## Hydrolysis:
· **First-order hydr half-life:**     Not expected to be significant
*Comment:*

· **Acid rate const $(M(H+)-hr)^{-1}$:**
*Comment:*

· **Base rate const $(M(OH-)-hr)^{-1}$:**
*Comment:*

# Toluene

CAS Registry Number: 108-88-3

## Half-lives:

**· Soil:**                          High:      528 hours          (22 days)
                                     Low:        96 hours          (4 days)
*Comment:* Scientific judgement based upon estimated aqueous aerobic biodegradation half-life.

**· Air:**                           High:      104 hours          (4.3 days)
                                     Low:        10 hours
*Comment:* Based upon photooxidation half-life in air.

**· Surface Water:**                 High:      528 hours          (22 days)
                                     Low:        96 hours          (4 days)
*Comment:* Scientific judgement based upon estimated aqueous aerobic biodegradation half-life.

**· Ground Water:**                  High:      672 hours          (4 weeks)
                                     Low:       168 hours          (7 days)
*Comment:* Scientific judgement based upon unacclimated grab sample data of aerobic soil from ground water aquifers (Wilson, JT et al. (1983A), Swindoll, CM et al. (1987)).

## Aqueous Biodegradation (unacclimated):

**· Aerobic half-life:**             High:      528 hours          (22 days)
                                     Low:        96 hours          (4 days)
*Comment:* Scientific judgement based upon an acclimated sea water die-away test (Wakeham, SG et al. (1983)).

**· Anaerobic half-life:**           High:     5040 hours          (30 weeks)
                                     Low:      1344 hours          (8 weeks)
*Comment:* Scientific judgement based upon anaerobic sediment grab sample data (high $t_{1/2}$) and anaerobic screening test data (Horowitz, A et al. (1982)).

**· Removal/secondary treatment:**   High:           75%
                                     Low:
*Comment:* Based upon % degraded in an 8 day period under anaerobic continuous flow conditions (Zeyer, J et al. (1986A)).

## Photolysis:

**· Atmos photol half-life:**        High:      No data
                                     Low:
*Comment:*

· **Max light absorption (nm):**   lambda max = 268, 264, 261, 259.5, 255, 253.5
*Comment:* Toluene at a concn of 1 g/L in methanol did not absorb UV light at wavelengths greater than 280 nm (Sadtler UV No. 155).

· **Aq photol half-life:**   High:   No data
   Low:
*Comment:*

## Photooxidation half-life:
· **Water:**   High:   1284 hours      (54 days)
   Low:   321 hours      (13 days)
*Comment:* Based upon measured rate data for hydroxyl radicals in aqueous solution (Dorfman, LM and Adams, GE (1973)).

· **Air:**   High:   104 hours      (4.3 days)
   Low:   10 hours
*Comment:* Based upon measured rate data for the vapor phase reaction with hydroxyl radicals in air (Atkinson, R (1985)).

## Reduction half-life:   High:   No data
   Low:
*Comment:*

## Hydrolysis:
· **First-order hydr half-life:**   No hydrolyzable groups
*Comment:* Rate constant at neutral pH is zero (Kollig, HP et al. (1987)).

· **Acid rate const (M(H+)-hr)$^{-1}$:**
*Comment:*

· **Base rate const (M(OH-)-hr)$^{-1}$:**
*Comment:*

# Chlorobenzene

<u>CAS Registry Number:</u>  108-90-7

<u>Half-lives:</u>
· **Soil:**                           High:    3600 hours              (150 days)
                                        Low:     1632 hours              (68 days)
*Comment:* Scientific judgement based upon estimated aqueous aerobic biodegradation half-life.

· **Air:**                            High:    729 hours               (30.4 days)
                                        Low:     72.9 hours              (3.0 days)
*Comment:* Based upon photooxidation half-life in air.

· **Surface Water:**                  High:    3600 hours              (150 days)
                                        Low:     1632 hours              (68 days)
*Comment:* Scientific judgement based upon unacclimated aerobic river die-away tests (low $t_{1/2}$: Hungspreugs, M et al. (1984), high $t_{1/2}$: Lee, RF and Ryan, C (1979)).

· **Ground Water:**                   High:    7200 hours              (300 days)
                                        Low:     3264 hours              (136 days)
*Comment:* Scientific judgement based upon estimated aqueous aerobic biodegradation half-life.

<u>Aqueous Biodegradation (unacclimated):</u>
· **Aerobic half-life:**              High:    3600 hours              (150 days)
                                        Low:     1632 hours              (68 days)
*Comment:* Scientific judgement based upon unacclimated aerobic river die-away tests (low $t_{1/2}$: Hungspreugs, M et al. (1984), high $t_{1/2}$: Lee, RF and Ryan, C (1979)).

· **Anaerobic half-life:**            High:    14400 hours             (600 days)
                                        Low:     6528 hours              (272 days)
*Comment:* Scientific judgement based upon estimated aqueous aerobic biodegradation half-life.

· **Removal/secondary treatment:**    High:    No data
                                        Low:
*Comment:*

<u>Photolysis:</u>
· **Atmos photol half-life:**         High:    No data
                                        Low:
*Comment:*

· **Max light absorption (nm):**         lambda max = 265, 215.5
*Comment:* No absorbance of UV light at wavelengths greater than 320 nm in iso-octane (Sadtler UV No. 16).

· **Aq photol half-life:**         High:      No data
                                   Low:

*Comment:*

## Photooxidation half-life:

· **Water:**         High:    62106 hours              (7.1 years)
                     Low:      1553 hours              (64.7 days)
*Comment:* Based upon measured rate data for hydroxyl radicals in aqueous solution (Dorfman, LM and Adams, GE (1973)).

· **Air:**         High:      729 hours              (30.4 days)
                   Low:      72.9 hours              (3.0 days)
*Comment:* Based upon measured rate data for the vapor phase reaction with hydroxyl radicals in air (Atkinson, R et al.(1985B)).

## Reduction half-life:         High:      No data
                                Low:

*Comment:*

## Hydrolysis:

· **First-order hydr half-life:**         >879 years
*Comment:* Scientific judgement based upon rate constant (<0.9 $M^{-1}$ $hr^{-1}$) extrapolated to pH 7 at 25 °C from 1% disappearance after 16 days at 85 °C and pH 9.7 (Ellington, JJ et al. (1988)).

· **Acid rate const $(M(H+)-hr)^{-1}$:**
*Comment:*

· **Base rate const $(M(OH-)-hr)^{-1}$:**         <0.9
*Comment:* Scientific judgement based upon 1% disappearance after 16 days at 85 °C and pH 9.7 (Ellington, JJ et al. (1988)).

# Phenol

<u>CAS Registry Number:</u>  108-95-2

<u>Half-lives:</u>
  · **Soil:**                              High:     240 hours              (10 days)
                                           Low:       24 hours              (1 day)
*Comment:*  Based upon aerobic soil die-away study data (Haider, K et al. (1974); Baker, MD and Mayfield, CI (1980)).


  · **Air:**                               High:     22.8 hours
                                           Low:      2.28 hours
*Comment:*  Based upon measured reaction rate constant for ·OH with phenol (Atkinson, R (1987A)).


  · **Surface Water:**                     High:     56.5 hours             (2.4 days)
                                           Low:       5.3 hours             (0.22 days)
*Comment:*  Scientific judgement based upon estimated aqueous aerobic biodegradation half-life and aqueous photolysis half-life.


  · **Ground Water:**                      High:     168 hours              (7 days)
                                           Low:       12 hours              (0.5 days)
*Comment:*  Scientific judgement based upon estimated aqueous aerobic biodegradation half-life.

<u>Aqueous Biodegradation (unacclimated):</u>
  · **Aerobic half-life:**                 High:      84 hours              (3.5 days)
                                           Low:        6 hours              (0.25 days)
*Comment:*  Scientific judgement based upon aerobic river die-away study data (high $t_{1/2}$) (Borighem, G and Vereecken, J (1978) and lake die-away study data (low $t_{1/2}$) (Rubin, HE and Alexander, M (1983)).


  · **Anaerobic half-life:**               High:     672 hours              (28 days)
                                           Low:      192 hours              (8 days)
*Comment:*  Scientific judgement based upon aqueous anaerobic screening studies (Boyd, SA et al. (1983); Healy, JB and Young, LY (1978)).


  · **Removal/secondary treatment:**       High:     99.9%
                                           Low:       90%
*Comment:*  Removal percentages based upon continuous activated sludge treatment simulator data (Stover, EL and Kincannon, DF (1983); Ludzack, FJ et al. (1961A)).

<u>Photolysis:</u>
  · **Atmos photol half-life:**            High:     173 hours              (7.2 days)

$$\text{Low:} \quad 46 \text{ hours} \quad (1.9 \text{ days})$$

*Comment:* Based upon reported half-lives for photolysis under sunlight for phenol in distilled water in summer (low $t_{1/2}$) and winter (high $t_{1/2}$) (Hwang, H et al. (1986)).

· **Max light absorption (nm):** lambda max = 269 nm (1% aqueous ethanol)
*Comment:* Drahanovsky, J and Vacek, Z (1971)).

· **Aq photol half-life:** High: 173 hours (7.2 days)
                                   Low: 46 hours (1.9 days)
*Comment:* Based upon reported half-lives for photolysis under sunlight for phenol in distilled water in summer (low $t_{1/2}$) and winter (high $t_{1/2}$) (Hwang, H et al. (1986)).

## Photooxidation half-life:
· **Water:** High: 3840 hours (160 days)
                  Low: 77 hours (3.2 days)
*Comment:* Scientific judgement based upon reported reaction rate constants for $RO_2\cdot$ with the phenol class (Mill, T and Mabey, W (1985)).

· **Air:** High: 22.8 hours
                Low: 2.28 hours
*Comment:* Based upon measured reaction rate constant for OH with phenol (Atkinson, R (1987A)).

## Reduction half-life:
High: No data
Low:

*Comment:*

## Hydrolysis:
· **First-order hydr half-life:** No hydrolyzable functions
*Comment:*

· **Acid rate const $(M(H+)\text{-hr})^{-1}$:**
*Comment:*

· **Base rate const $(M(OH-)\text{-hr})^{-1}$:**
*Comment:*

# 2-Methoxyethanol

CAS Registry Number: 109-86-4

**Half-lives:**
- **Soil:**                         High:     672 hours           (4 weeks)
                                    Low:      168 hours           (7 days)

*Comment:* Scientific judgement based upon estimated unacclimated aqueous aerobic biodegradation half-life.

- **Air:**                          High:     57 hours            (2.4 days)
                                    Low:      5.7 hours

*Comment:* Scientific judgement based upon estimated rate constant for reaction with hydroxyl radicals in air (Atkinson, R (1987A)).

- **Surface Water:**                High:     672 hours           (4 weeks)
                                    Low:      168 hours           (7 days)

*Comment:* Scientific judgement based upon estimated unacclimated aqueous aerobic biodegradation half-life.

- **Ground Water:**                 High:     1344 hours          (8 weeks)
                                    Low:      336 hours           (14 days)

*Comment:* Scientific judgement based upon estimated unacclimated aqueous aerobic biodegradation half-life.

**Aqueous Biodegradation (unacclimated):**
- **Aerobic half-life:**            High:     672 hours           (4 weeks)
                                    Low:      168 hours           (7 days)

*Comment:* Scientific judgement based upon unacclimated aerobic aqueous screening test data (Bridie, AL et al. (1979); Price, KS et al. (1974); Heukelekian, H and Rand, MC (1955)).

- **Anaerobic half-life:**          High:     2688 hours          (16 weeks)
                                    Low:      672 hours           (28 days)

*Comment:* Scientific judgement based upon estimated unacclimated aqueous aerobic biodegradation half-life.

- **Removal/secondary treatment:**  High:     65%
                                    Low:

*Comment:* Based upon biological oxygen demand results from an activated sludge dispersed seed aeration treatment simulator (Mills, EJ Jr and Stack, VT Jr (1954)).

**Photolysis:**
- **Atmos photol half-life:**       High:     No data

416

Low:

*Comment:*

· **Max light absorption (nm):**
*Comment:*

· **Aq photol half-life:**                     High:
                                               Low:

*Comment:*

## Photooxidation half-life:
· **Water:**                                   High: $4.7 \times 10^5$ hours        (54 years)
                                               Low:     7400 hours               (308 days)

*Comment:* Based upon measure rate constants for reaction with hydroxyl radicals in water (Anbar, M and Neta, P (1967); Dorfman, LM and Adams, GE (1973)).

· **Air:**                                     High:     57 hours               (2.4 days)
                                               Low:     5.7 hours

*Comment:* Scientific judgement based upon estimated rate constant for reaction with hydroxyl radicals in air (Atkinson, R (1987A)).

## Reduction half-life:                        High:     No data
                                               Low:

*Comment:*

## Hydrolysis:
· **First-order hydr half-life:**              No data
*Comment:*

· **Acid rate const $(M(H+)\text{-hr})^{-1}$:**
*Comment:*

· **Base rate const $(M(OH-)\text{-hr})^{-1}$:**
*Comment:*

# Furan

<u>CAS Registry Number:</u>  110-00-9

<u>Half-lives:</u>
  · **Soil:**                                  High:    672 hours          (4 weeks)
                                          Low:     168 hours          (7 days)
*Comment:* Scientific judgement based upon estimated unacclimated aqueous aerobic biodegradation half-life.

  · **Air:**                                   High:    4.72 hours
                                          Low:     0.477 hours
*Comment:* Based upon photooxidation half-life in air.

  · **Surface Water:**                  High:    672 hours          (4 weeks)
                                          Low:     13.8 hours
*Comment:* Scientific judgement based upon estimated unacclimated aqueous aerobic biodegradation half-life (high $t_{1/2}$) and estimated photooxidation half-life in water (low $t_{1/2}$).

  · **Ground Water:**                 High:    2688 hours        (16 weeks)
                                          Low:     672 hours          (28 days)
*Comment:* Scientific judgement based upon estimated unacclimated aqueous aerobic biodegradation half-life.

<u>Aqueous Biodegradation (unacclimated):</u>
  · **Aerobic half-life:**              High:    672 hours          (4 weeks)
                                          Low:     168 hours          (7 days)
*Comment:* Scientific judgement based upon river die-away test data for benzene (Vaishnav, DD and Babeu, L (1987)).

  · **Anaerobic half-life:**           High:    2688 hours        (16 weeks)
                                          Low:     672 hours          (28 days)
*Comment:* Scientific judgement based upon estimated unacclimated aqueous aerobic biodegradation half-life.

  · **Removal/secondary treatment:**    High:    No data
                                          Low:
*Comment:*

<u>Photolysis:</u>
  · **Atmos photol half-life:**        High:
                                          Low:
*Comment:*

**· Max light absorption (nm):**     No data
*Comment:*

**· Aq photol half-life:**     High:
                               Low:

*Comment:*

## Photooxidation half-life:
**· Water:**     High:     1375 hours          (57.3 days)
                 Low:      13.8 hours
*Comment:* Based upon measured rate constant for reaction with singlet oxygen in water (Mill, T and Mabey, W (1985)).

**· Air:**     High:     4.72 hours
               Low:      0.477 hours
*Comment:* Based upon measured rate constants for reaction with hydroxyl radical (Atkinson, R (1987A)) and ozone (Atkinson, R and Carter, WPL (1984)) in air.

## Reduction half-life:     High:     No data
                            Low:
*Comment:*

## Hydrolysis:
**· First-order hydr half-life:**
*Comment:*

**· Acid rate const $(M(H+)-hr)^{-1}$:**     No hydrolyzable groups
*Comment:*

**· Base rate const $(M(OH-)-hr)^{-1}$:**
*Comment:*

# 2-Ethoxyethanol

<u>CAS Registry Number:</u>  110-80-5

<u>Half-lives:</u>
  · **Soil:**                          High:     672 hours           (4 weeks)
                                        Low:      168 hours           (7 days)
*Comment:* Scientific judgement based upon estimated unacclimated aqueous aerobic biodegradation half-life.

  · **Air:**                           High:     53.5 hours          (2.23 days)
                                        Low:      5.35 hours
*Comment:* Based upon photooxidation half-life in air.

  · **Surface Water:**                 High:     672 hours           (4 weeks)
                                        Low:      168 hours           (7 days)
*Comment:* Scientific judgement based upon estimated unacclimated aqueous aerobic biodegradation half-life.

  · **Ground Water:**                  High:     1344 hours          (8 weeks)
                                        Low:      336 hours           (14 days)
*Comment:* Scientific judgement based upon estimated unacclimated aqueous aerobic biodegradation half-life.

<u>Aqueous Biodegradation (unacclimated):</u>
  · **Aerobic half-life**:             High:     672 hours           (4 weeks)
                                        Low:      168 hours           (7 days)
*Comment:* Scientific judgement based upon estimated unacclimated aqueous aerobic biodegradation screening test data (Bogan, RH and Sawyer, CN (1955)).

  · **Anaerobic half-life**:           High:     2688 hours          (16 weeks)
                                        Low:      672 hours           (28 days)
*Comment:* Scientific judgement based upon estimated unacclimated aqueous aerobic biodegradation half-life.

  · **Removal/secondary treatment:**   High:     85%
                                        Low:      83%
*Comment:* Removal percentages based upon biological oxygen demand results from an activated sludge dispersed seed aeration treatment simulator (low %) (Mills, EJ Jr and Stack, VT Jr (1954)) and data from a continuous activated sludge biological treatment simulator (Brown, JA Jr and Weintraub, M (1982)).

<u>Photolysis:</u>

· **Atmos photol half-life:**          High:
                                         Low:

*Comment:*

· **Max light absorption (nm):**    No data
*Comment:*

· **Aq photol half-life:**         High:
                                         Low:

*Comment:*

## Photooxidation half-life:
· **Water:**                    High: $3.9 \times 10^5$ hours         (44 years)
                                    Low:    9625 hours         (1.1 year)
*Comment:* Based upon measured rate constant for reaction with hydroxyl radical in water (Anbar, M and Neta, P (1967)).

· **Air:**                      High:    53.5 hours         (2.23 days)
                                    Low:    5.35 hours
*Comment:* Based upon measured rate constant for reaction with hydroxyl radical in air (Atkinson, R (1987A)).

## Reduction half-life:
                                        High:      No data
                                        Low:

*Comment:*

## Hydrolysis:
· **First-order hydr half-life:**
*Comment:*

· **Acid rate const $(M(H+)\text{-}hr)^{-1}$:**    No hydrolyzable groups
*Comment:*

· **Base rate const $(M(OH\text{-})\text{-}hr)^{-1}$:**
*Comment:*

# Cyclohexane

<u>CAS Registry Number:</u>  110-82-7

Half-lives:
  · **Soil:**                         High:     4320 hours              (6 months)
                                      Low:       672 hours              (4 weeks)
  *Comment:*  Scientific judgement based upon unacclimated grab sample of aerobic soil (high $t_{1/2}$:
  Haider, K et al. (1974)) and aerobic aqueous screening test data (Kawasaki, M (1980)).

  · **Air:**                          High:        87 hours              (3.6 days)
                                      Low:        8.7 hours
  *Comment:*  Based upon photooxidation half-life in air.

  · **Surface Water:**                High:     4320 hours              (6 months)
                                      Low:       672 hours              (4 weeks)
  *Comment:*  Scientific judgement based upon estimated unacclimated aqueous aerobic
  biodegradation half-life.

  · **Ground Water:**                 High:     8640 hours              (12 months)
                                      Low:      1344 hours              (8 weeks)
  *Comment:*  Scientific judgement based upon estimated unacclimated aqueous aerobic
  biodegradation half-life.

Aqueous Biodegradation (unacclimated):
  · **Aerobic half-life:**            High:     4032 hours              (6 months)
                                      Low:       672 hours              (4 weeks)
  *Comment:*  Scientific judgement based upon unacclimated grab sample of aerobic soil (high $t_{1/2}$:
  Haider, K et al. (1974)) and aerobic aqueous screening test data (Kawasaki, M (1980)).

  · **Anaerobic half-life:**          High:    16128 hours              (24 months)
                                      Low:      2688 hours              (16 weeks)
  *Comment:*  Scientific judgement based upon estimated unacclimated aqueous aerobic
  biodegradation half-life.

  · **Removal/secondary treatment:**  High:      No data
                                      Low:
  *Comment:*

Photolysis:
  · **Atmos photol half-life:**       High:
                                      Low:
  *Comment:*

· **Max light absorption (nm):**        No data
*Comment:*

· **Aq photol half-life:**        High:
                                  Low:
*Comment:*

## Photooxidation half-life:
· **Water:**        High:$6.9 \times 10^{10}$ hours    ($7.8 \times 10^{6}$ years)
                    Low:  $1.4 \times 10^{9}$ hours    ($1.6 \times 10^{5}$ years)
*Comment:* Based upon measured rate data for alkylperoxyl radicals in aqueous solution (Hendry, DG (1974)).

· **Air:**        High:    87 hours    (3.6 days)
                Low:    8.7 hours
*Comment:* Based upon measured rate data for the vapor phase reaction with hydroxyl radicals in air (Atkinson, R (1985)).

## Reduction half-life:
                             High:     No data
                             Low:
*Comment:*

## Hydrolysis:
· **First-order hydr half-life:**    No data
*Comment:*

· **Acid rate const (M(H+)-hr)$^{-1}$:**
*Comment:*

· **Base rate const (M(OH-)-hr)$^{-1}$:**
*Comment:*

# Pyridine

<u>CAS Registry Number:</u>  110-86-1

<u>Half-lives:</u>
  · **Soil:**                            High:     168 hours         (7 days)
                                          Low:      24 hours         (1 day)

*Comment:* Scientific judgement based upon unacclimated grab sample of aerobic soil (Sims, GK and Sommers, LE (1985)).

  · **Air:**                                High:    1284 hours     (53 days)
                                          Low:     128 hours      (5.3 days)

*Comment:* Based upon photooxidation half-life in air.

  · **Surface Water:**               High:     168 hours         (7 days)
                                            Low:      24 hours         (1 day)

*Comment:* Scientific judgement based upon estimated aqueous aerobic biodegradation half-life.

  · **Ground Water:**               High:     336 hours       (14 days)
                                            Low:      48 hours        (2 days)

*Comment:* Scientific judgement based upon estimated aqueous aerobic biodegradation half-life.

<u>Aqueous Biodegradation (unacclimated):</u>
  · **Aerobic half-life:**           High:     168 hours         (7 days)
                                          Low:      24 hours         (1 day)

*Comment:* Scientific judgement based upon unacclimated grab sample of aerobic soil (Sims, GK and Sommers, LE (1985)).

  · **Anaerobic half-life:**        High:     672 hours       (28 days)
                                          Low:     168 hours       (7 days)

*Comment:* Scientific judgement based upon anaerobic acclimated screening test data (Naik, MN et al. (1972)).

  · **Removal/secondary treatment:**    High:        99%
                                          Low:

*Comment:* Based upon % degraded in a 1 day period under acclimated aerobic continuous flow conditions (Gerike, P and Fischer WK (1979)).

<u>Photolysis:</u>
  · **Atmos photol half-life:**       High:     No data
                                          Low:

*Comment:*

**· Max light absorption (nm):**          lambda max 256.5
*Comment:* Pyridine in methanol did not absorb UV light at wavelengths greater than 290 nm (Sadtler UV No. 9).

**· Aq photol half-life:**          High:     No data
                                    Low:
*Comment:*

## Photooxidation half-life:
**· Water:**          High: $2.1 \times 10^5$ hours          (24.4 years)
                      Low: $5.4 \times 10^3$ hours          (14.7 years)
*Comment:* Based upon measured rate data for hydroxyl radicals in aqueous solution (Dorfman, LM and Adams, GE (1973)).

**· Air:**          High:     1284 hours          (53 days)
                    Low:      128 hours           (5.3 days)
*Comment:* Based upon measured rate data for the vapor phase reaction with hydroxyl radicals in air (Atkinson, R et al. (1987)).

## Reduction half-life:          High:     No data
                                 Low:
*Comment:*

## Hydrolysis:
**· First-order hydr half-life:**          No hydrolyzable groups
*Comment:* Rate constant at neutral pH is zero (Kollig, HP et al. (1987)).

**· Acid rate const $(M(H+)-hr)^{-1}$:**          0.0
*Comment:* Based upon measured rate data at acid pH (Kollig, HP et al. (1987)).

**· Base rate const $(M(OH-)-hr)^{-1}$:**          0.0
*Comment:* Based upon measured rate data at basic pH (Kollig, HP et al. (1987)).

# Diethanolamine

CAS Registry Number: 111-42-2

Half-lives:
· Soil:                    High:    168 hours         (7 days)
                           Low:     14.4 hours        (0.6 days)
Comment: Scientific judgement based upon estimated unacclimated aqueous aerobic biodegradation half-life.

· Air:                     High:    7.2 hours
                           Low:     0.72 hours
Comment: Scientific judgement based upon estimated rate constant for reaction with hydroxyl radicals in air (Atkinson, R (1987A)).

· Surface Water:           High:    168 hours         (7 days)
                           Low:     14.4 hours        (0.6 days)
Comment: Scientific judgement based upon estimated unacclimated aqueous aerobic biodegradation half-life.

· Ground Water:            High:    336 hours         (14 days)
                           Low:     28.8 hours        (1.2 days)
Comment: Scientific judgement based upon estimated unacclimated aqueous aerobic biodegradation half-life.

Aqueous Biodegradation (unacclimated):
· Aerobic half-life:       High:    168 hours         (7 days)
                           Low:     14.4 hours        (0.6 days)
Comment: Scientific judgement based upon measured half-life in surface water grab sample experiment (low $t_{1/2}$) (Boethling, RS and Alexander, M (1979)) and aqueous aerobic screening test data (high $t_{1/2}$) (Gerike, P and Fischer, WK (1979); Bridie, AL et al. (1979)).

· Anaerobic half-life:     High:    672 hours         (28 days)
                           Low:     57.6 hours        (2.4 days)
Comment: Scientific judgement based upon estimated unacclimated aqueous aerobic biodegradation half-life.

· Removal/secondary treatment:   High:    94%
                                 Low:     0%
Comment: Percent removal based upon activated sludge coupled unit treatment simulator (high removal percentage) (Gerike, P and Fischer, WK (1979)) and change in BOD in 24 hours in a domestic sewage semicontinuous treatment simulator (Mills, EJ Jr and Stack, VT Jr (1954)).

## Photolysis:
 · **Atmos photol half-life:**　　　　　High:　　　No data
 　　　　　　　　　　　　　　　　　　　Low:

 *Comment:*

 · **Max light absorption (nm):**
 *Comment:*

 · **Aq photol half-life:**　　　　　　　High:
 　　　　　　　　　　　　　　　　　　　Low:

 *Comment:*

## Photooxidation half-life:
 · **Water:**　　　　　　　　　　　　　High:　　　No data
 　　　　　　　　　　　　　　　　　　　Low:

 *Comment:*

 · **Air:**　　　　　　　　　　　　　　High:　　　7.2 hours
 　　　　　　　　　　　　　　　　　　　Low:　　　0.72 hours
 *Comment:*  Based upon estimated rate constant for reaction with hydroxyl radicals in air (Atkinson, R (1987A)).

## Reduction half-life:　　　　　　　　　High:　　　No data
 　　　　　　　　　　　　　　　　　　　Low:

 *Comment:*

## Hydrolysis:
 · **First-order hydr half-life:**　　　　No data
 *Comment:*

 · **Acid rate const $(M(H+)-hr)^{-1}$:**
 *Comment:*

 · **Base rate const $(M(OH-)-hr)^{-1}$:**
 *Comment:*

# Bis(2-chloroethyl) ether

<u>CAS Registry Number:</u>  111-44-4

<u>Half-lives:</u>
· **Soil:**                        High:    4320 hours            (6 months)
                                    Low:     672 hours            (4 weeks)
*Comment:* Scientific judgement based upon estimated unacclimated aerobic aqueous biodegradation half-life.

· **Air:**                         High:    96.5 hours           (4.02 days)
                                    Low:     9.65 hours
*Comment:* Scientific judgement based upon estimated photooxidation half-life in air.

· **Surface Water:**               High:    4320 hours            (6 months)
                                    Low:     672 hours            (4 weeks)
*Comment:* Scientific judgement based upon estimated unacclimated aerobic aqueous biodegradation half-life.

· **Ground Water:**                High:    8640 hours            (12 months)
                                    Low:     1344 hours           (8 weeks)
*Comment:* Scientific judgement based upon estimated unacclimated aerobic aqueous biodegradation half-life.

<u>Aqueous Biodegradation (unacclimated):</u>
· **Aerobic half-life**:           High:    4320 hours            (6 months)
                                    Low:     672 hours            (4 weeks)
*Comment:* Scientific judgement based upon river die-away test data (Ludzack, FJ and Ettinger, MB (1963); Doljido, JR (1979)).

· **Anaerobic half-life**:         High:    17280 hours           (24 months)
                                    Low:     2688 hours           (16 weeks)
*Comment:* Scientific judgement based upon estimated unacclimated aerobic aqueous biodegradation half-life.

· **Removal/secondary treatment:**  High:
                                     Low:
*Comment:*

<u>Photolysis:</u>
· **Atmos photol half-life:**       High:
                                    Low:
*Comment:*

· **Max light absorption (nm):**   No data
*Comment:*

· **Aq photol half-life:**   High:
   Low:

*Comment:*

**Photooxidation half-life:**
   · **Water:**   High:   No data
   Low:

*Comment:*

   · **Air:**   High:   96.5 hours   (4.02 days)
   Low:   9.65 hours
*Comment:* Scientific judgement based upon estimated rate constant for reaction with hydroxyl radical in air (Atkinson, R (1987A)).

**Reduction half-life:**   High:   No data
   Low:
*Comment:*

**Hydrolysis:**
   · **First-order hydr half-life:**   $1.93 \times 10^5$ hours   (22 years)
   *Comment:* Scientific judgement based upon neutral hydrolysis rate constant at 20°C which was extrapolated from data for hydrolysis in aqueous dioxane at 100°C (Mabey, WR et al. (1981)).

   · **Acid rate const $(M(H+)-hr)^{-1}$:**   No data
   *Comment:*

   · **Base rate const $(M(OH-)-hr)^{-1}$:**   No data
   *Comment:*

# Ethylene glycol, monobutyl ether

<u>CAS Registry Number:</u>  111-76-2

<u>Half-lives:</u>
- **· Soil:**

  | | High: | 672 hours | (4 weeks) |
  | | Low: | 168 hours | (7 days) |

  *Comment:*  Scientific judgement based upon estimated unacclimated aqueous aerobic biodegradation half-life.

- **· Air:**

  | | High: | 32.8 hours |
  | | Low: | 3.28 hours |

  *Comment:*  Scientific judgement based upon estimated photooxidation half-life in air.

- **· Surface Water:**

  | | High: | 672 hours | (4 weeks) |
  | | Low: | 168 hours | (7 days) |

  *Comment:*  Scientific judgement based upon estimated unacclimated aqueous aerobic biodegradation half-life.

- **· Ground Water:**

  | | High: | 1344 hours | (8 weeks) |
  | | Low: | 336 hours | (14 days) |

  *Comment:*  Scientific judgement based upon estimated unacclimated aqueous aerobic biodegradation half-life.

<u>Aqueous Biodegradation (unacclimated):</u>
- **· Aerobic half-life:**

  | | High: | 672 hours | (4 weeks) |
  | | Low: | 168 hours | (7 days) |

  *Comment:*  Scientific judgement based upon unacclimated aqueous aerobic biodegradation screening test data (Bridie, AL et al. (1979)).

- **· Anaerobic half-life:**

  | | High: | 2688 hours | (16 weeks) |
  | | Low: | 672 hours | (28 days) |

  *Comment:*  Scientific judgement based upon estimated unacclimated aqueous aerobic biodegradation half-life.

- **· Removal/secondary treatment:**

  | | High: | No data |
  | | Low: | |

  *Comment:*

<u>Photolysis:</u>
- **· Atmos photol half-life:**

  | | High: |
  | | Low: |

  *Comment:*

· **Max light absorption (nm):**  No data
*Comment:*

· **Aq photol half-life:**  High:
Low:

*Comment:*

**Photooxidation half-life:**
· **Water:**  High:  No data
Low:
*Comment:*

· **Air:**  High:  32.8 hours
Low:  3.28 hours
*Comment:* Scientific judgement based upon estimated rate constant for reaction with hydroxyl radical in air (Atkinson, R (1987A)).

**Reduction half-life:**  High:  No data
Low:
*Comment:*

**Hydrolysis:**
· **First-order hydr half-life:**
*Comment:*

· **Acid rate const (M(H+)-hr)$^{-1}$:**  No hydrolyzable groups
*Comment:*

· **Base rate const (M(OH-)-hr)$^{-1}$:**
*Comment:*

# Diethylene glycol, monoethyl ether

<u>CAS Registry Number:</u>  111-90-0

<u>Half-lives:</u>
  · **Soil:**                          High:     672 hours           (4 weeks)
                                       Low:      168 hours           (7 days)
  *Comment:* Scientific judgement based upon estimated unacclimated aqueous aerobic biodegradation half-life.

  · **Air:**                           High:     22.3 hours
                                       Low:      2.23 hours
  *Comment:* Scientific judgement based upon estimated photooxidation half-life in air.

  · **Surface Water:**                 High:     672 hours           (4 weeks)
                                       Low:      168 hours           (7 days)
  *Comment:* Scientific judgement based upon estimated unacclimated aqueous aerobic biodegradation half-life.

  · **Ground Water:**                  High:     1344 hours          (8 weeks)
                                       Low:      336 hours           (14 days)
  *Comment:* Scientific judgement based upon estimated unacclimated aqueous aerobic biodegradation half-life.

<u>Aqueous Biodegradation (unacclimated):</u>
  · **Aerobic half-life:**             High:     672 hours           (4 weeks)
                                       Low:      168 hours           (7 days)
  *Comment:* Scientific judgement based upon estimated unacclimated aqueous aerobic biodegradation screening test data (Bogan, RH and Sawyer, CN (1955)).

  · **Anaerobic half-life:**           High:     2688 hours          (16 weeks)
                                       Low:      672 hours           (28 days)
  *Comment:* Scientific judgement based upon estimated unacclimated aqueous aerobic biodegradation half-life.

  · **Removal/secondary treatment:**   High:     No data
                                       Low:
  *Comment:*

<u>Photolysis:</u>
  · **Atmos photol half-life:**        High:
                                       Low:
  *Comment:*

· **Max light absorption (nm):**      No data
*Comment:*

· **Aq photol half-life:**      High:
                                  Low:

*Comment:*

**Photooxidation half-life:**
· **Water:**      High:      No data
                                Low:
*Comment:*

· **Air:**      High:      22.3 hours
                                Low:      2.23 hours
*Comment:* Scientific judgement based upon estimated rate constant for reaction with hydroxyl radical in air (Atkinson, R (1987A)).

**Reduction half-life:**      High:      No data
                                Low:
*Comment:*

**Hydrolysis:**
· **First-order hydr half-life:**
*Comment:*

· **Acid rate const (M(H+)-hr)$^{-1}$:**      No hydrolyzable groups
*Comment:*

· **Base rate const (M(OH-)-hr)$^{-1}$:**
*Comment:*

# Propoxur

**CAS Registry Number:**  114-26-1

**Structure:**

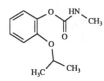

**Half-lives:**
 · **Soil:**                          High:     672 hours              (4 weeks)
                                      Low:       38 hours              (1.6 days)
*Comment:* Scientific judgement based upon estimated hydrolysis $t_{1/2}$ at pH 9.0 (low $t_{1/2}$) (Aly, DM and El-Dib, MA (1971A)) and estimated unacclimated aqueous aerobic biodegradation half-life (high $t_{1/2}$).

 · **Air:**                           High:     7.1 hours
                                      Low:      0.71 hours
*Comment:* Scientific judgement based upon estimated rate constant for reaction with hydroxyl radicals in air (Atkinson, R (1987A)).

 · **Surface Water:**                 High:     672 hours              (4 weeks)
                                      Low:       38 hours              (1.6 days)
*Comment:* Scientific judgement based upon estimated hydrolysis $t_{1/2}$ at pH 9.0 (low $t_{1/2}$) (Aly, DM and El-Dib, MA (1971A)) and estimated unacclimated aqueous aerobic biodegradation half-life (high $t_{1/2}$).

 · **Ground Water:**                  High:     1344 hours             (8 weeks)
                                      Low:        38 hours             (1.6 days)
*Comment:* Scientific judgement based upon estimated hydrolysis $t_{1/2}$ at pH 9.0 (low $t_{1/2}$) (Aly, DM and El-Dib, MA (1971A)) and estimated unacclimated aqueous aerobic biodegradation half-life (high $t_{1/2}$).

**Aqueous Biodegradation (unacclimated):**
 · **Aerobic half-life**:             High:     672 hours              (4 weeks)
                                      Low:      168 hours              (7 days)
*Comment:* Scientific judgement based upon unacclimated aqueous aerobic screening test data (Gummer, WD (1979); Kanazawa, J (1987)).

 · **Anaerobic half-life**:           High:     2688 hours             (16 weeks)
                                      Low:       672 hours             (28 days)
*Comment:* Scientific judgement based upon estimated unacclimated aqueous aerobic

biodegradation half-life.

· **Removal/secondary treatment:**     High:     No data
                                      Low:
*Comment:*

## Photolysis:
· **Atmos photol half-life:**     High:     87.9 hours     (3.7 days)
                                  Low:     62.5 hours     (2.6 days)
*Comment:* Based upon measured rate of photolysis on bean leaves in sunlight (low $t_{1/2}$) (Ivie, GW and Casida, JE (1971)) and in aqueous solution under simulated sunlight (high $t_{1/2}$) (Jensen-Korte, U et al. (1987)).

· **Max light absorption (nm):**     lambda max = 254 nm
*Comment:* (Aly, DM and El-Dib, MA (1971A)).

· **Aq photol half-life:**     High:     87.9 hours     (3.7 days)
                               Low:     62.5 hours     (2.6 days)
*Comment:* Based upon measured rate of photolysis on bean leaves in sunlight (low $t_{1/2}$) (Ivie, GW and Casida, JE (1971)) and in aqueous solution under simulated sunlight (high $t_{1/2}$) (Jensen-Korte, U et al. (1987)).

## Photooxidation half-life:
· **Water:**     High:     No data
                 Low:
*Comment:*

· **Air:**     High:     7.1 hours
               Low:     0.71 hours
*Comment:* Scientific judgement based upon estimated rate constant for reaction with hydroxyl radicals in air (Atkinson, R (1987A)).

## Reduction half-life:
                 High:     No data
                 Low:
*Comment:*

## Hydrolysis:
· **First-order hydr half-life:**     No data
*Comment:* Scientific judgement based upon measured base catalyzed hydrolysis rate constant ($t_{1/2}$ = 157 days at pH 7) (Aly, DM and El-Dib, MA (1971A)).

· **Acid rate const $(M(H+)-hr)^{-1}$:**     No data
*Comment:*

435

· **Base rate const (M(OH-)-hr)$^{-1}$:**     3.04X10$^1$ l/mole-sec
*Comment:*  (t$_{1/2}$ = 38 hours at pH 9) (Aly, DM and El-Dib, MA (1971A)).

# Azaserine

CAS Registry Number: 115-02-6

Structure:

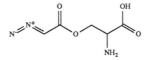

Half-lives:
- Soil:                        High:    1344 hours        (8 weeks)
                               Low:     196 hours         (8.2 days)
Comment: Scientific judgement based upon hydrolysis half-life for pH 5 and 25 °C (Ellington, JJ et al. (1988A)) and estimated aqueous aerobic biodegradation half-life.

- Air:                         High:    10.8 hours
                               Low:     1.1 hours
Comment: Scientific judgement based upon estimated photooxidation half-life in air.

- Surface Water:               High:    1344 hours        (8 weeks)
                               Low:     196 hours         (8.2 days)
Comment: Scientific judgement based upon hydrolysis half-life for pH 5 and 25 °C (Ellington, JJ et al. (1988A)) and estimated aqueous aerobic biodegradation half-life.

- Ground Water:                High:    1344 hours        (8 weeks)
                               Low:     196 hours         (8.2 days)
Comment: Scientific judgement based upon hydrolysis half-life for pH 5 and 25 °C (Ellington, JJ et al. (1988A)) and estimated aqueous aerobic biodegradation half-life.

Aqueous Biodegradation (unacclimated):
- Aerobic half-life:           High:    672 hours         (4 weeks)
                               Low:     168 hours         (1 week)
Comment: Scientific judgement.

- Anaerobic half-life:         High:    2688 hours        (16 weeks)
                               Low:     672 hours         (4 weeks)
Comment: Scientific judgement based upon estimated aqueous aerobic biodegradation half-life.

- Removal/secondary treatment: High:    No data
                               Low:
Comment:

## Photolysis:
· **Atmos photol half-life:**       High:
                                           Low:

*Comment:*

· **Max light absorption (nm):**     No data
*Comment:*

· **Aq photol half-life:**          High:
                                           Low:

*Comment:*

## Photooxidation half-life:
· **Water:**                       High:       No data
                                           Low:

*Comment:*

· **Air:**                           High:       10.8 hours
                                           Low:        1.1 hours
*Comment:* Scientific judgement based upon estimated rate data for hydroxyl radicals in air (Atkinson, R (1987A)).

## Reduction half-life:                  High:       No data
                                               Low:

*Comment:*

## Hydrolysis:
· **First-order hydr half-life:**           2376 hours           (99 days)
*Comment:* Based upon neutral rate constant ($k_N = 2.6 \times 10^{-4}$ $hr^{-1}$) at pH 7 and 25 °C (Ellington, JJ et al. (1988A)).

· **Acid rate const $(M(H+)\text{-}hr)^{-1}$:**     328
*Comment:* ($t_{1/2}$ = 8.2 days) Based upon neutral ($k_N = 2.6 \times 10^{-4}$ $hr^{-1}$) and acid rate constants (328 $M^{-1}$ $hr^{-1}$) at pH 5 and 25 °C (Ellington, JJ et al. (1988A)).

· **Base rate const $(M(OH\text{-})\text{-}hr)^{-1}$:**     6.8
*Comment:* ($t_{1/2}$ = 8.8 days) Based upon neutral ($k_N = 2.6 \times 10^{-4}$ $hr^{-1}$) and base rate constants (6.8 $M^{-1}$ $hr^{-1}$) at pH 9 and 25 °C (Ellington, JJ et al. (1988A)).

438

# Propylene

**CAS Registry Number:** 115-07-1

## Half-lives:
· **Soil:**                              High:     672 hours        (4 weeks)

                                         Low:      168 hours        (7 days)

*Comment:* Scientific judgement based upon estimated aerobic biodegradation half-lives.

· **Air:**                               High:     13.7 hours

                                         Low:      1.7 hours

*Comment:* Based upon measured photooxidation rate constants in air.

· **Surface Water:**                High:     672 hours        (4 weeks)

                                         Low:      168 hours        (7 days)

*Comment:* Scientific judgement based upon estimated aerobic biodegradation half-lives.

· **Ground Water:**               High:     1344 hours      (8 weeks)

                                         Low:      336 hours       (14 days)

*Comment:* Scientific judgement based upon estimated aerobic biodegradation half-lives.

## Aqueous Biodegradation (unacclimated):
· **Aerobic half-life:**            High:     672 hours        (4 weeks)

                                         Low:      168 hours        (7 days)

*Comment:* Scientific judgement.

· **Anaerobic half-life:**       High:     2688 hours      (16 weeks)

                                         Low:      672 hours       (28 days)

*Comment:* Scientific judgement based upon estimated aerobic biodegradation half-lives.

· **Removal/secondary treatment:**    High:     No data

                                         Low:

*Comment:*

## Photolysis:
· **Atmos photol half-life:**      High:     No data

                                         Low:

*Comment:*

· **Max light absorption (nm):**

*Comment:*

· **Aq photol half-life:**               High:

*Comment:*

## Photooxidation half-life:
  · **Water:**                      High:     43000 hours           (4.9 years)

                                          Low:      1070 hours            (44.5 days)

*Comment:* Based upon measured photooxidation rate constant in water with ·OH (Guesten, H et al. (1981)).

  · **Air:**                            High:     13.7 hours

                                            Low:       1.7 hours

*Comment:* Based upon measured photooxidation rate constants in air with ·OH and ozone (Atkinson, R (1985); Atkinson, R and Carter, WPL (1984)).

## Reduction half-life:
                                            High:     No data

                                            Low:

*Comment:*

## Hydrolysis:
  · **First-order hydr half-life:**            No hydrolyzable groups

*Comment:*

  · **Acid rate const $(M(H+)-hr)^{-1}$:**

*Comment:*

  · **Base rate const $(M(OH-)-hr)^{-1}$:**

*Comment:*

# Endosulfan

**CAS Registry Number:** 115-29-7

**Structure:**

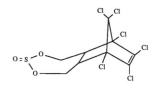

**Half-lives:**

· **Soil:**              High:    218 hours             (9.1 days)
                         Low:     4.5 hours

*Comment:* Scientific judgement based upon aqueous hydrolysis half-lives for alpha- and beta-endosulfan at pH 7 (low $t_{1/2}$) and 9 (high $t_{1/2}$) and 25° C, respectively (Ellington, JJ et al. (1988A)).

· **Air:**               High:    24.8 hours
                         Low:     2.5 hours

*Comment:* Scientific judgement based upon estimated photooxidation half-life in air for a deoxygenated endosulfan analog.

· **Surface Water:**     High:    218 hours             (9.1 days)
                         Low:     4.5 hours

*Comment:* Scientific judgement based upon aqueous hydrolysis half-lives for alpha- and beta-endosulfan at pH 7 (low $t_{1/2}$) and 9 (high $t_{1/2}$) and 25° C, respectively (Ellington, JJ et al. (1988A)).

· **Ground Water:**      High:    218 hours             (9.1 days)
                         Low:     4.5 hours

*Comment:* Scientific judgement based upon aqueous hydrolysis half-lives for alpha- and beta-endosulfan at pH 7 (low $t_{1/2}$) and 9 (high $t_{1/2}$) and 25° C, respectively (Ellington, JJ et al. (1988A)).

**Aqueous Biodegradation (unacclimated):**

· **Aerobic half-life:**     High:    336 hours             (14 days)
                             Low:     48 hours              (2 days)

*Comment:* Scientific judgement based upon unacclimated aerobic river die-away test data (high $t_{1/2}$: Eichelberger, JW and Lichtenberg, JJ (1971)) and reported soil grab sample data (low $t_{1/2}$: Bowman, MC et al. (1965)).

· **Anaerobic half-life:**   High:    1344 hours            (56 days)
                             Low:     192 hours             (8 days)

441

*Comment:* Scientific judgement based upon unacclimated aerobic biodegradation half-life.

· **Removal/secondary treatment:**  High: No data
Low:
*Comment:*

## Photolysis:
· **Atmos photol half-life:**  High: No data
Low:
*Comment:*

· **Max light absorption (nm):**  No data
*Comment:*

· **Aq photol half-life:**  High: No data
Low:
*Comment:*

## Photooxidation half-life:
· **Water:**  High: No data
Low:
*Comment:*

· **Air:**  High: 28.8 hours
Low: 2.5 hours
*Comment:* Scientific judgement based upon an estimated rate constant for the vapor phase reaction with hydroxyl radicals in air with a deoxygenated endosulfan analog (Atkinson, R (1987A)).

## Reduction half-life:  High: No data
Low:
*Comment:*

## Hydrolysis:
· **First-order hydr half-life:**  218 hours  (9.1 days)
*Comment:* Scientific judgement based upon neutral rate constant for alpha-endosulfan ($3.2 \times 10^{-3}$ $hr^{-1}$) at pH 7 and 25 °C (Ellington, JJ et al. (1988A)).

· **First-order hydr half-life:**  187 hours  (7.8 days)
*Comment:* Scientific judgement based upon neutral rate constant for beta-endosulfan ($3.7 \times 10^{-3}$ $hr^{-1}$) at pH 7 and 25 °C (Ellington, JJ et al. (1988A)).

· **Acid rate const $(M(H+)-hr)^{-1}$:**  $8.1 \times 10^{-3}$ $M^{-1}$ $hr^{-1}$
*Comment:* ($t_{1/2}$ = 9.1 days) For alpha-endosulfan at pH 5 and 25 °C (Ellington, JJ et al. (1988A)).

· **Acid rate const (M(H+)-hr)$^{-1}$:**      $7.4 \times 10^{-3}$ M$^{-1}$ hr$^{-1}$

*Comment:* (t$_{1/2}$ = 7.8 days) For beta-endosulfan at pH 5 and 25 °C (Ellington, JJ et al. (1988A)).

· **Base rate const (M(OH-)-hr)$^{-1}$:**      $1.0 \times 10^{4}$ M$^{-1}$ hr$^{-1}$

*Comment:*   (t$_{1/2}$ = 6.7 hours) Based upon a rate constant for alpha-endosulfan at pH 9 and 25 °C (Ellington, JJ et al. (1988A)).

· **Base rate const (M(OH-)-hr)$^{-1}$:**      $1.5 \times 10^{4}$ M$^{-1}$ hr$^{-1}$

*Comment:*   (t$_{1/2}$ = 4.5 hours) Based upon a rate constant for beta-endosulfan at pH 9 and 25 °C (Ellington, JJ et al. (1988A)).

# Aldicarb

CAS Registry Number: 116-06-3

Structure:

Half-lives:
 · Soil:                               High:    8664 hours         (361 days)
                                       Low:      480 hours         (20 days)
 Comment: Scientific judgement based upon unacclimated aerobic soil grab sample data (Ou, LT et al. (1985)).

 · Air:                                High:     9.5 hours
                                       Low:       1 hour
 Comment: Scientific judgement based upon estimated photooxidation half-life in air.

 · Surface Water:                      High:    8664 hours         (361 days)
                                       Low:      480 hours         (20 days)
 Comment: Scientific judgement based upon estimated aqueous aerobic biodegradation half-life.

 · Ground Water:                       High:   15240 hours         (635 days)
                                       Low:      960 hours         (40 days)
 Comment: Scientific judgement based upon estimated aqueous aerobic biodegradation half-life and anaerobic ground water grab sample data (high $t_{1/2}$: Miles, CJ and Delfino JJ (1985)).

Aqueous Biodegradation (unacclimated):
 · Aerobic half-life:                  High:    8664 hours         (361 days)
                                       Low:      480 hours         (20 days)
 Comment: Scientific judgement based upon unacclimated aerobic soil grab sample data (Ou, LT et al. (1985)).

 · Anaerobic half-life:                High:   15240 hours         (635 days)
                                       Low:     1488 hours         (62 days)
 Comment: Scientific judgement based upon anaerobic ground water grab sample data (Miles, CJ and Delfino JJ (1985)).

 · Removal/secondary treatment:        High:     No data
                                       Low:

444

*Comment:*

**Photolysis:**
   · **Atmos photol half-life:**             High:      No data
                                         Low:
   *Comment:*

   · **Max light absorption (nm):**     No data
   *Comment:*

   · **Aq photol half-life:**             High:      No data
                                         Low:
   *Comment:*

**Photooxidation half-life:**
   · **Water:**                       High:      No data
                                         Low:
   *Comment:*

   · **Air:**                          High:      9.5 hours
                                       Low:        1 hour
   *Comment:* Scientific judgement based upon an estimated rate constant for vapor phase reaction with hydroxyl radicals in air (Atkinson, R (1987A)).

**Reduction half-life:**              High:      No data
                                         Low:
   *Comment:*

**Hydrolysis:**
   · **First-order hydr half-life:**          4580 days        (12.5 years)
   *Comment:* Based upon a first order rate constant ($1.51 \times 10^{-4}$ day$^{-1}$) at pH 5.5 and 5 °C (Hansen, JL and Spiegel, MH (1983)).

   · **First-order hydr half-life:**          1320 hours        (55 days)
   *Comment:* Based upon a first order rate constant ($1.3 \times 10^{-2}$ day$^{-1}$) at pH 8.85 and 20 °C (Given, CJ and Dierberg, FE (1985)).

   · **Acid rate const (M(H+)-hr)$^{-1}$:**
   *Comment:*

   · **Base rate const (M(OH-)-hr)$^{-1}$:**
   *Comment:*

# 2-Aminoanthraquinone

<u>CAS Registry Number:</u>  117-79-3

<u>Structure:</u>

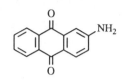

<u>Half-lives:</u>
- **Soil:**

|  | | |
|---|---|---|
| High: | 672 hours | (4 weeks) |
| Low: | 168 hours | (7 days) |

*Comment:* Scientific judgement based upon estimated aqueous aerobic biodegradation half-lives.

- **Air:**

|  | | |
|---|---|---|
| High: | 23 hours | |
| Low: | 2.3 hours | |

*Comment:* Based upon estimated photooxidation rate constant in air.

- **Surface Water:**

|  | | |
|---|---|---|
| High: | 672 hours | (4 weeks) |
| Low: | 62.4 hours | (2.6 days) |

*Comment:* The high value is scientific judgement based upon estimated aqueous aerobic biodegradation half-life. The low value is based upon aqueous photooxidation half-life.

- **Ground Water:**

|  | | |
|---|---|---|
| High: | 1344 hours | (8 weeks) |
| Low: | 336 hours | (14 days) |

*Comment:* Scientific judgement based upon estimated aqueous aerobic biodegradation half-lives.

<u>Aqueous Biodegradation (unacclimated):</u>
- **Aerobic half-life**:

|  | | |
|---|---|---|
| High: | 672 hours | (4 weeks) |
| Low: | 168 hours | (7 days) |

*Comment:* Scientific judgement based upon biological screening test data for anthraquinone (Kondo, M et al. (1988A); Rontani, JF et al. (1985A)).

- **Anaerobic half-life**:

|  | | |
|---|---|---|
| High: | 2688 hours | (16 weeks) |
| Low: | 672 hours | (28 days) |

*Comment:* Scientific judgement based upon estimated aqueous aerobic biodegradation half-lives.

- **Removal/secondary treatment:**

|  | |
|---|---|
| High: | No data |
| Low: | |

*Comment:*

## Photolysis:
· **Atmos photol half-life:** High: No data
Low:

*Comment:*


· **Max light absorption (nm):** lambda max = 323 nm
*Comment:* Absorption maxima in dioxane (Sadtler 3788 UV); overall absorption extends above 355 nm.


· **Aq photol half-life:** High: No data
Low:

*Comment:* No photodegradation was observed when an alkaline solution was irradiated with visible light (Kato, S et al. (1978)).


## Photooxidation half-life:
· **Water:** High: 3480 hours (145 days)
Low: 62.4 hours (2.6 days)

*Comment:* Scientific judgement based upon photooxidation rate constants with ·OH and $RO_2$· for the aromatic amine class (Mill, T and Mabey, W (1985); Guesten, H et al. (1981)).


· **Air:** High: 23 hours
Low: 2.3 hours

*Comment:* Based upon estimated photooxidation rate constant with ·OH in air (Atkinson, R (1987A)).


## Reduction half-life:
High: No data
Low:

*Comment:*


## Hydrolysis:
· **First-order hydr half-life:** No hydrolyzable groups
*Comment:*


· **Acid rate const $(M(H+)-hr)^{-1}$:**
*Comment:*


· **Base rate const $(M(OH-)-hr)^{-1}$:**
*Comment:*

447

# Bis-(2-ethylhexyl) phthalate

<u>CAS Registry Number:</u>  117-81-7

<u>Half-lives:</u>

| · Soil: | High: | 550 hours | (23 days) |
|---|---|---|---|
| | Low: | 120 hours | (5 days) |

*Comment:* Scientific judgement based upon unacclimated aqueous aerobic biodegradation half-life.

| · Air: | High: | 29 hours | |
|---|---|---|---|
| | Low: | 2.9 hours | |

*Comment:* Based upon estimated photooxidation half-life in air.

| · Surface Water: | High: | 550 hours | (23 days) |
|---|---|---|---|
| | Low: | 120 hours | (5 days) |

*Comment:* Based upon unacclimated aqueous aerobic biodegradation half-life.

| · Ground Water: | High: | 9336 hours | (389 days) |
|---|---|---|---|
| | Low: | 240 hours | (10 days) |

*Comment:* Scientific judgement based upon estimated unacclimated aqueous anaerobic (high $t_{1/2}$) and aerobic (low $t_{1/2}$) biodegradation half-life.

<u>Aqueous Biodegradation (unacclimated):</u>

| · Aerobic half-life: | High: | 550 hours | (23 days) |
|---|---|---|---|
| | Low: | 120 hours | (5 days) |

*Comment:* Based upon grab sample die-away test data and scientific judgement (Schouten, MJ et al. (1979); Johnson, BT and Lulves, W (1975)).

| · Anaerobic half-life: | High: | 9336 hours | (389 days) |
|---|---|---|---|
| | Low: | 980 hours | (42 days) |

*Comment:* Scientific judgement based upon anaerobic die-away test data (low $t_{1/2}$) (Shanker, R et al. (19859)) and anaerobic aqueous screening studies in which 0-9% theoretical methane production was observed in up to 70 days (Horowitz, A et al. (1982); Shelton, DR et al. (1984)).

| · Removal/secondary treatment: | High: | 91% | |
|---|---|---|---|
| | Low: | 70% | |

*Comment:* high: Graham, PR (1973); low: Saeger, VW and Tucker, ES (1976).

<u>Photolysis:</u>

| · Atmos photol half-life: | High: | 4800 hours | (200 days) |
|---|---|---|---|
| | Low: | 3500 hours | (144 days) |

*Comment:* Scientific judgement based upon measured rate of aqueous photolysis for dimethyl phthalate (Wolfe, NL et al. (1980A)).

· **Max light absorption (nm):**        No data
*Comment:*

· **Aq photol half-life:**        High:    4800 hours        (200 days)
                                    Low:     3500 hours        (144 days)
*Comment:* Scientific judgement based upon measured rate of aqueous photolysis for dimethyl phthalate (Wolfe, NL et al. (1980A)).

## Photooxidation half-life:
· **Water:**        High: $1.4 \times 10^4$ hours        (584 days)
                         Low:     1056 hours        (44 days)
*Comment:* Scientific judgement based upon measured rate data for reaction with peroxyl radicals in water (Wolfe, NL et al. (1980A)).

· **Air:**        High:     29 hours
                   Low:     2.9 hours
*Comment:* Based upon estimated rate constant for reaction with hydroxyl radicals in air (Atkinson, R (1987A)).

## Reduction half-life:
                                 High:     No data
                                 Low:
*Comment:*

## Hydrolysis:
· **First-order hydr half-life:**        No data ($t_{1/2}$ = 2,000 years at pH 7)
*Comment:* Scientific judgement based upon measured base catalyzed hydrolysis rate constant (Wolfe, NL et al. (1980B)).

· **Acid rate const (M(H+)-hr)$^{-1}$:**        No data

· **Base rate const (M(OH-)-hr)$^{-1}$:**        $1.1 \times 10^{-4}$ l/mole-sec ($t_{1/2}$ = 20 years at pH 9)
*Comment:* (Wolfe, NL et al. (1980B)).

# Di(n-octyl) phthalate

CAS Registry Number: 117-84-0

## Half-lives:
· **Soil:**                                High:    672 hours              (4 weeks)
                                           Low:     168 hours              (7 days)
*Comment:* Scientific judgement based upon estimated aqueous aerobic biodegradation half-life.

· **Air:**                                 High:    44.8 hours             (1.9 days)
                                           Low:     4.5 hours
*Comment:* Scientific judgement based upon estimated photooxidation half-life in air.

· **Surface Water:**                       High:    672 hours              (4 weeks)
                                           Low:     168 hours              (7 days)
*Comment:* Scientific judgement based upon estimated aqueous aerobic biodegradation half-life.

· **Ground Water:**                        High:    8760 hours             (1 year)
                                           Low:     336 hours              (14 days)
*Comment:* Scientific judgement based upon estimated aqueous aerobic (low $t_{1/2}$) and anaerobic (high $t_{1/2}$) biodegradation half-lives.

## Aqueous Biodegradation (unacclimated):
· **Aerobic half-life:**                   High:    672 hours              (4 weeks)
                                           Low:     168 hours              (7 days)
*Comment:* Scientific judgement based upon unacclimated (Sasaki, S (1978)) and acclimated aqueous screening test data (Sugatt, RH et al. (1984)).

· **Anaerobic half-life:**                 High:    8760 hours             (1 year)
                                           Low:     4320 hours             (6 months)
*Comment:* Scientific judgement based upon acclimated anaerobic screening test data (Shelton, DR et al. (1984); Horowitz, A et al. (1982)).

· **Removal/secondary treatment:**         High:    68.8%
                                           Low:     0%
*Comment:* Based upon % degraded under aerobic continuous flow conditions (Davis, EM et al. (1983); Petrasek, AC et al. (1983)).

## Photolysis:
· **Atmos photol half-life:**              High:    No data
                                           Low:
*Comment:*

**· Max light absorption (nm):**  lambda max = 274.5, 223
*Comment:* No absorbance occurs above 320 nm in cyclohexane (Sadtler UV No. 226).

**· Aq photol half-life:**  High:   No data
  Low:
*Comment:*

## Photooxidation half-life:
**· Water:**  High:   No data
  Low:
*Comment:*

**· Air:**  High:   44.8 hours   (1.9 days)
  Low:   4.5 hours
*Comment:* Scientific judgement based upon an estimated rate constant for vapor phase reaction with hydroxyl radicals in air (Atkinson, R (1987A)).

## Reduction half-life:  High:   No data
  Low:
*Comment:*

## Hydrolysis:
**· First-order hydr half-life:**  107 years
*Comment:* Based upon base rate constant ($k_B = 7.4$ M$^{-1}$ hr$^{-1}$) at 25 °C and pH of 7 (Ellington, JJ et al. (1987)).

**· Acid rate const (M(H+)-hr)$^{-1}$:**  No data
*Comment:*

**· Base rate const (M(OH-)-hr)$^{-1}$:**  7.4
*Comment:* At pH 9, $t_{1/2} = 1$ year, based upon rate constant at 25 °C (Ellington, JJ et al. (1987)).

# Hexachlorobenzene

<u>CAS Registry Number:</u>  118-74-1

<u>Half-lives:</u>
  · **Soil:**                          High:    50136 hours              (5.7 years)
                                        Low:     23256 hours              (2.7 years)
  *Comment:*  Scientific judgement based upon unacclimated aerobic soil grab sample data (Beck, J and Hansen, KE (1974)).

  · **Air:**                           High:    37530 hours              (4.2 years)
                                        Low:      3753 hours             (156.4 days)
  *Comment:*  Scientific judgement based upon estimated photooxidation half-life in air.

  · **Surface Water:**                 High:    50136 hours              (5.7 years)
                                        Low:     23256 hours              (2.7 years)
  *Comment:*  Scientific judgement based upon estimated unacclimated aqueous aerobic biodegradation half-life.

  · **Ground Water:**                  High:   100272 hours             (11.4 years)
                                        Low:     46512 hours              (5.3 years)
  *Comment:*  Scientific judgement based upon estimated unacclimated aqueous aerobic biodegradation half-life.

<u>Aqueous Biodegradation (unacclimated):</u>
  · **Aerobic half-life**:             High:    50136 hours              (5.7 years)
                                        Low:     23256 hours              (2.7 years)
  *Comment:*  Scientific judgement based upon unacclimated aerobic soil grab sample data (Beck, J and Hansen, KE (1974)).

  · **Anaerobic half-life**:           High:   200544 hours             (22.9 years)
                                        Low:     93024 hours             (10.6 years)
  *Comment:*  Scientific judgement based upon estimated unacclimated aqueous aerobic biodegradation half-life.

  · **Removal/secondary treatment:**   High:        No data
                                        Low:
  *Comment:*

<u>Photolysis:</u>
  · **Atmos photol half-life:**        High:        No data
                                        Low:
  *Comment:*

· **Max light absorption (nm):**      lambda max = 301.0, 291.0
*Comment:* In isooctane (IARC 20: 155 (1979)).

· **Aq photol half-life:**            High:     No data
                                      Low:

*Comment:*

**Photooxidation half-life:**
· **Water:**                          High:     No data
                                      Low:

*Comment:*

· **Air:**                            High:   37530 hours         (4.2 years)
                                      Low:     3753 hours         (156.4 days)
*Comment:* Scientific judgement based upon estimated rate constant for the vapor phase reaction with hydroxyl radicals in air (Atkinson, R (1987A)).

**Reduction half-life:**              High:     No data
                                      Low:

*Comment:*

**Hydrolysis:**
· **First-order hydr half-life:**     Not expected to be important
*Comment:* Scientific judgement based upon zero hydrolysis was observed after 13 days for pHs of 3, 7 and 11 at 85 °C (Ellington, JJ et al. (1987)).

· **Acid rate const (M(H+)-hr)$^{-1}$:**     Not expected to be important
*Comment:* Scientific judgement based upon zero hydrolysis was observed after 13 days at 85 °C for a pH of 3 (Ellington, JJ et al. (1987)).

· **Base rate const (M(OH-)-hr)$^{-1}$:**     Not expected to be important
*Comment:* Scientific judgement based upon zero hydrolysis was observed after 13 days at 85 °C for a pH of 11 (Ellington, JJ et al. (1987)).

# 2,4,6-Trinitrotoluene

<u>CAS Registry Number:</u>  118-96-7

<u>Half-lives:</u>

· **Soil:**                    High:    4320 hours           (6 months)

                               Low:      672 hours            (4 weeks)

*Comment:*  Scientific judgement based upon estimated unacclimated aqueous aerobic biodegradation half-life.

· **Air:**                     High:    11.3 hours

                               Low:      3.7 hours

*Comment:*  Based upon measured rate constant for sunlight photolysis in distilled water (Spanggord, RJ et al. (1980A)).

· **Surface Water:**           High:    1.28 hours

                               Low:     0.160 hours

*Comment:*  Based upon measured rate constant for sunlight photolysis and photooxidation in natural waters (Spanggord, RJ et al. (1980A)).

· **Ground Water:**            High:    8640 hours           (12 months)

                               Low:      672 hours            (4 weeks)

*Comment:*  Scientific judgement based upon estimated unacclimated aqueous anaerobic (low $t_{1/2}$) and aerobic (high $t_{1/2}$) biodegradation half-lives.

<u>Aqueous Biodegradation (unacclimated):</u>

· **Aerobic half-life**:       High:    4320 hours           (6 months)

                               Low:      672 hours            (4 weeks)

*Comment:*  Scientific judgement based upon aerobic natural water die-away test data (Spanggord, RJ et al. (1981)).  Reduction of the nitro group observed in these studies may not be due to biodegradation.

· **Anaerobic half-life**:     High:    4320 hours           (6 months)

                               Low:      672 hours            (4 weeks)

*Comment:*  Scientific judgement based upon anaerobic natural water die-away test data (Spanggord, RJ et al. (1981)).  Reduction of the nitro group observed in these studies may not be due to biodegradation.

· **Removal/secondary treatment:**    High:    90%

                               Low:     86%

*Comment:*  Removal percentages based upon data from a continuous activated sludge biological treatment simulator (Bringmann, G and Kuehn, R (1971)).

## Photolysis:
**· Atmos photol half-life:**  High: 11.3 hours
Low: 3.7 hours
*Comment:* Based upon measured rate constant for sunlight photolysis in distilled water (Spanggord, RJ et al. (1980A)).

**· Max light absorption (nm):**  lambda max not available
*Comment:* Absorption in water solution extends to >400 nm (Spanggord, RJ et al. (1980)).

**· Aq photol half-life:**  High: 11.3 hours
Low: 3.7 hours
*Comment:* Based upon measured rate constant for sunlight photolysis in distilled water (Spanggord, RJ et al. (1980A)).

## Photooxidation half-life:
**· Water:**  High: 1.45 hours
Low: 0.168 hours
*Comment:* Based upon measured rate constant for sunlight photolysis and photooxidation in natural waters corrected for rate of sunlight photolysis in distilled water (Spanggord, RJ et al. (1980A)).

**· Air:**  High: 4407 hours   (184 days)
Low: 441 hours   (18.4 days)
*Comment:* Scientific judgement based upon estimated rate constant for reaction with hydroxyl radical in air (Atkinson, R (1987A)).

## Reduction half-life:  High: No data
Low:
*Comment:*

## Hydrolysis:
**· First-order hydr half-life:**
*Comment:*

**· Acid rate const (M(H+)-hr)$^{-1}$:**  No hydrolyzable groups
*Comment:*

**· Base rate const (M(OH-)-hr)$^{-1}$:**
*Comment:*

# 3,3'-Dimethoxybenzidine

CAS Registry Number: 119-90-4

Half-lives:
- Soil:                          High:    4320 hours          (6 months)
                                 Low:      672 hours          (4 weeks)
Comment: Scientific judgement based upon estimated unacclimated aqueous aerobic biodegradation half-life.

- Air:                           High:    3.47 hours
                                 Low:     0.347 hours
Comment: Scientific judgement based upon estimated photooxidation half-life in air.

- Surface Water:                 High:    1740 hours          (72.5 days)
                                 Low:      31.2 hours         (1.3 days)
Comment: Scientific judgement based upon photooxidation half-life in water.

- Ground Water:                  High:    8640 hours          (12 months)
                                 Low:      672 hours          (4 weeks)
Comment: Scientific judgement based upon estimated unacclimated aqueous aerobic (high $t_{1/2}$) and anaerobic (low $t_{1/2}$) biodegradation half-life.

Aqueous Biodegradation (unacclimated):
- Aerobic half-life:             High:    4320 hours          (6 months)
                                 Low:      672 hours          (4 weeks)
Comment: Scientific judgement based upon unacclimated aqueous aerobic biodegradation screening test data (Kawasaki, M (1980); Brown, D and Labourer, P (1983A)).

- Anaerobic half-life:           High:    2688 hours          (16 weeks)
                                 Low:      672 hours          (28 days)
Comment: Scientific judgement based upon unacclimated aqueous anaerobic biodegradation screening test data (Brown, D and Hamburger, B (1987)).

- Removal/secondary treatment:   High:    No data
                                 Low:
Comment:

Photolysis:
- Atmos photol half-life:        High:    No data
                                 Low:
Comment:

· **Max light absorption (nm):**          lambda max 212, 302 nm (methanol)
*Comment:* Absorption continues up to 348 nm (Sadtler UV No. 17845).


· **Aq photol half-life:**          High:          No data
                                    Low:

*Comment:*


## Photooxidation half-life:
· **Water:**                        High:    1740 hours          (72.5 days)
                                    Low:     31.2 hours          (1.3 days)
*Comment:* Scientific judgement based upon estimated rate constants for reactions of
representative aromatic amines with $HO^.$ and $RO_2^.$ (Mill, T and Mabey, W (1985)).  It is assumed
that 3,3'-dimethoxybenzidine reacts twice as fast as aniline.


· **Air:**                          High:    3.47 hours
                                    Low:     0.347 hours
*Comment:* Scientific judgement based upon estimated rate constant for reaction with hydroxyl
radical in air (Atkinson, R (1987A)).


## Reduction half-life:                High:          No data
                                    Low:

*Comment:*


## Hydrolysis:
· **First-order hydr half-life:**
*Comment:*


· **Acid rate const $(M(H+)-hr)^{-1}$:**          No hydrolyzable groups
*Comment:*


· **Base rate const $(M(OH-)-hr)^{-1}$:**
*Comment:*

# 3,3'-Dimethylbenzidine

<u>CAS Registry Number:</u>  119-93-7

## Half-lives:
· **Soil:**                          High:      168 hours            (7 days)
                                     Low:        24 hours            (1 day)
*Comment:* Scientific judgement based upon estimated unacclimated aqueous aerobic biodegradation half-life.

· **Air:**                           High:      2.67 hours
                                     Low:       0.267 hours
*Comment:* Scientific judgement based upon estimated photooxidation half-life in air.

· **Surface Water:**                 High:      168 hours            (7 days)
                                     Low:        24 hours            (1 day)
*Comment:* Scientific judgement based upon estimated unacclimated aqueous aerobic biodegradation half-life.

· **Ground Water:**                  High:      336 hours            (14 days)
                                     Low:        48 hours            (2 days)
*Comment:* Scientific judgement based upon estimated unacclimated aqueous aerobic biodegradation half-life.

## Aqueous Biodegradation (unacclimated):
· **Aerobic half-life**:             High:      168 hours            (7 days)
                                     Low:        24 hours            (1 day)
*Comment:* Scientific judgement based upon unacclimated aqueous aerobic biodegradation screening test data (Baird, R et al. (1977)).

· **Anaerobic half-life**:           High:      672 hours            (28 days)
                                     Low:        96 hours            (4 days)
*Comment:* Scientific judgement based upon estimated unacclimated aqueous aerobic biodegradation half-life.

· **Removal/secondary treatment:**   High:      No data
                                     Low:
*Comment:*

## Photolysis:
· **Atmos photol half-life:**        High:
                                     Low:
*Comment:*

· **Max light absorption (nm):**          No data
*Comment:*

· **Aq photol half-life:**          High:
                                    Low:

*Comment:*

**Photooxidation half-life:**
    · **Water:**          High:    1740 hours          (72.5 days)
                          Low:     31.2 hours          (1.3 days)
*Comment:* Scientific judgement based upon estimated rate constants for reactions of representative aromatic amines with HO· and RO$_2$ (Mill, T and Mabey, W (1985)). It is assumed that 3,3'-dimethylbenzidine reacts twice as fast as aniline.

    · **Air:**          High:    2.67 hours
                        Low:     0.267 hours
*Comment:* Scientific judgement based upon estimated rate constant for reaction with hydroxyl radical in air (Atkinson, R (1987A)).

**Reduction half-life:**          High:          No data
                                  Low:
*Comment:*

**Hydrolysis:**
    · **First-order hydr half-life:**
*Comment:*

    · **Acid rate const (M(H+)-hr)$^{-1}$:**          No hydrolyzable groups
*Comment:*

    · **Base rate const (M(OH-)-hr)$^{-1}$:**
*Comment:*

# Anthracene

CAS Registry Number: 120-12-7

Half-lives:

· **Soil:**
    High:   11040 hours     (1.26 years)
    Low:    1200 hours      (50 days)

*Comment:* Based upon aerobic soil die-away test data (Coover, MP and Sims, RC (1987); Sims, RC (1990)).

· **Air:**
    High:   1.7 hours
    Low:   0.58 hours

*Comment:* Scientific judgement based upon photolysis half-life in water.

· **Surface Water:**
    High:   1.7 hours
    Low:   0.58 hours

*Comment:* Scientific judgement based upon photolysis half-life in water.

· **Ground Water:**
    High:   22080 hours     (2.52 years)
    Low:    2400 hours     (100 days)

*Comment:* Scientific judgement based upon estimated unacclimated aqueous aerobic biodegradation half-life.

Aqueous Biodegradation (unacclimated):

· **Aerobic half-life:**
    High:   11040 hours     (1.26 years)
    Low:    1200 hours      (50 days)

*Comment:* Based upon aerobic soil die-away test data (Coover, MP and Sims, RC (1987); Sims, RC (1990)).

· **Anaerobic half-life:**
    High:   44160 hours     (5.04 years)
    Low:    4800 hours     (200 days)

*Comment:* Scientific judgement based upon estimated unacclimated aqueous aerobic biodegradation half-life.

· **Removal/secondary treatment:**
    High:   97%
    Low:

*Comment:* Removal percentage based upon data from a continuous activated sludge biological treatment simulator (Petrasek, AC et al. (1983)).

Photolysis:

· **Atmos photol half-life:**
    High:   1.7 hours
    Low:   0.58 hours

*Comment:* Based upon measured aqueous photolysis rate constant for midday summer sunlight at

35°N latitude (low $t_{1/2}$) (Southworth, GR (1979)) and adjusted for approximate winter sunlight intensity (high $t_{1/2}$) (Lyman, WJ et al. (1982)).

· **Max light absorption (nm):**      lambda max = 251.5, 308, 323, 338, 354.5, 374.5
(methanol/ethanol mixture)
*Comment:* IARC (1983A).

· **Aq photol half-life:**         High:    1.7 hours
                       Low:    0.58 hours
*Comment:* Based upon measured aqueous photolysis rate constant for midday summer sunlight at 35°N latitude (low $t_{1/2}$) (Southworth, GR (1979)) and adjusted for approximate winter sunlight intensity (high $t_{1/2}$) (Lyman, WJ et al. (1982)).

## Photooxidation half-life:
· **Water:**                High:    38500 hours      (4.39 years)
                       Low:    1111 hours       (46.3 days)
*Comment:* Based upon measured rate constant for reaction with alkylperoxyl radical in water (Radding, SB et al. (1976)).

· **Air:**                 High:    5.01 hours
                       Low:    0.501 hours
*Comment:* Scientific judgement based upon estimated rate constant for reaction with hydroxyl radical in air (Atkinson, R (1987A)).

## Reduction half-life:              High:    No data
                       Low:
*Comment:*

## Hydrolysis:
· **First-order hydr half-life:**
*Comment:*

· **Acid rate const (M(H+)-hr)$^{-1}$:**    No hydrolyzable groups
*Comment:*

· **Base rate const (M(OH-)-hr)$^{-1}$:**
*Comment:*

# Isosafrole

**CAS Registry Number:** 120-58-1

**Structure:**

**Half-lives:**
  **· Soil:**                      High:     672 hours             (4 weeks)
                                   Low:      168 hours             (7 days)
  *Comment:* Scientific judgement based upon estimated aqueous aerobic biodegradation half-life.

  **· Air:**                       High:     2.9 hours
                                   Low:      0.3 hour              (16 minutes)
  *Comment:* Scientific judgement based upon estimated photooxidation half-life in air.

  **· Surface Water:**             High:     672 hours             (4 weeks)
                                   Low:      168 hours             (7 days)
  *Comment:* Scientific judgement based upon estimated aqueous aerobic biodegradation half-life.

  **· Ground Water:**              High:     1344 hours            (8 weeks)
                                   Low:      336 hours             (14 days)
  *Comment:* Scientific judgement based upon estimated aqueous aerobic biodegradation half-life.

**Aqueous Biodegradation (unacclimated):**
  **· Aerobic half-life:**         High:     672 hours             (4 weeks)
                                   Low:      168 hours             (7 days)
  *Comment:* Scientific judgement.

  **· Anaerobic half-life:**       High:     2688 hours            (16 weeks)
                                   Low:      672 hours             (28 days)
  *Comment:* Scientific judgement based upon estimated aqueous aerobic biodegradation half-life.

  **· Removal/secondary treatment:**  High:     No data
                                      Low:
  *Comment:*

**Photolysis:**
  **· Atmos photol half-life:**    High:

Low:

*Comment:*

· **Max light absorption (nm):**　　　　No data
*Comment:*

· **Aq photol half-life:**　　　　High:
　　　　　　　　　　　　　　　　Low:

*Comment:*

## Photooxidation half-life:
　· **Water:**　　　　　　　　　High:　　　No data
　　　　　　　　　　　　　　　Low:

*Comment:*

　· **Air:**　　　　　　　　　　High:　　　2.9 hours
　　　　　　　　　　　　　　　Low:　　　0.3 hours　　　　(16 minutes)
*Comment:* Scientific judgement based upon an estimated rate constant for vapor phase reaction with hydroxyl radicals and ozone in air (Atkinson, R (1987A); Atkinson, R and Carter, WPL (1984)).

## Reduction half-life:　　　　　　High:　　　No data
　　　　　　　　　　　　　　　　Low:
*Comment:*

## Hydrolysis:
　· **First-order hydr half-life:**　　No data
　*Comment:*

　· **Acid rate const (M(H+)-hr)$^{-1}$:**
　*Comment:*

　· **Base rate const (M(OH-)-hr)$^{-1}$:**
　*Comment:*

# Dimethyl terephthalate

**CAS Registry Number:** 120-61-6

**Half-lives:**
· **Soil:**             High:   672 hours       (4 weeks)
                        Low:    168 hours       (1 week)
*Comment:* Scientific judgement based upon estimated aqueous aerobic biodegradation half-life.

· **Air:**              High:   1118 hours      (46.6 days)
                        Low:    112 hours       (4.7 days)
*Comment:* Scientific judgement based upon estimated photooxidation half-life in air.

· **Surface Water:**    High:   672 hours       (4 weeks)
                        Low:    168 hours       (1 week)
*Comment:* Scientific judgement based upon estimated aqueous aerobic biodegradation half-life.

· **Ground Water:**     High:   8640 hours      (8 weeks)
                        Low:    336 hours       (2 weeks)
*Comment:* Scientific judgement based upon estimated aqueous aerobic biodegradation half-life.

**Aqueous Biodegradation (unacclimated):**
· **Aerobic half-life:**    High:   672 hours       (4 weeks)
                            Low:    168 hours       (1 week)
*Comment:* Scientific judgement based upon soil grab sample data for dimethyl phtalate (Russell, DJ et al. (1985)).

· **Anaerobic half-life:**  High:   2688 hours      (16 weeks)
                            Low:    672 hours       (4 weeks)
*Comment:* Scientific judgement based upon estimated aqueous aerobic biodegradation half-life.

· **Removal/secondary treatment:**  High:   No data
                                    Low:
*Comment:*

**Photolysis:**
· **Atmos photol half-life:**   High:
                                Low:
*Comment:*

· **Max light absorption (nm):**   No data
*Comment:*

464

**· Aq photol half-life:**          High:
                                     Low:

*Comment:*

## Photooxidation half-life:
**· Water:**                        High:       No data
                                     Low:

*Comment:*

**· Air:**                          High:       1118 hours          (46.6 days)
                                     Low:        112 hours           (4.7 days)
*Comment:*  Scientific judgement based upon estimated rate data for hydroxyl radicals in air
(Atkinson, R (1987A)).

## Reduction half-life:              High:       No data
                                     Low:

*Comment:*

## Hydrolysis:
**· First-order hydr half-life:**              7704 hours          (321 days)
*Comment:*  Based upon overall rate constant ($k_H = 2.5 \times 10^{-8}$ $s^{-1}$) at pH 7 and 25 °C (Mabey, W
and Mill, T (1978)).

**· Acid rate const $(M(H+)-hr)^{-1}$:**
*Comment:*

**· Base rate const $(M(OH-)-hr)^{-1}$:**
*Comment:*

# p-Cresidine

CAS Registry Number: 120-71-8

Structure:

Half-lives:
  · **Soil:**                High:   4320 hours        (6 months)
                             Low:     672 hours        (4 weeks)
  *Comment:* Scientific judgement based upon estimated aqueous aerobic biodegradation half-lives.

  · **Air:**                 High:    2.9 hours
                             Low:     0.29 hours
  *Comment:* Based upon estimated photooxidation rate constant in air.

  · **Surface Water:**       High:   4320 hours        (6 months)
                             Low:    62.4 hours        (2.6 days)
  *Comment:* High value based upon scientific judgement of estimated high aqueous biodegradation half-life. Low value based upon estimated low value for aqueous photooxidation.

  · **Ground Water:**        High:   8640 hours        (12 months)
                             Low:    1344 hours        (8 weeks)
  *Comment:* Scientific judgement based upon estimated aqueous aerobic biodegradation half-lives.

Aqueous Biodegradation (unacclimated):
  · **Aerobic half-life**:   High:   4320 hours        (6 months)
                             Low:     672 hours        (4 weeks)
  *Comment:* Scientific judgement based upon aerobic aqueous screening studies for 2-anisidine (Kondo, M et al. (1988A); Kool, HJ (1984)).

  · **Anaerobic half-life**: High:  17280 hours        (24 months)
                             Low:    2688 hours        (16 weeks)
  *Comment:* Scientific judgement based upon estimated aqueous aerobic biodegradation half-lives.

  · **Removal/secondary treatment:**   High:   No data
                                       Low:
  *Comment:*

Photolysis:

**· Atmos photol half-life:**  High:  No data
  Low:

*Comment:*

**· Max light absorption (nm):**
*Comment:*

**· Aq photol half-life:**  High:
  Low:

*Comment:*

## Photooxidation half-life:
  **· Water:**  High:  3480 hours  (145 days)
    Low:  62.4 hours  (2.6 days)
*Comment:* Scientific judgement based upon photooxidation rate constants with ·OH and RO$_2$·
for the aromatic amine class (Mill, T and Mabey, W (1985); Guesten, H et al. (1981)).

  **· Air:**  High:  2.9 hours
    Low:  0.29 hours
*Comment:* Based upon estimated photooxidation rate constant with ·OH in air (Atkinson, R
(1987A)).

## Reduction half-life:  High:  No data
  Low:

*Comment:*

## Hydrolysis:
  **· First-order hydr half-life:**  No hydrolyzable groups
*Comment:*

  **· Acid rate const (M(H+)-hr)$^{-1}$:**
*Comment:*

  **· Base rate const (M(OH-)-hr)$^{-1}$:**
*Comment:*

# Catechol

**CAS Registry Number:** 120-80-9

**Structure:**

OH

OH

**Half-lives:**
&middot; **Soil:**  High: 168 hours  (7 days)

Low: 24 hours  (1 day)

*Comment:* Scientific judgement based upon estimated aqueous aerobic biodegradation half-lives.

&middot; **Air:**  High: 26 hours

Low: 2.6 hours

*Comment:* Based upon estimated photooxidation half-lives in air.

&middot; **Surface Water:**  High: 168 hours  (7 days)

Low: 24 hours  (1 day)

*Comment:* Scientific judgement based upon the estimated aqueous aerobic biodegradation half-life.

&middot; **Ground Water:**  High: 336 hours  (14 days)

Low: 48 hours  (2 days)

*Comment:* Scientific judgement based upon estimated aqueous aerobic biodegradation half-lives.

**Aqueous Biodegradation (unacclimated):**
&middot; **Aerobic half-life:**  High: 168 hours  (7 days)

Low: 24 hours  (1 day)

*Comment:* Scientific judgement based upon aerobic biological screening test data (Gerike, P and Fischer, WK (1979); Heukelekian, H and Rand, MC (1955); Okey, RW and Bogan, RH (1965); Pitter, P (1976); Urushigawa, Y et al. (1983); Urushigawa, Y et al. (1984B)).

&middot; **Anaerobic half-life:**  High: 672 hours  (28 days)

Low: 96 hours  (4 days)

*Comment:* Scientific judgement based upon estimated aqueous aerobic biodegradation half-lives.

&middot; **Removal/secondary treatment:**  High: 98.8%

Low: 97.5%

*Comment:* Continuous biological treatment simulation (Chudoba, J et al. (1968)).

**Photolysis:**

**· Atmos photol half-life:**  High:  No data
  Low:

*Comment:*

**· Max light absorption (nm):**  lambda max = 313 nm
*Comment:* Absorption maxima in a basic aqueous solution (Perbet, G et al. (1979)) in which a portion of the catechol exists in ionized form. In methanol solution (no ionization), catechol does not absorb above 298 nm (Sadtler UV No. 1592).

**· Aq photol half-life:**  High:  No data
  Low:

*Comment:*

**Photooxidation half-life:**
  **· Water:**  High:  3840 hours  (160 days)
    Low:  77 hours  (3.2 days)
*Comment:* Scientific judgement based upon reported reaction rate constants for $RO_2\cdot$ with the phenol class (Mill, T and Mabey, W (1985)).

**· Air:**  High:  26 hours
  Low:  2.6 hours
*Comment:* Based upon estimated reaction rate constant with $\cdot OH$ (Atkinson, R (1987A)).

**Reduction half-life:**  High:  No data
  Low:

*Comment:*

**Hydrolysis:**
  **· First-order hydr half-life:**  No hydrolyzable groups
  *Comment:*

  **· Acid rate const $(M(H+)-hr)^{-1}$:**
  *Comment:*

  **· Base rate const $(M(OH-)-hr)^{-1}$:**
  *Comment:*

# 1,2,4-Trichlorobenzene

<u>CAS Registry Number:</u>  120-82-1

<u>Half-lives:</u>

· **Soil:**                              High:    4320 hours              (6 months)
                                          Low:      672 hours              (4 weeks)
*Comment:*  Scientific judgement based upon unacclimated aerobic soil grab sample data (low $t_{1/2}$:
Haider, K et al. (1981), high $t_{1/2}$: Marinucci, AC and Bartha, R (1979)).

· **Air:**                               High:    1284 hours              (53.5 days)
                                          Low:      128.4 hours           (5.4 days)
*Comment:*  Based upon photooxidation half-life in air.

· **Surface Water:**               High:    4320 hours              (6 months)
                                          Low:      672 hours              (4 weeks)
*Comment:*  Scientific judgement based upon estimated unacclimated aqueous aerobic
biodegradation half-life.

· **Ground Water:**                High:    8640 hours              (12 months)
                                          Low:      1344 hours            (8 weeks)
*Comment:*  Scientific judgement based upon estimated unacclimated aqueous aerobic
biodegradation half-life.

<u>Aqueous Biodegradation (unacclimated):</u>

· **Aerobic half-life:**           High:    4320 hours              (6 months)
                                          Low:      672 hours              (4 weeks)
*Comment:*  Scientific judgement based upon unacclimated aerobic soil grab sample data (low $t_{1/2}$:
Haider, K et al. (1981), high $t_{1/2}$: Marinucci, AC and Bartha, R (1979)).

· **Anaerobic half-life:**         High:    17280 hours            (24 months)
                                          Low:      2688 hours            (16 weeks)
*Comment:*  Scientific judgement based upon estimated unacclimated aqueous aerobic
biodegradation half-life.

· **Removal/secondary treatment:**   High:    No data
                                          Low:
*Comment:*

<u>Photolysis:</u>

· **Atmos photol half-life:**      High:    No data
                                          Low:
*Comment:*

· **Max light absorption (nm):**          lambda max = 286.0, 278.0
*Comment:* No absorbance of UV light at wavelengths greater than 300 nm in cyclohexane
(Sadtler UV No. 1277).

· **Aq photol half-life:**          High:     No data
                                    Low:
*Comment:*

## Photooxidation half-life:
· **Water:**                        High:     No data
                                    Low:
*Comment:*

· **Air:**                          High:   1284 hours              (53.5 days)
                                    Low:    128.4 hours             (5.4 days)
*Comment:* Based upon measured rate data for the vapor phase reaction with hydroxyl radicals in
air (Atkinson, R et al.(1985B)).

## Reduction half-life:
                                    High:     No data
                                    Low:
*Comment:*

## Hydrolysis:
· **First-order hydr half-life:**          29784 hours             (3.4 years)
*Comment:* Scientific judgement based upon overall rate constant ($2.3 \times 10^{-5}$ $hr^{-1}$) at pH 7 and 25
°C (Ellington, JJ et al. (1988)).

· **Acid rate const $(M(H+)-hr)^{-1}$:**          No data
*Comment:*

· **Base rate const $(M(OH-)-hr)^{-1}$:**          No data
*Comment:*

# 2,4-Dichlorophenol

<u>CAS Registry Number:</u>  120-83-2

<u>Half-lives:</u>
  · **Soil:**                               High:    1680 hours         (70 days)
                                            Low:      176 hours         (7.32 days)
*Comment:*  Based upon aerobic soil die-away test data (low $t_{1/2}$: Baker, MD et al. (1980); high $t_{1/2}$: Haider, K et al. (1974)).

  · **Air:**                                High:    212 hours          (8.83 days)
                                            Low:     21.2 hours
*Comment:*  Scientific judgement based upon estimated photooxidation half-life in air.

  · **Surface Water:**                      High:      3 hours
                                            Low:      0.8 hours
*Comment:*  Based upon measured rate of photolysis in distilled water under sunlight in summer (low $t_{1/2}$) and winter (high $t_{1/2}$) (Hwang, H et al. (1986)).

  · **Ground Water:**                       High:    1032 hours         (43 days)
                                            Low:      133 hours         (5.54 days)
*Comment:*  Based upon unacclimated aqueous aerobic (low $t_{1/2}$) and anaerobic (high $t_{1/2}$) biodegradation half-lives.

<u>Aqueous Biodegradation (unacclimated):</u>
  · **Aerobic half-life:**                  High:    199 hours          (8.3 days)
                                            Low:     66.7 hours         (2.78 days)
*Comment:*  Based upon aerobic lake die-away test data (Aly, OM and Faust, SD (1964)).

  · **Anaerobic half-life:**                High:    1032 hours         (43 days)
                                            Low:     324 hours          (13.5 days)
*Comment:*  Based upon anaerobic lake die-away test data (Aly, OM and Faust, SD (1964)).

  · **Removal/secondary treatment:**        High:     95.2%
                                            Low:      84.5%
*Comment:*  Removal percentages based upon data from a continuous activated sludge biological treatment simulator (Stover, EL and Kincannon, DF (1983)).

<u>Photolysis:</u>
  · **Atmos photol half-life:**             High:      3 hours
                                            Low:      0.8 hours
*Comment:*  Based upon measured rate of photolysis in distilled water under sunlight in summer (low $t_{1/2}$) and winter (high $t_{1/2}$) (Hwang, H et al. (1986)).

· **Max light absorption (nm):** lambda max = 228.5, 220.5, 246, 287, 308 nm (methanol)
*Comment:* Significant absorption continues up to approximately 310 nm (Sadtler UV No. 9786).

· **Aq photol half-life:**     High:     3 hours
                               Low:     0.8 hours
*Comment:* Based upon measured rate of photolysis in distilled water under sunlight in summer (low $t_{1/2}$) and winter (high $t_{1/2}$) (Hwang, H et al. (1986)).

## Photooxidation half-life:
· **Water:**     High:     3840 hours     (160 days)
                 Low:      13 hours
*Comment:* Scientific judgement based upon reported reaction rate constants for $RO_2 \cdot$ with the phenol class (high $t_{1/2}$) (Mill, T and Mabey, W (1985)) and the measured reaction rate constant for reaction with singlet oxygen (Scully, FE Jr and Hoigne, J (1987)) combined with the rate constant for reaction of the phenol class with $RO_2 \cdot$ (Mill, T and Mabey, W (1985)).

· **Air:**     High:     212 hours     (8.83 days)
               Low:      21.2 hours
*Comment:* Scientific judgement based upon estimated rate constant for reaction with hydroxyl radical in air (Atkinson, R (1987A)).

## Reduction half-life:
                High:     No data
                Low:
*Comment:*

## Hydrolysis:
· **First-order hydr half-life:**
*Comment:*

· **Acid rate const $(M(H+)-hr)^{-1}$:**     No hydrolyzable groups
*Comment:*

· **Base rate const $(M(OH-)-hr)^{-1}$:**
*Comment:*

# 2,4-Dinitrotoluene

CAS Registry Number: 121-14-2

Half-lives:
· Soil:                            High:    4320 hours              (6 months)
                                   Low:     672 hours              (4 weeks)
Comment: Scientific judgement based upon estimated unacclimated aqueous aerobic
biodegradation half-life.

· Air:                             High:    72 hours
                                   Low:     23 hours
Comment: Based upon estimated photolysis half-life in air.

· Surface Water:                   High:    33 hours
                                   Low:      3 hours
Comment: Based upon photooxidation half-life in natural water.

· Ground Water:                    High:    8640 hours             (12 months)
                                   Low:       48 hours             (2 days)
Comment: Scientific judgement based upon estimated unacclimated aqueous anaerobic (low $t_{1/2}$)
and aerobic (high $t_{1/2}$) biodegradation half-lives.

Aqueous Biodegradation (unacclimated):
· Aerobic half-life:               High:    4320 hours             (6 months)
                                   Low:     672 hours              (4 weeks)
Comment: Scientific judgement based upon aerobic natural water die-away test data (Spanggord,
RJ et al. (1981)). Reduction of the nitro group observed in these studies may not be due to
biodegradation.

· Anaerobic half-life:             High:    240 hours              (10 days)
                                   Low:      48 hours              (2 days)
Comment: Scientific judgement based upon anaerobic natural water die-away test data
(Spanggord, RJ et al. (1980)). Reduction of the nitro group observed in these studies may not be
due to biodegradation

· Removal/secondary treatment:     High:    No data
                                   Low:
Comment:

Photolysis:
· Atmos photol half-life:          High:    72 hours
                                   Low:     23 hours

*Comment:* Scientific judgement based upon measured photolysis rates in water (Mill, T and Mabey, W (1985); Simmons, MS and Zepp, RG (1986)).

· **Max light absorption (nm):**     No data
*Comment:*

· **Aq photol half-life:**          High:      72 hours
                              Low:       23 hours
*Comment:* Based upon measured photolysis rates in water (Mill, T and Mabey, W (1985); Simmons, MS and Zepp, RG (1986)).

**Photooxidation half-life:**
· **Water:**                      High:      33 hours
                              Low:        3 hours
*Comment:* Based upon measured photooxidation rates in natural waters (Spanggord, RJ et al. (1980); Simmons, MS and Zepp, RG (1986)).

· **Air:**                        High:    2840 hours        (118 days)
                              Low:      284 hours        (11.8 days)
*Comment:* Based upon estimated rate constant for reaction with hydroxyl radicals in air (Atkinson, R (1987A)).

**Reduction half-life:**            High:      No data
                              Low:
*Comment:*

**Hydrolysis:**
· **First-order hydr half-life:**
*Comment:*

· **Acid rate const (M(H+)-hr)$^{-1}$:**     No hydrolyzable groups
*Comment:*

· **Base rate const (M(OH-)-hr)$^{-1}$:**
*Comment:*

# N,N-Dimethylaniline

CAS Registry Number:  121-69-7

## Half-lives:

· **Soil:**

|  | High: | 4320 hours | (6 months) |
|---|---|---|---|
|  | Low: | 672 hours | (4 weeks) |

*Comment:*  Scientific judgement based upon estimated unacclimated aqueous aerobic biodegradation half-life.

· **Air:**

|  | High: | 21 hours |
|---|---|---|
|  | Low: | 2.7 hours |

*Comment:*  Based upon photooxidation half-life in air.

· **Surface Water:**

|  | High: | 1925 hours | (80.2 days) |
|---|---|---|---|
|  | Low: | 19.3 hours | (0.8 days) |

*Comment:*  Based upon photooxidation half-life in water.

· **Ground Water:**

|  | High: | 8640 hours | (12 months) |
|---|---|---|---|
|  | Low: | 1344 hours | (8 weeks) |

*Comment:*  Scientific judgement based upon estimated unacclimated aqueous aerobic biodegradation half-life.

## Aqueous Biodegradation (unacclimated):

· **Aerobic half-life:**

|  | High: | 4320 hours | (6 months) |
|---|---|---|---|
|  | Low: | 672 hours | (4 weeks) |

*Comment:*  Scientific judgement based upon unacclimated aqueous screening test data (low $t_{1/2}$: Kitano, M (1978); high $t_{1/2}$: Baird, R et al. (1977))

· **Anaerobic half-life:**

|  | High: | 17280 hours | (24 months) |
|---|---|---|---|
|  | Low: | 2880 hours | (16 weeks) |

*Comment:*  Scientific judgement based upon estimated unacclimated aqueous aerobic biodegradation half-life.

· **Removal/secondary treatment:**

|  | High: | No data |
|---|---|---|
|  | Low: |  |

*Comment:*

## Photolysis:

· **Atmos photol half-life:**

|  | High: |
|---|---|
|  | Low: |

*Comment:*

· **Max light absorption (nm):**      No data
*Comment:*

· **Aq photol half-life:**      High:
      Low:

*Comment:*

## Photooxidation half-life:
· **Water:**      High:   1925 hours    (80.2 days)
      Low:    19.3 hours    (0.8 days)
*Comment:* Based upon measured rate data for singlet oxygen in aqueous solution (Haag, WR, Hoigne, J (1985A)).

· **Air:**      High:    21 hours
      Low:    2.7 hours
*Comment:* Based upon measured rate data for hydroxyl radicals and ozone (Atkinson, R et al. (1987A)).

## Reduction half-life:
      High:   No data
      Low:

*Comment:*

## Hydrolysis:
· **First-order hydr half-life:**
*Comment:*

· **Acid rate const (M(H+)-hr)$^{-1}$:**      No hydrolyzable groups
*Comment:*

· **Base rate const (M(OH-)-hr)$^{-1}$:**
*Comment:*

# Malathion

CAS Registry Number: 121-75-5

Structure:

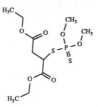

Half-lives:
  · Soil:                                  High:    168 hours            (7 days)
                                           Low:      72 hours            (3 days)
  Comment: Scientific judgement based upon unacclimated aerobic soil grab sample data (Walker, WW and Stojanovic, BJ (1973)).

  · Air:                                   High:     9.8 hours
                                           Low:      1.0 hour
  Comment: Scientific judgement based upon estimated photooxidation half-life in air.

  · Surface Water:                         High:    1236 hours           (51.5 days)
                                           Low:      100 hours           (4.2 days)
  Comment: Scientific judgement based upon unacclimated aerobic river die-away test data (low $t_{1/2}$: Eichelberger, JW and Lichtenberg, JJ (1971)) and estuarine water grab sample data (high $t_{1/2}$: Walker, WW (1978)).

  · Ground Water:                          High:    2472 hours           (103 days)
                                           Low:      200 hours           (8.4 days)
  Comment: Scientific judgement based upon estimated aqueous aerobic biodegradation half-life.

Aqueous Biodegradation (unacclimated):
  · Aerobic half-life:                     High:    1236 hours           (51.5 days)
                                           Low:      100 hours           (4.2 days)
  Comment: Scientific judgement based upon unacclimated aerobic river die-away test data (low $t_{1/2}$: Eichelberger, JW and Lichtenberg, JJ (1971)) and estuarine water grab sample data (high $t_{1/2}$: Walker, WW (1978)).

  · Anaerobic half-life:                   High:    4944 hours           (206 days)
                                           Low:      400 hours           (16.8 days)
  Comment: Scientific judgement based upon unacclimated aerobic biodegradation half-life.

  · Removal/secondary treatment:           High:     No data

Low:
*Comment:*

## Photolysis:
· **Atmos photol half-life:**    High:  20000 hours        (833 days)
                                 Low:      990 hours        (41.3 days)
*Comment:* Scientific judgement based upon aqueous photolysis data (low $t_{1/2}$: Wolfe, NL et al. (1977A); high $t_{1/2}$: Wolfe, NL et al. (1976)).

· **Max light absorption (nm):**    No data
*Comment:*

· **Aq photol half-life:**    High:  20000 hours        (833 days)
                              Low:      990 hours        (41.3 days)
*Comment:* Based upon aqueous photolysis rate constant at pH 6 for wavelengths greater than 290 nm (low $t_{1/2}$: Wolfe, NL et al. (1977A)) and for summer sunlight 30° N (high $t_{1/2}$: Wolfe, NL et al. (1976)).

## Photooxidation half-life:
· **Water:**    High:    No data
                Low:
*Comment:*

· **Air:**    High:    9.8 hours
              Low:     1.0 hour
*Comment:* Scientific judgement based upon an estimated rate constant for vapor phase reaction with hydroxyl radicals in air (Atkinson, R (1987A)).

## Reduction half-life:    High:    No data
                           Low:
*Comment:*

## Hydrolysis:
· **First-order hydr half-life:**    $7.7 \times 10^{-4}$ hours  (8.8 years)
*Comment:* Scientific judgement based upon reported rate constant ($2.5 \times 10^{-2}$ $M^{-1}$ $s^{-1}$) at pH 7 and 0 °C (Wolfe, NL et al. (1977A)).

· **Acid rate const $(M(H+)-hr)^{-1}$:**
*Comment:*

· **Base rate const $(M(OH-)-hr)^{-1}$:**    1.4 $M^{-1}$ $s^{-1}$
*Comment:* ($t_{1/2}$ = 14 hr) Scientific judgement based upon rate constant at pH 9 and 27 °C (Wolfe, NL et al. (1977A)).

# Diphenylamine

CAS Registry Number: 122-39-4

Half-lives:
- Soil:                          High:   672 hours          (4 weeks)
                                 Low:    168 hours          (7 days)

Comment: Scientific judgement based upon estimated unacclimated aqueous aerobic biodegradation half-life.

- Air:                           High:   2.47 hours
                                 Low:    0.247 hours

Comment: Scientific judgement based upon estimated photooxidation half-life in air.

- Surface Water:                 High:   672 hours          (4 weeks)
                                 Low:    31 hours           (1.3 days)

Comment: Scientific judgement based upon estimated unacclimated aqueous aerobic biodegradation half-life (high $t_{1/2}$) and photooxidation half-life in water (low $t_{1/2}$).

- Ground Water:                  High:   1344 hours         (8 weeks)
                                 Low:    336 hours          (14 days)

Comment: Scientific judgement based upon estimated unacclimated aqueous aerobic biodegradation half-life.

Aqueous Biodegradation (unacclimated):
- Aerobic half-life:             High:   672 hours          (4 weeks)
                                 Low:    168 hours          (7 days)

Comment: Scientific judgement based upon aqueous aerobic biodegradation screening test data (Malaney, GW (1960)).

- Anaerobic half-life:           High:   2688 hours         (16 weeks)
                                 Low:    672 hours          (28 days)

Comment: Scientific judgement based upon estimated unacclimated aqueous aerobic biodegradation half-life.

- Removal/secondary treatment:   High:   99%
                                 Low:    65%

Comment: Removal percentage based upon data from a continuous synthetic sewage biological treatment simulator (Gardner, AM et al. (1982)).

Photolysis:
- Atmos photol half-life:        High:   No data
                                 Low:

*Comment:*

· **Max light absorption (nm):**     lambda max = 281.5 nm (cyclohexane)
*Comment:*  Significant absorption extends to 330 nm (Sadtler UV   No. 30).

· **Aq photol half-life:**          High:     No data
                                    Low:
*Comment:*

**Photooxidation half-life:**
  · **Water:**                      High:     1740 hours          (72 days)
                                    Low:        31 hours          (1.3 days)
*Comment:*  Scientific judgement based upon photooxidation rate constants with ·OH and $RO_2$·
for the aromatic amine class (Mill, T and Mabey, W (1985); Guesten, H et al. (1981)).

  · **Air:**                        High:     2.47 hours
                                    Low:     0.247 hours
*Comment:*  Scientific judgement based upon estimated rate constant for reaction with hydroxyl
radical in air (Atkinson, R (1978)).

**Reduction half-life:**            High:     No data
                                    Low:
*Comment:*

**Hydrolysis:**
  · **First-order hydr half-life:**
*Comment:*

  · **Acid rate const $(M(H+)-hr)^{-1}$:**     No hydrolyzable groups
*Comment:*

  · **Base rate const $(M(OH-)-hr)^{-1}$:**
*Comment:*

# Hydrazobenzene

__CAS Registry Number:__  122-66-7

__Structure:__

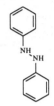

__Half-lives:__
   · __Soil:__                          High:      4320 hours          (6 months)
                                       Low:        672 hours           (4 weeks)
       *Comment:*  Scientific judgement based upon estimated aqueous aerobic biodegradation half-life.


   · __Air:__                           High:      3.0 hours
                                       Low:        0.3 hours           (18 minutes)
       *Comment:*  Scientific judgement based upon estimated photooxidation half-life in air.


   · __Surface Water:__                 High:      1740 hours          (72 days)
                                       Low:         31 hours           (1.3 days)
       *Comment:*  Scientific judgement based upon photooxidation rate constants with ·OH and $RO_2$·
       for the aromatic amine class (Mill, T and Mabey, W (1985); Guesten, H et al. (1981)). It is
       assumed that hydrazobenzene reacts twice as fast as aniline.


   · __Ground Water:__                  High:      8640 hours          (12 months)
                                       Low:       1344 hours           (8 weeks)
       *Comment:*  Scientific judgement based upon estimated unacclimated aqueous aerobic
       biodegradation half-life.

__Aqueous Biodegradation (unacclimated):__
   · __Aerobic half-life:__             High:      4320 hours          (6 months)
                                       Low:        672 hours           (4 weeks)
       *Comment:*  Scientific judgement based upon acclimated aerobic aqueous screening test data
       (Malaney, GW (1960)).


   · __Anaerobic half-life:__           High:     17280 hours          (24 months)
                                       Low:       2880 hours           (16 weeks)
       *Comment:*  Scientific judgement based upon estimated unacclimated aqueous aerobic
       biodegradation half-life.


   · __Removal/secondary treatment:__   High:      No data

Low:

*Comment:*

## Photolysis:
· **Atmos photol half-life:**   High:   No data
                                Low:

*Comment:*

· **Max light absorption (nm):**   lambda max = 316.0, 229.5
*Comment:*  In methanol (Sadtler UV No. 4611).

· **Aq photol half-life:**   High:   No data
                             Low:

*Comment:*

## Photooxidation half-life:
· **Water:**   High:   1740 hours   (72 days)
               Low:    31 hours     (1.3 days)
*Comment:*  Scientific judgement based upon photooxidation rate constants with ·OH and $RO_2$·
for the aromatic amine class (Mill, T and Mabey, W (1985); Guesten, H et al. (1981)). It is
assumed that hydrazobenzene reacts twice as fast as aniline.

· **Air:**   High:   3.0 hours
             Low:    0.3 hours     (18 minutes)
*Comment:*  Scientific judgement based upon estimated rate data for hydroxyl radicals in air
(Atkinson, R (1987A)).

## Reduction half-life:
                    High:   No data
                    Low:

*Comment:*

## Hydrolysis:
· **First-order hydr half-life:**   No data
*Comment:*

· **Acid rate const $(M(H+)-hr)^{-1}$:**
*Comment:*

· **Base rate const $(M(OH-)-hr)^{-1}$:**
*Comment:*

# Hydroquinone

**CAS Registry Number:** 123-31-9

**Half-lives:**
   · **Soil:**                           High:      168 hours              (7 days)
                                          Low:       24 hours              (1 day)
   *Comment:* Scientific judgement based upon estimated unacclimated aqueous aerobic biodegradation half-life.

   · **Air:**                            High:      26.1 hours
                                         Low:       2.6 hours
   *Comment:* Scientific judgement based upon estimated photooxidation half-life in air.

   · **Surface Water:**                  High:      19.3 hours
                                         Low:       0.39 hours            (23 minutes)
   *Comment:* Based upon photooxidation half-life in water.

   · **Ground Water:**                   High:      336 hours             (14 days)
                                         Low:       48 hours              (2 days)
   *Comment:* Scientific judgement based upon estimated unacclimated aqueous aerobic biodegradation half-life.

**Aqueous Biodegradation (unacclimated):**
   · **Aerobic half-life:**              High:      168 hours             (7 days)
                                         Low:       24 hours              (1 day)
   *Comment:* Scientific judgement based upon aqueous screening test data (Belly, R and Goodhue, CT (1976); Gerike, P and Fischer, WK (1979), Ludzack FJ and Ettinger, MB (1960)).

   · **Anaerobic half-life:**            High:      672 hours             (28 days)
                                         Low:       96 hours              (4 days)
   *Comment:* Scientific judgement based upon estimated unacclimated aqueous aerobic biodegradation half-life.

   · **Removal/secondary treatment:**    High:      No data
                                         Low:
   *Comment:*

**Photolysis:**
   · **Atmos photol half-life:**         High:      No data
                                         Low:
   *Comment:*

484

· **Max light absorption (nm):**        lambda max = 293.5

*Comment:* In methanol at a concn of 0.1 g/L (Sadtler UV No. 60).

· **Aq photol half-life:**        High:      No data

                                        Low:

*Comment:*

## Photooxidation half-life:

· **Water:**        High:      19.3 hours

                         Low:      0.39 hours        (23 minutes)

*Comment:* Scientific judgement based upon measured rate data for alkylperoxyl radicals in aqueous solution (Mill, T (1982)).

· **Air:**        High:      26.1 hours

                         Low:      2.6 hours

*Comment:* Scientific judgement based upon an estimated rate constant for vapor phase reaction with hydroxyl radicals in air (Atkinson, R (1987A)).

## Reduction half-life:        High:      No data

                                        Low:

*Comment:*

## Hydrolysis:

· **First-order hydr half-life:**

*Comment:*

· **Acid rate const (M(H+)-hr)$^{-1}$:**      No hydrolyzable groups

*Comment:*

· **Base rate const (M(OH-)-hr)$^{-1}$:**

*Comment:*

# Propionaldehyde

CAS Registry Number: 123-38-6

## Half-lives:
**· Soil:**                           High:       168 hours              (7 days)
                        Low:        24 hours               (1 day)
*Comment:* Scientific judgement based upon estimated aqueous aerobic biodegradation half-lives.

**· Air:**                            High:       33 hours
                        Low:        3.3 hours
*Comment:* Based upon measured photooxidation half-lives in air.

**· Surface Water:**                  High:       168 hours              (7 days)
                        Low:        24 hours               (1 day)
*Comment:* Scientific judgement based upon estimated aqueous aerobic biodegradation half-lives.

**· Ground Water:**                   High:       336 hours              (14 days)
                        Low:        48 hours               (2 days)
*Comment:* Scientific judgement based upon estimated aqueous aerobic biodegradation half-lives.

## Aqueous Biodegradation (unacclimated):
**· Aerobic half-life:**              High:       168 hours              (7 days)
                        Low:        24 hours               (1 day)
*Comment:* Scientific judgement based upon aerobic biological screening test data (Dore, M et al. (1975); Urano, K and Kato, Z (1986); Gerhold, RM and Malaney, GW (1966)).

**· Anaerobic half-life:**            High:       672 hours              (28 days)
                        Low:        96 hours               (4 days)
*Comment:* Scientific judgement based upon estimated aqueous aerobic biodegradation half-lives.

**· Removal/secondary treatment:**    High:       95%
                        Low:        60%
*Comment:* Semi-continuous biological treatment simulator (Hatfield, R (1957)).

## Photolysis:
**· Atmos photol half-life:**         High:       No data
                        Low:
*Comment:*

**· Max light absorption (nm):**
*Comment:*

**· Aq photol half-life:**          High:

                                        Low:

*Comment:*

## Photooxidation half-life:

**· Water:**          High:      No data

                                         Low:

*Comment:*

**· Air:**          High:      33 hours

                                         Low:       3.3 hours

*Comment:* Based upon measured reaction rate constant with ·OH (Atkinson, R (1987)).

## Reduction half-life:          High:      No data

                                         Low:

*Comment:*

## Hydrolysis:

**· First-order hydr half-life:**      No hydrolyzable groups

*Comment:*

**· Acid rate const $(M(H+)-hr)^{-1}$:**

*Comment:*

**· Base rate const $(M(OH-)-hr)^{-1}$:**

*Comment:*

# Butyraldehyde

**CAS Registry Number:** 123-72-8

**Half-lives:**
  · **Soil:**                              High:    168 hours        (7 days)
                                       Low:     24 hours         (1 day)
  *Comment:* Scientific judgement based upon estimated aqueous aerobic biodegradation half-lives.

  · **Air:**                               High:    28 hours
                                       Low:     2.8 hours
  *Comment:* Based upon measured photooxidation half-lives in air.

  · **Surface Water:**                High:    168 hours        (7 days)
                                       Low:     24 hours         (1 day)
  *Comment:* Scientific judgement based upon estimated aqueous aerobic biodegradation half-lives.

  · **Ground Water:**               High:    336 hours       (14 days)
                                       Low:     48 days         (2 days)
  *Comment:* Scientific judgement based upon estimated aqueous aerobic biodegradation half-lives.

**Aqueous Biodegradation (unacclimated):**
  · **Aerobic half-life:**             High:    168 hours        (7 days)
                                       Low:     24 hours         (1 day)
  *Comment:* Scientific judgement based upon aerobic biological screening test data (Dore, M et al. (1975); Urano, K and Kato, Z (1986); Lamb, CB and Jenkins, GF (1952); Heukelekian, H and Rand, MC (1955)).

  · **Anaerobic half-life:**         High:    672 hours       (28 days)
                                       Low:     96 hours         (4 days)
  *Comment:* Scientific judgement based upon estimated aqueous aerobic biodegradation half-lives.

  · **Removal/secondary treatment:**    High:    No data
                                       Low:
  *Comment:*

**Photolysis:**
  · **Atmos photol half-life:**       High:    No data
                                       Low:
  *Comment:*

  · **Max light absorption (nm):**
  *Comment:*

· **Aq photol half-life:**                    High:
                                              Low:

*Comment:*

**Photooxidation half-life:**
· **Water:**                                  High: 167,000 hours          (19 years)
                                              Low:    2750 hours          (114 days)
*Comment:* Based upon measured reaction rate constant with ·OH in water (Anbar, M and Neta, P (1967); Dorfman, LM and Adams, GE (1973)).

· **Air:**                                    High:        28 hours
                                              Low:        2.8 hours
*Comment:* Based upon measured reaction rate constant with ·OH in air (Atkinson, R (1987)).

**Reduction half-life:**                      High:        No data
                                              Low:

*Comment:*

**Hydrolysis:**
· **First-order hydr half-life:**             No hydrolyzable groups
*Comment:*

· **Acid rate const (M(H+)-hr)$^{-1}$:**
*Comment:*

· **Base rate const (M(OH-)-hr)$^{-1}$:**
*Comment:*

# Crotonaldehyde (trans)

**CAS Registry Number:** 123-73-9

**Structure:**

$$\underset{H}{}\overset{O}{\underset{}{\diagup\!\!\!\diagdown}}\;CH_3$$

**Half-lives:**

- **Soil:**

  High: 168 hours (7 days)
  Low: 24 hours (1 day)

  *Comment:* Scientific judgement based upon unacclimated aqueous aerobic biodegradation half-life.

- **Air:**

  High: 17.9 hours
  Low: 1.94 hours

  *Comment:* Scientific judgement based upon estimated photooxidation half-life in air.

- **Surface Water:**

  High: 168 hours (7 days)
  Low: 24 hours (1 day)

  *Comment:* Scientific judgement based upon unacclimated aqueous aerobic biodegradation half-life.

- **Ground Water:**

  High: 336 hours (14 days)
  Low: 48 hours (2 days)

  *Comment:* Scientific judgement based upon unacclimated aqueous aerobic biodegradation half-life.

**Aqueous Biodegradation (unacclimated):**

- **Aerobic half-life:**

  High: 168 hours (7 days)
  Low: 24 hours (1 day)

  *Comment:* Scientific judgement based upon unacclimated aqueous aerobic biodegradation screening test data for the aldehyde class (Heukelekian, H and Rand, MC (1955); Gerhold, RM and Malaney, GW (1966)).

- **Anaerobic half-life:**

  High: 672 hours (28 days)
  Low: 96 hours (4 days)

  *Comment:* Scientific judgement based upon unacclimated aqueous aerobic biodegradation half-life.

- **Removal/secondary treatment:** High: No data

490

                                                    Low:
*Comment:*

## Photolysis:
   · **Atmos photol half-life:**      High:
                                              Low:
*Comment:*

   · **Max light absorption (nm):**   No data
*Comment:*

   · **Aq photol half-life:**         High:
                                              Low:
*Comment:*

## Photooxidation half-life:
   · **Water:**                     High:      No data
                                              Low:
*Comment:*

   · **Air:**                       High:      17.9 hours
                                            Low:      1.94 hours
*Comment:* Scientific judgement based upon estimated rate constants for reaction with hydroxy radical (Atkinson, R (1987A)) and ozone (Atkinson, R and Carter, WPL (1984)).

## Reduction half-life:                High:      No data
                                            Low:
*Comment:*

## Hydrolysis:
   · **First-order hydr half-life:**   No data
*Comment:*

   · **Acid rate const (M(H+)-hr)$^{-1}$:**
*Comment:*

   · **Base rate const (M(OH-)-hr)$^{-1}$:**
*Comment:*

# 1,4-Dioxane

CAS Registry Number: 123-91-1

## Half-lives:
- **Soil:**                                   High:    4320 hours          (6 months)
                                              Low:     672 hours           (4 weeks)
*Comment:* Scientific judgement based upon estimated unacclimated aqueous aerobic biodegradation half-life.

- **Air:**                                    High:    81 hours            (3.4 days)
                                              Low:     8.1 hours           (0.34 days)
*Comment:* Based upon photooxidation half-life in air.

- **Surface Water:**                          High:    4320 hours          (6 months)
                                              Low:     672 hours           (4 weeks)
*Comment:* Based upon estimated unacclimated aqueous aerobic biodegradation half-life.

- **Ground Water:**                           High:    8640 hours          (12 months)
                                              Low:     1344 hours          (8 weeks)
*Comment:* Scientific judgement based upon estimated unacclimated aqueous aerobic biodegradation half-life.

## Aqueous Biodegradation (unacclimated):
- **Aerobic half-life:**                      High:    4320 hours          (6 months)
                                              Low:     672 hours           (4 weeks)
*Comment:* Scientific judgement based upon unacclimated aerobic aqueous screening test data which confirmed resistance to biodegradation (Kawasaki, M (1980); Sasaki, S (1978)).

- **Anaerobic half-life:**                    High:    17280 hours         (24 months)
                                              Low:     2688 hours          (16 weeks)
*Comment:* Scientific judgement based upon estimated aerobic biodegradation half-life.

- **Removal/secondary treatment:**            High:    No data
                                              Low:
*Comment:*

## Photolysis:
- **Atmos photol half-life:**                 High:  Will not directly photolyze
                                              Low:
*Comment:*

- **Max light absorption (nm):**              No data

492

*Comment:*

**· Aq photol half-life:**        High:

                                        Low:

*Comment:*

## Photooxidation half-life:

**· Water:**        High:    $8 \times 10^4$ hours      (9.1 years)

                Low:    1608 hours       (67 days)

*Comment:* Based upon measured rates for reaction with hydroxyl radicals in water (Dorfman, LM and Adams, GE (1973); Anbar, M and Neta, P (1967)).

**· Air:**        High:    81 hours      (3.4 days)

           Low:    8.1 hours     (0.34 days)

*Comment:* Based upon measured rate constant for reaction of 1,3,5-trioxane with hydroxyl radicals in air (Atkinson, R (1987A)).

## Reduction half-life:        High: No reducible groups

                                      Low:

*Comment:*

## Hydrolysis:

**· First-order hydr half-life:**

*Comment:*

**· Acid rate const $(M(H+)\text{-}hr)^{-1}$:**     No hydrolyzable groups

*Comment:*

**· Base rate const $(M(OH\text{-})\text{-}hr)^{-1}$:**

*Comment:*

# Dimethylamine

CAS Registry Number: **CAS Registry Number:** 124-40-3

## Half-lives:
  · **Soil:**                           High:      336 hours              (14 days)
                                        Low:       86 hours               (3.6 days)
*Comment:* Based upon soil die-away test data (Greene, S et al. (1981); Tate, RL III and Alexander, M (1976)).

  · **Air:**                            High:      9.20 hours
                                        Low:       0.892 hours
*Comment:* Based upon photooxidation half-life in air.

  · **Surface Water:**                  High:      79 hours               (3.3 days)
                                        Low:       2 hours
*Comment:* Scientific judgement based upon estimated unacclimated aqueous aerobic biodegradation half-life.

  · **Ground Water:**                   High:      158 hours              (6.58 days)
                                        Low:       4 hours
*Comment:* Scientific judgement based upon estimated unacclimated aqueous aerobic biodegradation half-life.

## Aqueous Biodegradation (unacclimated):
  · **Aerobic half-life:**              High:      79 hours               (3.3 days)
                                        Low:       2 hours
*Comment:* Based upon river die-away test data (Digeronimo, MJ et al. (1979); Dojlido, JR (1979)).

  · **Anaerobic half-life:**            High:      316 hours              (13.2 days)
                                        Low:       8 hours
*Comment:* Scientific judgement based upon estimated unacclimated aqueous aerobic biodegradation half-life.

  · **Removal/secondary treatment:**    High:      100%
                                        Low:       93%
*Comment:* Removal percentages based upon data from semicontinuous activated sludge and continuous sewage biological treatment simulators (Dojlido, JR (1979); Greene, S et al. (1981)).

## Photolysis:
  · **Atmos photol half-life:**         High:
                                        Low:

*Comment:*

**· Max light absorption (nm):**        No data
*Comment:*

**· Aq photol half-life:**        High:
                                    Low:

*Comment:*

**Photooxidation half-life:**
   **· Water:**        High:        No data
                                    Low:

*Comment:*

   **· Air:**        High:        9.20 hours
                                    Low:     0.892 hours
*Comment:* Based upon measured rate constants for reaction with hydroxyl radical (Atkinson, R (1985)) and ozone (Tuazon, EC et al. (1978)) in air.

**Reduction half-life:**        High:        No data
                                    Low:

*Comment:*

**Hydrolysis:**
   **· First-order hydr half-life:**
*Comment:*

   **· Acid rate const (M(H+)-hr)$^{-1}$:**        No hydrolyzable groups
*Comment:*

   **· Base rate const (M(OH-)-hr)$^{-1}$:**
*Comment:*

# Dibromochloromethane

**CAS Registry Number:** 124-48-1

**Half-lives:**

  · **Soil:**                          High:     4320 hours             (6 months)

                                       Low:      672 hours              (4 weeks)

*Comment:* Scientific judgement based upon estimated aqueous aerobic biodegradation half-life.

  · **Air:**                           High:   10252 hours         (1.2 years)

                                         Low:    1025 hours         (42.7 days)

*Comment:* Scientific judgement based upon estimated photooxidation half-life in air.

  · **Surface Water:**             High:     4320 hours             (6 months)

                                       Low:      672 hours              (4 weeks)

*Comment:* Scientific judgement based upon estimated aqueous aerobic biodegradation half-life.

  · **Ground Water:**            High:     4320 hours             (6 months)

                                       Low:      336 hours              (14 days)

*Comment:* Scientific judgement based upon estimated aqueous aerobic and anaerobic biodegradation half-lives.

**Aqueous Biodegradation (unacclimated):**

  · **Aerobic half-life:**          High:     4320 hours             (6 months)

                                       Low:      672 hours              (4 weeks)

*Comment:* Scientific judgement based upon aerobic screening test data (Tabak, HH et al. (1981A)).

  · **Anaerobic half-life:**       High:     4320 hours             (6 months)

                                       Low:      672 hours              (4 weeks)

*Comment:* Scientific judgement based upon unacclimated anaerobic screening test data (Bouwer, EJ and McCarty, PL (1983), Bouwer, EJ, et al. (1981)).

  · **Removal/secondary treatment:**     High:          99%

                                              Low:

*Comment:* Based upon % degraded under aerobic continuous flow conditions (Bouwer, EJ and McCarty, PL (1983)).

**Photolysis:**

  · **Atmos photol half-life:**        High:     No data

                                           Low:

*Comment:*

· **Max light absorption (nm):**        No data
*Comment:*

· **Aq photol half-life:**        High:        No data
                                  Low:
*Comment:*

**Photooxidation half-life:**
· **Water:**                      High:        No data
                                  Low:
*Comment:*

· **Air:**                        High:    10252 hours        (1.2 years)
                                  Low:     1025 hours         (42.7 days)
*Comment:* Scientific judgement based upon an estimated rate constant for the vapor phase
reaction with hydroxyl radicals in air (Atkinson, R (1987A)).

**Reduction half-life:**          High:        No data
                                  Low:
*Comment:*

**Hydrolysis:**
· **First-order hydr half-life:**        275 years
*Comment:* Scientific judgement based upon reported rate constant ($8.0 \times 10^{-11}$ s$^{-1}$) at pH 7 and 25
°C (Mabey, W and Mill, T (1978)).

· **Acid rate const (M(H+)-hr)$^{-1}$:**
*Comment:*

· **Base rate const (M(OH-)-hr)$^{-1}$:**
*Comment:*

# Tris (2,3-dibromopropyl) phosphate

**CAS Registry Number:** 126-72-7

**Half-lives:**
- **Soil:** High: 168 hours (7 days)
  Low: 24 hours (1 day)

  *Comment:* Scientific judgement based upon estimated unacclimated aqueous aerobic biodegradation half-life.

- **Air:** High: 9.6 hours
  Low: 0.96 hours (57 minutes)

  *Comment:* Scientific judgement based upon estimated photooxidation half-life in air.

- **Surface Water:** High: 168 hours (7 days)
  Low: 24 hours (1 day)

  *Comment:* Scientific judgement based upon estimated unacclimated aqueous aerobic biodegradation half-life.

- **Ground Water:** High: 336 hours (14 days)
  Low: 48 hours (2 days)

  *Comment:* Scientific judgement based upon estimated unacclimated aqueous aerobic biodegradation half-life.

**Aqueous Biodegradation (unacclimated):**
- **Aerobic half-life:** High: 168 hours (7 days)
  Low: 24 hours (1 day)

  *Comment:* Scientific judgement based upon unacclimated aqueous screening test data (Alvarez, GH et al. (1982)).

- **Anaerobic half-life:** High: 672 hours (28 days)
  Low: 96 hours (4 days)

  *Comment:* Scientific judgement based upon unacclimated aerobic biodegradation half-life.

- **Removal/secondary treatment:** High: No data
  Low:

  *Comment:*

**Photolysis:**
- **Atmos photol half-life:** High: No data
  Low:

  *Comment:*

**· Max light absorption (nm):**          No data
*Comment:*

**· Aq photol half-life:**          High:          No data
                                    Low:
*Comment:*

**Photooxidation half-life:**
   **· Water:**          High:          No data
                         Low:
*Comment:*

   **· Air:**          High:          9.6 hours
                       Low:          0.96 hours          (57 minutes)
*Comment:* Scientific judgement based upon an estimated rate constant for the vapor phase reaction with hydroxyl radicals in air (Atkinson, R (1987A)).

**Reduction half-life:**          High:          No data
                                  Low:
*Comment:*

**Hydrolysis:**
   **· First-order hydr half-life:**          No data
*Comment:*

   **· Acid rate const (M(H+)-hr)$^{-1}$:**
*Comment:*

   **· Base rate const (M(OH-)-hr)$^{-1}$:**
*Comment:*

# Chloroprene

<u>CAS Registry Number:</u>  126-99-8

<u>Half-lives:</u>
- **Soil:**                      High:    4320 hours          (6 months)
                                   Low:      672 hours           (4 weeks)
*Comment:*  Scientific judgement based upon estimated unacclimated aqueous aerobic biodegradation half-life.

- **Air:**                       High:    27.8 hours
                                   Low:      2.9 hours
*Comment:*  Scientific judgement based upon estimated rate constants for reaction with hydroxyl radicals and ozone in air (Atkinson, R (1987A); Atkinson, R and Carter, WPL (1984)).

- **Surface Water:**             High:    4320 hours          (6 months)
                                   Low:      672 hours           (4 weeks)
*Comment:*  Scientific judgement based upon estimated unacclimated aqueous aerobic biodegradation half-life.

- **Ground Water:**              High:    8640 hours          (12 months)
                                   Low:     1344 hours          (8 weeks)
*Comment:*  Scientific judgement based upon estimated unacclimated aqueous aerobic biodegradation half-life.

<u>Aqueous Biodegradation (unacclimated):</u>
- **Aerobic half-life:**         High:    4320 hours          (6 months)
                                   Low:      672 hours           (4 weeks)
*Comment:*  Scientific judgement based upon aqueous aerobic screening test data for vinyl chloride (Freitag, D et al. (1984A); Helfgott, TB et al. (1977)).

- **Anaerobic half-life:**       High:    17280 hours         (24 months)
                                   Low:     2688 hours          (16 weeks)
*Comment:*  Scientific judgement based upon estimated unacclimated aqueous aerobic biodegradation half-life.

- **Removal/secondary treatment:**    High:    No data
                                         Low:
*Comment:*

<u>Photolysis:</u>
- **Atmos photol half-life:**    High:    No data
                                   Low:

500

*Comment:*

**· Max light absorption (nm):**
*Comment:*

**· Aq photol half-life:**                    High:
                                              Low:
*Comment:*

**Photooxidation half-life:**
  **· Water:**                        High:        No data
                                                Low:
*Comment:*

  **· Air:**                          High:        27.8 hours
                                                Low:         2.9 hours
*Comment:* Based upon estimated rate constants for reaction with hydroxyl radicals and ozone in air (Atkinson, R (1987A); Atkinson, R and Carter, WPL (1984)).

**Reduction half-life:**                      High:         No data
                                              Low:
*Comment:*

**Hydrolysis:**
  **· First-order hydr half-life:**    No data
*Comment:*

  **· Acid rate const (M(H+)-hr)$^{-1}$:**
*Comment:*

  **· Base rate const (M(OH-)-hr)$^{-1}$:**
*Comment:*

# Tetrachloroethylene

**CAS Registry Number:** 127-18-4

**Half-lives:**

| | | | |
|---|---|---|---|
| · **Soil:** | High: | 8640 hours | (1 year) |
| | Low: | 4320 hours | (6 months) |

*Comment:* Scientific judgement based upon estimated aqueous aerobic biodegradation half-life.

| | | | |
|---|---|---|---|
| · **Air:** | High: | 3843 hours | (160 days) |
| | Low: | 384 hours | (16 days) |

*Comment:* Based upon photooxidation half-life in air.

| | | | |
|---|---|---|---|
| · **Surface Water:** | High: | 8640 hours | (1 year) |
| | Low: | 4320 hours | (6 months) |

*Comment:* Scientific judgement based upon aerobic river die-away test data (Mudder, TI (1981)) and saltwater grab sample data (Jensen, S and Rosenberg, R (1975)).

| | | | |
|---|---|---|---|
| · **Ground Water:** | High: | 17280 hours | (2 years) |
| | Low: | 8640 hours | (12 months) |

*Comment:* Scientific judgement based upon estimated aqueous aerobic biodegradation half-life.

**Aqueous Biodegradation (unacclimated):**

| | | | |
|---|---|---|---|
| · **Aerobic half-life:** | High: | 8640 hours | (1 year) |
| | Low: | 4320 hours | (6 months) |

*Comment:* Scientific judgement based upon aerobic river die-away test data (Mudder, TI (1981)) and saltwater grab sample data (Jensen, S and Rosenberg, R (1975)).

| | | | |
|---|---|---|---|
| · **Anaerobic half-life:** | High: | 39672 hours | (4.5 years) |
| | Low: | 2352 hours | (98 days) |

*Comment:* Scientific judgement based upon anaerobic screening test data (Bouwer, EJ et al. (1981)).

| | | | |
|---|---|---|---|
| · **Removal/secondary treatment:** | High: | 86% | |
| | Low: | | |

*Comment:* Based upon % degraded under anaerobic continuous flow conditions (Bouwer, EJ and McCarty, PL (1983)).

**Photolysis:**

| | | |
|---|---|---|
| · **Atmos photol half-life:** | High: | No data |
| | Low: | |

*Comment:*

· **Max light absorption (nm):**   No data
*Comment:* Does not absorb UV light at wavelengths greater than 260 nm (Dahlberg, JA (1969)).

· **Aq photol half-life:**     High:  No data
             Low:
*Comment:*

## Photooxidation half-life:
· **Water:**         High:  No data
             Low:
*Comment:*

· **Air:**          High:  3843 hours   (160 days)
             Low:   384 hours    (16 days)
*Comment:* Based upon measured rate data for the vapor phase reaction with hydroxyl radicals in air (Atkinson, R (1985)).

## Reduction half-life:       High:  No data
             Low:
*Comment:*

## Hydrolysis:
· **First-order hydr half-life:**   No hydrolyzable groups
*Comment:* Rate constant at pH 3, 7, and 11 is zero (Kollig, HP et al. (1987A)).

· **Acid rate const $(M(H+)-hr)^{-1}$:**
*Comment:*

· **Base rate const $(M(OH-)-hr)^{-1}$:**
*Comment:*

# C.I. Vat Yellow 4

<u>CAS Registry Number:</u>  128-66-5

<u>Structure:</u>

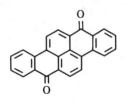

<u>Half-lives:</u>
  **· Soil:**                        High:    4320 hours            (6 months)
                                     Low:      672 hours            (4 weeks)
  *Comment:*  Scientific judgement based upon estimated unacclimated aqueous aerobic biodegradation half-life.

  **· Air:**                         High:     157 hours            (6.5 days)
                                     Low:      15.7 hours           (0.65 days)
  *Comment:*  Scientific judgement based upon estimated rate constants for reaction with hydroxyl radicals in air (Atkinson, R (1987A)).

  **· Surface Water:**               High:    4320 hours            (6 months)
                                     Low:      672 hours            (4 weeks)
  *Comment:*  Scientific judgement based upon estimated unacclimated aqueous aerobic biodegradation half-life.

  **· Ground Water:**                High:    8640 hours            (12 months)
                                     Low:     1344 hours            (8 weeks)
  *Comment:*  Scientific judgement based upon estimated unacclimated aqueous aerobic biodegradation half-life.

<u>Aqueous Biodegradation (unacclimated):</u>
  **· Aerobic half-life:**           High:    4320 hours            (6 months)
                                     Low:      672 hours            (4 weeks)
  *Comment:*  Scientific judgement based upon slow biodegradation for polycyclic aromatic compounds.

  **· Anaerobic half-life:**         High:   17280 hours            (24 months)
                                     Low:     2688 hours            (16 weeks)
  *Comment:*  Scientific judgement based upon estimated unacclimated aqueous aerobic biodegradation half-life.

**· Removal/secondary treatment:**       High:       No data
                                          Low:

*Comment:*

## Photolysis:
**· Atmos photol half-life:**             High:       No data
                                          Low:

*Comment:*

**· Max light absorption (nm):**
*Comment:*

**· Aq photol half-life:**                High:
                                          Low:

*Comment:*

## Photooxidation half-life:
**· Water:**                              High:       No data
                                          Low:

*Comment:*

**· Air:**                                High:       157 hours        (6.5 days)
                                          Low:        15.7 hours       (0.65 days)
*Comment:* Scientific judgement based upon estimated rate constant for reaction with hydroxyl radicals in air (Atkinson, R (1987A)).

## Reduction half-life:
                                          High:       No data
                                          Low:

*Comment:*

## Hydrolysis:
**· First-order hydr half-life:**         No data
*Comment:*

**· Acid rate const (M(H+)-hr)$^{-1}$:**
*Comment:*

**· Base rate const (M(OH-)-hr)$^{-1}$:**
*Comment:*

# Pyrene

**CAS Registry Number:** 129-00-0

**Structure:**

**Half-lives:**

· **Soil:**           High:    45600 hours        (5.2 years)

                      Low:      5040 hours        (210 days)

*Comment:* Based upon aerobic soil die-away test data at 10-30°C (Coover, MP and Sims, RCC (1987); Sims, RC (1990)).

· **Air:**            High:    2.04 hours

                      Low:     0.68 hours

*Comment:* Scientific judgement based upon estimated sunlight photolysis half-life in water.

· **Surface Water:**  High:    2.04 hours

                      Low:     0.68 hours

*Comment:* Scientific judgement based upon estimated sunlight photolysis half-life in water.

· **Ground Water:**   High:    91200 hours       (10.4 years)

                      Low:     10080 hours       (1.15 years)

*Comment:* Scientific judgement based upon estimated unacclimated aqueous aerobic biodegradation half-life.

**Aqueous Biodegradation (unacclimated):**

· **Aerobic half-life:**      High:    45600 hours        (5.2 years)

                              Low:      5040 hours        (210 days)

*Comment:* Based upon aerobic soil die-away test data at 10-30°C (Coover, MP and Sims, RCC (1987); Sims, RC (1990)).

· **Anaerobic half-life:**    High:   182400 hours        (20.8 years)

                              Low:     20160 hours        (2.3 years)

*Comment:* Scientific judgement based upon estimated unacclimated aqueous aerobic biodegradation half-life.

· **Removal/secondary treatment:**    High:

                                       Low:         0%

*Comment:* Removal percentage based upon data from a continuous activated sludge biological

506

treatment simulator (Petrasek, AC et al. (1983)).

## Photolysis:
· **Atmos photol half-life:**  High: 2.04 hours
Low: 0.68 hours

*Comment:* Based upon measured aqueous photolysis quantum yields and calculated for midday summer sunlight at 40°N latitude (low $t_{1/2}$) (Zepp, RG and Schlotzhauer, PF (1979)) and adjusted for approximate winter sunlight intensity (high $t_{1/2}$) (Lyman, WJ et al. (1982)).

· **Max light absorption (nm):**  lambda max = 230.5, 241, 251, 261.5, 272, 292, 305, 318, 333.5, 351.5, 356, 362, 371.5 (methanol/ethanol mixture)
*Comment:* IARC (1983A).

· **Aq photol half-life:**  High: 2.04 hours
Low: 0.68 hours

*Comment:* Based upon measured aqueous photolysis quantum yields and calculated for midday summer sunlight at 40°N latitude (low $t_{1/2}$) (Zepp, RG and Schlotzhauer, PF (1979)) and adjusted for approximate winter sunlight intensity (high $t_{1/2}$) (Lyman, WJ et al. (1982)).

## Photooxidation half-life:
· **Water:**  High: No data
Low:

*Comment:*

· **Air:**  High: 8.02 hours
Low: 0.802 hours

*Comment:* Scientific judgement based upon estimated rate constant for reaction with hydroxyl radical in air (Atkinson, R (1987A)).

## Reduction half-life:  High: No data
Low:

*Comment:*

## Hydrolysis:
· **First-order hydr half-life:**
*Comment:*

· **Acid rate const $(M(H+)-hr)^{-1}$:**  No hydrolyzable groups
*Comment:*

· **Base rate const $(M(OH-)-hr)^{-1}$:**
*Comment:*

# Dimethyl phthalate

CAS Registry Number: 131-11-3

## Half-lives:

· **Soil:**                          High:    168 hours              (7 days)
                                       Low:     24 hours               (1 day)
*Comment:* Scientific judgement based upon unacclimated aerobic river die-away test data (Hattori, Y et al. (1975)) and acclimated aerobic soil grab sample data (Russell, DJ et al. (1985)).

· **Air:**                           High:    1118 hours             (46.6 days)
                                       Low:     112 hours              (4.7 days)
*Comment:* Scientific judgement based upon estimated photooxidation half-life in air.

· **Surface Water:**                 High:    168 hours              (7 days)
                                       Low:     24 hours               (1 day)
*Comment:* Scientific judgement based upon unacclimated river die-away test data (Hattori, Y et al. (1975)).

· **Ground Water:**                  High:    336 hours              (14 days)
                                       Low:     48 hours               (2 days)
*Comment:* Scientific judgement based upon estimated unacclimated aqueous aerobic biodegradation half-life.

## Aqueous Biodegradation (unacclimated):

· **Aerobic half-life:**             High:    168 hours              (7 days)
                                       Low:     24 hours               (1 day)
*Comment:* Scientific judgement based upon unacclimated aerobic river die-away test data (Hattori, Y et al. (1975)) and acclimated aerobic soil grab sample data (Russell, DJ et al. (1985)).

· **Anaerobic half-life:**           High:    672 hours              (28 days)
                                       Low:     96 hours               (4 days)
*Comment:* Scientific judgement based upon unacclimated aerobic biodegradation half-life.

· **Removal/secondary treatment:**   High:           96%
                                       Low:
*Comment:* Based upon % removal under aerobic continuous flow conditions (Petrasek, AC et al. (1983)).

## Photolysis:

· **Atmos photol half-life:**        High:    No data
                                       Low:

*Comment:*

· **Max light absorption (nm):**        No data
*Comment:*


· **Aq photol half-life:**        High:        No data
                              Low:
*Comment:*


**Photooxidation half-life:**
  · **Water:**                    High:        No data
                              Low:
*Comment:*


· **Air:**                        High:        1118 hours                (46.6 days)
                              Low:        112 hours                (4.7 days)
*Comment:* Scientific judgement based upon estimated rate data for hydroxyl radicals in air (Atkinson, R (1987A)).


**Reduction half-life:**            High:        No data
                              Low:
*Comment:*


**Hydrolysis:**
  · **First-order hydr half-life:**            27903 hours            (1163 days)
*Comment:* Based upon second order rate constant ($k_B = 6.9 \times 10^{-2}$ $M^{-1}$ $s^{-1}$) at pH 7 and 30 °C (Wolfe, NL et al. (1980B)).


· **Acid rate const (M(H+)-hr)$^{-1}$:**
*Comment:*


· **Base rate const (M(OH-)-hr)$^{-1}$:**        111.6 $M^{-1}$ $hr^{-1}$
*Comment:* ($t_{1/2}$ = 1.2 days) Based upon second order rate constant at pH 9 and 18 °C (Wolfe, NL et al. (1980B)).

# Dibenzofuran

CAS Registry Number: 132-64-9

Half-lives:
- Soil:                              High:     672 hours          (4 weeks)
                                     Low:      168 hours          (7 days)

Comment: Scientific judgement based upon aerobic acclimated and unacclimated ground water die-away test data (Lee, MD et al. (1984); Ward, CH et al. (1986)) and aerobic subsurface-soil microcosm screening study data (Lee, MD et al. (1984)).

- Air:                               High:      19 hours
                                     Low:       1.9 hours

Comment: Scientific judgement based upon estimated rate constants for reaction with hydroxyl radicals in air (Atkinson, R (1987A)).

- Surface Water:                     High:     672 hours          (4 weeks)
                                     Low:      168 hours          (7 days)

Comment: Scientific judgement based upon aerobic acclimated and unacclimated ground water die-away test data (Lee, MD et al. (1984); Ward, CH et al. (1986)).

- Ground Water:                      High:     835 hours          (35 days)
                                     Low:      205 hours          (8.5 days)

Comment: Scientific judgement based upon aerobic acclimated and unacclimated ground water die-away test data (Lee, MD et al. (1984); Ward, CH et al. (1986)) and aerobic subsurface-soil microcosm screening study data (Lee, MD et al. (1984)).

Aqueous Biodegradation (unacclimated):
- Aerobic half-life:                 High:     672 hours          (4 weeks)
                                     Low:      168 hours          (7 days)

Comment: Scientific judgement based upon aerobic acclimated and unacclimated ground water die-away test data (Lee, MD et al. (1984); Ward, CH et al. (1986)).

- Anaerobic half-life:               High:     2688 hours         (16 weeks)
                                     Low:      672 hours          (28 days)

Comment: Scientific judgement based upon estimated unacclimated aqueous aerobic biodegradation half-life.

- Removal/secondary treatment:       High:     No data
                                     Low:

Comment:

Photolysis:

510

· **Atmos photol half-life:**                 High:      No data
                                              Low:

*Comment:*

· **Max light absorption (nm):**
*Comment:*

· **Aq photol half-life:**                    High:
                                              Low:

*Comment:*

**Photooxidation half-life:**
· **Water:**                                  High:      No data
                                              Low:

*Comment:*

· **Air:**                                    High:      19 hours
                                              Low:       1.9 hours

*Comment:*  Based upon estimated rate constant for reaction with hydroxyl radicals in air
(Atkinson, R (1987A)).

**Reduction half-life:**                      High:      No data
                                              Low:

*Comment:*

**Hydrolysis:**
· **First-order hydr half-life:**        No data
*Comment:*

· **Acid rate const (M(H+)-hr)$^{-1}$:**
*Comment:*

· **Base rate const (M(OH-)-hr)$^{-1}$:**
*Comment:*

# Captan

**CAS Registry Number:** 133-06-2

**Structure:**

## Half-lives:

**· Soil:**

| | High: | 1440 hours | (60 days) |
|---|---|---|---|
| | Low: | 48 hours | (2 days) |

*Comment:* Scientific judgement based upon unacclimated (low $t_{1/2}$: Agnihotri, VP (1970)) and acclimated soil grab sample data (high $t_{1/2}$: Foschi, S et al. (1970)).

**· Air:**

| | High: | 32 hours |
|---|---|---|
| | Low: | 3.2 hours |

*Comment:* Scientific judgement based upon estimated photooxidation half-life in air.

**· Surface Water:**

| | High: | 10.3 hours |
|---|---|---|
| | Low: | 10.5 minutes |

*Comment:* Scientific judgement based upon aqueous hydrolysis half-lives for pH 5.2 (high $t_{1/2}$) and 8.3 (low $t_{1/2}$) at 28 °C (Wolfe, NL et al. (1976)).

**· Ground Water:**

| | High: | 10.3 hours |
|---|---|---|
| | Low: | 10.5 minutes |

*Comment:* Scientific judgement based upon aqueous hydrolysis half-lives for pH 5.2 (high $t_{1/2}$) and 8.3 (low $t_{1/2}$) at 28 °C (Wolfe, NL et al. (1976)).

## Aqueous Biodegradation (unacclimated):

**· Aerobic half-life:**

| | High: | 1440 hours | (60 days) |
|---|---|---|---|
| | Low: | 48 hours | (2 days) |

*Comment:* Scientific judgement based upon unacclimated (low $t_{1/2}$: Agnihotri, VP (1970)) and acclimated soil grab sample data (high $t_{1/2}$: Foschi, S et al. (1970)).

**· Anaerobic half-life:**

| | High: | 5760 hours | (240 days) |
|---|---|---|---|
| | Low: | 192 hours | (8 days) |

*Comment:* Scientific judgement based upon unacclimated aerobic biodegradation half-life.

**· Removal/secondary treatment:**

| | High: | No data |
|---|---|---|
| | Low: | |

*Comment:*

512

**Photolysis:**
  · **Atmos photol half-life:**          High:        No data
                                         Low:

  *Comment:*

  · **Max light absorption (nm):**        No data
  *Comment:*

  · **Aq photol half-life:**             High:        No data
                                         Low:

  *Comment:*

**Photooxidation half-life:**
  · **Water:**                           High:        No data
                                         Low:

  *Comment:*

  · **Air:**                             High:        32 hours
                                         Low:         3.2 hours
  *Comment:* Scientific judgement based upon an estimated rate constant for the vapor phase reaction with hydroxyl radicals in air (Atkinson, R (1987A)).

**Reduction half-life:**                 High:        No data
                                         Low:

  *Comment:*

**Hydrolysis:**
  · **First-order hydr half-life:**      1.8 hours
  *Comment:* Scientific judgement based upon first order rate constant ($6.5 \times 10^{-3}$ $s^{-1}$) at pH 7.1 and 28 °C (Wolfe, NL et al. (1976)).

  · **Acid rate const (M(H+)-hr)$^{-1}$:**      $1.87 \times 10^{-5}$ $s^{-1}$
  *Comment:* ($t_{1/2}$ = 10.3 hours) Based upon first order rate constant at pH 5.2 and 28 °C (Wolfe, NL et al. (1976)).

  · **Base rate const (M(OH-)-hr)$^{-1}$:**      $1.1 \times 10^{-3}$ $s^{-1}$
  *Comment:* ($t_{1/2}$ = 10.5 minutes) Based upon first order rate constant at pH 8.3 and 28 °C (Wolfe, NL et al. (1976)).

# alpha-Naphthylamine

<u>CAS Registry Number:</u>  134-32-7

<u>Half-lives:</u>
  · **Soil:**                          High:    4320 hours         (6 months)
                                        Low:      672 hours         (4 weeks)
*Comment:* Scientific judgement based upon slow biodegradation observed in an aerobic soil die-away test study (Graveel, JG et al. (1986)) and aerobic activated sludge screening tests (Pitter, P (1976)).

  · **Air:**                           High:    2.92 hours
                                        Low:     0.292 hours
*Comment:* Scientific judgement based  upon estimated photooxidation half-life in air.

  · **Surface Water:**                 High:    3480 hours         (145 days)
                                        Low:     62.4 hours         (2.6 days)
*Comment:* Scientific judgement based upon estimated rate constants for reactions of representative aromatic amines with ·OH and $RO_2$· in water (Mill, T and Mabey, W (1985)).

  · **Ground Water:**                  High:    8640 hours         (12 months)
                                        Low:     1344 hours         (8 weeks)
*Comment:* Scientific judgement based upon slow biodegradation observed in an aerobic soil die-away test study (Graveel, JG et al. (1986)) and aerobic activated sludge screening tests (Pitter, P (1976)).

<u>Aqueous Biodegradation (unacclimated):</u>
  · **Aerobic half-life:**             High:    4320 hours         (6 months)
                                        Low:      672 hours         (4 weeks)
*Comment:* Scientific judgement based upon slow biodegradation observed in an aerobic soil die-away test study (Graveel, JG et al. (1986)) and aerobic activated sludge screening tests (Pitter, P (1976)).

  · **Anaerobic half-life:**           High:    17280 hours        (24 months)
                                        Low:     2688 hours         (16 weeks)
*Comment:* Scientific judgement based upon slow biodegradation observed in an aerobic soil die-away test study (Graveel,JG et al. (1986)) and aerobic activated sludge screening tests (Pitter, P (1976)).

  · **Removal/secondary treatment:**   High:    No data
                                        Low:
*Comment:*

514

## Photolysis:
**· Atmos photol half-life:**        High:      No data

                                         Low:

*Comment:*

**· Max light absorption (nm):**      lambda max = 243, 318 nm (cyclohexane)

*Comment:* Significant absorption up to approximately 355 nm (Sadtler UV No. 2545).

**· Aq photol half-life:**            High:      No data

                                         Low:

*Comment:*

## Photooxidation half-life:
**· Water:**                       High:     3480 hours            (145 days)

                                         Low:      62.4 hours            (2.6 days)

*Comment:* Scientific judgement based upon estimated rate constants for reactions of represntative aromatic amines with ·OH and $RO_2$· (Mill, T and Mabey, W (1985)).

**· Air:**                               High:     2.92 hours

                                         Low:      0.292 hours

*Comment:* Scientific judgement based upon estimated rate constant for reaction with hydroxyl radical in air (Atkinson, R (1987A)).

## Reduction half-life:
                                           High:     No data

                                           Low:

*Comment:*

## Hydrolysis:
**· First-order hydr half-life:**

*Comment:*

**· Acid rate const $(M(H+)-hr)^{-1}$:**     No hydrolyzable groups

*Comment:*

**· Base rate const $(M(OH-)-hr)^{-1}$:**

*Comment:*

# Cupferron

CAS Registry Number: 135-20-6

Structure:

Half-lives:
  · Soil:                          High:     4320 hours         (6 months)
                                   Low:       672 hours         (4 weeks)
  *Comment:* Scientific judgement based upon estimated unacclimated aqueous aerobic
  biodegradation half-life.

  · Air:                           High:     14.0 hours
                                   Low:      1.40 hours
  *Comment:* Scientific judgement based upon estimated photooxidation half-life in air.

  · Surface Water:                 High:     4320 hours         (6 months)
                                   Low:       672 hours         (4 weeks)
  *Comment:* Scientific judgement based upon estimated unacclimated aqueous rates predicted for
  the nitrosamine class.

  · Ground Water:                  High:     8640 hours         (12 months)
                                   Low:      1344 hours         (8 weeks)
  *Comment:* Scientific judgement based upon estimated unacclimated aqueous aerobic
  biodegradation half-life.

Aqueous Biodegradation (unacclimated):
  · Aerobic half-life:             High:     4320 hours         (6 months)
                                   Low:       672 hours         (4 weeks)
  *Comment:* Scientific judgement based upon estimated unacclimated aqueous rates predicted for
  the nitrosamine class.

  · Anaerobic half-life:           High:    17280 hours         (24 months)
                                   Low:      2688 hours         (16 weeks)
  *Comment:* Scientific judgement based upon estimated unacclimated aqueous aerobic
  biodegradation half-life.

  · Removal/secondary treatment:   High:      No data
                                   Low:

  *Comment:*

**Photolysis:**
  · **Atmos photol half-life:**           High:      No data
                                              Low:
    *Comment:*

  · **Max light absorption (nm):**     No data
    *Comment:*

  · **Aq photol half-life:**          High:      No data
                                           Low:
    *Comment:*

**Photooxidation half-life:**
  · **Water:**                    High:      No data
                                           Low:
    *Comment:*

  · **Air:**                      High:    14.0 hours
                                         Low:     1.40 hours
    *Comment:*  Scientific judgement based upon estimated rate constant for reaction with hydroxyl radicals in air (Atkinson, R (1987A)).

**Reduction half-life:**              High:      No data
                                          Low:
    *Comment:*

**Hydrolysis:**
  · **First-order hydr half-life:**
    *Comment:*

  · **Acid rate const (M(H+)-hr)$^{-1}$:**     No data
    *Comment:*

  · **Base rate const (M(OH-)-hr)$^{-1}$:**
    *Comment:*

# Nitrilotriacetic acid

**CAS Registry Number:** 139-13-9

**Half-lives:**

· **Soil:** High: 672 hours (4 weeks)

Low: 72 hours (3 days)

*Comment:* Scientific judgement based upon measured half-lives from aerobic grab soil sample studies (low $t_{1/2}$) (Tabatabai, MA and Bremner, JM (1975)) and estimated unacclimated aqueous aerobic biodegradation half-life (high $t_{1/2}$).

· **Air:** High: 8.1 hours

Low: 0.81 hours

*Comment:* Scientific judgement based upon estimated rate constants for reaction with hydroxyl radicals in air (Atkinson, R (1987A)).

· **Surface Water:** High: 672 hours (4 weeks)

Low: 168 hours (7 days)

*Comment:* Scientific judgement based upon estimated unacclimated aqueous aerobic biodegradation half-life.

· **Ground Water:** High: 1344 hours (8 weeks)

Low: 336 hours (14 days)

*Comment:* Scientific judgement based upon estimated unacclimated aqueous aerobic biodegradation half-life.

**Aqueous Biodegradation (unacclimated):**

· **Aerobic half-life:** High: 672 hours (4 weeks)

Low: 168 hours (7 days)

*Comment:* Scientific judgement based upon aqueous aerobic grab sample study data which indicate rapid biodegradation following acclimation (Larson, RJ and Davidson, DH (1982); Larson, RJ et al. (1981); Thompson, JE and Duthie, JR (1968)).

· **Anaerobic half-life:** High: 2688 hours (16 weeks)

Low: 672 hours (28 days)

*Comment:* Scientific judgement based upon estimated unacclimated aqueous aerobic biodegradation half-life.

· **Removal/secondary treatment:** High:

Low:

*Comment:*

**Photolysis:**

· **Atmos photol half-life:**　　　　　　High:　　No data
　　　　　　　　　　　　　　　　　　　　Low:

*Comment:*

· **Max light absorption (nm):**
*Comment:*

· **Aq photol half-life:**　　　　　　　High:
　　　　　　　　　　　　　　　　　　　Low:

*Comment:*

**Photooxidation half-life:**
　· **Water:**　　　　　　　　　　　　High:　　No data
　　　　　　　　　　　　　　　　　　　Low:

*Comment:*

· **Air:**　　　　　　　　　　　　　　High:　　8.1 hours
　　　　　　　　　　　　　　　　　　　Low:　　0.81 hours
*Comment:* Based upon estimated rate constant for reaction with hydroxyl radicals (Atkinson, R (1987A)).

**Reduction half-life:**　　　　　　　　High:　　No data
　　　　　　　　　　　　　　　　　　　Low:

*Comment:*

**Hydrolysis:**
　· **First-order hydr half-life:**　　　No data
*Comment:*

· **Acid rate const (M(H+)-hr)$^{-1}$:**
*Comment:*

· **Base rate const (M(OH-)-hr)$^{-1}$:**
*Comment:*

# 4,4'-Thiodianiline

CAS Registry Number: 139-65-1

Structure:

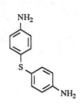

Half-lives:
**· Soil:**                                  High:      672 hours          (4 weeks)
                                             Low:       168 hours          (7 days)
*Comment:* Scientific judgement based upon estimated aqueous aerobic biodegradation half-life.

**· Air:**                                   High:      3.5 hours
                                             Low:       0.35 hours
*Comment:* Based upon estimated photooxidation half-lives in air.

**· Surface Water:**                         High:      672 hours          (4 weeks)
                                             Low:       31.2 hours
*Comment:* High value is scientific judgement based upon the high value estimated for aqueous aerobic biodegradation. The low value is based upon estimated photooxidation in water.

**· Ground Water:**                          High:      1344 hours         (8 weeks)
                                             Low:       336 hours          (14 days)
*Comment:* Scientific judgement based upon estimated aqueous aerobic biodegradation half-life.

Aqueous Biodegradation (unacclimated):
**· Aerobic half-life:**                     High:      672 hours          (4 weeks)
                                             Low:       168 hours          (7 days)
*Comment:* Scientific judgement.

**· Anaerobic half-life:**                   High:      2688 hours         (16 weeks)
                                             Low:       672 hours          (28 days)
*Comment:* Scientific judgement based upon estimated aqueous aerobic biodegradation half-life.

**· Removal/secondary treatment:**           High:      No data
                                             Low:
*Comment:*

Photolysis:

520

· **Atmos photol half-life:**        High:     No data

                                           Low:

*Comment:*

· **Max light absorption (nm):**     lambda max = 264 nm

*Comment:* Absorption maxima in methanol solution (Sadtler 6007 UV); UV absorption extends to approximately 350 nm.

· **Aq photol half-life:**          High:     No data

                                           Low:

*Comment:*

## Photooxidation half-life:

· **Water:**                    High:     1740 hours         (72.5 days)

                                           Low:      31.2 hours

*Comment:* Scientific judgement based upon reaction rates of ·OH and $RO_2$· with the aromatic amine class (Mill, T and Mabey, W (1985); Guesten, H et al. (1981)). It is assumed that thioaniline reacts twice as fast as aniline.

· **Air:**                          High:     3.5 hours

                                           Low:      0.35 hours

*Comment:* Based upon estimated reaction rate constant with ·OH (Atkinson, R (1987A)).

## Reduction half-life:              High:     No data

                                         Low:

*Comment:*

## Hydrolysis:

· **First-order hydr half-life:**     No hydrolyzable groups

*Comment:*

· **Acid rate const $(M(H+)-hr)^{-1}$:**

*Comment:*

· **Base rate const $(M(OH-)-hr)^{-1}$:**

*Comment:*

# Ethyl acrylate

**CAS Registry Number:** 140-88-5

**Half-lives:**

· **Soil:**                                  High:    168 hours              (7 days)

                                             Low:      24 hours              (1 day)

*Comment:* Scientific judgment based upon estimated unacclimated aqueous aerobic biodegradation half-life.

· **Air:**                                   High:    22.7 hours

                                             Low:      2.37 hours

*Comment:* Scientific judgement based upon estimated photooxidation rate constant in air.

· **Surface Water:**                         High:    168 hours              (7 days)

                                             Low:      24 hours              (1 day)

*Comment:* Scientific judgment based upon estimated unacclimated aqueous aerobic biodegradation half-life.

· **Ground Water:**                          High:    336 hours             (14 days)

                                             Low:      48 hours              (2 days)

*Comment:* Scientific judgment based upon estimated unacclimated aqueous aerobic biodegradation half-life.

**Aqueous Biodegradation (unacclimated):**

· **Aerobic half-life:**                     High:    168 hours              (7 days)

                                             Low:      24 hours              (1 day)

*Comment:* Scientific judgement based upon aqueous aerobic screening test data (Price, KS et al. (1974); Sasaki, S (1978)).

· **Anaerobic half-life:**                   High:    672 hours             (28 days)

                                             Low:      96 hours              (4 days)

*Comment:* Scientific judgment based upon estimated unacclimated aqueous aerobic biodegradation half-life.

· **Removal/secondary treatment:**           High:    No data

                                             Low:

*Comment:*

**Photolysis:**

· **Atmos photol half-life:**                High:    No data

                                             Low:

*Comment:*

**· Max light absorption (nm):** lambda max = 196, 240 nm (ethanol); 196, 239 nm (methanol) (Brunn, J et al. (1976)).
*Comment:*

**· Aq photol half-life:** High: No data
Low:
*Comment:*

**Photooxidation half-life:**
**· Water:** High: No data
Low:
*Comment:*

**· Air:** High: 22.7 hours
Low: 2.37 hours
*Comment:* Scientific judgement based upon estimated rate constants for reaction with hydroxyl radical (Atkinson, R (1987A)) and ozone (Atkinson, R and Carter, WPL (1984)).

**Reduction half-life:** High: No data
Low:
*Comment:*

**Hydrolysis:**
**· First-order hydr half-life:** $2.47 \times 10^4$ hours (2.8 years)
*Comment:* ($t_{1/2}$ at pH 7) Based upon acid and base catalyzed hydrolysis rate constants (Mabey, W and Mill, T (1978)).

**· Acid rate const $(M(H+)-hr)^{-1}$:** $1.2 \times 10^{-6}$ $M^{-1}$ $s^{-1}$
*Comment:* ($2.14 \times 10^6$ hours (244 years) $t_{1/2}$ at pH 5) Based upon acid and base catalyzed hydrolysis rate constants (Mabey, W and Mill, T (1978)).

**· Base rate const $(M(OH-)-hr)^{-1}$:** 0.078 $M^{-1}$ $s^{-1}$
*Comment:* (247 hours (10.3 days) $t_{1/2}$ at pH 9) Based upon acid and base catalyzed hydrolysis rate constants (Mabey, W and Mill, T (1978)).

# Butyl acrylate

**CAS Registry Number:** 141-32-2

**Half-lives:**

  · **Soil:**                            High:    168 hours        (7 days)

                                   Low:     24 hours         (1 day)

*Comment:* Scientific judgement based upon estimated aqueous aerobic biodegradation half-lives.

  · **Air:**                            High:    23 hours

                                   Low:     2.3 hours

*Comment:* Based upon estimated photooxidation rate constant in air.

  · **Surface Water:**            High:    168 hours        (7 days)

                                   Low:     24 hours         (1 day)

*Comment:* Scientific judgement based upon estimated aqueous aerobic biodegradation half-lives.

  · **Ground Water:**             High:    336 hours        (14 days)

                                     Low:     48 hours         (2 days)

*Comment:* Scientific judgement based upon estimated aqueous aerobic biodegradation half-lives.

**Aqueous Biodegradation (unacclimated):**

  · **Aerobic half-life:**          High:    168 hours        (7 days)

                                   Low:     24 hours         (1 day)

*Comment:* Scientific judgement based upon aerobic biological screening studies (Kondo, M et al. (1988); Sasaki, S (1978)).

  · **Anaerobic half-life:**       High:    672 hours        (28 days)

                                   Low:     96 hours         (4 days)

*Comment:* Scientific judgement based upon estimated aqueous aerobic biodegradation half-lives.

  · **Removal/secondary treatment:**    High:    No data

                                   Low:

*Comment:*

**Photolysis:**

  · **Atmos photol half-life:**      High:    No data

                                     Low:

*Comment:*

  · **Max light absorption (nm):**

*Comment:*

**· Aq photol half-life:**　　　　　　　　High:
　　　　　　　　　　　　　　　　　　　　　Low:

*Comment:*

**Photooxidation half-life:**
　**· Water:**　　　　　　　　　　　　High:　　　No data
　　　　　　　　　　　　　　　　　　　Low:

*Comment:*

　**· Air:**　　　　　　　　　　　　　　High:　　　23 hours
　　　　　　　　　　　　　　　　　　　Low:　　　2.3 hours
*Comment:* Based upon estimated photooxidation rate constant with ·OH (Atkinson, R (1987A)).

**Reduction half-life:**　　　　　　　　High:　　　No data
　　　　　　　　　　　　　　　　　　　Low:

*Comment:*

**Hydrolysis:**
　**· First-order hydr half-life:**　　　　30,700 hours　　　　　(3.5 years)
*Comment:* Scientific judgement based upon reported half-life for ethyl acrylate at 25°C and pH 7 (Mabey, W and Mill, T (1978)).

　**· Acid rate const (M(H+)-hr)$^{-1}$:**　　0.00432
*Comment:* Value reported for ethyl acrylate at 25°C (Mabey, W and Mill, T (1978)).

　**· Base rate const (M(OH-)-hr)$^{-1}$:**　　280
*Comment:* Value reported for ethyl acrylate at 25°C (Mabey, W and Mill, T (1978)). At 25°C and pH 9, the half-life would be 248 hours (10.3 days).

# Ethyl acetate

CAS Registry Number: 141-78-6

## Half-lives:
· **Soil:**           High:    168 hours        (7 days)
                      Low:      24 hours        (1 day)
*Comment:* Scientific judgement based upon unacclimated aqueous aerobic biodegradation half-life.

· **Air:**            High:    353 hours        (14.7 days)
                      Low:     35.3 hours       (1.47 days)
*Comment:* Based upon photooxidation half-life in air.

· **Surface Water:**  High:    168 hours        (7 days)
                      Low:      24 hours        (1 day)
*Comment:* Scientific judgement based upon unacclimated aqueous aerobic biodegradation half-life.

· **Ground Water:**   High:    336 hours        (14 days)
                      Low:      48 hours        (2 days)
*Comment:* Scientific judgement based upon unacclimated aqueous aerobic biodegradation half-life.

## Aqueous Biodegradation (unacclimated):
· **Aerobic half-life:**    High:    168 hours        (7 days)
                           Low:      24 hours        (1 day)
*Comment:* Scientific judgement based upon unacclimated aqueous aerobic biodegradation screening test data (Heukelekian, H and Rand, MC (1955); Price, KS et al. (1974)).

· **Anaerobic half-life:**  High:    672 hours        (28 days)
                           Low:      96 hours        (4 days)
*Comment:* Scientific judgement based upon unacclimated aqueous aerobic biodegradation half-life.

· **Removal/secondary treatment:**  High:      96%
                                    Low:      99.9%
*Comment:* Removal percentages based upon data from a continuous activated sludge biological treatment simulator (Stover, EL and Kincannon, DF (1983)).

## Photolysis:
· **Atmos photol half-life:**    High:
                                 Low:

*Comment:*

**· Max light absorption (nm):**      No data
*Comment:*

**· Aq photol half-life:**      High:
                                     Low:

*Comment:*

## Photooxidation half-life:
**· Water:**      High: $9.6 \times 10^5$ hours      (110 years)
                        Low:   24090 hours        (2.75 years)
*Comment:* Based upon measured rate constant for reaction with hydroxyl radical in water (Dorfman, LM and Adams, GE (1973)).

**· Air:**      High:    353 hours        (14.7 days)
                     Low:    35.3 hours        (1.47 days)
*Comment:* Based upon measured rate constant for reaction with hydroxyl radical in air (Atkinson, R (1985)).

## Reduction half-life:
                                    High:      No data
                                    Low:
*Comment:*

## Hydrolysis:
**· First-order hydr half-life:**      $1.77 \times 10^4$ hours      (2.02 years)
*Comment:* ($t_{1/2}$ at pH 7 and 20°C) Based upon measured rate constants for acid and base catalyzed and neutral hydrolysis (Mabey, W and Mill, T (1978)).

**· Acid rate const $(M(H+)\text{-hr})^{-1}$:**      $3.05 \times 10^{-8}$
*Comment:* ($t_{1/2}$ = $1.42 \times 10^4$ hours (1.62 years) at pH 5 and 20°C) Based upon measured rate constants for acid and base catalyzed and neutral hydrolysis (Mabey, W and Mill, T (1978)).

**· Base rate const $(M(OH-)\text{-hr})^{-1}$:**      $2.99 \times 10^{-5}$
*Comment:* ($t_{1/2}$ = 178 hours (7.42 days) at pH 9 and 20°C) Based upon measured rate constants for acid and base catalyzed and neutral hydrolysis (Mabey, W and Mill, T (1978)).

# Kepone

**CAS Registry Number:** 143-50-0

**Structure:**

**Half-lives:**

· **Soil:** High: 17280 hours (2 years)

Low: 7488 hours (312 days)

*Comment:* Scientific judgement based upon aerobic aquatic microcosm study (soil and water grab samples) (Portier, RJ (1985)).

· **Air:** High: $4.2 \times 10^7$ hours (200 years)

Low: 438000 hours (50 years)

*Comment:* Scientific judgement based upon estimated half-lives for loss of trichlorofluoromethane from the troposphere to the stratosphere where direct photolysis can occur. Kepone is fully chlorinated and therefore will have a very slow rate in the troposphere (Brice, KA et al. (1982); Penkett, SA et al. (1980)).

· **Surface Water:** High: 17280 hours (2 years)

Low: 7488 hours (312 days)

*Comment:* Scientific judgement based upon aerobic aquatic microcosm study (soil and water grab samples) (Portier, RJ (1985)).

· **Ground Water:** High: 34560 hours (4 years)

Low: 14976 hours (624 days)

*Comment:* Scientific judgement based upon estimated aqueous aerobic biodegradation half-life.

**Aqueous Biodegradation (unacclimated):**

· **Aerobic half-life:** High: 17280 hours (2 years)

Low: 7488 hours (312 days)

*Comment:* Scientific judgement based upon aerobic aquatic microcosm study (soil and water grab samples) (Portier, RJ (1985)).

· **Anaerobic half-life:** High: 69120 hours (8 years)

Low: 29952 hours (1248 days)

*Comment:* Scientific judgement based upon unacclimated aerobic biodegradation half-life.

**· Removal/secondary treatment:**     High:     No data
                                        Low:

*Comment:*

## Photolysis:
**· Atmos photol half-life:**           High:     No data
                                        Low:

*Comment:*

**· Max light absorption (nm):**        No data
*Comment:*

**· Aq photol half-life:**              High:     No data
                                        Low:

*Comment:*

## Photooxidation half-life:
**· Water:**                            High:     No data
                                        Low:

*Comment:*

**· Air:**                              High:     Indefinite
                                        Low:
*Comment:* Scientific judgement based upon a rate constant of zero for vapor phase reaction with hydroxyl radicals in air (Atkinson, R (1987A)).

## Reduction half-life:                 High:     No data
                                        Low:

*Comment:*

## Hydrolysis:
**· First-order hydr half-life:**       No hydrolyzable groups
*Comment:*

**· Acid rate const (M(H+)-hr)$^{-1}$:**
*Comment:*

**· Base rate const (M(OH-)-hr)$^{-1}$:**
*Comment:*

# Phenylalanine mustard

CAS Registry Number: 148-82-3

Structure:

## Half-lives:
- **· Soil:** High: 24 hours
  Low: 4.62 hours
  *Comment:* Scientific judgement based upon estimated hydrolysis half-life in water.

- **· Air:** High: 3.90 hours
  Low: 0.390 hours
  *Comment:* Scientific judgement based upon estimated photooxidation half-life in air.

- **· Surface Water:** High: 24 hours
  Low: 4.62 hours
  *Comment:* Scientific judgement based upon estimated hydrolysis half-life in water.

- **· Ground Water:** High: 24 hours
  Low: 4.62 hours
  *Comment:* Scientific judgement based upon estimated hydrolysis half-life in water.

## Aqueous Biodegradation (unacclimated):
- **· Aerobic half-life:** High: 672 hours (4 weeks)
  Low: 168 hours (7 days)
  *Comment:* Scientific judgement.

- **· Anaerobic half-life:** High: 2688 hours (16 weeks)
  Low: 672 hours (28 days)
  *Comment:* Scientific judgement based upon unacclimated aqueous aerobic biodegradation half-life.

- **· Removal/secondary treatment:** High: No data
  Low:
  *Comment:*

## Photolysis:
- **· Atmos photol half-life:** High:
  Low:

*Comment:*

· **Max light absorption (nm):**  No data
*Comment:*

· **Aq photol half-life:**  High:
Low:

*Comment:*

## Photooxidation half-life:
· **Water:**  High: 1740 hours  (72 days)
Low: 31 hours  (1.3 days)

*Comment:* Scientific judgement based upon photooxidation rate constants with ·OH and $RO_2$· for the aromatic amine class (Mill, T and Mabey, W (1985); Guesten, H et al. (1981)).

· **Air:**  High: 3.90 hours
Low: 0.390 hours

*Comment:* Scientific judgement based upon estimated rate constant for reaction with hydroxyl radical in air (Atkinson, R (1987A)).

## Reduction half-life:  High: No data
Low:

*Comment:*

## Hydrolysis:
· **First-order hydr half-life:**  High: 24 hours
Low: 4.62 hours

*Comment:* Low value is based upon measured neutral hydrolysis rate constant for phenylalanine mustard at pH 5-9 (Ellington, JJ et al. (1987)) and high value is based upon scientific judgement of experimentally reported reaction kinetics of nitrogen mustard and similar homologs in dilute aqueous solutions at various temperatures (Cohen, B et al. (1948); Bartlett, PD et al. (1947)).

· **Acid rate const (M(H+)-hr)$^{-1}$:**  No data
*Comment:*

· **Base rate const (M(OH-)-hr)$^{-1}$:**  No data
*Comment:*

# Aziridine

CAS Registry Number: 151-56-4

Structure:

Half-lives:
- **· Soil:**                                     High:     672 hours              (4 weeks)
                                                  Low:      168 hours              (1 week)
  *Comment:* Scientific judgement based upon estimated aqueous aerobic biodegradation half-life.

- **· Air:**                                      High:     105 hours              (4.4 days)
                                                  Low:      10.5 hours
  *Comment:* Based upon photooxidation half-life in air.

- **· Surface Water:**                            High:     672 hours              (4 weeks)
                                                  Low:      168 hours              (1 week)
  *Comment:* Scientific judgement based upon estimated aqueous aerobic biodegradation half-life.

- **· Ground Water:**                             High:     8640 hours             (8 weeks)
                                                  Low:      336 hours              (2 weeks)
  *Comment:* Scientific judgement based upon estimated aqueous aerobic biodegradation half-life.

Aqueous Biodegradation (unacclimated):
- **· Aerobic half-life:**                        High:     672 hours              (4 weeks)
                                                  Low:      168 hours              (1 week)
  *Comment:* Scientific judgement.

- **· Anaerobic half-life:**                      High:     2688 hours             (16 weeks)
                                                  Low:      672 hours              (4 weeks)
  *Comment:* Scientific judgement based upon estimated aqueous aerobic biodegradation half-life.

- **· Removal/secondary treatment:**              High:        No data
                                                  Low:
  *Comment:*

Photolysis:
- **· Atmos photol half-life:**                   High:
                                                  Low:
  *Comment:*

- **· Max light absorption (nm):**                No data

*Comment:*

**· Aq photol half-life:**                    High:
                                              Low:
*Comment:*

**Photooxidation half-life:**
   **· Water:**                               High:         No data
                                              Low:
*Comment:*

   **· Air:**                                 High:         105 hours              (4.4 days)
                                              Low:          10.5 hours
*Comment:*  Based upon measured rate constant for hydroxyl radicals in air (Atkinson, R (1985)).

**Reduction half-life:**                      High:         No data
                                              Low:
*Comment:*

**Hydrolysis:**
   **· First-order hydr half-life:**                        3696 hours             (154 days)
*Comment:*  Based upon overall rate constant ($k_H = 5.2 \times 10^{-8}$ s$^{-1}$) at pH 7 and 25 °C (Mabey, W and Mill, T (1978)).

   **· Acid rate const (M(H+)-hr)$^{-1}$:**
*Comment:*

   **· Base rate const (M(OH-)-hr)$^{-1}$:**
*Comment:*

# p-Nitrosodiphenylamine

CAS Registry Number:  156-10-5

## Half-lives:
· **Soil:**                                       High:      4320 hours              (6 months)

Low:       672 hours               (4 weeks)

*Comment:*  Scientific judgement based upon estimated unacclimated aqueous aerobic biodegradation half-life.

· **Air:**                                         High:      2.5 hours

Low:       0.25 hours

*Comment:*  Scientific judgement based upon estimated photooxidation half-life in air.

· **Surface Water:**                          High:      3480 hours              (145 days)

Low:       62.4 hours              (2.6 days)

*Comment:*  Scientific judgement based upon estimated rate constants for reactions of representative aromatic amines with $\cdot$OH and $RO_2\cdot$ (Mill, T and Mabey, W (1985)).

· **Ground Water:**                          High:      8640 hours              (12 months)

Low:       1344 hours              (8 weeks)

*Comment:*  Scientific judgement based upon estimated unacclimated aqueous aerobic biodegradation half-life.

## Aqueous Biodegradation (unacclimated):
· **Aerobic half-life**:                       High:      4320 hours              (6 months)

Low:       672 hours               (4 weeks)

*Comment:*  Scientific judgement based upon estimated unacclimated aqueous rates predicted for the nitrosamine class.

· **Anaerobic half-life**:                     High:   17280 hours              (24 months)

Low:     2688 hours               (16 weeks)

*Comment:*  Scientific judgement based upon estimated unacclimated aqueous aerobic biodegradation half-life.

· **Removal/secondary treatment:**       High:      No data

Low:

*Comment:*

## Photolysis:
· **Atmos photol half-life:**              High:      No data

Low:

*Comment:*

· **Max light absorption (nm):**          No data
*Comment:*

· **Aq photol half-life:**          High:          No data
                                    Low:
*Comment:*

**Photooxidation half-life:**
· **Water:**          High:          3480 hours          (145 days)
                      Low:          62.4 hours          (2.6 days)
*Comment:* Scientific judgement based upon estimated rate constants for reactions of representative aromatic amines with ·OH and $RO_2$· (Mill, T and Mabey, W (1985)).

· **Air:**          High:          2.5 hours
                    Low:          0.25 hours
*Comment:* Scientific judgement based upon estimated rate constant for reaction with hydroxyl radicals in air (Atkinson, R (1987A)).

**Reduction half-life:**          High:          No data
                                  Low:
*Comment:*

**Hydrolysis:**
· **First-order hydr half-life:**
*Comment:*

· **Acid rate const $(M(H+)-hr)^{-1}$:**          No data
*Comment:*

· **Base rate const $(M(OH-)-hr)^{-1}$:**
*Comment:*

535

# Cyanamide, calcium salt

**CAS Registry Number:** 156-62-7

**Structure:**

$$H_2N\!\!=\!\!\!=\!\!\!=N$$

$$Ca^{++}$$

## Half-lives:
  · **Soil:**                                High:    672 hours        (4 weeks)
                                          Low:     168 hours        (1 week)
*Comment:* Scientific judgement based upon estimated aqueous aerobic biodegradation half-life.

  · **Air:**                                 High:    32 hours        (1.3 days)
                                          Low:     3.2 hours
*Comment:* Scientific judgement based upon estimated photooxidation half-life of hydrogen cyanamide in air.

  · **Surface Water:**                 High:    672 hours        (4 weeks)
                                        Low:     168 hours        (1 week)
*Comment:* Scientific judgement based upon estimated aqueous aerobic biodegradation half-life.

  · **Ground Water:**                High:    1344 hours      (8 weeks)
                                          Low:     336 hours       (2 weeks)
*Comment:* Scientific judgement based upon estimated aqueous aerobic biodegradation half-life.

## Aqueous Biodegradation (unacclimated):
  · **Aerobic half-life:**             High:    672 hours        (4 weeks)
                                        Low:     168 hours        (1 week)
*Comment:* Scientific judgement based upon acclimated aerobic screening test data for hydrocyanic acid (Pettet, AJ and Mills, EV (1954)).

  · **Anaerobic half-life:**           High:    2688 hours      (16 weeks)
                                          Low:     672 hours       (4 weeks)
*Comment:* Scientific judgement based upon estimated aqueous aerobic biodegradation half-life.

  · **Removal/secondary treatment:**    High:    No data
                                          Low:
*Comment:*

## Photolysis:
  · **Atmos photol half-life:**        High:    No data
                                          Low:

*Comment:*

· **Max light absorption (nm):**     No data
*Comment:*

· **Aq photol half-life:**           High:      No data
                                     Low:
*Comment:*

**Photooxidation half-life:**
  · **Water:**                       High:      No data
                                     Low:

*Comment:*

  · **Air:**                         High:      32 hours            (1.3 days)
                                     Low:       3.2 hours
*Comment:* Scientific judgement based upon estimated rate constant for the vapor phase reaction of hydrogen cyanamide with hydroxyl radicals in air (Atkinson, R (1987A)).

**Reduction half-life:**             High:      No data
                                     Low:
*Comment:*

**Hydrolysis:**
  · **First-order hydr half-life:**  No data
*Comment:*

  · **Acid rate const (M(H+)-hr)$^{-1}$:**
*Comment:*

  · **Base rate const (M(OH-)-hr)$^{-1}$:**
*Comment:*

# 1,2,7,8-Dibenzopyrene

**CAS Registry Number:** 189-55-9

**Structure:**

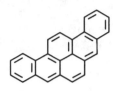

**Half-lives:**
  · **Soil:**                                  High:     8664 hours              (361 days)
                                                        Low:      5568 hours              (232 days)
  *Comment:* Based upon aerobic soil die-away test data (Sims, RC (1990)).

  · **Air:**                                   High:     3.21 hours
                                                        Low:      0.321 hours
  *Comment:* Scientific judgement based upon estimated photooxidation half-life in air.

  · **Surface Water:**                   High:     8664 hours              (361 days)
                                                        Low:      5568 hours              (232 days)
  *Comment:* Based upon aerobic soil die-away test data (Sims, RC (1990)).

  · **Ground Water:**                    High:     17328 hours             (1.98 years)
                                                        Low:      11136 hours             (1.27 years)
  *Comment:* Scientific judgement based upon estimated unacclimated aqueous aerobic biodegradation half-life.

**Aqueous Biodegradation (unacclimated):**
  · **Aerobic half-life:**                High:     8664 hours              (361 days)
                                                        Low:      5568 hours              (232 days)
  *Comment:* Based upon aerobic soil die-away test data (Sims, RC (1990)).

  · **Anaerobic half-life:**             High:     34656 hours             (3.96 years)
                                                        Low:      22272 hours             (2.54 years)
  *Comment:* Scientific judgement based upon estimated unacclimated aqueous aerobic biodegradation half-life.

  · **Removal/secondary treatment:**     High:     No data
                                                        Low:
  *Comment:*

**Photolysis:**

**· Atmos photol half-life:**        High:      No data
                                      Low:

*Comment:*

**· Max light absorption (nm):**    lambda max = 222, 242, 272 nm (ethanol).
*Comment:* IARC (1983A).

**· Aq photol half-life:**          High:      No data
                                      Low:

*Comment:*

**Photooxidation half-life:**
   **· Water:**                 High:      No data
                                      Low:

*Comment:*

   **· Air:**                    High:    3.21 hours
                                      Low:    0.321 hours
*Comment:* Scientific judgement based upon estimated rate constant for reaction with hydroxyl radical in air (Atkinson, R (1987A)).

**Reduction half-life:**          High:      No data
                                      Low:

*Comment:*

**Hydrolysis:**
   **· First-order hydr half-life:**
   *Comment:*

   **· Acid rate const (M(H+)-hr)$^{-1}$:**    No hydrolyzable groups
   *Comment:*

   **· Base rate const (M(OH-)-hr)$^{-1}$:**
   *Comment:*

# Benzo(ghi)perylene

<u>CAS Registry Number:</u>  191-24-2

<u>Structure:</u>

<u>Half-lives:</u>
  · **Soil:**                                  High:    15600 hours              (650 days)
                                               Low:     14160 hours              (590 days)
  *Comment:*  Based upon aerobic soil die-away test data at 10-30°C (Coover, MP and Sims, RCC (1987)).

  · **Air:**                                   High:     3.21 hours
                                               Low:      0.321 hours
  *Comment:*  Scientific judgement based upon estimated photooxidation half-life in air.

  · **Surface Water:**                         High:    15600 hours              (650 days)
                                               Low:     14160 hours              (590 days)
  *Comment:*  Based upon aerobic soil die-away test data at 10-30°C (Coover, MP and Sims, RCC (1987)).

  · **Ground Water:**                          High:    31200 hours              (3.6 years)
                                               Low:     28320 hours              (3.2 years)
  *Comment:*  Based upon aerobic soil die-away test data at 10-30°C (Coover, MP and Sims, RCC (1987)).

<u>Aqueous Biodegradation (unacclimated):</u>
  · **Aerobic half-life:**                     High:    15600 hours              (650 days)
                                               Low:     14160 hours              (590 days)
  *Comment:*  Based upon aerobic soil die-away test data at 10-30°C (Coover, MP and Sims, RCC (1987)).

  · **Anaerobic half-life:**                   High:    62400 hours              (7.1 years)
                                               Low:     56640 hours              (6.5 years)
  *Comment:*  Based upon aerobic soil die-away test data at 10-30°C (Coover, MP and Sims, RCC (1987)).

  · **Removal/secondary treatment:**           High:       No data

Low:
*Comment:*

## Photolysis:
· **Atmos photol half-life:**        High:    No data
Low:
*Comment:*

· **Max light absorption (nm):**    lambda max = 222, 252, 275, 288, 299, 311, 324, 328, 344, 362, 383, 392, 396 nm (cyclohexane)
*Comment:* IARC (1983A).

· **Aq photol half-life:**           High:    No data
Low:
*Comment:*

## Photooxidation half-life:
· **Water:**                         High:    No data
Low:
*Comment:*

· **Air:**                           High:    3.21 hours
Low:    0.321 hours
*Comment:* Scientific judgement based upon estimated rate constant for reaction with hydroxyl radical in air (Atkinson, R (1987A)).

## Reduction half-life:                High:    No data
Low:
*Comment:*

## Hydrolysis:
· **First-order hydr half-life:**
*Comment:*

· **Acid rate const $(M(H+)-hr)^{-1}$:**     No hydrolyzable groups
*Comment:*

· **Base rate const $(M(OH-)-hr)^{-1}$:**
*Comment:*

541

# Indeno(1,2,3-cd)pyrene

**CAS Registry Number:** 193-39-5

**Structure:**

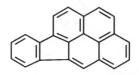

**Half-lives:**
  · **Soil:**                          High:    17520 hours          (2.0 years)
                                      Low:     14400 hours          (1.64 years)
*Comment:* Based upon aerobic soil die-away test data (Coover, MP and Sims, RC (1987)).

  · **Air:**                           High:       6.29 hours
                                     Low:       0.629 hours
*Comment:* Scientific judgement based upon estimated photooxidation half-life in air.

  · **Surface Water:**               High:      6000 hours          (250 days)
                                      Low:      3000 hours          (125 days)
*Comment:* Scientific judgement based upon estimated aqueous photolysis half-life.

  · **Ground Water:**              High:    35040 hours          (4.0 years)
                                      Low:     28800 hours          (3.29 years)
*Comment:* Scientific judgement based upon estimated unacclimated aqueous aerobic biodegradation half-life.

## Aqueous Biodegradation (unacclimated):
  · **Aerobic half-life:**            High:    17520 hours          (2.0 years)
                                      Low:     14400 hours          (1.64 years)
*Comment:* Based upon aerobic soil die-away test data (Coover, MP and Sims, RC (1987)).

  · **Anaerobic half-life:**          High:    70080 hours          (8.0 years)
                                      Low:     57600 hours          (6.58 years)
*Comment:* Scientific judgement based upon estimated unacclimated aqueous aerobic biodegradation half-life.

  · **Removal/secondary treatment:**     High:       No data
                                        Low:
*Comment:*

## Photolysis:
  · **Atmos photol half-life:**        High:      6000 hours          (250 days)

Low:     3000 hours          (125 days)

*Comment:* Scientific judgement based upon measured rate of photolysis for photolysis in heptane solution under November sunlight (Muel, B and Saguem, S (1985)) and adjusted for relative sunlight intensity in the summer (Lyman, WJ et al. (1982)).

· **Max light absorption (nm):**          lambda max = 242, 249, 274, 290, 301, 314, 340, 358, 376, 382, 407, 426, 446, 453, 460 nm (cyclohexane)
*Comment:* IARC (1983A).

· **Aq photol half-life:**          High:     6000 hours          (250 days)
                                    Low:      3000 hours          (125 days)
*Comment:* Scientific judgement based upon measured rate of photolysis for photolysis in heptane solution under November sunlight (Muel, B and Saguem, S (1985)).

## Photooxidation half-life:

· **Water:**          High:     No data
                      Low:

*Comment:*

· **Air:**          High:     6.29 hours
                    Low:      0.629 hours
*Comment:* Scientific judgement based upon estimated rate constant for reaction with hydroxyl radical in air (Atkinson, R (1987A)).

## Reduction half-life:          High:     No data
                                 Low:

*Comment:*

## Hydrolysis:

· **First-order hydr half-life:**
*Comment:*

· **Acid rate const $(M(H+)-hr)^{-1}$:**          No hydrolyzable groups
*Comment:*

· **Base rate const $(M(OH-)-hr)^{-1}$:**
*Comment:*

# Benzo(b)fluoranthene

**CAS Registry Number:** 205-99-2

**Structure:**

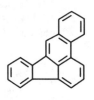

**Half-lives:**

· **Soil:**                          High:   14640 hours          (1.67 years)
                                     Low:     8640 hours          (360 days)
*Comment:* Based upon aerobic soil die-away test data (Coover, MP and Sims, RC (1987)).

· **Air:**                           High:   14.3 hours
                                     Low:     1.43 hours
*Comment:* Scientific judgement based upon estimated photooxidation half-life in air.

· **Surface Water:**                 High:   720 hours            (30 days)
                                     Low:     8.7 hours
*Comment:* Scientific judgement based upon estimated aqueous photolysis half-life.

· **Ground Water:**                  High:   29280 hours          (3.34 years)
                                     Low:   17280 hours          (1.97 years)
*Comment:* Scientific judgement based upon estimated unacclimated aqueous aerobic
biodegradation half-life.

**Aqueous Biodegradation (unacclimated):**

· **Aerobic half-life:**             High:   14640 hours          (1.67 years)
                                     Low:     8640 hours          (360 days)
*Comment:* Based upon aerobic soil die-away test data (Coover, MP and Sims, RC (1987)).

· **Anaerobic half-life:**           High:   58560 hours          (6.68 years)
                                     Low:   34560 hours          (3.95 years)
*Comment:* Scientific judgement based upon estimated unacclimated aqueous aerobic
biodegradation half-life.

· **Removal/secondary treatment:**   High:   No data
                                     Low:
*Comment:*

**Photolysis:**

· **Atmos photol half-life:**  High:  720 hours  (30 days)
   Low:  8.7 hours
*Comment:* Scientific judgement based upon measured rate of photolysis in heptane irradiated with light >290 nm (Lane, DA and Katz, M (1977); Muel, B and Saguem, S (1985)).

· **Max light absorption (nm):**  lambda max = 255, 275, 288, 300, 340, 349, 367 nm (cyclohexane).
*Comment:* IARC (1983A).

· **Aq photol half-life:**  High:  720 hours  (30 days)
   Low:  8.7 hours
*Comment:* Scientific judgement based upon measured rate of photolysis in heptane irradiated with light >290 nm (Lane, DA and Katz, M (1977); Muel, B and Saguem, S (1985)).

**Photooxidation half-life:**
· **Water:**  High:  No data
   Low:
*Comment:*

· **Air:**  High:  14.3 hours
   Low:  1.43 hours
*Comment:* Scientific judgement based upon estimated rate constant for reaction with hydroxyl radical in air (Atkinson, R (1987A)).

**Reduction half-life:**  High:  No data
   Low:
*Comment:*

**Hydrolysis:**
· **First-order hydr half-life:**
*Comment:*

· **Acid rate const $(M(H+)-hr)^{-1}$:**  No hydrolyzable groups
*Comment:*

· **Base rate const $(M(OH-)-hr)^{-1}$:**
*Comment:*

# Fluoranthene

**CAS Registry Number:** 206-44-0

**Structure:**

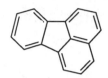

**Half-lives:**

· **Soil:**        High:   10560 hours        (440 days)
                   Low:     3360 hours        (140 days)

*Comment:* Based upon aerobic soil die-away test data at 10-30°C (Coover, MP and Sims, RCC (1987)).

· **Air:**        High:   20.2 hours
                  Low:    2.02 hours

*Comment:* Scientific judgement based upon estimated sunlight photolysis half-life in water.

· **Surface Water:**        High:   63 hours        (2.6 days)
                           Low:    21 hours

*Comment:* Based upon photolysis half-life in water.

· **Ground Water:**        High:   21120 hours        (2.41 years)
                          Low:     6720 hours        (280 days)

*Comment:* Scientific judgement based upon estimated unacclimated aqueous aerobic biodegradation half-life.

**Aqueous Biodegradation (unacclimated):**

· **Aerobic half-life:**        High:   10560 hours        (440 days)
                               Low:     3360 hours        (140 days)

*Comment:* Based upon aerobic soil die-away test data at 10-30°C (Coover, MP and Sims, RCC (1987)).

· **Anaerobic half-life:**        High:   42240 hours        (4.82 years)
                                 Low:    13440 hours        (1.53 years)

*Comment:* Scientific judgement based upon estimated unacclimated aqueous aerobic biodegradation half-life.

· **Removal/secondary treatment:**        High:
                                         Low:        0%

*Comment:* Removal percentage based upon data from a continuous activated sludge biological treatment simulator (Petrasek, AC et al. (1983)).

## Photolysis:
· **Atmos photol half-life:**        High:     63 hours         (2.6 days)
                                      Low:     21 hours
*Comment:* Based upon measured sunlight photolysis rate constant in water adjusted for midday summer sunlight at 40°N latitude in (low $t_{1/2}$) (Zepp, RG and Schlotzhauer, PF (1979)) and adjusted for approximate winter sunlight intensity (high $t_{1/2}$) (Lyman, WJ et al. (1982)).

· **Max light absorption (nm):**     lambda max = 235, 252, 262, 270, 275, 281, 286, 306, 322, 339, 357 nm (cyclohexane).
*Comment:* IARC (1983A).

· **Aq photol half-life:**           High:     63 hours         (2.6 days)
                                        Low:     21 hours
*Comment:* Based upon measured sunlight photolysis rate constant in water adjusted for midday summer sunlight at 40°N latitude in (low $t_{1/2}$) (Zepp, RG and Schlotzhauer, PF (1979)) and adjusted for approximate winter sunlight intensity (high $t_{1/2}$) (Lyman, WJ et al. (1982)).

## Photooxidation half-life:
· **Water:**                       High:     No data
                                        Low:
*Comment:*

· **Air:**                         High:     20.2 hours
                                        Low:     2.02 hours
*Comment:* Scientific judgement based upon estimated rate constant for reaction with hydroxyl radical in air (Atkinson, R (1987A)).

## Reduction half-life:                 High:     No data
                                        Low:
*Comment:*

## Hydrolysis:
· **First-order hydr half-life:**
*Comment:*

· **Acid rate const $(M(H+)-hr)^{-1}$:**     No hydrolyzable groups
*Comment:*

· **Base rate const $(M(OH-)-hr)^{-1}$:**
*Comment:*

# Benzo(k)fluoranthene

**CAS Registry Number:** 207-08-9

**Structure:**

**Half-lives:**

**· Soil:**  High: 51360 hours (5.86 years)
Low: 21840 hours (2.49 years)
*Comment:* Based upon aerobic soil die-away test data (Coover, MP and Sims, RC (1987); Bossert, I et al. (1984)).

**· Air:**  High: 11.0 hours
Low: 1.10 hours
*Comment:* Scientific judgement based upon estimated photooxidation half-life in air.

**· Surface Water:**  High: 499 hours
Low: 3.8 hours
*Comment:* Scientific judgement based upon photolysis half-life in water.

**· Ground Water:**  High: 102720 hours (11.7 years)
Low: 42680 hours (4.99 years)
*Comment:* Scientific judgement based upon estimated unacclimated aqueous aerobic biodegradation half-life.

**Aqueous Biodegradation (unacclimated):**

**· Aerobic half-life:**  High: 51360 hours (5.86 years)
Low: 21840 hours (2.49 years)
*Comment:* Based upon aerobic soil die-away test data (Coover, MP and Sims, RC (1987); Bossert, I et al. (1984)).

**· Anaerobic half-life:**  High: 205440 hours (23.5 years)
Low: 87360 hours (9.97 years)
*Comment:* Scientific judgement based upon estimated unacclimated aqueous aerobic biodegradation half-life.

**· Removal/secondary treatment:**  High: No data
Low:
*Comment:*

548

## Photolysis:
**· Atmos photol half-life:**  High: 499 hours
    Low: 3.8 hours
*Comment:* Scientific judgement based upon measured rate of photolysis in heptane under November sunlight (high $t_{1/2}$) (Muel, B and Saguem, S (1985)) and the above data adjusted by ratio of sunlight photolysis half-lives for benz(a)anthracene in water vs. heptane (low $t_{1/2}$) (Smith, JH et al. (1978); Muel, B and Saguem, S (1985)).

**· Max light absorption (nm):**  lambda max = 245, 266, 282, 295, 306, 358, 370, 378, 400 nm (cyclohexane).
*Comment:* IARC (1983A).

**· Aq photol half-life:**  High: 499 hours
    Low: 3.8 hours
*Comment:* Scientific judgement based upon measured rate of photolysis in heptane under November sunlight (high $t_{1/2}$) (Muel, B and Saguem, S (1985)) and the above data adjusted by ratio of sunlight photolysis half-lives for benz(a)anthracene in water vs. heptane (low $t_{1/2}$) (Smith, JH et al. (1978); Muel, B and Saguem, S (1985)).

## Photooxidation half-life:
**· Water:**  High: No data
    Low:

*Comment:*

**· Air:**  High: 11.0 hours
    Low: 1.10 hours
*Comment:* Scientific judgement based upon estimated rate constant for reaction with hydroxyl radical in air (Atkinson, R (1987A)).

## Reduction half-life:
    High: No data
    Low:

*Comment:*

## Hydrolysis:
**· First-order hydr half-life:**
*Comment:*

**· Acid rate const $(M(H+)-hr)^{-1}$:**  No hydrolyzable groups
*Comment:*

**· Base rate const $(M(OH-)-hr)^{-1}$:**
*Comment:*

549

# Acenaphthylene

CAS Registry Number:  208-96-8

Structure:

Half-lives:
 · Soil:                           High:    1440 hours              (60 days)
                                   Low:     1020 hours              (42.5 days)
 *Comment:*  Scientific judgement based upon soil column study data (Kincannon, DF and Lin, YS (1985)).

 · Air:                            High:    1.27 hours
                                   Low:     0.191 hours
 *Comment:*  Scientific judgement based upon estimated photooxidation half-life in air.

 · Surface Water:                  High:    1440 hours              (60 days)
                                   Low:     1020 hours              (42.5 days)
 *Comment:*  Scientific judgement based upon estimated unacclimated aqueous aerobic biodegradation half-life.

 · Ground Water:                   High:    2880 hours              (120 days)
                                   Low:     2040 hours              (85 days)
 *Comment:*  Scientific judgement based upon estimated unacclimated aqueous aerobic biodegradation half-life.

Aqueous Biodegradation (unacclimated):
 · Aerobic half-life:              High:    1440 hours              (60 days)
                                   Low:     1020 hours              (42.5 days)
 *Comment:*  Scientific judgement based upon soil column study data (Kincannon, DF and Lin, YS (1985)).

 · Anaerobic half-life:            High:    5760 hours              (240 days)
                                   Low:     4080 hours              (170 days)
 *Comment:*  Scientific judgement based upon estimated unacclimated aqueous aerobic biodegradation half-life.

 · Removal/secondary treatment:    High:       No data
                                   Low:
 *Comment:*

## Photolysis:
· **Atmos photol half-life:**  High:    No data
Low:

*Comment:*

· **Max light absorption (nm):**  lambda max approximately: 230, 272, 283, 310, 323, 340, 345, 417, 445, 475 nm (Mabey,WR et al. (1981)).
*Comment:*

· **Aq photol half-life:**  High:    No data
Low:

*Comment:*

## Photooxidation half-life:
· **Water:**  High:    No data
Low:

*Comment:*

· **Air:**  High:    1.27 hours
Low:    0.191 hours

*Comment:* Scientific judgement based upon estimated rate constants for reaction in air with hydroxyl radical (Atkinson, R (1987A)) and ozone (Atkinson, R and Carter, WPL (1984)).

## Reduction half-life:  High:    No data
Low:

*Comment:*

## Hydrolysis:
· **First-order hydr half-life:**
*Comment:*

· **Acid rate const $(M(H+)-hr)^{-1}$:**  No hydrolyzable groups
*Comment:*

· **Base rate const $(M(OH-)-hr)^{-1}$:**
*Comment:*

# Chrysene

<u>CAS Registry Number:</u>  218-01-9

<u>Structure:</u>

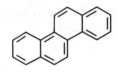

## <u>Half-lives:</u>
- **· Soil:**                             High:    24000 hours          (2.72 years)
                                          Low:      8904 hours          (1.02 years)
*Comment:* Based upon aerobic soil die-away test data (Coover, MP and Sims, RCC (1987); Sims, RC (1990)).

- **· Air:**                              High:     8.02 hours
                                          Low:      0.802 hours
*Comment:* Scientific judgement based upon estimated photooxidation half-life in air.

- **· Surface Water:**                    High:       13 hours
                                          Low:       4.4 hours
*Comment:* Scientific judgement based upon sunlight photolysis half-life in water.

- **· Ground Water:**                     High:    48000 hours          (5.48 years)
                                          Low:     17808 hours          (2.04 years)
*Comment:* Scientific judgement based upon estimated unacclimated aqueous aerobic biodegradation half-life.

## <u>Aqueous Biodegradation (unacclimated):</u>
- **· Aerobic half-life:**                High:    24000 hours          (2.72 years)
                                          Low:      8904 hours          (1.02 years)
*Comment:* Based upon aerobic soil die-away test data (Coover, MP and Sims, RCC (1987); Sims, RC (1990)).

- **· Anaerobic half-life:**              High:    96000 hours          (11.0 years)
                                          Low:     35616 hours          (4.06 years)
*Comment:* Scientific judgement based upon estimated unacclimated aqueous aerobic biodegradation half-life.

- **· Removal/secondary treatment:**      High:
                                          Low:        9%
*Comment:* Removal percentage based upon data from a continuous activated sludge biological treatment simulator (Petrasek, AC et al. (1983)).

## Photolysis:
· **Atmos photol half-life:**              High:     13 hours
                                          Low:     4.4 hours

*Comment:* Based upon measured aqueous photolysis quantum yields and calculated for midday summer sunlight at 40°N latitude (low $t_{1/2}$) (Zepp, RG and Schlotzhauer, PF (1979)) and adjusted for approximate winter sunlight intensity (Lyman, WJ et al. (1982)).

· **Max light absorption (nm):**     lambda max = 344, 351, 360 (ethanol); 220, 240, 250, 257, 293, 305, 318, 342, 350, 359 nm (cyclohexane).
*Comment:* Ethanol spectrum: Radding, SB et al. (1976); cyclohexane spectrum: IARC (1983A).

· **Aq photol half-life:**               High:     13 hours
                                          Low:     4.4 hours

*Comment:* Based upon measured aqueous photolysis quantum yields and calculated for midday summer sunlight at 40°N latitude (low $t_{1/2}$) (Zepp, RG and Schlotzhauer, PF (1979)) and adjusted for approximate winter sunlight intensity (Lyman, WJ et al. (1982)).

## Photooxidation half-life:
· **Water:**                            High:     No data
                                            Low:

*Comment:*

· **Air:**                                High:     8.02 hours
                                          Low:     0.802 hours

*Comment:* Scientific judgement based upon estimated rate constant for reaction with hydroxyl radical in air (Atkinson, R (1987A)).

## Reduction half-life:                   High:     No data
                                            Low:

*Comment:*

## Hydrolysis:
· **First-order hydr half-life:**
*Comment:*

· **Acid rate const $(M(H+)-hr)^{-1}$:**     No hydrolyzable groups
*Comment:*

· **Base rate const $(M(OH-)-hr)^{-1}$:**
*Comment:*

# Benz(c)acridine

CAS Registry Number:  225-51-4

Structure:

Half-lives:

· **Soil:**                               High:     8760 hours              (1 year)
                                          Low:      4320 hours              (6 months)
*Comment:*  Based upon aerobic soil die-away test data for benz(a)anthracene (Coover, MP and Sims, RCC (1987); Groenewegen, D and Stolp, H (1976)).

· **Air:**                                High:     22.9 hours
                                          Low:      2.29 hours
*Comment:*  Scientific judgement based upon estimated photooxidation half-life in air.

· **Surface Water:**                      High:     8760 hours              (1 year)
                                          Low:      4320 hours              (6 months)
*Comment:*  Based upon aerobic soil die-away test data for benz(a)anthracene (Coover, MP and Sims, RCC (1987); Groenewegen, D and Stolp, H (1976)).

· **Ground Water:**                       High:     17520 hours             (2 years)
                                          Low:      8640 hours              (12 months)
*Comment:*  Scientific judgement based upon estimated unacclimated aqueous aerobic biodegradation half-life.

Aqueous Biodegradation (unacclimated):

· **Aerobic half-life**:                  High:     8760 hours              (1 year)
                                          Low:      4320 hours              (6 months)
*Comment:*  Based upon aerobic soil die-away test data for benz(a)anthracene (Coover, MP and Sims, RCC (1987); Groenewegen, D and Stolp, H (1976)).

· **Anaerobic half-life**:                High:     35040 hours             (4 years)
                                          Low:      17280 hours             (24 months)
*Comment:*  Scientific judgement based upon estimated unacclimated aqueous aerobic biodegradation half-life.

· **Removal/secondary treatment:**        High:     No data
                                          Low:

*Comment:*

## Photolysis:
· **Atmos photol half-life:**          High:     No data
                                                Low:
*Comment:*

· **Max light absorption (nm):**     lambda max = 204, 219, 222, 232, 266, 273, 284, 315, 330, 338, 346, 355, 364, 374, 383 nm (cyclohexane).
*Comment:* IARC (1983A).

· **Aq photol half-life:**          High:     No data
                                                Low:
*Comment:*

## Photooxidation half-life:
· **Water:**          High:     No data
                                                Low:
*Comment:*

· **Air:**          High:     22.9 hours
                                                Low:     2.29 hours
*Comment:* Scientific judgement based upon estimated rate constant for reaction with hydroxyl radical in air (Atkinson, R (1987A)).

## Reduction half-life:
                                              High:     No data
                                              Low:
*Comment:*

## Hydrolysis:
· **First-order hydr half-life:**
*Comment:*

· **Acid rate const $(M(H+)-hr)^{-1}$:**     No hydrolyzable groups
*Comment:*

· **Base rate const $(M(OH-)-hr)^{-1}$:**
*Comment:*

# Methyl parathion

CAS Registry Number:  298-00-0

Structure:

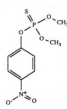

Half-lives:
· Soil:

| | High: | 8640 hours | (1 year) |
| --- | --- | --- | --- |
| | Low: | 240 hours | (10 days) |

*Comment:* Scientific judgement based upon unacclimated aerobic soil grab sample data (Davidson, JM et al. (1980)). Degrades rapidly at concn <100 ppm (Sanborn, JR et al. (1977)). Persistent at high application rates (high $t_{1/2}$: Davidson, JM et al. (1980), Butler, LC et al. (1981)).

· Air:

| | High: | 10.5 hours |
| --- | --- | --- |
| | Low: | 1.0 hour |

*Comment:* Scientific judgement based upon estimated photooxidation half-life in air.

· Surface Water:

| | High: | 912 hours | (38 days) |
| --- | --- | --- | --- |
| | Low: | 192 hours | (8 days) |

*Comment:* Based upon aqueous photolysis half-lives for summer and winter sunlight (Smith, JH et al. (1978)).

· Ground Water:

| | High: | 1680 hours | (70 days) |
| --- | --- | --- | --- |
| | Low: | 24 hours | (1 day) |

*Comment:* Scientific judgement based upon estimated aqueous aerobic and anaerobic biodegradation half-lives.

Aqueous Biodegradation (unacclimated):
· Aerobic half-life:

| | High: | 1680 hours | (70 days) |
| --- | --- | --- | --- |
| | Low: | 360 hours | (15 days) |

*Comment:* Scientific judgement based upon an unacclimated aerobic river die-away test data (low $t_{1/2}$: Bourquin, AW et al. (1979); high $t_{1/2}$: Spain, JC et al. (1980)).

· Anaerobic half-life:

| | High: | 168 hours | (7 days) |
| --- | --- | --- | --- |
| | Low: | 24 hours | (1 day) |

*Comment:* Scientific judgement based upon unacclimated anaerobic soil and sediment grab sample data (Adhya, TK et al. (1981A); Wolfe NL et al. (1986)).

· **Removal/secondary treatment:**  High: No data  
                                    Low:  

*Comment:*

## Photolysis:
· **Atmos photol half-life:**  High: 912 hours  (38 days)  
                               Low:  192 hours  (8 days)  
*Comment:* Scientific judgement based upon aqueous photolysis data (Smith, JH et al. (1978)).

· **Max light absorption (nm):**  lambda max = 400.0, 370.0, 340.0, 320.0, 305.0 and 297.5  
*Comment:* No absorbance above 410 nm (Smith, JH et al. (1978)).

· **Aq photol half-life:**  High: 912 hours  (38 days)  
                            Low:  192 hours  (8 days)  
*Comment:* Based upon aqueous photolysis rate constants at pH 5-8 for summer and winter sunlight (Smith, JH et al. (1978)).

## Photooxidation half-life:
· **Water:**  High: No data  
              Low:  

*Comment:*

· **Air:**  High: 10.5 hours  
            Low:  1.0 hour  
*Comment:* Scientific judgement based upon an estimated rate constant for vapor phase reaction with hydroxyl radicals in air (Atkinson, R (1987A)).

## Reduction half-life:
                                    High: No data  
                                    Low:  

*Comment:*

## Hydrolysis:
· **First-order hydr half-life:**  1728 hours  (72 days)  
*Comment:* Scientific judgement based upon reported rate constant ($1.1 \times 10^{-7}$ s$^{-1}$) at pH 7 and 25 °C (Mabey, W and Mill, T (1978)).

· **Acid rate const (M(H+)-hr)$^{-1}$:**  
*Comment:*

· **Base rate const (M(OH-)-hr)$^{-1}$:**  $1.637 \times 10^{-6}$ s$^{-1}$  
*Comment:* ($t_{1/2}$ = 45 days) Scientific judgement based upon an overall rate constant at pH 9 and 25 °C (Smith, JH et al. (1978)).

# Disulfoton

**CAS Registry Number:** 298-04-4

**Structure:**

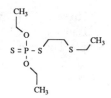

**Half-lives:**
- **Soil:**

| | High: | 504 hours | (21 days) |
|---|---|---|---|
| | Low: | 72 hours | (3 days) |

*Comment:* Scientific judgement based upon aerobic soil field data (high $t_{1/2}$: Szeto, SY et al. (1983)) and reported half-lives for soil (Domsch, KH (1984)).

- **Air:**

| | High: | 4.8 hours | |
|---|---|---|---|
| | Low: | 0.5 hour | (29 minutes) |

*Comment:* Scientific judgement based upon estimated photooxidation half-life in air.

- **Surface Water:**

| | High: | 504 hours | (21 days) |
|---|---|---|---|
| | Low: | 72 hours | (3 days) |

*Comment:* Scientific judgement based upon unacclimated aerobic biodegradation half-life.

- **Ground Water:**

| | High: | 1008 hours | (42 days) |
|---|---|---|---|
| | Low: | 144 hours | (6 days) |

*Comment:* Scientific judgement based upon estimated aqueous aerobic biodegradation half-life.

**Aqueous Biodegradation (unacclimated):**
- **Aerobic half-life:**

| | High: | 504 hours | (21 days) |
|---|---|---|---|
| | Low: | 72 hours | (3 days) |

*Comment:* Scientific judgement based upon aerobic soil field data (high $t_{1/2}$: Szeto, SY et al. (1983)) and reported half-lives for soil (Domsch, KH (1984)).

- **Anaerobic half-life:**

| | High: | 2016 hours | (84 days) |
|---|---|---|---|
| | Low: | 288 hours | (12 days) |

*Comment:* Scientific judgement based upon unacclimated aerobic biodegradation half-life.

- **Removal/secondary treatment:**

| | High: | No data |
|---|---|---|
| | Low: | |

*Comment:*

## Photolysis:
· **Atmos photol half-life:**          High:      No data
                                         Low:

*Comment:*

· **Max light absorption (nm):**     No data
*Comment:*

· **Aq photol half-life:**           High:      No data
                                         Low:

*Comment:*

## Photooxidation half-life:
· **Water:**                      High:      No data
                                         Low:

*Comment:*

· **Air:**                         High:      4.8 hours
                                         Low:      0.5 hour        (29 minutes)
*Comment:* Scientific judgement based upon an estimated rate constant for vapor phase reaction with hydroxyl radicals in air (Atkinson, R (1987A)).

## Reduction half-life:
                                         High:      No data
                                       Low:

*Comment:*

## Hydrolysis:
· **First-order hydr half-life:**          2475 hours        (103 days)
*Comment:* Scientific judgement based upon overall rate constant ($2.8 \times 10^{-2}$ hr$^{-1}$) at pH 7 and 25 °C (Ellington, JJ et al. (1987A)).

· **Acid rate const (M(H+)-hr)$^{-1}$:**
*Comment:*

· **Base rate const (M(OH-)-hr)$^{-1}$:**     5.99 M$^{-1}$ hr$^{-1}$
*Comment:* ($t_{1/2}$ = 85 days) Scientific judgement based upon a rate constant at pH 9 and 25 °C (Ellington, JJ et al. (1987A)).

# Hydrazine

<u>CAS Registry Number:</u>  302-01-2

<u>Half-lives:</u>

· **Soil:**  High: 168 hours  (7 days)

Low: 24 hours  (1 day)

*Comment:* Scientific judgement based upon unacclimated aqueous aerobic biodegradation half-life.

· **Air:**  High: 5.57 hours

Low: 0.674 hours

*Comment:* Based upon photooxidation half-life in air.

· **Surface Water:**  High: 168 hours  (7 days)

Low: 24 hours  (1 day)

*Comment:* Scientific judgement based upon unacclimated aqueous aerobic biodegradation half-life.

· **Ground Water:**  High: 336 hours  (14 days)

Low: 48 hours  (2 days)

*Comment:* Scientific judgement based upon unacclimated aqueous aerobic biodegradation half-life.

<u>Aqueous Biodegradation (unacclimated):</u>

· **Aerobic half-life:**  High: 168 hours  (7 days)

Low: 24 hours  (1 day)

*Comment:* Scientific judgement.

· **Anaerobic half-life:**  High: 672 hours  (28 days)

Low: 96 hours  (4 days)

*Comment:* Scientific judgement based upon unacclimated aqueous aerobic biodegradation half-life.

· **Removal/secondary treatment:**  High: 100%

Low: 69%

*Comment:* Removal percentages based upon data from a continuous activated sludge biological treatment simulator (MacNaughton, MG and Farmwald, JA (1979)).

<u>Photolysis:</u>

· **Atmos photol half-life:**  High:

Low:

*Comment:*

**· Max light absorption (nm):**          No data
*Comment:*

**· Aq photol half-life:**          High:
                                    Low:

*Comment:*

**Photooxidation half-life:**
  **· Water:**          High:          No data
                        Low:

*Comment:*

  **· Air:**          High:          5.57 hours
                      Low:          0.674 hours
*Comment:*  Based upon measured rate constants for reaction with hydroxyl radical (Harris, GW et al. (1979)) and ozone (Atkinson, R and Carter, WPL (1984)) in air.

**Reduction half-life:**          High:          No data
                                  Low:

*Comment:*

**Hydrolysis:**
  **· First-order hydr half-life:**
*Comment:*

  **· Acid rate const (M(H+)-hr)$^{-1}$:**          No hydrolyzable groups
*Comment:*

  **· Base rate const (M(OH-)-hr)$^{-1}$:**
*Comment:*

# Lasiocarpine

CAS Registry Number: 303-34-4

Structure:

## Half-lives:
· **Soil:**                          High:     672 hours          (4 weeks)
                                     Low:      168 hours          (7 days)
*Comment:* Scientific judgement based upon unacclimated aqueous aerobic biodegradation half-life.

· **Air:**                           High:     1.97 hours
                                     Low:      0.233 hours
*Comment:* Scientific judgement based upon estimated photooxidation half-life in air.

· **Surface Water:**                 High:     672 hours          (4 weeks)
                                     Low:      168 hours          (7 days)
*Comment:* Scientific judgement based upon unacclimated aqueous aerobic biodegradation half-life.

· **Ground Water:**                  High:     1344 hours         (8 weeks)
                                     Low:      336 hours          (14 days)
*Comment:* Scientific judgement based upon unacclimated aqueous aerobic biodegradation half-life.

## Aqueous Biodegradation (unacclimated):
· **Aerobic half-life**:             High:     672 hours          (4 weeks)
                                     Low:      168 hours          (7 days)
*Comment:*

· **Anaerobic half-life**:           High:     2688 hours         (16 weeks)
                                     Low:      672 hours          (28 days)
*Comment:* Scientific judgement based upon unacclimated aqueous aerobic biodegradation half-life.

· **Removal/secondary treatment:**   High:     No data
                                     Low:

*Comment:*

## Photolysis:
· **Atmos photol half-life:**         High:
                                      Low:
*Comment:*

· **Max light absorption (nm):**    No data
*Comment:*

· **Aq photol half-life:**          High:
                                      Low:
*Comment:*

## Photooxidation half-life:
· **Water:**          High:     No data
                                    Low:
*Comment:*

· **Air:**          High:     1.97 hours
                                    Low:     0.233 hours
*Comment:* Scientific judgement based upon estimated rate constant for reaction with hydroxyl radical (Atkinson, R (1987A)) and ozone (Atkinson, R and Carter, WPL (1984)) in air.

## Reduction half-life:
                                    High:     No data
                                    Low:
*Comment:*

## Hydrolysis:
· **First-order hydr half-life:**    $1.39 \times 10^4$ hours (1.58 years)
*Comment:* ($t_{1/2}$ at pH 5-7) Based upon measured rate of neutral and base catalyzed hydrolysis in water (Ellington, JJ et al. (1987)).

· **Acid rate const (M(H+)-hr)$^{-1}$:**    No data
*Comment:*

· **Base rate const (M(OH-)-hr)$^{-1}$:**    $9.8$ M$^{-1}$ hr$^{-1}$
*Comment:* ($t_{1/2}$ at pH 9 = 4704 hours (196 days)) Based upon measured rate of neutral and base catalyzed hydrolysis in water (Ellington, JJ et al. (1987)).

# Aldrin

**CAS Registry Number:** 309-00-2

**Structure:**

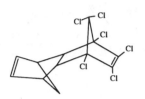

**Half-lives:**
  · **Soil:**                 High:    14200 hours         (1.6 years)
                                      Low:      504 hours           (3 weeks)
*Comment:* Scientific judgement based upon unacclimated aerobic river die-away test data (low $t_{1/2}$: Eichelberger, JW and Lichtenberg, JJ (1971)) and soil field test data (high $t_{1/2}$: Lichtenstein, EP et al. (1971)).

  · **Air:**                  High:    9.1 hours
                                      Low:      0.9 hours          (55 minutes)
*Comment:* Scientific judgement based upon estimated photooxidation half-life in air.

  · **Surface Water:**        High:    14200 hours         (1.6 years)
                                      Low:      504 hours           (3 weeks)
*Comment:* Scientific judgement based upon unacclimated aerobic river die-away test data (low $t_{1/2}$: Eichelberger, JW and Lichtenberg, JJ (1971)) and soil field test data (high $t_{1/2}$: Lichtenstein, EP et al. (1971)).

  · **Ground Water:**       High:    28400 hours         (3.2 years)
                                      Low:      24 hours            (1 day)
*Comment:* Scientific judgement based upon estimated aqueous anaerobic and aerobic biodegradation half-lives.

**Aqueous Biodegradation (unacclimated):**
  · **Aerobic half-life:**      High:    14200 hours         (1.6 years)
                                      Low:      504 hours           (3 weeks)
*Comment:* Scientific judgement based upon unacclimated aerobic river die-away test data (low $t_{1/2}$: Eichelberger, JW and Lichtenberg, JJ (1971)) and soil field test data (high $t_{1/2}$: Lichtenstein, EP et al. (1971))

  · **Anaerobic half-life:**    High:    168 hours           (7 days)
                                      Low:      24 hours            (1 day)
*Comment:* Scientific judgement based upon soil and freshwater mud grab sample data (Maule, A et al. (1987)).

· **Removal/secondary treatment:**      High:      No data
                                         Low:

*Comment:*

## Photolysis:

· **Atmos photol half-life:**      High:      No data
                                   Low:

*Comment:*

· **Max light absorption (nm):**      lambda max approximately 211 nm

*Comment:* In hexane, aldrin does not absorb UV light at wavelengths greater than 260 nm (Gore, RC et al. (1971)).

· **Aq photol half-life:**      High:      No data
                                Low:

*Comment:*

## Photooxidation half-life:

· **Water:**      High:      No data
                  Low:

*Comment:*

· **Air:**      High:      9.1 hours
                Low:       0.9 hours          (55 minutes)

*Comment:* Scientific judgement based upon an estimated rate constant for vapor phase reaction with hydroxyl radicals in air (Atkinson, R (1987A)).

## Reduction half-life:      High:      No data
                             Low:

*Comment:*

## Hydrolysis:

· **First-order hydr half-life:**      760 days

*Comment:* Based upon a first order rate constant ($3.8 \times 10^{-5}$ hr$^{-1}$) at pH 7 and 25 °C (Ellington, JJ et al. (1987A)).

· **Acid rate const (M(H+)-hr)$^{-1}$:**
*Comment:*

· **Base rate const (M(OH-)-hr)$^{-1}$:**
*Comment:*

# alpha-Hexachlorocyclohexane

__CAS Registry Number:__ 319-84-6

__Half-lives:__
  · __Soil:__                                  High:     3240 hours          (135 days)

| | High: | 3240 hours | (135 days) |
|---|---|---|---|
| | Low: | 330 hours | (13.8 days) |

_Comment:_ Based upon hydrolysis half-life for gamma-hexachlorocyclohexane (low $t_{1/2}$) (Ellington, JJ et al. (1987)) and data from one aerobic soil die-away study data for one soil (high $t_{1/2}$) (Macrae, IC et al. (1984)).

  · __Air:__

| | High: | 92.4 hours | (3.85 days) |
|---|---|---|---|
| | Low: | 9.24 hours | |

_Comment:_ Scientific judgement based upon estimated photooxidation half-life in air.

  · __Surface Water:__

| | High: | 3240 hours | (135 days) |
|---|---|---|---|
| | Low: | 330 hours | (13.8 days) |

_Comment:_ Based upon hydrolysis half-life for gamma-hexachlorocyclohexane (low $t_{1/2}$) (Ellington, JJ et al. (1987)) and data from one aerobic soil die-away study data for one soil (high $t_{1/2}$) (Macrae, IC et al. (1984)).

  · __Ground Water:__

| | High: | 6480 hours | (270 days) |
|---|---|---|---|
| | Low: | 330 hours | (13.8 days) |

_Comment:_ Based upon hydrolysis half-life for gamma-hexachlorocyclohexane (low $t_{1/2}$) (Ellington, JJ et al. (1987)) and data from one aerobic soil die-away study data for one soil (high $t_{1/2}$) (Macrae, IC et al. (1984)).

__Aqueous Biodegradation (unacclimated):__
  · __Aerobic half-life:__

| | High: | 3240 hours | (135 days) |
|---|---|---|---|
| | Low: | 1920 hours | (80 days) |

_Comment:_ Based upon data from one aerobic soil die-away study data for one soil (Macrae, IC et al. (1984)).

  · __Anaerobic half-life:__

| | High: | 960 hours | (40 days) |
|---|---|---|---|
| | Low: | 168 hours | (7 days) |

_Comment:_ Based upon anaerobic soil die-away study data (high $t_{1/2}$: Macrae, IC et al. (1984); low $t_{1/2}$: Castro, TF and Yoshida, T (1974)).

  · __Removal/secondary treatment:__

| | High: | No data |
|---|---|---|
| | Low: | |

_Comment:_

__Photolysis:__

· **Atmos photol half-life:**    High:
                                 Low:
*Comment:*

· **Max light absorption (nm):**    No data
*Comment:*

· **Aq photol half-life:**    High:
                              Low:
*Comment:*

## Photooxidation half-life:
   · **Water:**    High:    No data
                   Low:
*Comment:*

   · **Air:**    High:    92.4 hours    (3.85 days)
                 Low:     9.24 hours
*Comment:* Scientific judgement based upon estimated rate constant for reaction with hydroxyl radical in air (Atkinson, R (1987A)).

## Reduction half-life:
                 High:    No data
                 Low:
*Comment:*

## Hydrolysis:
   · **First-order hydr half-life:**    4957 hours    (207 days)
*Comment:* ($t_{1/2}$ at pH 7 and 25°C) Based upon measured neutral and base catalyzed hydrolysis rate constants for gamma-hexachlorocyclohexane (Ellington, JJ et al. (1987)).

   · **Acid rate const (M(H+)-hr)$^{-1}$:**    No data
*Comment:* ($t_{1/2}$ = 5765 hours (240 days) at pH 5 and 25°C) Based upon measured neutral and base catalyzed hydrolysis rate constants for gamma-hexachlorocyclohexane (Ellington, JJ et al. (1987)).

   · **Base rate const (M(OH-)-hr)$^{-1}$:**    198
*Comment:* ($t_{1/2}$ = 330 hours (13.8 days) at pH 9 and 25°C) Based upon measured neutral and base catalyzed hydrolysis rate constants for gamma-hexachlorocyclohexane (Ellington, JJ et al. (1987)).

# beta-Hexachlorocyclohexane

**CAS Registry Number:** 319-85-7

**Half-lives:**

· **Soil:** High: 2976 hours (124 days)

Low: 330 hours (13.8 days)

*Comment:* Based upon hydrolysis half-life for gamma-hexachlorocyclohexane (low $t_{1/2}$) (Ellington, JJ et al. (1987)) and data from one aerobic soil die-away study data for one soil (high $t_{1/2}$) (Macrae, IC et al. (1984)).

· **Air:** High: 92.4 hours (3.85 days)

Low: 9.24 hours

*Comment:* Scientific judgement based upon estimated photooxidation half-life in air.

· **Surface Water:** High: 2976 hours (124 days)

Low: 330 hours (13.8 days)

*Comment:* Based upon hydrolysis half-life for gamma-hexachlorocyclohexane (low $t_{1/2}$) (Ellington, JJ et al. (1987)) and data from one aerobic soil die-away study data for one soil (high $t_{1/2}$) (Macrae, IC et al. (1984)).

· **Ground Water:** High: 5952 hours (248 days)

Low: 330 hours (13.8 days)

*Comment:* Based upon hydrolysis half-life for gamma-hexachlorocyclohexane (low $t_{1/2}$) (Ellington, JJ et al. (1987)) and data from one aerobic soil die-away study data for one soil (high $t_{1/2}$) (Macrae, IC et al. (1984)).

**Aqueous Biodegradation (unacclimated):**

· **Aerobic half-life:** High: 2976 hours (124 days)

Low: 1440 hours (60 days)

*Comment:* Based upon data from one aerobic soil die-away study data for one soil (Macrae, IC et al. (1984)).

· **Anaerobic half-life:** High: 2256 hours (94 hours)

Low: 720 hours (30 days)

*Comment:* Based upon anaerobic soil die-away study data (high $t_{1/2}$: Macrae, IC et al. (1984); low $t_{1/2}$: Zhang, S et al. (1982)).

· **Removal/secondary treatment:** High: No data

Low:

*Comment:*

**Photolysis:**

· **Atmos photol half-life:** High:
Low:

*Comment:*

· **Max light absorption (nm):** No data
*Comment:*

· **Aq photol half-life:** High:
Low:

*Comment:*

## Photooxidation half-life:
· **Water:** High: No data
Low:

*Comment:*

· **Air:** High: 92.4 hours (3.85 days)
Low: 9.24 hours
*Comment:* Scientific judgement based upon estimated rate constant for reaction with hydroxyl radical in air (Atkinson, R (1987A)).

## Reduction half-life:
High: No data
Low:

*Comment:*

## Hydrolysis:
· **First-order hydr half-life:** 4957 hours (207 days)
*Comment:* ($t_{1/2}$ at pH 7 and 25°C) Based upon measured neutral and base catalyzed hydrolysis rate constants for gamma-hexachlorocyclohexane (Ellington, JJ et al. (1987)).

· **Acid rate const $(M(H+)-hr)^{-1}$:** No data
*Comment:* ($t_{1/2}$ = 5765 hours (240 days) at pH 5 and 25°C) Based upon measured neutral and base catalyzed hydrolysis rate constants for gamma-hexachlorocyclohexane (Ellington, JJ et al. (1987)).

· **Base rate const $(M(OH-)-hr)^{-1}$:** 198
*Comment:* ($t_{1/2}$ = 330 hours (13.8 days) at pH 9 and 25°C) Based upon measured neutral and base catalyzed hydrolysis rate constants for gamma-hexachlorocyclohexane (Ellington, JJ et al. (1987)).

# delta-Hexachlorocyclohexane

<u>**CAS Registry Number:**</u> 319-86-8

<u>**Half-lives:**</u>

· **Soil:**  High: 2400 hours  (100 days)
  Low: 330 hours  (13.8 days)

*Comment:* Based upon hydrolysis half-life for gamma-hexachlorocyclohexane (low $t_{1/2}$) (Ellington, JJ et al. (1987)) and data from one aerobic soil die-away study data for one soil (high $t_{1/2}$) (Macrae, IC et al. (1984)).

· **Air:**  High: 92.4 hours  (3.85 days)
  Low: 9.24 hours

*Comment:* Scientific judgement based upon estimated photooxidation half-life in air.

· **Surface Water:**  High: 2400 hours  (100 days)
  Low: 330 hours  (13.8 days)

*Comment:* Based upon hydrolysis half-life for gamma-hexachlorocyclohexane (low $t_{1/2}$) (Ellington, JJ et al. (1987)) and data from one aerobic soil die-away study data for one soil (high $t_{1/2}$) (Macrae, IC et al. (1984)).

· **Ground Water:**  High: 4800 hours  (200 days)
  Low: 330 hours  (13.8 days)

*Comment:* Based upon hydrolysis half-life for gamma-hexachlorocyclohexane (low $t_{1/2}$) (Ellington, JJ et al. (1987)) and data from one aerobic soil die-away study data for one soil (high $t_{1/2}$) (Macrae, IC et al. (1984)).

<u>**Aqueous Biodegradation (unacclimated):**</u>

· **Aerobic half-life:**  High: 2400 hours  (100 days)
  Low: 960 hours  (40 days)

*Comment:* Based upon data from one aerobic soil die-away study data for one soil (Macrae, IC et al. (1984)).

· **Anaerobic half-life:**  High: 2400 hours  (100 days)
  Low: 720 hours  (30 days)

*Comment:* Based upon anaerobic soil die-away study data (high $t_{1/2}$: Macrae, IC et al. (1984); low $t_{1/2}$: Zhang, S et al. (1982)).

· **Removal/secondary treatment:**  High: No data
  Low:

*Comment:*

<u>**Photolysis:**</u>

· **Atmos photol half-life:**        High:

                                               Low:

*Comment:*

· **Max light absorption (nm):**    No data

*Comment:*

· **Aq photol half-life:**           High:

                                               Low:

*Comment:*

**Photooxidation half-life:**

  · **Water:**                    High:       No data

                                               Low:

*Comment:*

  · **Air:**                       High:       92.4 hours          (3.85 days)

                                             Low:        9.24 hours

*Comment:* Scientific judgement based upon estimated rate constant for reaction with hydroxyl radical in air (Atkinson, R (1987A)).

**Reduction half-life:**               High:       No data

                                             Low:

*Comment:*

**Hydrolysis:**

  · **First-order hydr half-life:**        4957 hours         (207 days)

*Comment:* ($t_{1/2}$ at pH 7 and 25°C) Based upon measured neutral and base catalyzed hydrolysis rate constants for gamma-hexachlorocyclohexane (Ellington, JJ et al. (1987)).

  · **Acid rate const (M(H+)-hr)$^{-1}$:**    No data

*Comment:* ($t_{1/2}$ = 5765 hours (240 days) at pH 5 and 25°C) Based upon measured neutral and base catalyzed hydrolysis rate constants for gamma-hexachlorocyclohexane (Ellington, JJ et al. (1987)).

  · **Base rate const (M(OH-)-hr)$^{-1}$:**    198

*Comment:* ($t_{1/2}$ = 330 hours (13.8 days) at pH 9 and 25°C) Based upon measured neutral and base catalyzed hydrolysis rate constants for gamma-hexachlorocyclohexane (Ellington, JJ et al. (1987)).

# Linuron

**CAS Registry Number:** 330-55-2

**Structure:**

**Half-lives:**

&middot; **Soil:**                       High:    4272 hours          (178 days)

Low:    672 hours          (28 days)

*Comment:* Based upon soil die-away test data (low $t_{1/2}$: Walker, A and Zimdahl, RL (1981); high $t_{1/2}$: Walker, A (1978)).

&middot; **Air:**                        High:    4.90 hours

Low:    0.490 hours

*Comment:* Scientific judgement based upon estimated photooxidation half-life in air.

&middot; **Surface Water:**             High:    4272 hours          (178 days)

Low:    672 hours          (28 days)

*Comment:* Scientific judgement based upon estimated unacclimated aqueous aerobic biodegradation half-life.

&middot; **Ground Water:**              High:    8544 hours          (356 days)

Low:    1344 hours          (56 days)

*Comment:* Scientific judgement based upon estimated unacclimated aqueous aerobic biodegradation half-life.

**Aqueous Biodegradation (unacclimated):**

&middot; **Aerobic half-life:**          High:    4272 hours          (178 days)

Low:    672 hours          (28 days)

*Comment:* Based upon soil die-away test data (low $t_{1/2}$: Walker, A and Zimdahl, RL (1981); high $t_{1/2}$: Walker, A (1978)).

&middot; **Anaerobic half-life:**        High:    17088 hours         (1.95 years)

Low:    2688 hours          (162 days)

*Comment:* Scientific judgement based upon estimated unacclimated aqueous aerobic biodegradation half-life.

&middot; **Removal/secondary treatment:**    High:       No data

Low:

*Comment:*

## Photolysis:
　· **Atmos photol half-life:**        High:    4032 hours        (168 days)

                                                   Low:    1344 hours        (56 days)

*Comment:* Scientific judgement based upon measured rate constant for summer sunlight photolysis in distilled water (low $t_{1/2}$) Rosen, JD et al. (1969)) and adjusted for relative winter sunlight intensity (high $t_{1/2}$) (Lyman, WJ et al. (1982)).

　· **Max light absorption (nm):**        No data
*Comment:*

　· **Aq photol half-life:**        High:    4032 hours        (168 days)

                                               Low:    1344 hours        (56 days)

*Comment:* Scientific judgement based upon measured rate constant for summer sunlight photolysis in distilled water (low $t_{1/2}$) (Rosen, JD et al. (1969)) and adjusted for relative winter sunlight intensity (high $t_{1/2}$) (Lyman, WJ et al. (1982)).

## Photooxidation half-life:
　· **Water:**        High:    No data

                 Low:

*Comment:*

　· **Air:**        High:    4.90 hours

                 Low:    0.490 hours

*Comment:* Scientific judgement based upon estimated rate constant for reaction with hydroxyl radicals in air (Atkinson, R (1987A)).

## Reduction half-life:        High:    No data

                 Low:

*Comment:*

## Hydrolysis:
　· **First-order hydr half-life:**        No data
*Comment:*

　· **Acid rate const (M(H+)-hr)$^{-1}$:**
*Comment:*

　· **Base rate const (M(OH-)-hr)$^{-1}$:**
*Comment:*

# Auramine

CAS Registry Number: 492-80-8

Structure:

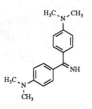

## Half-lives:
· **Soil:**

| | | |
|---|---|---|
| High: | 4320 hours | (6 months) |
| Low: | 672 hours | (4 weeks) |

*Comment:* Scientific judgement based upon estimated unacclimated aqueous aerobic biodegradation half-life.

· **Air:**

| | |
|---|---|
| High: | 1.97 hours |
| Low: | 0.197 hours |

*Comment:* Scientific judgement based upon estimated photooxidation half-life in air.

· **Surface Water:**

| | | |
|---|---|---|
| High: | 4320 hours | (6 months) |
| Low: | 672 hours | (4 weeks) |

*Comment:* Scientific judgement based upon estimated unacclimated aqueous aerobic biodegradation half-life.

· **Ground Water:**

| | | |
|---|---|---|
| High: | 8640 hours | (12 months) |
| Low: | 1344 hours | (8 weeks) |

*Comment:* Scientific judgement based upon estimated unacclimated aqueous aerobic biodegradation half-life.

## Aqueous Biodegradation (unacclimated):
· **Aerobic half-life:**

| | | |
|---|---|---|
| High: | 4320 hours | (6 months) |
| Low: | 672 hours | (4 weeks) |

*Comment:* Scientific judgement based upon unacclimated aqueous screening test data (Saito, T et al. (1984)).

· **Anaerobic half-life:**

| | | |
|---|---|---|
| High: | 17280 hours | (24 months) |
| Low: | 2880 hours | (16 weeks) |

*Comment:* Scientific judgement based upon estimated unacclimated aqueous aerobic biodegradation half-life.

· **Removal/secondary treatment:**  High:  No data

*Comment:*

## Photolysis:
· **Atmos photol half-life:**        High:
                                          Low:

*Comment:*

· **Max light absorption (nm):**      No data
*Comment:*

· **Aq photol half-life:**            High:
                                          Low:

*Comment:*

Water:                                     High:
                                          Low:

*Comment:*

· **Air:**                                 High:     1.97 hours
                                          Low:     0.197 hours
*Comment:* Scientific judgement based upon estimated rate constant for reaction with hydroxyl radical in air (Atkinson, R (1987A)).

## Reduction half-life:                         High:            No data
                                                               Low:

*Comment:*

## Hydrolysis:
· **First-order hydr half-life:**
*Comment:*

· **Acid rate const $(M(H+)-hr)^{-1}$:**     No data
*Comment:*

· **Base rate const $(M(OH-)-hr)^{-1}$:**
*Comment:*

# Mustard gas

**CAS Registry Number:** 505-60-2

**Structure:**

Cl~~~S~~~Cl

**Half-lives:**
  **· Soil:**                          High:      0.26 hours
                                       Low:       0.065 hours
  *Comment:* Scientific judgement based upon hydrolysis half-life in water.

  **· Air:**                           High:      186 hours            (7.8 days)
                                       Low:       18.6 hours
  *Comment:* Scientific judgement based upon estimated photooxidation half-life in air.

  **· Surface Water:**                 High:      0.26 hours
                                       Low:       0.065 hours
  *Comment:* Scientific judgement based upon hydrolysis half-life in water.

  **· Ground Water:**                  High:      0.26 hours
                                       Low:       0.065 hours
  *Comment:* Scientific judgement based upon hydrolysis half-life in water.

**Aqueous Biodegradation (unacclimated):**
  **· Aerobic half-life:**             High:      672 hours            (4 weeks)
                                       Low:       168 hours            (7 days)
  *Comment:* Scientific judgement.

  **· Anaerobic half-life:**           High:      2688 hours           (16 weeks)
                                       Low:       672 hours            (28 days)
  *Comment:* Scientific judgement.

  **· Removal/secondary treatment:**   High:      No data
                                       Low:
  *Comment:*

**Photolysis:**
  **· Atmos photol half-life:**        High:
                                       Low:

*Comment:*

· **Max light absorption (nm):**       No data
*Comment:*

· **Aq photol half-life:**       High:
                                      Low:

*Comment:*

## Photooxidation half-life:
· **Water:**       High:       No data
                                      Low:

*Comment:*

· **Air:**       High:       186 hours       (7.8 days)
                                      Low:       18.6 hours
*Comment:* Scientific judgement based upon estimated rate constant for reaction with hydroxyl radical in air (Atkinson, R (1987A)).

## Reduction half-life:       High:       No data
                                      Low:
*Comment:*

## Hydrolysis:
· **First-order hydr half-life:**       High:       0.26 hours
                                      Low:       0.065 hours
*Comment:* ($t_{1/2}$ at pH 5-9) Based upon hydrolysis half-life in distilled water at 20°C and 25°C (Small, MJ (1984)).

· **Acid rate const $(M(H+)-hr)^{-1}$:**
*Comment:*

· **Base rate const $(M(OH-)-hr)^{-1}$:**
*Comment:*

# Chlorobenzilate

**CAS Registry Number:** 510-15-6

**Structure:**

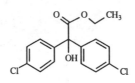

**Half-lives:**
· **Soil:**

| | High: | 840 hours | (5 weeks) |
|---|---|---|---|
| | Low: | 168 hours | (1 week) |

*Comment:* Scientific judgement based upon unacclimated soil grab sample data (Wheeler, WB et al. (1973)).

· **Air:**

| | High: | 130 hours | |
|---|---|---|---|
| | Low: | 13 hours | |

*Comment:* Scientific judgement based upon estimated photooxidation half-life in air.

· **Surface Water:**

| | High: | 840 hours | (5 weeks) |
|---|---|---|---|
| | Low: | 168 hours | (1 week) |

*Comment:* Scientific judgement based upon estimated aqueous aerobic biodegradation half-life.

· **Ground Water:**

| | High: | 1680 hours | (10 weeks) |
|---|---|---|---|
| | Low: | 336 hours | (2 weeks) |

*Comment:* Scientific judgement based upon estimated aqueous aerobic biodegradation half-life.

**Aqueous Biodegradation (unacclimated):**
· **Aerobic half-life:**

| | High: | 840 hours | (5 weeks) |
|---|---|---|---|
| | Low: | 168 hours | (1 week) |

*Comment:* Scientific judgement based upon unacclimated soil grab sample data (Wheeler, WB et al. (1973)).

· **Anaerobic half-life:**

| | High: | 2688 hours | (16 weeks) |
|---|---|---|---|
| | Low: | 672 hours | (4 weeks) |

*Comment:* Scientific judgement based upon estimated aqueous aerobic biodegradation half-life.

· **Removal/secondary treatment:**

| | High: | No data |
|---|---|---|
| | Low: | |

*Comment:*

**Photolysis:**

· **Atmos photol half-life:**          High:          No data
                                        Low:

*Comment:*

· **Max light absorption (nm):**          lambda max = 275, 265, 258 240, 230, 197
*Comment:*  No absorbance above 290 nm in hexane (Gore, RC et al. (1971)).

· **Aq photol half-life:**          High:          No data
                                     Low:

*Comment:*

## Photooxidation half-life:
· **Water:**          High:          No data
                      Low:

*Comment:*

· **Air:**          High:          130 hours
                    Low:          13 hours
*Comment:*  Scientific judgement based upon estimated rate data for hydroxyl radicals in air
(Atkinson, R (1987A)).

## Reduction half-life:          High:          No data
                                 Low:

*Comment:*

## Hydrolysis:
· **First-order hydr half-life:**          No data
*Comment:*

· **Acid rate const $(M(H+)\text{-}hr)^{-1}$:**
*Comment:*

· **Base rate const $(M(OH\text{-})\text{-}hr)^{-1}$:**
*Comment:*

579

# 2-Chloroacetophenone

<u>CAS Registry Number:</u>  532-27-4

<u>Half-lives:</u>
  · **Soil:**  High:   672 hours   (4 weeks)

| | | | |
|---|---|---|---|
| · **Soil:** | High: | 672 hours | (4 weeks) |
| | Low: | 168 hours | (7 days) |

*Comment:* Scientific judgement based upon estimated aerobic biodegradation half-life.

| | | | |
|---|---|---|---|
| · **Air:** | High: | 739 hours | (30.8 days) |
| | Low: | 73.9 hours | (3.08 days) |

*Comment:* Scientific judgement based upon estimated photooxidation half-life in air.

| | | | |
|---|---|---|---|
| · **Surface Water:** | High: | 672 hours | (4 weeks) |
| | Low: | 168 hours | (7 days) |

*Comment:* Scientific judgement based upon estimated aerobic biodegradation half-life.

| | | | |
|---|---|---|---|
| · **Ground Water:** | High: | 1344 hours | (8 weeks) |
| | Low: | 336 hours | (14 days) |

*Comment:* Scientific judgement based upon estimated aerobic biodegradation half-life.

<u>Aqueous Biodegradation (unacclimated):</u>

| | | | |
|---|---|---|---|
| · **Aerobic half-life:** | High: | 672 hours | (4 weeks) |
| | Low: | 168 hours | (7 days) |

*Comment:* Scientific judgement based upon river die-away test data for acetophenone (Ludzack, FJ and Ettinger, MB (1963)).

| | | | |
|---|---|---|---|
| · **Anaerobic half-life:** | High: | 2688 hours | (16 weeks) |
| | Low: | 672 hours | (28 days) |

*Comment:* Scientific judgement based upon estimated aerobic biodegradation half-life.

| | | | |
|---|---|---|---|
| · **Removal/secondary treatment:** | High: | No data | |
| | Low: | | |

*Comment:*

<u>Photolysis:</u>

| | | |
|---|---|---|
| · **Atmos photol half-life:** | High: | |
| | Low: | |

*Comment:*

| | |
|---|---|
| · **Max light absorption (nm):** | No data |

*Comment:*

**· Aq photol half-life:**               High:
                                           Low:

*Comment:*

## Photooxidation half-life:
**· Water:**                       High:      No data
                                           Low:

*Comment:*

**· Air:**                          High:    739 hours         (30.8 days)
                                         Low:     73.9 hours        (3.08 days)
*Comment:* Scientific judgement based upon estimated rate constant for reaction with hydroxyl radicals in air (Atkinson, R (1987A)).

## Reduction half-life:
                                             High: No reducible groups
                                           Low:

*Comment:*

## Hydrolysis:
**· First-order hydr half-life:**       Expected to be extremely slow
*Comment:* Scientific judgement based upon measured first-order hydrolysis rate constant for monochloroacetone (Osterman-Golkar, S (1984)).

**· Acid rate const $(M(H+)-hr)^{-1}$:**     No data
*Comment:*

**· Base rate const $(M(OH-)-hr)^{-1}$:**     No data
*Comment:*

# 4,6-Dinitro-o-cresol

CAS Registry Number: 534-52-1

Half-lives:
  · Soil:                              High:    504 hours          (21 days)
                                       Low:     168 hours          (7 days)
  Comment: Scientific judgement based upon data from a soil die-away study (Kincannon, DF and
  Lin, YS (1985)).

  · Air:                               High:    3098 hours         (129 days)
                                       Low:     310 hours          (12.9 days)
  Comment: Scientific judgement based upon estimated photooxidation half-life in air.

  · Surface Water:                     High:    504 hours          (21 days)
                                       Low:     77 hours           (3.2 days)
  Comment: Scientific judgement based upon data from a soil die-away study (high $t_{1/2}$)
  (Kincannon, DF and Lin, YS (1985)) and photooxidation half-life in water (low $t_{1/2}$).

  · Ground Water:                      High:    1008 hours         (42 days)
                                       Low:     68 hours           (2.8 days)
  Comment: Scientific judgement based upon estimated aqueous aerobic biodegradation half-life
  (high $t_{1/2}$) and estimated aqueous anaerobic biodegradation half-life (low $t_{1/2}$).

Aqueous Biodegradation (unacclimated):
  · Aerobic half-life:                 High:    504 hours          (21 days)
                                       Low:     168 hours          (7 days)
  Comment: Scientific judgement based upon data from a soil die-away study (Kincannon, DF and
  Lin, YS (1985)).

  · Anaerobic half-life:               High:    170 hours          (7.1 days)
                                       Low:     68 hours           (2.8 days)
  Comment: Based upon anaerobic flooded soil die-away tests for 2,4-dinitrophenol (Sudhakar-
  Barik, R and Sethunathan, N (1978A)).

  · Removal/secondary treatment:       High:    No data
                                       Low:
  Comment:

Photolysis:
  · Atmos photol half-life:            High:    No data
                                       Low:
  Comment:

· **Max light absorption (nm):**     lambda max = 265 nm (dioxane); lambda
max = 372 nm (aqueous NaOH)
*Comment:* Absorption extends to >400 nm in dioxane (Callahan, MA et al. (1979A)).


· **Aq photol half-life:**           High:     No data
                                     Low:
*Comment:*


## Photooxidation half-life:
· **Water:**                         High:     3840 hours          (160 days)
                                     Low:      77 hours            (3.2 days)
*Comment:* Scientific judgement based upon reported reaction rate constants for $RO_2 \cdot$ with the phenol class (Mill, T and Mabey, W (1985)).


· **Air:**                           High:     3098 hours          (129 days)
                                     Low:      310 hours           (12.9 days)
*Comment:* Scientific judgement based upon estimated rate constant for reaction with hydroxyl radical in air (Atkinson, R (1987A)).


## Reduction half-life:               High:     No data
                                     Low:
*Comment:*


## Hydrolysis:
· **First-order hydr half-life:**
*Comment:*


· **Acid rate const (M(H+)-hr)$^{-1}$:**     No hydrolyzable groups
*Comment:*


· **Base rate const (M(OH-)-hr)$^{-1}$:**
*Comment:*

# 1,2-Dichloroethylene

<u>CAS Registry Number:</u>  540-59-0

<u>Half-lives:</u>
· **Soil:**                                    High:     4320 hours              (6 months)
                                               Low:       672 hours              (4 weeks)
*Comment:*  Scientific judgement based upon estimated unacclimated aqueous aerobic biodegradation half-life.

· **Air:**                                     High:      286 hours              (11.9 days)
                                               Low:      25.2 hours              (1.1 days)
*Comment:*  Scientific judgement based upon estimated photooxidation half-life in air.

· **Surface Water:**                           High:     4320 hours              (6 months)
                                               Low:       672 hours              (4 weeks)
*Comment:*  Scientific judgement based upon estimated unacclimated aqueous aerobic biodegradation half-life.

· **Ground Water:**                            High:  69,000 hours              (95 months)
                                               Low:     1344 hours              (8 weeks)
*Comment:*  Scientific judgement based upon estimated unacclimated aqueous aerobic biodegradation half-life (low $t_{1/2}$) and an estimated half-life for anaerobic biodegradation of cis- and trans-1,2-dichloroethylene from a ground water field study of chlorinated ethylenes (Silka, LR and Wallen, DA (1988)).

<u>Aqueous Biodegradation (unacclimated):</u>
· **Aerobic half-life:**                       High:     4320 hours              (6 months)
                                               Low:       672 hours              (4 weeks)
*Comment:*  Scientific judgement based upon unacclimated aerobic aqueous screening test data (Tabak, HH et al. (1981)).

· **Anaerobic half-life:**                     High:    17280 hours              (24 months)
                                               Low:      2688 hours              (16 weeks)
*Comment:*  Scientific judgement based upon estimated unacclimated aqueous aerobic biodegradation half-life.

· **Removal/secondary treatment:**             High:         No data
                                               Low:

*Comment:*

<u>Photolysis:</u>
· **Atmos photol half-life:**                  High:         No data

Low:

*Comment:*

**· Max light absorption (nm):**     Max light absorption (nm):  lambda max ≈195 nm (cis-1,2-dichloroethylene); lambda max = 200 nm (trans-1,2-dichloroethylene)
*Comment:* No absorption above = 240 nm for both the cis and trans isomers (Dahlberg, JA (1969)).

**· Aq photol half-life:**          High:     No data
                                Low:
*Comment:*

## Photooxidation half-life:
**· Water:**                        High:     No data
                                Low:
*Comment:*

**· Air:**                          High:     286 hours          (11.9 days)
                                Low:      25.2 hours         (1.1 days)
*Comment:* Scientific judgement based upon estimated rate constants for reaction of cis-1,2-dichloroethylene (high $t_{1/2}$) and trans-1,2-dichloroethylene (low $t_{1/2}$) with hydroxyl radicals in air (Atkinson, R 1987A)).

## Reduction half-life:
                                High:  No reducible groups
                                Low:
*Comment:*

## Hydrolysis:
**· First-order hydr half-life:**
*Comment:*

**· Acid rate const $(M(H+)-hr)^{-1}$:**     No hydrolyzable groups
*Comment:*

**· Base rate const $(M(OH-)-hr)^{-1}$:**
*Comment:*

# 1,2-Dimethyl hydrazine

<u>CAS Registry Number:</u>  540-73-8

<u>Half-lives:</u>

· **Soil:**  High:  672 hours  (4 weeks)

  Low:  168 hours  (1 week)

  *Comment:*  Scientific judgement based upon estimated aqueous aerobic biodegradation half-life.

· **Air:**  High:  5.2 hours

  Low:  0.5 hours  (31 minutes)

  *Comment:*  Scientific judgement based upon estimated photooxidation half-life in air.

· **Surface Water:**  High:  672 hours  (4 weeks)

  Low:  168 hours  (1 week)

  *Comment:*  Scientific judgement based upon estimated aqueous aerobic biodegradation half-life.

· **Ground Water:**  High:  8640 hours  (8 weeks)

  Low:  336 hours  (2 weeks)

  *Comment:*  Scientific judgement based upon estimated aqueous aerobic biodegradation half-life.

<u>Aqueous Biodegradation (unacclimated):</u>

· **Aerobic half-life:**  High:  672 hours  (4 weeks)

  Low:  168 hours  (1 week)

  *Comment:*  Scientific judgement based upon freshwater grab sample data for 1,1-dimethyl hydrazine (Braun, BA and Zirrolli, JA (1983).

· **Anaerobic half-life:**  High:  2688 hours  (16 weeks)

  Low:  672 hours  (4 weeks)

  *Comment:*  Scientific judgement based upon estimated aqueous aerobic biodegradation half-life.

· **Removal/secondary treatment:**  High:  No data

  Low:

  *Comment:*

<u>Photolysis:</u>

· **Atmos photol half-life:**  High:

  Low:

  *Comment:*

· **Max light absorption (nm):**  No data

  *Comment:*

586

· **Aq photol half-life:**          High:
                                            Low:

*Comment:*

## Photooxidation half-life:

· **Water:**                  High:      No data
                                            Low:

*Comment:*

· **Air:**                     High:      5.2 hours
                                          Low:      0.5 hours        (31 minutes)
*Comment:* Scientific judgement based upon estimated rate data for hydroxyl radicals in air (Atkinson, R (1987A)).

## Reduction half-life:            High:      No data
                                          Low:

*Comment:*

## Hydrolysis:

· **First-order hydr half-life:**      No data
*Comment:*

· **Acid rate const $(M(H+)-hr)^{-1}$:**
*Comment:*

· **Base rate const $(M(OH-)-hr)^{-1}$:**
*Comment:*

# Ethyl chloroformate

**CAS Registry Number:** 541-41-3

## Half-lives:

| | | | |
|---|---|---|---|
| **· Soil:** | High: | 0.89 hours | (53 minutes) |
| | Low: | 0.55 hours | (33 minutes) |

*Comment:* Scientific judgement based upon hydrolysis half-lives at 24.525°C (low $t_{1/2}$) and 19.8°C (high $t_{1/2}$) (Queen, A (1967)).

| | | | |
|---|---|---|---|
| **· Air:** | High: | 445 hours | (18.5 days) |
| | Low: | 44.5 hours | (1.85 days) |

*Comment:* Scientific judgement based upon estimated photooxidation half-life in air.

| | | | |
|---|---|---|---|
| **· Surface Water:** | High: | 0.89 hours | (53 minutes) |
| | Low: | 0.55 hours | (33 minutes) |

*Comment:* Scientific judgement based upon hydrolysis half-lives at 24.525°C (low $t_{1/2}$) and 19.8°C (high $t_{1/2}$) (Queen, A (1967)).

| | | | |
|---|---|---|---|
| **· Ground Water:** | High: | 0.89 hours | (53 minutes) |
| | Low: | 0.55 hours | (33 minutes) |

*Comment:* Scientific judgement based upon hydrolysis half-lives at 24.525°C (low $t_{1/2}$) and 19.8°C (high $t_{1/2}$) (Queen, A (1967)).

## Aqueous Biodegradation (unacclimated):

| | | | |
|---|---|---|---|
| **· Aerobic half-life**: | High: | 672 hours | (4 weeks) |
| | Low: | 168 hours | (7 days) |

*Comment:* Scientific judgement.

| | | | |
|---|---|---|---|
| **· Anaerobic half-life**: | High: | 2688 hours | (16 weeks) |
| | Low: | 672 hours | (28 days) |

*Comment:* Scientific judgement based upon estimated unacclimated aqueous aerobic biodegradation half-life.

| | | |
|---|---|---|
| **· Removal/secondary treatment:** | High: | No data |
| | Low: | |

*Comment:* No data

## Photolysis:

| | |
|---|---|
| **· Atmos photol half-life:** | High: |
| | Low: |

*Comment:*

**· Max light absorption (nm):**          No data
*Comment:*

**· Aq photol half-life:**          High:
                                    Low:

*Comment:*

**Photooxidation half-life:**
**· Water:**          High:     No data
                      Low:
*Comment:*

**· Air:**          High:     445 hours          (18.5 days)
                    Low:      44.5 hours         (1.85 days)
*Comment:* Scientific judgement based upon estimated rate constant for reaction with hydroxyl radicals in air (Atkinson, R (1987A)).

**Reduction half-life:**          High:     No data
                                  Low:
*Comment:*

**Hydrolysis:**
**· First-order hydr half-life:**          0.55 hours     (33 minutes) at 24.525°C
                                           0.89 hours     (53 minutes) at 19.8°C
*Comment:* Based upon measured hydrolysis half-life in water (Queen, A (1967)).

**· Acid rate const (M(H+)-hr)$^{-1}$:**          No data
*Comment:*

**· Base rate const (M(OH-)-hr)$^{-1}$:**          No data
*Comment:*

# m-Dichlorobenzene

CAS Registry Number: 541-73-1

Half-lives:
  · Soil:                         High:   4320 hours        (6 months)
                                  Low:    672 hours         (4 weeks)
  Comment: Scientific judgement based upon unacclimated aerobic screening test data (Canton, JH et al. (1985)) and aerobic soil grab sample data (Haider, K et al. (1981)).

  · Air:                          High:   891 hours         (37.1 days)
                                  Low:    89.1 hours        (3.7 days)
  Comment: Based upon photooxidation half-life in air.

  · Surface Water:                High:   4320 hours        (6 months)
                                  Low:    672 hours         (4 weeks)
  Comment: Scientific judgement based upon estimated unacclimated aqueous aerobic biodegradation half-life.

  · Ground Water:                 High:   8640 hours        (12 months)
                                  Low:    1344 hours        (8 weeks)
  Comment: Scientific judgement based upon estimated unacclimated aqueous aerobic biodegradation half-life.

Aqueous Biodegradation (unacclimated):
  · Aerobic half-life:            High:   4320 hours        (6 months)
                                  Low:    672 hours         (4 weeks)
  Comment: Scientific judgement based upon unacclimated aerobic screening test data (Canton, JH et al. (1985)) and aerobic soil grab sample data (Haider, K et al. (1981)).

  · Anaerobic half-life:          High:   17280 hours       (24 months)
                                  Low:    2880 hours        (16 weeks)
  Comment: Scientific judgement based upon estimated unacclimated aqueous aerobic biodegradation half-life.

  · Removal/secondary treatment:  High:   99.5%
                                  Low:
  Comment: Based upon % degraded in a 6 day period under aerobic continuous flow conditions (may include stripping) (Kinncannon, DF et al. (1983)).

Photolysis:
  · Atmos photol half-life:       High:   No data
                                  Low:

590

*Comment:*

· **Max light absorption (nm):**          lambda max = 278.0, 270.0, 263.0, 256.0, 250.0, 216.0
*Comment:* No absorbance of UV light at wavelengths greater than 290 nm in methanol (Sadtler UV No. 1671).

· **Aq photol half-life:**          High:          No data
                                    Low:
*Comment:*

**Photooxidation half-life:**
   · **Water:**                     High:          No data
                                    Low:
*Comment:*

   · **Air:**                       High:      891 hours          (37.1 days)
                                    Low:       89.1 hours         (3.7 days)
*Comment:* Based upon measured rate data for the vapor phase reaction with hydroxyl radicals in air (Atkinson, R et al.(1985B)).

**Reduction half-life:**            High:          No data
                                    Low:
*Comment:*

**Hydrolysis:**
   · **First-order hydr half-life:**          >879 years
*Comment:* Scientific judgement based upon rate constant ($<0.9$ $M^{-1}$ $hr^{-1}$) extrapolated to pH 7 at 25 °C from 1% disappearance after 16 days at 85 °C and pH 9.7 (Ellington, JJ et al. (1988)).

   · **Acid rate const $(M(H+)-hr)^{-1}$:**
*Comment:*

   · **Base rate const $(M(OH-)-hr)^{-1}$:**          <0.9
*Comment:* Scientific judgement based upon 1% disappearance after 16 days at 85 °C and pH 9.7 (Ellington, JJ et al. (1988)).

# 1,3-Dichloropropene

<u>**CAS Registry Number:**</u>  542-75-6

<u>**Half-lives:**</u>
| | | | |
|---|---|---|---|
| · **Soil:** | High: | 271 hours | (11.3 days) |
| | Low: | 133 hours | (5.5 days) |

*Comment:* Based upon measured rate of hydrolysis in water at pH 7 at 25°C (Mill, T et al. (1985)) and at pH 5 and 20°C (McCall, PJ (1987)). Rate of hydrolysis is independent of pH over the range pH 5 to 10 (McCall, PJ (1987)).

| | | | |
|---|---|---|---|
| · **Air:** | High: | 80.3 hours | (3.35 days) |
| | Low: | 4.66 hours | |

*Comment:*  Based upon measured rate constants for reaction with hydroxyl radical (Atkinson, R and Carter, WPL (1984)) and ozone (Pitts, JN et al. (1984)) for cis-1,3-dichloropropene (high $t_{1/2}$) and trans-1,3-dichloropropene (low $t_{1/2}$).

| | | | |
|---|---|---|---|
| · **Surface Water:** | High: | 271 hours | (11.3 days) |
| | Low: | 133 hours | (5.5 days) |

*Comment:* Based upon measured rate of hydrolysis at pH 7 at 25°C (Mill, T et al. (1985)) and at pH 5 and 20°C (McCall, PJ (1987)). Rate of hydrolysis is independent of pH over the range pH 5 to 10 (McCall, PJ (1987)).

| | | | |
|---|---|---|---|
| · **Ground Water:** | High: | 271 hours | (11.3 days) |
| | Low: | 133 hours | (5.5 days) |

*Comment:* Based upon measured rate of hydrolysis at pH 7 at 25°C (Mill, T et al. (1985)) and at pH 5 and 20°C (McCall, PJ (1987)). Rate of hydrolysis is independent of pH over the range pH 5 to 10 (McCall, PJ (1987)).

<u>**Aqueous Biodegradation (unacclimated):**</u>
| | | | |
|---|---|---|---|
| · **Aerobic half-life:** | High: | 672 hours | (4 weeks) |
| | Low: | 168 hours | (7 days) |

*Comment:*  Scientific judgement based upon unacclimated aqueous aerobic biodegradation screening studies (Krijgsheld, KR and Van der Gen, A (1986A); Tabak, HH et al. (1981)).

| | | | |
|---|---|---|---|
| · **Anaerobic half-life:** | High: | 2688 hours | (16 weeks) |
| | Low: | 672 hours | (28 days) |

*Comment:*  Scientific judgement based upon estimated unacclimated aqueous aerobic biodegradation half-life.

| | | | |
|---|---|---|---|
| · **Removal/secondary treatment:** | High: | No data | |
| | Low: | | |

*Comment:*

## Photolysis:

**· Atmos photol half-life:**   High:
                                Low:

*Comment:*

**· Max light absorption (nm):**   No data
*Comment:*

**· Aq photol half-life:**   High:
                             Low:

*Comment:*

## Photooxidation half-life:

**· Water:**   High:   No data
               Low:

*Comment:*

**· Air:**   High:   80.3 hours   (3.35 days)
             Low:   4.66 hours

*Comment:* Based upon measured rate constants for reaction with hydroxyl radical (Atkinson, R and Carter, WPL (1984)) and ozone (Pitts, JN et al. (1984)) for cis-1,3-dichloropropene (high $t_{1/2}$) and trans-1,3-dichloropropene (low $t_{1/2}$).

## Reduction half-life:   High:   No data
                          Low:

*Comment:*

## Hydrolysis:

**· First-order hydr half-life:**   High:   271 hours   (11.3 days)
                                    Low:   133 hours   (5.5 days)

*Comment:* ($t_{1/2}$ at pH 5, 7, and 9) Based upon measured rate of hydrolysis at pH 7 at 25°C (Mill, T et al. (1985)) and at pH 7 and 20°C (McCall, PJ (1987)). Rate of hydrolysis is independent of pH over the range pH 5 to 10 (McCall, PJ (1987)).

**· Acid rate const (M(H+)-hr)$^{-1}$:**   No data
*Comment:*

**· Base rate const (M(OH-)-hr)$^{-1}$:**   No data
*Comment:*

593

# Bis-(chloromethyl) ether

CAS Registry Number: 542-88-1

Half-lives:
· Soil:                              High:    0.106 hours
                                     Low:    0.0106 hours
*Comment:* Scientific judgement based upon estimated hydrolysis half-life in water.

· Air:                               High:    1.96 hours
                                     Low:    0.196 hours
*Comment:* Scientific judgement based upon estimated photooxidation half-life in air.

· Surface Water:                     High:    0.106 hours
                                     Low:    0.0106 hours
*Comment:* Scientific judgement based upon estimated hydrolysis half-life in water.

· Ground Water:                      High:    0.106 hours
                                     Low:    0.0106 hours
*Comment:* Scientific judgement based upon estimated hydrolysis half-life in water.

Aqueous Biodegradation (unacclimated):
· Aerobic half-life:                 High:    672 hours        (4 weeks)
                                     Low:    168 hours         (7 days)
*Comment:* Scientific judgement.

· Anaerobic half-life:               High:    2688 hours       (16 weeks)
                                     Low:    672 hours         (28 days)
*Comment:* Scientific judgement based upon unacclimated aqueous aerobic biodegradation half-life.

· Removal/secondary treatment:       High:        100%
                                     Low:
*Comment:* Removal percentage based upon data from a continuous activated sludge biological treatment simulator (Patterson, JW and Kodukala, PS (1981)) and scientific judgement.

Photolysis:
· Atmos photol half-life:            High:
                                     Low:
*Comment:*

· Max light absorption (nm):         No data
*Comment:*

**· Aq photol half-life:**  High:
                           Low:

*Comment:*


## Photooxidation half-life:
**· Water:**  High:    No data
              Low:

*Comment:*


**· Air:**  High:    1.96 hours
            Low:    0.196 hours

*Comment:*  Scientific judgement based upon estimated rate constant for reaction with hydroxyl radical in air (Atkinson, R (1987A)).


## Reduction half-life:
High:    No data
Low:

*Comment:*


## Hydrolysis:
**· First-order hydr half-life:**  High:    0.106 hours
                                   Low:    0.0106 hours

*Comment:*  ($t_{1/2}$ at pH 5-9) Low $t_{1/2}$ based upon measured hydrolysis rate constant (Tou, JC et al. (1974)) and high $t_{1/2}$ based upon scientific judgement.


**· Acid rate const $(M(H+)-hr)^{-1}$:**
*Comment:*


**· Base rate const $(M(OH-)-hr)^{-1}$:**
*Comment:*

# C.I. Basic Green 4

CAS Registry Number: 569-64-2

Structure:

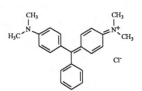

**Half-lives:**
 · **Soil:**                         High:    4320 hours          (6 months)
                                     Low:      672 hours          (4 weeks)
   *Comment:* Scientific judgement based upon estimated aqueous aerobic biodegradation half-lives.

 · **Air:**                          High:    0.265 hours
                                     Low:    0.0389 hours
   *Comment:* Scientific judgement based upon estimated photooxidation half-life in air.

 · **Surface Water:**                High:    4320 hours          (6 months)
                                     Low:      672 hours          (4 weeks)
   *Comment:* Scientific judgement based upon estimated aqueous aerobic biodegradation half-life.

 · **Ground Water:**                 High:    8640 hours          (12 months)
                                     Low:     1344 hours          (8 weeks)
   *Comment:* Scientific judgement based upon estimated aqueous aerobic biodegradation half-lives.

**Aqueous Biodegradation (unacclimated):**
 · **Aerobic half-life**:            High:    4320 hours          (6 months)
                                     Low:      672 hours          (4 weeks)
   *Comment:* Scientific judgement.

 · **Anaerobic half-life**:          High:   17280 hours          (24 months)
                                     Low:     2688 hours          (16 weeks)
   *Comment:* Scientific judgement based upon estimated aqueous aerobic biodegradation half-lives.

 · **Removal/secondary treatment:**  High:     No data
                                     Low:
   *Comment:*

**Photolysis:**
 · **Atmos photol half-life:**       High:
                                     Low:

*Comment:*

· **Max light absorption (nm):**        No data
*Comment:*

· **Aq photol half-life:**        High:
                                        Low:

*Comment:*

**Photooxidation half-life:**
  · **Water:**                         High:       No data
                                        Low:

*Comment:*

  · **Air:**                            High:    0.265 hours
                                        Low:    0.0389 hours
*Comment:* Scientific judgement based upon estimated rate constants for reaction with hydroxyl radical (Atkinson, R (1987A)) and ozone (Atkinson, R and Carter, WPL (1984)) in air.

**Reduction half-life:**                  High:       No data
                                        Low:

*Comment:*

**Hydrolysis:**
  · **First-order hydr half-life:**
*Comment:*

  · **Acid rate const (M(H+)-hr)$^{-1}$:**        No data
*Comment:*

  · **Base rate const (M(OH-)-hr)$^{-1}$:**
*Comment:*

# Bromoethylene

**CAS Registry Number:** 593-60-2

**Half-lives:**
 · Soil:                              High:    4320 hours          (6 months)
                                      Low:     672 hours           (4 weeks)
 *Comment:* Scientific judgement based upon estimated unacclimated aqueous aerobic biodegradation half-life.

 · **Air:**                           High:    94 hours
                                      Low:     9.4 hours
 *Comment:* Based upon photooxidation half-life in air.

 · **Surface Water:**                 High:    4320 hours          (6 months)
                                      Low:     672 hours           (4 weeks)
 *Comment:* Scientific judgement based upon estimated unacclimated aqueous aerobic biodegradation half-life.

 · **Ground Water:**                  High:    69000 hours         (95 months)
                                      Low:     1344 hours          (8 weeks)
 *Comment:* Scientific judgement based upon estimated unacclimated aqueous aerobic biodegradation half-life (low $t_{1/2}$) and an estimated half-life for anaerobic biodegradation of vinyl chloride from a ground water field study of chlorinated ethenes (Silka, LR and Wallen, DA (1988)).

**Aqueous Biodegradation (unacclimated):**
 · **Aerobic half-life:**             High:    4320 hours          (6 months)
                                      Low:     672 hours           (4 weeks)
 *Comment:* Scientific judgement based upon aqueous screening test data for vinyl chloride (Freitag, D et al. (1984A); Helfgott, TB et al. (1977)).

 · **Anaerobic half-life:**           High:    17280 hours         (24 months)
                                      Low:     2880 hours          (4 months)
 *Comment:* Scientific judgement based upon estimated unacclimated aqueous aerobic biodegradation half-life.

 · **Removal/secondary treatment:**   High:    No data
                                      Low:
 *Comment:*

**Photolysis:**
 · **Atmos photol half-life:**        High:

*Comment:*

· **Max light absorption (nm):**     No data
*Comment:*

· **Aq photol half-life:**     High:
                               Low:

*Comment:*

## Photooxidation half-life:
· **Water:**     High:     No data
                 Low:

*Comment:*

· **Air:**     High:     94 hours          (3.9 days)
               Low:      9.4 hours
*Comment:* Based upon measured rate constant for reaction with hydroxyl radicals in air (Atkinson, R (1985))

## Reduction half-life:     High:     No data
                            Low:

*Comment:*

## Hydrolysis:
· **First-order hydr half-life:**
*Comment:*

· **Acid rate const $(M(H+)-hr)^{-1}$:**     No hydrolyzable groups
*Comment:*

· **Base rate const $(M(OH-)-hr)^{-1}$:**
*Comment:*

# 2,3-Dinitrotoluene

CAS Registry Number: 602-01-7

## Half-lives:

· **Soil:**

| | High: | 4320 hours | (6 months) |
|---|---|---|---|
| | Low: | 672 hours | (4 weeks) |

*Comment:* Scientific judgement based upon estimated unacclimated aqueous aerobic biodegradation half-life.

· **Air:**

| | High: | 50 hours | (2.08 days) |
|---|---|---|---|
| | Low: | 45 hours | (1.88 days) |

*Comment:* Based upon estimated photolysis half-life in air.

· **Surface Water:**

| | High: | 50 hours | (2.08 days) |
|---|---|---|---|
| | Low: | 45 hours | (1.88 days) |

*Comment:* Based upon photooxidation half-life in natural water.

· **Ground Water:**

| | High: | 8640 hours | (12 months) |
|---|---|---|---|
| | Low: | 48 hours | (2 days) |

*Comment:* Scientific judgement based upon estimated unacclimated aqueous anaerobic biodegradation half-life for 2,4-dinitrotoluene (low $t_{1/2}$) and estimated unacclimated aqueous aerobic biodegradation half-life for 2,3-dinitrotoluene (high $t_{1/2}$).

## Aqueous Biodegradation (unacclimated):

· **Aerobic half-life**:

| | High: | 4320 hours | (6 months) |
|---|---|---|---|
| | Low: | 672 hours | (4 weeks) |

*Comment:* Scientific judgement based upon aerobic natural water die-away test data (Spanggord, RJ et al. (1981)). Reduction of the nitro group observed in these studies may not be due to biodegradation.

· **Anaerobic half-life**:

| | High: | 300 hours | (13 days) |
|---|---|---|---|
| | Low: | 48 hours | (2 days) |

*Comment:* Scientific judgement based upon anaerobic natural water die-away test data for 2,4-dinitrotoluene (Spanggord, RJ et al. (1980)). Reduction of the nitro group observed in these studies may not be due to biodegradation

· **Removal/secondary treatment**:

| | High: | No data |
|---|---|---|
| | Low: | |

*Comment:*

## Photolysis:

· **Atmos photol half-life:**

| | High: | 72 hours | (3 days) |
|---|---|---|---|

Low:      24 hours        (1 day)

*Comment:* Scientific judgement based upon measured photolysis rates in distilled water (Simmons, MS and Zepp, RG (1986); Mill, T and Mabey, W (1985)).

**· Max light absorption (nm):**     No data
*Comment:*

**· Aq photol half-life:**     High:    72 hours     (3 days)
                                  Low:    24 hours     (1 day)
*Comment:* Scientific judgement based upon measured photolysis rates in distilled water (Simmons, MS and Zepp, RG (1986); Mill, T and Mabey, W (1985)).

**Photooxidation half-life:**
  **· Water:**     High:    No data
                     Low:
*Comment:*

  **· Air:**     High:    3197 hours    (133 days)
                     Low:    320 hours    (13.3 days)
*Comment:* Scientific judgement based upon estimated rate constant for reaction with hydroxyl radical in air (Atkinson, R (1987A)).

**Reduction half-life:**     High:    No data
                     Low:
*Comment:*

**Hydrolysis:**
  **· First-order hydr half-life:**
*Comment:*

  **· Acid rate const (M(H+)-hr)$^{-1}$:**     No hydrolyzable groups
*Comment:*

  **· Base rate const (M(OH-)-hr)$^{-1}$:**
*Comment:*

# 2,6-Dinitrotoluene

<u>CAS Registry Number:</u>  606-20-2

<u>Half-lives:</u>
·  **Soil:**                                           High:     4320 hours                (6 months)
                                                        Low:       672 hours                 (4 weeks)
*Comment:*  Scientific judgement based upon estimated unacclimated aqueous aerobic biodegradation half-life.

·  **Air:**                                            High:       25 hours
                                                        Low:        17 hours
*Comment:*  Based upon estimated photolysis half-life in air.

·  **Surface Water:**                                  High:       17 hours
                                                        Low:         2 hours
*Comment:*  Based upon photooxidation half-life in natural water.

·  **Ground Water:**                                   High:     8640 hours                (12 months)
                                                        Low:        48 hours                 (2 days)
*Comment:*  Scientific judgement based upon estimated unacclimated aqueous anaerobic biodegradation half-life for 2,4-dinitrotoluene (low $t_{1/2}$) and estimated unacclimated aqueous aerobic biodegradation half-life for 2,6-dinitrotoluene (high $t_{1/2}$).

<u>Aqueous Biodegradation (unacclimated):</u>
·  **Aerobic half-life:**                              High:     4320 hours                (6 months)
                                                        Low:       672 hours                 (4 weeks)
*Comment:*  Scientific judgement based upon aerobic natural water die-away test data (Spanggord, RJ et al. (1981)).  Reduction of the nitro group observed in these studies may not be due to biodegradation.

·  **Anaerobic half-life:**                            High:      300 hours                (13 days)
                                                        Low:        48 hours                 (2 days)
*Comment:*  Scientific judgement based upon anaerobic natural water die-away test data for 2,4-dinitrotoluene (Spanggord, RJ et al. (1980)).  Reduction of the nitro group observed in these studies may not be due to biodegradation.

·  **Removal/secondary treatment:**                    High:
                                                        Low:
*Comment:*

<u>Photolysis:</u>
·  **Atmos photol half-life:**                         High:       25 hours

602

Low:       17 hours

*Comment:* Scientific judgement based upon measured photolysis rates in water (Simmons, MS and Zepp, RG (1986); Mill, T and Mabey, W (1985)).

· **Max light absorption (nm):**       No data
*Comment:*

· **Aq photol half-life:**       High:       25 hours
                                 Low:        17 hours
*Comment:* Based upon measured photolysis rates in water (Simmons, MS and Zepp, RG (1986); Mill, T and Mabey, W (1985)).

## Photooxidation half-life:
· **Water:**       High:       17 hours
                   Low:         2 hours
*Comment:* Based upon measured photooxidation rates in natural waters (Simmons, MS and Zepp, RG (1986)).

· **Air:**       High:       2840 hours       (118 days)
                 Low:         284 hours        (11.8 days)
*Comment:* Based upon estimated rate constant for reaction with hydroxyl radicals in air (Atkinson, R (1987A)).

## Reduction half-life:       High:       No data
                              Low:

*Comment:*

## Hydrolysis:
· **First-order hydr half-life:**
*Comment:*

· **Acid rate const (M(H+)-hr)$^{-1}$:**       No hydrolyzable groups
*Comment:*

· **Base rate const (M(OH-)-hr)$^{-1}$:**
*Comment:*

# Pentachlorobenzene

**CAS Registry Number:** 608-93-5

**Half-lives:**

| | | | |
|---|---|---|---|
| · **Soil:** | High: | 8280 hours | (345 days) |
| | Low: | 4656 hours | (194 days) |

*Comment:* Scientific judgement based upon unacclimated aerobic soil grab sample data (Beck, J and Hansen, KE (1974)).

| | | | |
|---|---|---|---|
| · **Air:** | High: | 10877 hours | (1.2 years) |
| | Low: | 1088 hours | (45.3 days) |

*Comment:* Scientific judgement based upon estimated photooxidation half-life in air.

| | | | |
|---|---|---|---|
| · **Surface Water:** | High: | 8280 hours | (345 days) |
| | Low: | 4656 hours | (194 days) |

*Comment:* Scientific judgement based upon estimated unacclimated aqueous aerobic biodegradation half-life.

| | | | |
|---|---|---|---|
| · **Ground Water:** | High: | 16560 hours | (690 days) |
| | Low: | 9312 hours | (388 days) |

*Comment:* Scientific judgement based upon estimated unacclimated aqueous aerobic biodegradation half-life.

**Aqueous Biodegradation (unacclimated):**

| | | | |
|---|---|---|---|
| · **Aerobic half-life:** | High: | 8280 hours | (345 days) |
| | Low: | 4656 hours | (194 days) |

*Comment:* Scientific judgement based upon unacclimated aerobic soil grab sample data (Beck, J and Hansen, KE (1974)).

| | | | |
|---|---|---|---|
| · **Anaerobic half-life:** | High: | 33120 hours | (1380 days) |
| | Low: | 18624 hours | (776 days) |

*Comment:* Scientific judgement based upon estimated unacclimated aqueous aerobic biodegradation half-life.

| | | |
|---|---|---|
| · **Removal/secondary treatment:** | High: | No data |
| | Low: | |

*Comment:*

**Photolysis:**

| | | |
|---|---|---|
| · **Atmos photol half-life:** | High: | No data |
| | Low: | |

*Comment:*

· **Max light absorption (nm):**     No data
*Comment:*

· **Aq photol half-life:**           High:     No data
                                     Low:
*Comment:*

**Photooxidation half-life:**
  · **Water:**                       High:     No data
                                     Low:
*Comment:*

· **Air:**                           High:   10877 hours        (1.2 years)
                                     Low:    1088 hours         (45.3 days)
*Comment:* Scientific judgement based upon estimated rate constant for the vapor phase reaction with hydroxyl radicals in air (Atkinson, R (1987A)).

**Reduction half-life:**             High:     No data
                                     Low:
*Comment:*

**Hydrolysis:**
  · **First-order hydr half-life:**     >879 years
*Comment:* Scientific judgement based upon rate constant ($<0.9$ $M^{-1}$ $hr^{-1}$) extrapolated to pH 7 at 25 °C from 1% disappearance after 16 days at 85 °C and pH 9.7 (Ellington, JJ et al. (1988)).

· **Acid rate const $(M(H+)-hr)^{-1}$:**
*Comment:*

· **Base rate const $(M(OH-)-hr)^{-1}$:**     <0.9
*Comment:* Scientific judgement based upon 1% disappearance after 16 days at 85 °C and pH 9.7 (Ellington, JJ et al. (1988)).

# 3,4-Dinitrotoluene

<u>CAS Registry Number:</u>  610-39-9

<u>Half-lives:</u>
- **Soil:**                                   High:    4320 hours              (6 months)
                                              Low:      672 hours              (4 weeks)
*Comment:*  Scientific judgement based upon estimated unacclimated aqueous aerobic biodegradation half-life.

- **Air:**                                    High:     326 hours              (13.6 days)
                                              Low:      319 hours              (13.3 days)
*Comment:*  Based upon estimated photolysis half-life in air.

- **Surface Water:**                          High:     326 hours              (13.6 days)
                                              Low:      319 hours              (13.3 days)
*Comment:*  Based upon photooxidation half-life in natural water.

- **Ground Water:**                           High:    8640 hours              (12 months)
                                              Low:       48 hours              (2 days)
*Comment:*  Scientific judgement based upon estimated unacclimated aqueous anaerobic biodegradation half-life for 2,4-dinitrotoluene (low $t_{1/2}$) and estimated unacclimated aqueous aerobic biodegradation half-life for 3,4-dinitrotoluene (high $t_{1/2}$).

<u>Aqueous Biodegradation (unacclimated):</u>
- **Aerobic half-life**:                      High:    4320 hours              (6 months)
                                              Low:      672 hours              (4 weeks)
*Comment:*  Scientific judgement based upon aerobic natural water die-away test data (Spanggord, RJ et al. (1981)).  Reduction of the nitro group observed in these studies may not be due to biodegradation.

- **Anaerobic half-life**:                    High:     300 hours              (13 days)
                                              Low:       48 hours              (2 days)
*Comment:*  Scientific judgement based upon anaerobic natural water die-away test data for 2,4-dinitrotoluene (Spanggord, RJ et al. (1980)).  Reduction of the nitro group observed in these studies may not be due to biodegradation

- **Removal/secondary treatment:**            High:    No data
                                              Low:
*Comment:*

<u>Photolysis:</u>
- **Atmos photol half-life:**                 High:     480 hours              (20 days)

606

|  | Low: | 160 hours | (6.7 days) |

*Comment:* Scientific judgement based upon measured photolysis rates in water (Simmons, MS and Zepp, RG (1986)).

**· Max light absorption (nm):**     No data
*Comment:*

**· Aq photol half-life:**     High:     480 hours     (20 days)
                              Low:     160 hours     (6.7 days)
*Comment:* Based upon measured photolysis rates in water (Simmons, MS and Zepp, RG (1986)).

## Photooxidation half-life:

**· Water:**     High:     No data
                Low:
*Comment:*

**· Air:**     High:     3197 hours     (133 days)
              Low:     320 hours     (13.3 days)
*Comment:* Scientific judgement based upon estimated rate constant for reaction with hydroxyl radical in air (Atkinson, R (1987A)).

## Reduction half-life:     High:     No data
                          Low:
*Comment:*

## Hydrolysis:

**· First-order hydr half-life:**
*Comment:*

**· Acid rate const $(M(H+)-hr)^{-1}$:**     No hydrolyzable groups
*Comment:*

**· Base rate const $(M(OH-)-hr)^{-1}$:**
*Comment:*

# Ethyl N-methyl-N-nitrosocarbamate

CAS Registry Number: 615-53-2

Half-lives:
· Soil:                          High:      24 hours
                                 Low:       12 hours
Comment: Based upon hydrolysis half-lives at pH 9 (low $t_{1/2}$) and pH 5 and 7 (high $t_{1/2}$) (Ellington, JJ et al. (1987)).

· Air:                           High:      223 hours
                                 Low:       22.3 hours
Comment: Scientific judgement based upon estimated photooxidation half-life in air.

· Surface Water:                 High:      24 hours
                                 Low:       12 hours
Comment: Based upon hydrolysis half-lives at pH 9 (low $t_{1/2}$) and pH 5 and 7 (high $t_{1/2}$) (Ellington, JJ et al. (1987)).

· Ground Water:                  High:      24 hours
                                 Low:       12 hours
Comment: Based upon hydrolysis half-lives at pH 9 (low $t_{1/2}$) and pH 5 and 7 (high $t_{1/2}$) (Ellington, JJ et al. (1987)).

Aqueous Biodegradation (unacclimated):
· Aerobic half-life:             High:      168 hours      (7 days)
                                 Low:       24 hours       (1 day)
Comment: Scientific judgement.

· Anaerobic half-life:           High:      672 hours      (28 days)
                                 Low:       96 hours       (4 days)
Comment: Scientific judgement.

· Removal/secondary treatment:   High:      No data
                                 Low:
Comment:

Photolysis:
· Atmos photol half-life:        High:
                                 Low:
Comment:

· Max light absorption (nm):     No data

608

*Comment:*

**· Aq photol half-life:**                         High:
                                                   Low:
*Comment:*

**Photooxidation half-life:**
   **· Water:**                                    High:          No data
                                                   Low:
   *Comment:*

   **· Air:**                                       High:          223 hours
                                                    Low:           22.3 hours
*Comment:* Scientific judgement based upon estimated rate constant for reaction with hydroxyl radical in air (Atkinson, R (1987A)).

**Reduction half-life:**                            High:          No data
                                                    Low:
   *Comment:*

**Hydrolysis:**
   **· First-order hydr half-life:**                24 hours
   *Comment:* ($t_{1/2}$ at pH 7 and 25°C) Based upon measured acid and base catalyzed and neutral hydrolysis rate constants (Ellington, JJ et al. (1987)).

   **· Acid rate const (M(H+)-hr)$^{-1}$:**          9.5
   *Comment:* ($t_{1/2}$ = 24 hours at pH 5 and 25°C) Based upon measured acid and base catalyzed and neutral hydrolysis rate constants (Ellington, JJ et al. (1987)).

   **· Base rate const (M(OH-)-hr)$^{-1}$:**         $2.9 \times 10^3$
   *Comment:* ($t_{1/2}$ = 12 hours at pH 9 and 25°C) Based upon measured acid and base catalyzed and neutral hydrolysis rate constants (Ellington, JJ et al. (1987)).

# 2,5-Dinitrotoluene

<u>CAS Registry Number:</u> 619-15-8

| · **Soil:** | High: | 4320 hours | (6 months) |
| | Low: | 672 hours | (4 weeks) |

*Comment:* Scientific judgement based upon estimated unacclimated aqueous aerobic biodegradation half-life.

| · **Air:** | High: | 2.78 hours | |
| | Low: | 1.85 hours | |

*Comment:* Scientific judgement based upon estimated photolysis half-life in air.

| · **Surface Water:** | High: | 2.78 hours | |
| | Low: | 1.85 hours | |

*Comment:* Based upon photolysis half-life in water.

| · **Ground Water:** | High: | 8640 hours | (12 months) |
| | Low: | 48 hours | (2 days) |

*Comment:* Scientific judgement based upon estimated unacclimated aqueous anaerobic biodegradation half-life for 2,4-dinitrotoluene (low $t_{1/2}$) and estimated unacclimated aqueous aerobic biodegradation half-life for 2,5-dinitrotoluene (high $t_{1/2}$).

<u>Aqueous Biodegradation (unacclimated):</u>

| · **Aerobic half-life:** | High: | 4320 hours | (6 months) |
| | Low: | 672 hours | (4 weeks) |

*Comment:* Scientific judgement based upon aerobic natural water die-away test data (Spanggord, RJ et al. (1981)). Reduction of the nitro group observed in these studies may not be due to biodegradation.

| · **Anaerobic half-life:** | High: | 300 hours | (13 days) |
| | Low: | 48 hours | (2 days) |

*Comment:* Scientific judgement based upon anaerobic natural water die-away test data for 2,4-dinitrotoluene (Spanggord, RJ et al. (1980)). Reduction of the nitro group observed in these studies may not be due to biodegradation

| · **Removal/secondary treatment:** | High: | No data | |
| | Low: | | |

*Comment:*

<u>Photolysis:</u>

| · **Atmos photol half-life:** | High: | 2.8 hours | |
| | Low: | 0.92 hours | |

*Comment:* Scientific judgement based upon measured photolysis rates in distilled water (Simmons, MS and Zepp, RG (1986)).

**· Max light absorption (nm):**     No data
*Comment:*

**· Aq photol half-life:**          High:     2.8 hours
                                     Low:      0.92 hours
*Comment:* Based upon measured photolysis rates in distilled water (Simmons, MS and Zepp, RG (1986)).

**Photooxidation half-life:**
  **· Water:**                        High:     No data
                                      Low:
*Comment:*

  **· Air:**                          High:     3197 hours        (133 days)
                                      Low:      320 hours         (13.3 days)
*Comment:* Scientific judgement based upon estimated rate constant for reaction with hydroxyl radical in air (Atkinson, R (1987A)).

**Reduction half-life:**              High:     No data
                                      Low:
*Comment:*

**Hydrolysis:**
  **· First-order hydr half-life:**
*Comment:*

  **· Acid rate const (M(H+)-hr)$^{-1}$:**     No hydrolyzable groups
*Comment:*

  **· Base rate const (M(OH-)-hr)$^{-1}$:**
*Comment:*

# N-Nitrosodipropylamine

CAS Registry Number: 621-64-7

## Half-lives:

· **Soil:**  High:   4320 hours        (6 months)
        Low:    504 hours         (21 days)

*Comment:* Based upon aerobic soil die-away data (high $t_{1/2}$: Tate, RL III and Alexander, M (1975); low $t_{1/2}$: Oliver, JE et al. (1979)).

· **Air:**  High:    1 hour
        Low:    0.17 hours

*Comment:* Based upon measured rate of photolysis in the vapor phase under sunlight (Oliver, JE (1981)).

· **Surface Water:**  High:    1 hour
        Low:    0.17 hours

*Comment:* Based upon measured rate of photolysis in the vapor phase under sunlight (Oliver, JE (1981)).

· **Ground Water:**  High:   8640 hours        (12 months)
        Low:    1008 hours        (42 days)

*Comment:* Scientific judgement based upon estimated unacclimated aqueous aerobic biodegradation half-lives.

## Aqueous Biodegradation (unacclimated):

· **Aerobic half-life:**  High:   4320 hours        (6 months)
        Low:    504 hours         (21 days)

*Comment:* Based upon aerobic soil die-away data (high $t_{1/2}$: Tate, RL III and Alexander, M (1975); low $t_{1/2}$: Oliver, JE et al. (1979)).

· **Anaerobic half-life:**  High:   17280 hours       (24 months)
        Low:    2016 hours        (84 days)

*Comment:* Scientific judgement based upon estimated unacclimated aqueous aerobic biodegradation half-lives.

· **Removal/secondary treatment:**  High:    No data
        Low:

*Comment:*

## Photolysis:

· **Atmos photol half-life:**  High:    1 hour
        Low:    0.17 hours

*Comment:* Based upon measured rate of photolysis in the vapor phase under sunlight (Oliver, JE (1981)).

· **Max light absorption (nm):**       lambda max = 233, 339 nm (water)
*Comment:* IARC (1978).

· **Aq photol half-life:**       High:      1 hour
                                 Low:     0.17 hours
*Comment:* Based upon measured rate of photolysis in the vapor phase under sunlight (Oliver, JE (1981)).

## Photooxidation half-life:
· **Water:**       High:     No data
               Low:

*Comment:*

· **Air:**       High:     26.6 hours
            Low:     2.66 hours
*Comment:* Scientific judgement based upon estimated rate constant for reaction with hydroxyl radical in air (Atkinson, R (1987A)).

## Reduction half-life:       High:     No data
            Low:

*Comment:*

## Hydrolysis:
· **First-order hydr half-life:**     Not expected to be significant
*Comment:* IARC (1978).

· **Acid rate const $(M(H+)-hr)^{-1}$:**     No data
*Comment:*

· **Base rate const $(M(OH-)-hr)^{-1}$:**     No data
*Comment:*

# Methylisocyanate

<u>CAS Registry Number:</u> 624-83-9

<u>Half-lives:</u>
- **· Soil:**
  - High: 0.326 hours
  - Low: 0.144 hours

  *Comment:* Based upon hydrolysis half-lives.

- **· Air:**
  - High: 18.6 hours
  - Low: 1.86 hours

  *Comment:* Scientific judgement based upon estimated photooxidation half-life in air.

- **· Surface Water:**
  - High: 0.326 hours
  - Low: 0.144 hours

  *Comment:* Based upon hydrolysis half-lives.

- **· Ground Water:**
  - High: 0.326 hours
  - Low: 0.144 hours

  *Comment:* Based upon hydrolysis half-lives.

<u>Aqueous Biodegradation (unacclimated):</u>
- **· Aerobic half-life:**
  - High: 672 hours (4 weeks)
  - Low: 168 hours (7 days)

  *Comment:* Scientific judgement.

- **· Anaerobic half-life:**
  - High: 2688 hours (16 weeks)
  - Low: 672 hours (28 days)

  *Comment:* Scientific judgement based upon estimated unacclimated aqueous aerobic biodegradation half-life.

- **· Removal/secondary treatment:**
  - High: No data
  - Low:

  *Comment:*

<u>Photolysis:</u>
- **· Atmos photol half-life:**
  - High:
  - Low:

  *Comment:*

- **· Max light absorption (nm):** No data

  *Comment:*

**· Aq photol half-life:**　　　　　　　High:
　　　　　　　　　　　　　　　　　　　　Low:

*Comment:*

**Photooxidation half-life:**
　**· Water:**　　　　　　　　　　　　High:　　　No data
　　　　　　　　　　　　　　　　　　　　Low:
*Comment:*

　**· Air:**　　　　　　　　　　　　　　High:　　　18.6 hours
　　　　　　　　　　　　　　　　　　　　Low:　　　1.86 hours
*Comment:* Scientific judgement based upon estimated rate constant for reaction with hydroxyl radical in air (Atkinson, R (1987A)).

**Reduction half-life:**　　　　　　　　High:　　　No data
　　　　　　　　　　　　　　　　　　　　Low:
*Comment:*

**Hydrolysis:**
　**· First-order hydr half-life:**　　　High:　　0.326 hours at 15°C
　　　　　　　　　　　　　　　　　　　　Low:　　　0.144 hours at 25°C
*Comment:* ($t_{1/2}$ at pH 5-9) Based upon measured neutral hydrolysis rate constants (Castro, EA et al. (1985)).

　**· Acid rate const (M(H+)-hr)$^{-1}$:**
*Comment:*

　**· Base rate const (M(OH-)-hr)$^{-1}$:**　　No data
*Comment:*

# 1,1,1,2-Tetrachloroethane

**CAS Registry Number:** 630-20-6

**Half-lives:**

· **Soil:**                         High:    1604 hours            (66.8 days)
                                     Low:      16 hours
*Comment:* Scientific judgement based upon hydrolysis half-lives at pH 7 and 9 (Mabey, WR et al. (1983A)).

· **Air:**                          High:    22361 hours         (2.6 years)
                                       Low:      2236 hours           (93 days)
*Comment:* Scientific judgement based upon estimated photooxidation half-life in air.

· **Surface Water:**             High:    1604 hours            (66.8 days)
                                     Low:      16 hours
*Comment:* Scientific judgement based upon hydrolysis half-lives at pH 7 and 9 (Mabey, WR et al. (1983A)).

· **Ground Water:**             High:    1604 hours            (66.8 days)
                                     Low:      16 hours
*Comment:* Scientific judgement based upon hydrolysis half-lives at pH 7 and 9 (Mabey, WR et al. (1983A)).

**Aqueous Biodegradation (unacclimated):**

· **Aerobic half-life:**          High:    672 hours             (4 weeks)
                                     Low:      4320 hours          (6 months)
*Comment:* Scientific judgement based upon acclimated river die-away rate data for 1,1,2,2-tetrachloroethane (Mudder, T (1981)); and unacclimated sea water (Pearson, CR and McConnell, G (1975)) and sub-soil grab sample data from a ground water aquifer (Wilson, JT et al. (1983)) for 1,1,1-trichloroethane.

· **Anaerobic half-life:**      High:    2688 hours         (16 weeks)
                                     Low:      17280 hours       (6 months)
*Comment:* Scientific judgement based upon aerobic biodegradation half-life.

· **Removal/secondary treatment:**     High:     No data
                                       Low:
*Comment:*

**Photolysis:**

· **Atmos photol half-life:**        High:
                                       Low:

*Comment:*

· **Max light absorption (nm):**　　　　No data
*Comment:*

· **Aq photol half-life:**　　　　　　High:
　　　　　　　　　　　　　　　　　Low:

*Comment:*

**Photooxidation half-life:**
　· **Water:**　　　　　　　　　　High:　　No data
　　　　　　　　　　　　　　　　　Low:

*Comment:*

　· **Air:**　　　　　　　　　　　High:　　22361 hours　　　　　(2.6 years)
　　　　　　　　　　　　　　　　　Low:　　2236 hours　　　　　(93 days)
*Comment:* Scientific judgement based upon an estimated rate constant for vapor phase reaction with hydroxyl radicals in air (Atkinson, R (1987A)).

**Reduction half-life:**　　　　　　High:　　No data
　　　　　　　　　　　　　　　　　Low:

*Comment:*

**Hydrolysis:**
　· **First-order hydr half-life:**　　　　1604 hours　　　　(66.8 days)
*Comment:* Based upon rate constant ($k_B = 1.2$ M$^{-1}$ s$^{-1}$) for base reaction at 25 °C and pH of 7 (Mabey, WR et al. (1983A)).

　· **Acid rate const (M(H+)-hr)$^{-1}$:**　　No data
*Comment:*

　· **Base rate const (M(OH-)-hr)$^{-1}$:**　　$k_B = 4320$
*Comment:* ($t_{1/2}$= 16 hours) Rate constant for base reaction at 25 °C and pH of 9 (Mabey, WR et al. (1983A)).

# N-Methyl-N-nitrosourea

**CAS Registry Number:** 684-93-5

**Half-lives:**

   · **Soil:**
                    High:     3.5 hours
                              Low:     0.013 hours

*Comment:* Based upon hydrolysis half-lives for N-nitroso-N-ethylurea at pH 9 (low $t_{1/2}$) and pH 5 (high $t_{1/2}$) (Ellington, JJ et al. (1987)).

   · **Air:**
                    High:     5.28 hours
                              Low:     0.528 hours

*Comment:* Scientific judgement based upon estimated photooxidation half-life in air.

   · **Surface Water:**
               High:     3.5 hours
                           Low:     0.013 hours

*Comment:* Based upon hydrolysis half-lives for N-nitroso-N-ethylurea at pH 9 (low $t_{1/2}$) and pH 5 (high $t_{1/2}$) (Ellington, JJ et al. (1987)).

   · **Ground Water:**
              High:     3.5 hours
                         Low:     0.013 hours

*Comment:* Based upon hydrolysis half-lives for N-nitroso-N-ethylurea at pH 9 (low $t_{1/2}$) and pH 5 (high $t_{1/2}$) (Ellington, JJ et al. (1987)).

**Aqueous Biodegradation (unacclimated):**

   · **Aerobic half-life:**
         High:    4320 hours        (6 months)
                   Low:     672 hours         (4 weeks)

*Comment:* Based upon aerobic soil die-away data for N-nitrosodimethylamine and N-nitrosodipropylamine (high $t_{1/2}$: Tate, RL III and Alexander, M (1975); low $t_{1/2}$: Oliver, JE et al. (1979)).

   · **Anaerobic half-life:**
        High:   17280 hours     (24 months)
                  Low:    2688 hours      (16 weeks)

*Comment:* Scientific judgement based upon estimated unacclimated aqueous aerobic biodegradation half-lives.

   · **Removal/secondary treatment:**
         High:    No data
                Low:

*Comment:*

**Photolysis:**

   · **Atmos photol half-life:**
         High:    No data

Comment:                                       Low:

· **Max light absorption (nm):**                lambda max = 231 nm (water)
*Comment:* IARC (1978).

· **Aq photol half-life:**                      High:        No data
                                                Low:
*Comment:*

**Photooxidation half-life:**
· **Water:**                                    High:        No data
                                                Low:
*Comment:*

· **Air:**                                       High:        5.28 hours
                                                Low:         0.528 hours
*Comment:* Scientific judgement based upon estimated rate constant for reaction with hydroxyl radical in air (Atkinson, R (1985)).

**Reduction half-life:**                        High:        No data
                                                Low:
*Comment:*

**Hydrolysis:**
· **First-order hydr half-life:**        0.96 hours
*Comment:* ($t_{1/2}$ at pH 7 and 25°C) Based upon measured acid and base catalyzed and neutral hydrolysis rate constants for N-nitroso-N-ethylurea (Ellington, JJ et al. (1987)).

· **Acid rate const (M(H+)-hr)$^{-1}$:**        63
*Comment:* ($t_{1/2}$ = 3.5 hours at pH 5 and 25°C) Based upon measured acid and base catalyzed and neutral hydrolysis rate constants for N-nitroso-N-ethylurea (Ellington, JJ et al. (1987)).

· **Base rate const (M(OH-)-hr)$^{-1}$:**        $5.3 \times 10^6$
*Comment:* ($t_{1/2}$ = 0.013 hours at pH 9 and 25°C) Based upon measured acid and base catalyzed and neutral hydrolysis rate constants for N-nitroso-N-ethylurea (Ellington, JJ et al. (1987)).

# N-Nitroso-N-ethylurea

CAS Registry Number: 759-73-9

Half-lives:
 · Soil:                                   High:      3.5 hours
                                           Low:      0.013 hours
Comment: Based upon hydrolysis half-lives at pH 9 (low $t_{1/2}$) and pH 5 (high $t_{1/2}$) (Ellington, JJ et al. (1987)).

 · Air:                                    High:      4.99 hours
                                           Low:      0.499 hours
Comment: Scientific judgement based upon estimated photooxidation half-life in air.

 · Surface Water:                          High:      3.5 hours
                                           Low:      0.013 hours
Comment: Based upon hydrolysis half-lives at pH 9 (low $t_{1/2}$) and pH 5 (high $t_{1/2}$) (Ellington, JJ et al. (1987)).

 · Ground Water:                           High:      3.5 hours
                                           Low:      0.013 hours
Comment: Based upon hydrolysis half-lives at pH 9 (low $t_{1/2}$) and pH 5 (high $t_{1/2}$) (Ellington, JJ et al. (1987)).

Aqueous Biodegradation (unacclimated):
 · Aerobic half-life:              High:      4320 hours          (6 months)
                                   Low:       672 hours          (4 weeks)
Comment: Scientific judgement.

 · Anaerobic half-life:            High:     17280 hours          (24 months)
                                   Low:      2688 hours          (16 weeks)
Comment: Scientific judgement.

 · Removal/secondary treatment:    High:      No data
                                   Low:
Comment:

Photolysis:
 · Atmos photol half-life:         High:      No data
                                   Low:
Comment:

 · Max light absorption (nm):      lambda max = 233 nm (water)

*Comment:* IARC (1978).

· **Aq photol half-life:**　　　　　　　　High:　　　No data
　　　　　　　　　　　　　　　　　　　　　　Low:

*Comment:*

## Photooxidation half-life:
· **Water:**　　　　　　　　　　　　　　High:　　　No data
　　　　　　　　　　　　　　　　　　　　　　Low:

*Comment:*

· **Air:**　　　　　　　　　　　　　　　High:　　4.99 hours
　　　　　　　　　　　　　　　　　　　　　　Low:　　0.499 hours

*Comment:* Scientific judgement based upon estimated rate constant for reaction with hydroxyl radical in air (Atkinson, R (1987A)).

## Reduction half-life:　　　　　　　　　High:　　　No data
　　　　　　　　　　　　　　　　　　　　　　Low:

*Comment:*

## Hydrolysis:
· **First-order hydr half-life:**　　　　0.96 hours

*Comment:* ($t_{1/2}$ at pH 7 and 25°C) Based upon measured acid and base catalyzed and neutral hydrolysis rate constants (Ellington, JJ et al. (1987)).

· **Acid rate const (M(H+)-hr)$^{-1}$:**　　　63

*Comment:* ($t_{1/2}$ = 3.5 hours at pH 5 and 25°C) Based upon measured acid and base catalyzed and neutral hydrolysis rate constants (Ellington, JJ et al. (1987)).

· **Base rate const (M(OH-)-hr)$^{-1}$:**　　　$5.3 \times 10^6$

*Comment:* ($t_{1/2}$ = 0.013 hours at pH 9 and 25°C) Based upon measured acid and base catalyzed and neutral hydrolysis rate constants (Ellington, JJ et al. (1987)).

# Glycidylaldehyde

CAS Registry Number: 765-34-4

Structure:

$$\triangle\!\!\!\!\!/ \overset{O}{\underset{CH}{\overset{\|}{\ }}}$$

Half-lives:
  · Soil:                     High:    672 hours       (4 weeks)
                              Low:     168 hours       (1 week)
  *Comment:* Scientific judgement based upon estimated aqueous aerobic biodegradation half-life.

  · Air:                      High:     30 hours
                              Low:       3 hours
  *Comment:* Scientific judgement based upon estimated photooxidation half-life in air.

  · Surface Water:            High:    672 hours       (4 weeks)
                              Low:     168 hours       (1 week)
  *Comment:* Scientific judgement based upon estimated aqueous aerobic biodegradation half-life.

  · Ground Water:             High:    8640 hours      (8 weeks)
                              Low:     336 hours       (2 weeks)
  *Comment:* Scientific judgement based upon estimated aqueous aerobic biodegradation half-life.

Aqueous Biodegradation (unacclimated):
  · Aerobic half-life:        High:    672 hours       (4 weeks)
                              Low:     168 hours       (1 week)
  *Comment:* Scientific judgement.

  · Anaerobic half-life:      High:    2688 hours      (16 weeks)
                              Low:     672 hours       (4 weeks)
  *Comment:* Scientific judgement based upon estimated aqueous aerobic biodegradation half-life.

  · Removal/secondary treatment:   High:    No data
                                   Low:
  *Comment:*

Photolysis:
  · Atmos photol half-life:   High:
                              Low:

*Comment:*

**· Max light absorption (nm):**          No data
*Comment:*

**· Aq photol half-life:**          High:
                                    Low:

*Comment:*

## Photooxidation half-life:
**· Water:**                        High:          No data
                                    Low:

*Comment:*

**· Air:**                          High:          30 hours
                                    Low:           3 hours
*Comment:* Scientific judgement based upon estimated rate data for hydroxyl radicals in air
(Atkinson, R (1987A)).

## Reduction half-life:               High:          No data
                                    Low:
*Comment:*

## Hydrolysis:
**· First-order hydr half-life:**          No data
*Comment:*

**· Acid rate const $(M(H+)-hr)^{-1}$:**
*Comment:*

**· Base rate const $(M(OH-)-hr)^{-1}$:**
*Comment:*

# C.I. Solvent Yellow 14

<u>CAS Registry Number:</u>  842-07-9

<u>Structure:</u>

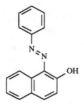

<u>Half-lives:</u>
  · **Soil:**                    High:     672 hours          (4 weeks)
                                 Low:      168 hours          (7 days)
  *Comment:*  Scientific judgement based upon estimated aqueous aerobic biodegradation half-life.

  · **Air:**                     High:     32.6 hours         (1.36 days)
                                 Low:      3.26 hours
  *Comment:*  Scientific judgement based upon estimated photooxidation half-life in air.

  · **Surface Water:**           High:     640 hours          (26.7 days)
                                 Low:      168 hours          (7 days)
  *Comment:*  Scientific judgement based upon estimated aqueous photolysis half-life (high $t_{1/2}$) and estimated aerobic biodegradation half-life (low $t_{1/2}$).

  · **Ground Water:**            High:     1344 hours         (8 weeks)
                                 Low:      336 hours          (14 days)
  *Comment:*  Scientific judgement based upon estimated aqueous aerobic biodegradation half-life.

<u>Aqueous Biodegradation (unacclimated):</u>
  · **Aerobic half-life:**       High:     672 hours          (4 weeks)
                                 Low:      168 hours          (7 days)
  *Comment:*  Scientific judgement based upon aerobic aqueous screening test data for 4-aminoazobenzene (Urushigawa, Y and Yonezawa, Y (1977)).

  · **Anaerobic half-life:**     High:     2688 hours         (16 weeks)
                                 Low:      672 hours          (28 days)
  *Comment:*  Scientific judgement based upon estimated aqueous aerobic biodegradation half-life.

  · **Removal/secondary treatment:**   High:    No data
                                       Low:
  *Comment:*

## Photolysis:

· **Atmos photol half-life:**   High:   640 hours      (26.7 days)
                                Low:    213 hours      (8.9 days)

*Comment:* Scientific judgement based upon the estimated photolysis half-life for distilled water solution in borosilicate glass tubes irradiated with light with a 313 nm max output under conditions closely approximating winter sunlight (high $t_{1/2}$) (Haag, WR and Mill, T (1987)); low $t_{1/2}$ based upon the experimental direct photolysis data and approximate summer sunlight intensity (Haag, WR and Mill, T (1987); Lyman, WJ et al. (1982)).

· **Max light absorption (nm):**   No data
*Comment:*

· **Aq photol half-life:**   High:   640 hours      (26.7 days)
                             Low:    213 hours      (8.9 days)

*Comment:* Scientific judgement based upon the estimated direct photolysis half-life for distilled water solutions in borosilicate glass tubes irradiated with light with a 313 nm max output under conditions closely approximating winter sunlight intensity (high $t_{1/2}$) (Haag, WR and Mill, T (1987)); low $t_{1/2}$ based upon the experimental direct photolysis data and approximate summer sunlight intensity (Haag, WR and Mill, T (1987); Lyman, WJ et al. (1982)).

## Photooxidation half-life:

· **Water:**   High:   28,000 hours      (3.2 years)
               Low:    283 hours         (12 days)

*Comment:* Scientific judgement based upon the estimated rate constant for reaction with singlet oxygen which is derived from data from an indirect photolysis study in humic acid solutions (synthetic natural water); the rate constant has been corrected for the rate of direct photolysis (Haag, WR and Mill, T (1985); Haag, WR and Mill, T (1987)).

· **Air:**   High:   32.6 hours      (1.36 days)
             Low:    3.26 hours

*Comment:* Scientific judgement based upon estimated rate constant for reaction with hydroxyl radicals in air (Atkinson, R (1987A)).

## Reduction half-life:   High:   No data
                          Low:

*Comment:*

## Hydrolysis:

· **First-order hydr half-life:**
*Comment:*

· **Acid rate const (M(H+)-hr)$^{-1}$:**   No data
*Comment:*

· **Base rate const $(M(OH-)-hr)^{-1}$:**
*Comment:*

# Dibutylnitrosamine

**CAS Registry Number:** 924-16-3

**Half-lives:**
　**· Soil:**　　　　　　　　　　　　High:　　4320 hours　　　　　(6 months)
　　　　　　　　　　　　　　　　　　　Low:　　　504 hours　　　　　(21 days)
*Comment:* Based upon aerobic soil die-away data for N-nitrosodimethylamine and N-nitrosodipropylamine (high $t_{1/2}$: Tate, RL III and Alexander, M (1975); low $t_{1/2}$: Oliver, JE et al. (1979)).

　**· Air:**　　　　　　　　　　　　　High:　　4.37 hours
　　　　　　　　　　　　　　　　　　　Low:　　0.437 hours
*Comment:* Scientific judgement based upon estimated photooxidation half-life in air.

　**· Surface Water:**　　　　　　　　High:　　　8 hours
　　　　　　　　　　　　　　　　　　　Low:　　　4 hours
*Comment:* Based upon measured rates for N-nitrosodiethylamine for sunlight photolysis in water (high $t_{1/2}$: Oliver, JE (1981); low $t_{1/2}$: Zhang, Z et al. (1983)).

　**· Ground Water:**　　　　　　　　High:　　8640 hours　　　　　(12 months)
　　　　　　　　　　　　　　　　　　　Low:　　1008 hours　　　　　(42 days)
*Comment:* Based upon aerobic soil die-away data for N-nitrosodimethylamine and N-nitrosodipropylamine (high $t_{1/2}$: Tate, RL III and Alexander, M (1975); low $t_{1/2}$: Oliver, JE et al. (1979)).

**Aqueous Biodegradation (unacclimated):**
　**· Aerobic half-life:**　　　　　　High:　　4320 hours　　　　　(6 months)
　　　　　　　　　　　　　　　　　　　Low:　　　504 hours　　　　　(21 days)
*Comment:* Based upon aerobic soil die-away data for N-nitrosodimethylamine and N-nitrosodipropylamine (high $t_{1/2}$: Tate, RL III and Alexander, M (1975); low $t_{1/2}$: Oliver, JE et al. (1979)).

　**· Anaerobic half-life:**　　　　　High:　　17280 hours　　　　(24 months)
　　　　　　　　　　　　　　　　　　　Low:　　2016 hours　　　　　(84 days)
*Comment:* Scientific judgement based upon estimated unacclimated aqueous aerobic biodegradation half-lives.

　**· Removal/secondary treatment:**　High:　　No data
　　　　　　　　　　　　　　　　　　　Low:
*Comment:*

**Photolysis:**

· **Atmos photol half-life:**       High:       8 hours
                                     Low:        4 hours
*Comment:* Scientific judgement based upon measured rates for N-nitrosodiethylamine for sunlight photolysis in water (high $t_{1/2}$: Oliver, JE (1981); low $t_{1/2}$: Zhang, Z et al. (1983)).

· **Max light absorption (nm):**       lambda max = 233, 347 nm (water)
*Comment:* IARC (1978).

· **Aq photol half-life:**       High:       8 hours
                                 Low:        4 hours
*Comment:* Based upon measured rates for N-nitrosodiethylamine for sunlight photolysis in water (high $t_{1/2}$: Oliver, JE (1981); low $t_{1/2}$: Zhang, Z et al. (1983)).

## Photooxidation half-life:
· **Water:**       High:       No data
                   Low:

*Comment:*

· **Air:**       High:       4.37 hours
                 Low:        0.437 hours
*Comment:* Scientific judgement based upon estimated rate constant for reaction with hydroxyl radical in air (Atkinson, R (1985)).

## Reduction half-life:       High:       No data
                              Low:

*Comment:*

## Hydrolysis:
· **First-order hydr half-life:**       Not expected to be significant
*Comment:* IARC (1978).

· **Acid rate const $(M(H+)\text{-hr})^{-1}$:**       No data
*Comment:*

· **Base rate const $(M(OH-)\text{-hr})^{-1}$:**       No data
*Comment:*

# N-Nitrosopyrrolidine

**CAS Registry Number:** 930-55-2

**Structure:**

**Half-lives:**

· **Soil:**

|  | High: | 4320 hours | (6 months) |
|---|---|---|---|
|  | Low: | 672 hours | (4 weeks) |

*Comment:* Based upon aerobic soil die-away data for N-nitrosodimethylamine and N-nitrosodipropylamine (high $t_{1/2}$: Tate, RL III and Alexander, M (1975); low $t_{1/2}$: Oliver, JE et al. (1979)).

· **Air:**

|  | High: | 32.9 hours |
|---|---|---|
|  | Low: | 3.29 hours |

*Comment:* Scientific judgement based upon estimated photooxidation half-life in air.

· **Surface Water:**

|  | High: | 4320 hours | (6 months) |
|---|---|---|---|
|  | Low: | 672 hours | (4 weeks) |

*Comment:* Based upon aerobic soil die-away data for N-nitrosodimethylamine and N-nitrosodipropylamine (high $t_{1/2}$: Tate, RL III and Alexander, M (1975); low $t_{1/2}$: Oliver, JE et al. (1979)).

· **Ground Water:**

|  | High: | 8640 hours | (12 months) |
|---|---|---|---|
|  | Low: | 1344 hours | (8 weeks) |

*Comment:* Based upon aerobic soil die-away data for N-nitrosodimethylamine and N-nitrosodipropylamine (high $t_{1/2}$: Tate, RL III and Alexander, M (1975); low $t_{1/2}$: Oliver, JE et al. (1979)).

**Aqueous Biodegradation (unacclimated):**

· **Aerobic half-life:**

|  | High: | 4320 hours | (6 months) |
|---|---|---|---|
|  | Low: | 672 hours | (4 weeks) |

*Comment:* Based upon aerobic soil die-away data for N-nitrosodimethylamine and N-nitrosodipropylamine (high $t_{1/2}$: Tate, RL III and Alexander, M (1975); low $t_{1/2}$: Oliver, JE et al. (1979)).

· **Anaerobic half-life:**

|  | High: | 17280 hours | (24 months) |
|---|---|---|---|
|  | Low: | 2688 hours | (16 weeks) |

*Comment:* Scientific judgement based upon estimated unacclimated aqueous aerobic biodegradation half-lives.

    **· Removal/secondary treatment:**        High:      No data
                                                                   Low:
    *Comment:*

## Photolysis:
    **· Atmos photol half-life:**             High:
                                                                     Low:
    *Comment:*

    **· Max light absorption (nm):**
    *Comment:*

    **· Aq photol half-life:**                High:
                                                                     Low:
    *Comment:*

## Photooxidation half-life:
    **· Water:**                             High:
                                                                        Low:
    *Comment:*

    **· Air:**                                 High:      32.9 hours
                                                                     Low:      3.29 hours
    *Comment:* Scientific judgement based upon estimated rate constant for reaction with hydroxyl radical in air (Atkinson, R (1985)).

## Reduction half-life:
                                              High:      No data
                                              Low:
    *Comment:*

## Hydrolysis:
    **· First-order hydr half-life:**           Not expected to be significant
    *Comment:* IARC (1978).

    **· Acid rate const (M(H+)-hr)$^{-1}$:**        No data
    *Comment:*

    **· Base rate const (M(OH-)-hr)$^{-1}$:**      No data
    *Comment:*

# Heptachlor epoxide

CAS Registry Number:  1024-57-3

Structure:

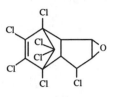

Half-lives:

· **Soil:**                                    High:    13248 hours            (552 days)
                                                 Low:       792 hours            (33 days)
*Comment:* Scientific judgement based upon aerobic soil grab sample data (Bowman, MC et al. (1965)).

· **Air:**                                     High:        60 hours            (2.5 days)
                                                 Low:          6 hours
*Comment:* Scientific judgement based upon estimated photooxidation half-life in air.

· **Surface Water:**                    High:    13248 hours            (552 days)
                                                 Low:       792 hours            (33 days)
*Comment:* Scientific judgement based upon estimated unacclimated aqueous aerobic biodegradation half-life.

· **Ground Water:**                    High:    26496 hours            (1104 days)
                                                 Low:         24 hours            (1 day)
*Comment:* Scientific judgement based upon estimated unacclimated aqueous aerobic and anaerobic biodegradation half-lives.

Aqueous Biodegradation (unacclimated):

· **Aerobic half-life:**                High:    13248 hours            (552 days)
                                                 Low:       792 hours            (33 days)
*Comment:* Scientific judgement based upon aerobic soil grab sample data (Bowman, MC et al. (1965)).

· **Anaerobic half-life:**             High:       168 hours            (7 days)
                                                 Low:         24 hours            (1 day)
*Comment:* Scientific judgement based upon soil and freshwater mud grab sample data (Maule, A et al. (1987)).

· **Removal/secondary treatment:**    High:        No data
                                                           Low:

631

*Comment:*

**Photolysis:**
  · **Atmos photol half-life:**         High:      No data
                                     Low:
  *Comment:*

  · **Max light absorption (nm):**      No data
  *Comment:*

  · **Aq photol half-life:**            High:      No data
                                       Low:
  *Comment:*

**Photooxidation half-life:**
  · **Water:**                        High:      No data
                                       Low:
  *Comment:*

  · **Air:**                           High:      60 hours           (2.5 days)
                                       Low:       6 hours
  *Comment:* Scientific judgement based upon an estimated rate constant for vapor phase reaction with hydroxyl radicals in air (Atkinson, R (1987A)).

**Reduction half-life:**               High:      No data
                                       Low:
  *Comment:*

**Hydrolysis:**
  · **First-order hydr half-life:**       Not expected to be important
  *Comment:* Based upon a reported rate constant of zero (Kollig, HP et al. (1987A)).

  · **Acid rate const (M(H+)-hr)$^{-1}$:**
  *Comment:*

  · **Base rate const (M(OH-)-hr)$^{-1}$:**
  *Comment:*

# N-Nitrosodiethanolamine

**CAS Registry Number:** 1116-54-7

**Half-lives:**

· **Soil:**                  High:    4320 hours         (6 months)
                                       Low:     120 hours          (5 days)
*Comment:* Scientific judgement based upon estimated unacclimated aqueous aerobic biodegradation half-lives.

· **Air:**                   High:    22.1 hours
                                       Low:     2.21 hours
*Comment:* Scientific judgement based upon estimated photooxidation half-life in air.

· **Surface Water:**       High:    4320 hours         (6 months)
                                       Low:     120 hours          (5 days)
*Comment:* Scientific judgement based upon estimated unacclimated aqueous aerobic biodegradation half-lives.

· **Ground Water:**       High:    8640 hours         (12 months)
                                       Low:     240 hours          (10 days)
*Comment:* Scientific judgement based upon estimated unacclimated aqueous aerobic biodegradation half-lives.

**Aqueous Biodegradation (unacclimated):**

· **Aerobic half-life:**     High:    4320 hours         (6 months)
                                       Low:     120 hours          (5 days)
*Comment:* Based upon aerobic lake die-away data (Yordy, JR and Alexander, M (1980)).

· **Anaerobic half-life:**   High:   17280 hours       (24 months)
                                       Low:     480 hours          (20 days)
*Comment:* Scientific judgement based upon estimated unacclimated aqueous aerobic biodegradation half-lives.

· **Removal/secondary treatment:**     High:    No data
                                           Low:
*Comment:*

**Photolysis:**

· **Atmos photol half-life:**     High:    No data
                                           Low:
*Comment:*

**· Max light absorption (nm):**         lambda max = 234, 345 nm (water)
*Comment:* IARC (1978).

**· Aq photol half-life:**              High:      No data
                                        Low:

*Comment:*

**Photooxidation half-life:**
**· Water:**                            High:      No data
                                        Low:
*Comment:*

**· Air:**                              High:      22.1 hours
                                        Low:       2.21 hours
*Comment:* Scientific judgement based upon estimated rate constant for reaction with hydroxyl radical in air (Atkinson, R (1987A)).

**Reduction half-life:**                High:      No data
                                        Low:
*Comment:*

**Hydrolysis:**
**· First-order hydr half-life:**       Not expected to be significant
*Comment:* IARC (1978).

**· Acid rate const $(M(H+)-hr)^{-1}$:**   No data
*Comment:*

**· Base rate const $(M(OH-)-hr)^{-1}$:**  No data
*Comment:*

# Propane sultone

<u>CAS Registry Number:</u> 1120-71-4

<u>Structure:</u>

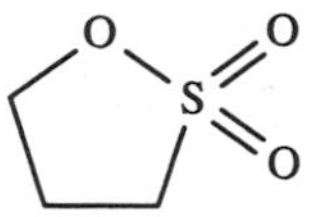

<u>Half-lives:</u>

**· Soil:** High: 672 hours (4 weeks)
Low: 8.5 hours

*Comment:* Scientific judgement based upon neutral hydrolysis rate constant in water at 25°C (low $t_{1/2}$) (Ellington, JJ et al. (1987)) and estimated unacclimated aerobic aqueous biodegradation half-life (high $t_{1/2}$).

**· Air:** High: 39.7 hours (1.65 days)
Low: 3.97 hours

*Comment:* Scientific judgment based upon estimated photooxidation half-life in air.

**· Surface Water:** High: 672 hours (4 weeks)
Low: 8.5 hours

*Comment:* Scientific judgement based upon neutral hydrolysis rate constant in water at 25°C (low $t_{1/2}$) and estimated unacclimated aerobic aqueous biodegradation half-life (high $t_{1/2}$).

**· Ground Water:** High: 1344 hours (8 weeks)
Low: 8.5 hours

*Comment:* Scientific judgement based upon neutral hydrolysis rate constant in water at 25°C (low $t_{1/2}$) and estimated unacclimated aerobic aqueous biodegradation half-life (high $t_{1/2}$).

<u>Aqueous Biodegradation (unacclimated):</u>

**· Aerobic half-life:** High: 672 hours (4 weeks)
Low: 168 hours (7 days)

*Comment:* Scientific judgement.

**· Anaerobic half-life:** High: 2688 hours (16 weeks)
Low: 672 hours (28 days)

*Comment:* Scientific judgement based upon estimated unacclimated aerobic aqueous biodegradation half-life.

**· Removal/secondary treatment:** High: No data
Low:

*Comment:*

**Photolysis:**
   **· Atmos photol half-life:**           High:
                                           Low:
      *Comment:*

   **· Max light absorption (nm):**        No data
      *Comment:*

   **· Aq photol half-life:**              High:
                                           Low:

      *Comment:*

**Photooxidation half-life:**
   **· Water:**                            High:      No data
                                           Low:
      *Comment:*

   **· Air:**                              High:      39.7 hours          (1.65 days)
                                           Low:       3.97 hours
   *Comment:* Scientific judgement based upon estimated rate constant for reaction with hydroxyl radical in air (Atkinson, R (1987A)).

**Reduction half-life:**                   High:      No data
                                           Low:
      *Comment:*

**Hydrolysis:**
   **· First-order hydr half-life:**       8.5 hours ($t_{1/2}$ at pH 7)
   *Comment:* Based upon neutral hydrolysis rate constant at 25°C (Ellington, JJ et al. (1987)).

   **· Acid rate const (M(H+)-hr)$^{-1}$:**      No data
   *Comment:*

   **· Base rate const (M(OH-)-hr)$^{-1}$:**      No data
   *Comment:*

# Aflatoxin B1

**CAS Registry Number:** 1162-65-8

**Structure:**

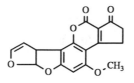

**Half-lives:**

**· Soil:**
      High: 672 hours      (4 weeks)
      Low: 168 hours      (7 days)
*Comment:* Scientific judgement based upon estimated unacclimated aqueous aerobic biodegradation half-life.

**· Air:**
      High: 0.935 hours
      Low: 0.109 hours
*Comment:* Based upon photooxidation half-life in air.

**· Surface Water:**
      High: 672 hours      (4 weeks)
      Low: 168 hours      (7 days)
*Comment:* Scientific judgement based upon estimated unacclimated aqueous aerobic biodegradation half-life.

**· Ground Water:**
      High: 1344 hours      (8 weeks)
      Low: 336 hours      (14 days)
*Comment:* Scientific judgement based upon estimated unacclimated aqueous aerobic biodegradation half-life.

**Aqueous Biodegradation (unacclimated):**

**· Aerobic half-life:**
      High: 672 hours      (4 weeks)
      Low: 168 hours      (7 days)
*Comment:* Scientific judgement based upon aqueous aerobic biodegradation screening test data (Price, KS et al. (1974)).

**· Anaerobic half-life:**
      High: 2688 hours      (16 weeks)
      Low: 672 hours      (28 days)
*Comment:* Scientific judgement based upon estimated unacclimated aqueous aerobic biodegradation half-life.

**· Removal/secondary treatment:**
      High:      No data
      Low:

*Comment:*

## Photolysis:
  · **Atmos photol half-life:**          High:
                                         Low:
  *Comment:*

  · **Max light absorption (nm):**    No data
  *Comment:*

  · **Aq photol half-life:**         High:
                                         Low:
  *Comment:*

## Photooxidation half-life:
  · **Water:**                  High:     No data
                                         Low:
  *Comment:*

  · **Air:**                   High:    0.935 hours
                                         Low:    0.109 hours
  *Comment:* Scientific judgement based upon estimated rate constants for reaction with hydroxyl radical (Atkinson, R (1987A)) and ozone (Atkinson, R and Carter, WPL (1984)) in air.

## Reduction half-life:               High:    No data
                                         Low:
  *Comment:*

## Hydrolysis:
  · **First-order hydr half-life:**
  *Comment:*
  · **Acid rate const (M(H+)-hr)$^{-1}$:**    No data
  *Comment:*

  · **Base rate const (M(OH-)-hr)$^{-1}$:**
  *Comment:*

# Decabromophenyl ether

**CAS Registry Number:** 1163-19-5

**Half-lives:**

· **Soil:**                              High:    8760 hours          (1 year)
                                         Low:     4320 hours          (6 months)
*Comment:* Scientific judgement based upon unacclimated aqueous aerobic biodegradation half-life.

· **Air:**                               High:    3686 hours          (154 days)
                                         Low:      369 hours          (15.4 days)
*Comment:* Scientific judgement based upon estimated photooxidation half-life in air.

· **Surface Water:**                     High:    8760 hours          (1 year)
                                         Low:     4320 hours          (6 months)
*Comment:* Scientific judgement based upon unacclimated aqueous aerobic biodegradation half-life.

· **Ground Water:**                      High:   17520 hours          (2 years)
                                         Low:     8640 hours          (12 months)
*Comment:* Scientific judgement based upon unacclimated aqueous aerobic biodegradation half-life.

**Aqueous Biodegradation (unacclimated):**

· **Aerobic half-life:**                 High:    8760 hours          (1 year)
                                         Low:     4320 hours          (6 months)
*Comment:* Scientific judgement based upon aqueous aerobic biodegradation screening test data (Sasaki, S (1978)).

· **Anaerobic half-life:**               High:   35040 hours          (4 years)
                                         Low:    17280 hours          (24 months)
*Comment:* Scientific judgement based upon unacclimated aqueous aerobic biodegradation half-life.

· **Removal/secondary treatment:**       High:      No data
                                         Low:
*Comment:*

**Photolysis:**

· **Atmos photol half-life:**            High:      No data
                                         Low:
*Comment:*

639

**· Max light absorption (nm):**     No data
*Comment:*

**· Aq photol half-life:**           High:     No data
                                     Low:
*Comment:*

**Photooxidation half-life:**
  **· Water:**                       High:     No data
                                     Low:
*Comment:*

  **· Air:**                         High:   3686 hours        (154 days)
                                     Low:     369 hours        (15.4 days)
*Comment:* Scientific judgement based upon estimated rate constant for reaction with hydroxyl radical in air (Atkinson, R (1987A)).

**Reduction half-life:**             High:     No data
                                     Low:
*Comment:*

**Hydrolysis:**
  **· First-order hydr half-life:**
*Comment:*

  **· Acid rate const (M(H+)-hr)$^{-1}$:**     No hydrolyzable groups
*Comment:*

  **· Base rate const (M(OH-)-hr)$^{-1}$:**
*Comment:*

# Cresol(s)

**CAS Registry Number:** 1319-77-3

**Half-lives:**

· **Soil:**                                      High:       696 hours                (29 days)
                                                 Low:            1 hour
*Comment:* Scientific judgement based upon estimated unacclimated aqueous aerobic biodegradation half-life.

· **Air:**                                       High:         16 hours
                                                 Low:          1.1 hours
*Comment:* Based upon photooxidation half-life in air.

· **Surface Water:**                             High:       696 hours                (29 days)
                                                 Low:            1 hour
*Comment:* Scientific judgement based upon estimated unacclimated aqueous aerobic biodegradation half-life.

· **Ground Water:**                              High:      1176 hours                (49 days)
                                                 Low:            2 hours
*Comment:* Scientific judgement based upon estimated unacclimated aqueous aerobic biodegradation half-life (low $t_{1/2}$) and aqueous anaerobic half-life (high $t_{1/2}$).

**Aqueous Biodegradation (unacclimated):**

· **Aerobic half-life:**                         High:       696 hours                (29 days)
                                                 Low:            1 hour
*Comment:* Scientific judgement based upon unacclimated marine water grab sample data for meta and para isomers (high $t_{1/2}$: Pfaender, FK and Bartholomew, GW (1982), low $t_{1/2}$: Van Veld, PA and Spain, JC (1983)).

· **Anaerobic half-life:**                       High:      1176 hours                (49 days)
                                                 Low:        240 hours                (10 days)
*Comment:* Scientific judgement based upon anaerobic screening test data for meta and para isomers (low $t_{1/2}$: Boyd, SA et al. (1983), high $t_{1/2}$: Shelton, DR and Tiedje, JM (1981)).

· **Removal/secondary treatment:**               High:               99%
                                                 Low:
*Comment:* Based upon % of p-cresol degraded under aerobic continuous flow conditions (Chudoba, J et al. (1968)).

**Photolysis:**

· **Atmos photol half-life:**                    High:       No data

Low:

*Comment:*

**· Max light absorption (nm):**          lambda max ortho: 282.0, 272.5, 238.0,
214.0; meta: 273.0; para: 279.0
*Comment:* Ortho and meta isomers absorb very little UV light at wavelengths greater than 290
nm in methanol (Sadtler UV No. 15, 259 and 622).

**· Aq photol half-life:**          High:     No data
                                    Low:

*Comment:*

**Photooxidation half-life:**
**· Water:**          High:     No data
                      Low:

*Comment:*

**· Air:**          High:     16 hours
                    Low:     1.1 hours
*Comment:* Based upon measured rate data for the vapor phase reaction of the ortho and meta
isomers with hydroxyl radicals in air (Atkinson, R (1985)).

**Reduction half-life:**          High:     No data
                                  Low:

*Comment:*

**Hydrolysis:**
**· First-order hydr half-life:**          No hydrolyzable groups
*Comment:* Rate constant at neutral pH is zero for ortho, meta and para isomers (Kollig, HP et al.
(1987A)).

**· Acid rate const (M(H+)-hr)$^{-1}$:**          0.0
*Comment:* Based upon measured rate data for ortho, meta and para isomers at acid pH (Kollig,
HP et al. (1987A)).

**· Base rate const (M(OH-)-hr)$^{-1}$:**          0.0
*Comment:* Based upon measured rate data for ortho, meta and para isomers at basic pH (Kollig,
HP et al. (1987A)).

# Xylenes

**CAS Registry Number:** 1330-20-7

**Half-lives:**
  · **Soil:**                                High:    672 hours           (4 weeks)
                                            Low:     168 hours           (1 week)
*Comment:* Scientific judgement based upon estimated aqueous aerobic biodegradation half-life.

  · **Air:**                                  High:     44 hours           (1.8 days)
                                            Low:     2.6 hours
*Comment:* Based upon photooxidation half-life in air.

  · **Surface Water:**                  High:    672 hours           (4 weeks)
                                            Low:     168 hours           (1 week)
*Comment:* Scientific judgement based upon estimated aqueous aerobic biodegradation half-life.

  · **Ground Water:**                 High:    8640 hours         (12 months)
                                            Low:     336 hours           (2 weeks)
*Comment:* Scientific judgement based upon estimated aqueous aerobic and anaerobic biodegradation half-lives.

**Aqueous Biodegradation (unacclimated):**
  · **Aerobic half-life:**              High:    672 hours           (4 weeks)
                                            Low:     168 hours           (1 week)
*Comment:* Scientific judgement based upon soil column study simulating an aerobic river/ground water infiltration system (high $t_{1/2}$: Kuhn, EP et al. (1985)) and aqueous screening test data (Bridie, AL et al. (1979)) for ortho, meta and para isomers.

  · **Anaerobic half-life:**           High:    8640 hours         (12 months)
                                            Low:     4320 hours         (6 months)
*Comment:* Scientific judgement based upon acclimated grab sample data for anaerobic soil from a ground water aquifer receiving landfill leachate (Wilson, BH et al. (1986)) and a soil column study simulating an anaerobic river/ground water infiltration system (Kuhn, EP et al. (1985)) for the ortho, meta and para isomers.

  · **Removal/secondary treatment:**     High:    No data
                                            Low:
*Comment:*

**Photolysis:**
  · **Atmos photol half-life:**          High:    No data
                                            Low:

*Comment:*

· **Max light absorption (nm):**          lambda max ortho: 269.5, 262; meta: 277, 268, 265; para: 274.5, 211.5

*Comment:*  Orto isomer does not absorb UV above 310 nm in methanol (Sadtler UV No. 7). Meta and para isomers do not absorb UV above 290 nm in cyclohexane (Sadtler UV No. 317 and 609).

· **Aq photol half-life:**          High:          No data
                                     Low:

*Comment:*

## Photooxidation half-life:
· **Water:**          High: $2.7 \times 10^8$ hours          (31397 years)
                      Low: $3.9 \times 10^5$ hours          (43 years)

*Comment:*  Scientific judgement based upon estimated rate data for o-xylene with alkylperoxyl radicals in aqueous solution (Hendry, DG et al. (1974)).

· **Air:**          High:          44 hours          (1.8 days)
                    Low:          2.6 hours

*Comment:*  Scientific judgement based upon measured rate data for vapor phase reaction of the ortho, meta, and para isomers with hydroxyl radicals in air (Atkinson, R (1985)).

## Reduction half-life:
                    High:          No data
                    Low:

*Comment:*

## Hydrolysis:
· **First-order hydr half-life:**          No hydrolyzable groups
*Comment:*

· **Acid rate const $(M(H+)-hr)^{-1}$:**
*Comment:*

· **Base rate const $(M(OH-)-hr)^{-1}$:**
*Comment:*

# Hexachloronaphthalene

**CAS Registry Number:** 1335-87-1

## Half-lives:
· **Soil:** High: 8760 hours (1 year)

Low: 4320 hours (6 months)

*Comment:* Scientific judgement based upon estimated unacclimated aqueous aerobic biodegradation half-life.

· **Air:** High: 3270 hours (136 days)

Low: 327 hours (13.6 days)

*Comment:* Scientific judgement based upon estimated photooxidation half-life in air.

· **Surface Water:** High: 8760 hours (1 year)

Low: 4320 hours (6 months)

*Comment:* Scientific judgement based upon estimated unacclimated aqueous aerobic biodegradation half-life.

· **Ground Water:** High: 17520 hours (2 years)

Low: 8640 hours (12 months)

*Comment:* Scientific judgement based upon estimated unacclimated aqueous aerobic biodegradation half-life.

## Aqueous Biodegradation (unacclimated):
· **Aerobic half-life:** High: 8760 hours (1 year)

Low: 4320 hours (6 months)

*Comment:* Scientific judgement based upon unacclimated aerobic soil grab sample data for pentachlorobenzene (Beck, J and Hansen, KE (1974)).

· **Anaerobic half-life:** High: 35040 hours (4 years)

Low: 17280 hours (24 months)

*Comment:* Scientific judgement based upon estimated unacclimated aqueous aerobic biodegradation half-life.

· **Removal/secondary treatment:** High: No data

Low:

*Comment:*

## Photolysis:
· **Atmos photol half-life:** High:

Low:

*Comment:*

· **Max light absorption (nm):**        No data
*Comment:*

· **Aq photol half-life:**        High:
                                    Low:

*Comment:*

**Photooxidation half-life:**
  · **Water:**        High:      No data
                             Low:

*Comment:*

  · **Air:**        High:    3270 hours        (136 days)
                     Low:     327 hours         (13.6 days)

*Comment:* Scientific judgement based upon estimated rate constant for reaction with hydroxyl radical in air (Atkinson, R (1987A)).

**Reduction half-life:**        High:      No data
                                  Low:

*Comment:*

**Hydrolysis:**
  · **First-order hydr half-life:**        No data
  *Comment:*

  · **Acid rate const $(M(H+)\text{-}hr)^{-1}$:**
  *Comment:*

  · **Base rate const $(M(OH\text{-})\text{-}hr)^{-1}$:**
  *Comment:*

# Methyl ethyl ketone peroxide

**CAS Registry Number:** 1338-23-4

**Half-lives:**

  · **Soil:**                                      High:    672 hours        (4 weeks)

                                                  Low:     168 hours        (7 days)

*Comment:* Scientific judgement based upon unacclimated aqueous aerobic biodegradation half-life.

  · **Air:**                                        High:    9.13 hours

                                                  Low:     0.913 hours

*Comment:* Scientific judgement based upon estimated photooxidation half-life in air.

  · **Surface Water:**                      High:    672 hours        (4 weeks)

                                                  Low:     168 hours        (7 days)

*Comment:* Scientific judgement based upon unacclimated aqueous aerobic biodegradation half-life.

  · **Ground Water:**                      High:    1344 hours      (8 weeks)

                                                  Low:     336 hours       (14 days)

*Comment:* Scientific judgement based upon unacclimated aqueous aerobic biodegradation half-life.

**Aqueous Biodegradation (unacclimated):**

  · **Aerobic half-life:**                  High:    672 hours        (4 weeks)

                                                  Low:     168 hours        (7 days)

*Comment:* Scientific judgement.

  · **Anaerobic half-life:**               High:    2688 hours      (16 weeks)

                                                  Low:     672 hours       (28 days)

*Comment:* Scientific judgement based upon unacclimated aqueous aerobic biodegradation half-life.

  · **Removal/secondary treatment:**      High:    No data

                                                  Low:

*Comment:*

**Photolysis:**

  · **Atmos photol half-life:**         High:

                                                  Low:

*Comment:*

**· Max light absorption (nm):**      No data
*Comment:*

**· Aq photol half-life:**      High:
                                       Low:

*Comment:*

## Photooxidation half-life:
**· Water:**      High:      No data
                                       Low:

*Comment:*

**· Air:**      High:      9.13 hours
                                       Low:      0.913 hours
*Comment:* Scientific judgement based upon estimated rate constant for reaction with hydroxyl radical in air (Atkinson, R (1987A)).

## Reduction half-life:
                                       High:      No data
                                       Low:
*Comment:*

## Hydrolysis:
**· First-order hydr half-life:**      No data
*Comment:*

**· Acid rate const $(M(H+)-hr)^{-1}$:**
*Comment:*

**· Base rate const $(M(OH-)-hr)^{-1}$:**
*Comment:*

# Diepoxybutane

**CAS Registry Number:** 1464-53-5

**Half-lives:**
  · **Soil:**                              High:     138 hours        (5.75 days)
                                       Low:      92 hours        (3.83 days)
    *Comment:* Scientific judgement based upon hydrolysis half-life in water.

  · **Air:**                                High:     750 hours        (31.3 days)
                                       Low:     75.0 hours     (3.13 days)
    *Comment:* Scientific judgement based upon estimated photooxidation half-life in air.

  · **Surface Water:**                High:     138 hours        (5.75 days)
                                       Low:      92 hours        (3.83 days)
    *Comment:* Scientific judgement based upon hydrolysis half-life in water.

  · **Ground Water:**               High:     138 hours        (5.75 days)
                                       Low:      92 hours        (3.83 days)
    *Comment:* Scientific judgement based upon hydrolysis half-life in water.

**Aqueous Biodegradation (unacclimated):**
  · **Aerobic half-life:**            High:     672 hours        (4 weeks)
                                       Low:     168 hours        (7 days)
    *Comment:* Scientific judgement.

  · **Anaerobic half-life:**         High:    2688 hours     (16 weeks)
                                       Low:     672 hours        (28 days)
    *Comment:* Scientific judgement based upon estimated aerobic biodegradation half-life.

  · **Removal/secondary treatment:**    High:     No data
                                       Low:
    *Comment:*

**Photolysis:**
  · **Atmos photol half-life:**        High:
                                       Low:
    *Comment:*

  · **Max light absorption (nm):**    No data
    *Comment:*

  · **Aq photol half-life:**          High:

Comment:                                          Low:

## Photooxidation half-life:
  · **Water:**                                    High:     No data
                                                  Low:
Comment:

  · **Air:**                                      High:     750 hours        (31.3 days)
                                                  Low:      75.0 hours       (3.13 days)
Comment:  Scientific judgement based upon estimated rate constant for reaction with hydroxyl radical in air (Atkinson, R (1987A)).

## Reduction half-life:                           High:     No data
                                                  Low:
Comment:

## Hydrolysis:
  · **First-order hydr half-life:**              High:     138 hours        (5.75 days)
                                                  Low:      92 hours         (3.83 days)
Comment:  ($t_{1/2}$ at pH 5-9) Based upon measured neutral hydrolysis rate constants in water (Bogyo, DA et al. (1980)).

  · **Acid rate const $(M(H+)-hr)^{-1}$:**        No data
Comment:

  · **Base rate const $(M(OH-)-hr)^{-1}$:**       No data
Comment:

# N,N'-Diethyl hydrazine

**CAS Registry Number:** 1615-80-1

**Half-lives:**

· **Soil:**　　　　　　　　　　　　High:　672 hours　　　　(4 weeks)
　　　　　　　　　　　　　　　　　Low:　168 hours　　　　(1 week)
*Comment:* Scientific judgement based upon estimated aqueous aerobic biodegradation half-life.

· **Air:**　　　　　　　　　　　　High:　4.7 hours
　　　　　　　　　　　　　　　　　Low:　0.5 hours　　　　(28 minutes)
*Comment:* Scientific judgement based upon estimated photooxidation half-life in air.

· **Surface Water:**　　　　　　　High:　672 hours　　　　(4 weeks)
　　　　　　　　　　　　　　　　　Low:　168 hours　　　　(1 week)
*Comment:* Scientific judgement based upon estimated aqueous aerobic biodegradation half-life.

· **Ground Water:**　　　　　　　High:　8640 hours　　　(8 weeks)
　　　　　　　　　　　　　　　　　Low:　336 hours　　　　(2 weeks)
*Comment:* Scientific judgement based upon estimated aqueous aerobic biodegradation half-life.

**Aqueous Biodegradation (unacclimated):**

· **Aerobic half-life:**　　　　　High:　672 hours　　　　(4 weeks)
　　　　　　　　　　　　　　　　　Low:　168 hours　　　　(1 week)
*Comment:* Scientific judgement based upon freshwater grab sample data for 1,1-dimethyl hydrazine (Braun, BA and Zirrolli, JA (1983).

· **Anaerobic half-life:**　　　　High:　2688 hours　　　(16 weeks)
　　　　　　　　　　　　　　　　　Low:　672 hours　　　　(4 weeks)
*Comment:* Scientific judgement based upon estimated aqueous aerobic biodegradation half-life.

· **Removal/secondary treatment:**　High:　No data
　　　　　　　　　　　　　　　　　Low:
*Comment:*

**Photolysis:**

· **Atmos photol half-life:**　　High:
　　　　　　　　　　　　　　　　　Low:
*Comment:*

· **Max light absorption (nm):**　No data
*Comment:*

**· Aq photol half-life:**          High:
                                         Low:

*Comment:*

## Photooxidation half-life:
**· Water:**                  High:      No data
                                         Low:

*Comment:*

**· Air:**                     High:      4.7 hours
                                         Low:       0.5 hours        (28 minutes)
*Comment:* Scientific judgement based upon estimated rate data for hydroxyl radicals in air (Atkinson, R (1987A)).

## Reduction half-life:          High:      No data
                                         Low:

*Comment:*

## Hydrolysis:
**· First-order hydr half-life:**      No data
*Comment:*

**· Acid rate const $(M(H+)-hr)^{-1}$:**
*Comment:*

**· Base rate const $(M(OH-)-hr)^{-1}$:**
*Comment:*

# Methyl t-butyl ether

**CAS Registry Number:** 1634-04-4

**Half-lives:**
  · **Soil:**                                   High:     4320 hours          (6 months)
                                                Low:      672 hours           (4 weeks)
  *Comment:* Scientific judgement based upon estimated aerobic biodegradation half-life.

  · **Air:**                                    High:     265 hours           (11 days)
                                                Low:      20.7 hours
  *Comment:* Based upon measured photooxidation rate constants in air (low $t_{1/2}$: Wallington, TJ et al. (1988); high $t_{1/2}$: Atkinson, R (1985)).

  · **Surface Water:**                          High:     4320 hours          (6 months)
                                                Low:      672 hours           (4 weeks)
  *Comment:* Scientific judgement based upon estimated aerobic biodegradation half-life.

  · **Ground Water:**                           High:     8640 hours          (12 months)
                                                Low:      1344 hours          (8 weeks)
  *Comment:* Scientific judgement based upon estimated aerobic biodegradation half-life.

**Aqueous Biodegradation (unacclimated):**
  · **Aerobic half-life:**                      High:     4320 hours          (6 months)
                                                Low:      672 hours           (4 weeks)
  *Comment:* Scientific judgement based upon unacclimated aerobic aqueous screening test data (Fujiwara, Y et al. (1984)).

  · **Anaerobic half-life:**                    High:     17280 hours         (24 months)
                                                Low:      2688 hours          (16 weeks)
  *Comment:* Scientific judgement based upon estimated aerobic biodegradation half-life.

  · **Removal/secondary treatment:**            High:         85%
                                                Low:
  *Comment:* Removal percentage based upon a conventional activated sludge treatment system data (Vanluin, AB and Teurlinckx, LVM (1987)).

**Photolysis:**
  · **Atmos photol half-life:**                 High:
                                                Low:
  *Comment:*

  · **Max light absorption (nm):**              No data

*Comment:*

· **Aq photol half-life:**  High:
  Low:
*Comment:*

**Photooxidation half-life:**
  · **Water:**  High:  No data
    Low:
*Comment:*

  · **Air:**  High:  265 hours  (11 days)
    Low:  20.7 hours
*Comment:* Based upon measured rate constants for reaction with hydroxyl radicals in air (low $t_{1/2}$: Wallington, TJ et al. (1988); high $t_{1/2}$: Atkinson, R (1985)).

**Reduction half-life:**  High:  No reducible groups
  Low:
*Comment:*

**Hydrolysis:**
  · **First-order hydr half-life:**
*Comment:*

  · **Acid rate const $(M(H+)-hr)^{-1}$:**  No hydrolyzable groups
*Comment:*

  · **Base rate const $(M(OH-)-hr)^{-1}$:**
*Comment:*

# Bromoxynil octanoate

**CAS Registry Number:** 1689-99-2

**Structure:**

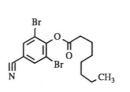

## Half-lives:

· **Soil:**
    High:    528 hours    (22 days)
    Low:    0.667 hours

*Comment:* Based upon aerobic soil die-away test data (high $t_{1/2}$) (Collins, RF (1973)) and estimated hydrolysis half-life (low $t_{1/2}$).

· **Air:**
    High:    63.7 hours    (2.65 days)
    Low:    6.37 hours

*Comment:* Scientific judgement based upon estimated photooxidation half-life in air.

· **Surface Water:**
    High:    528 hours    (22 days)
    Low:    0.667 hours

*Comment:* Scientific judgement based upon estimated unacclimated aerobic aqueous biodegradation half-life (high $t_{1/2}$) and estimated hydrolysis half-life (low $t_{1/2}$).

· **Ground Water:**
    High:    1056 hours    (44 days)
    Low:    0.667 hours

*Comment:* Scientific judgement based upon estimated unacclimated aerobic aqueous biodegradation half-life and estimated hydrolysis half-life (low $t_{1/2}$).

## Aqueous Biodegradation (unacclimated):

· **Aerobic half-life:**
    High:    528 hours    (22 days)
    Low:    168 hours    (7 days)

*Comment:* Based upon aerobic soil die-away test data (Collins, RF (1973)).

· **Anaerobic half-life:**
    High:    2112 hours    (88 days)
    Low:    672 hours    (28 days)

*Comment:* Scientific judgement based upon estimated unacclimated aerobic aqueous biodegradation half-life.

· **Removal/secondary treatment:**
    High:    No data
    Low:

*Comment:*

**Photolysis:**
- **Atmos photol half-life:**            High:
                                             Low:

    *Comment:*

- **Max light absorption (nm):**    No data
    *Comment:*

- **Aq photol half-life:**           High:
                                             Low:

    *Comment:*

**Photooxidation half-life:**
- **Water:**                      High:     No data
                                             Low:

    *Comment:*

- **Air:**                         High:    63.7 hours         (2.65 days)
                                           Low:     6.37 hours
    *Comment:* Scientific judgement based upon estimated rate constant for reaction with hydroxyl radical in air (Atkinson, R (1987A)).

**Reduction half-life:**              High:    No data
                                           Low:
    *Comment:*

**Hydrolysis:**
- **First-order hydr half-life:**          66.7 hours        (2.78 days)
    *Comment:* ($t_{1/2}$ at pH 7) Scientific judgement based upon estimated rate constant for base catalyzed hydrolysis (PCHYDRO; PCGEMS Graphical Exposure Modeling System USEPA (1987)).

- **Acid rate const $(M(H+)\text{-}hr)^{-1}$:**    No data
    *Comment:*

- **Base rate const $(M(OH\text{-})\text{-}hr)^{-1}$:**    $1.038 \times 10^5$
    *Comment:* ($t_{1/2}$ = 0.667 hours at pH 9) Scientific judgement based upon estimated rate constant for base catalyzed hydrolysis (PCHYDRO; PCGEMS Graphical Exposure Modeling System USEPA (1987)).

656

# 2,3,7,8-TCDD (Dioxin)

**CAS Registry Number:** 1746-01-6

## Half-lives:
· **Soil:**  High:  14160 hours  (1.62 years)
  Low:  10032 hours  (1.15 years)
*Comment:* Based upon soil die-away test data for two soils (Kearney, PC et al. (1971)).

· **Air:**  High:  223 hours  (9.3 days)
  Low:  22.3 hours
*Comment:* Scientific judgement based upon estimated photooxidation half-life in air.

· **Surface Water:**  High:  14160 hours  (1.62 years)
  Low:  10032 hours  (1.15 years)
*Comment:* Scientific judgement based upon estimated unacclimated aqueous aerobic biodegradation half-life.

· **Ground Water:**  High:  28320 hours  (3.23 years)
  Low:  20064 hours  (2.29 years)
*Comment:* Scientific judgement based upon estimated unacclimated aqueous aerobic biodegradation half-life.

## Aqueous Biodegradation (unacclimated):
· **Aerobic half-life:**  High:  14160 hours  (1.62 years)
  Low:  10032 hours  (1.15 years)
*Comment:* Based upon soil die-away test data (low $t_{1/2}$) (Kearney, PC et al. (1971)) and lake water and sediment die-away data (Ward, CT and Matsumura, F (1978)).

· **Anaerobic half-life:**  High:  56640 hours  (6.45 years)
  Low:  40128 hours  (4.58 years)
*Comment:* Scientific judgement based upon estimated unacclimated aqueous aerobic biodegradation half-life.

· **Removal/secondary treatment:**  High:  No data
  Low:
*Comment:*

## Photolysis:
· **Atmos photol half-life:**  High:  81 hours  (3.4 days)
  Low:  27 hours  (1.1 days)
*Comment:* Scientific judgement based upon measured rate constant for photolysis in a 90:10 mixture of distilled water and acetonitrile under summer sunlight (low $t_{1/2}$) (Dulin, D et al. (1986))

and adjusted for relative winter sunlight intensity (high $t_{1/2}$) (Lyman, WJ et al. (1982)). Scientific judgement based upon a very measured high soil adsorption coefficient for 2,3,7,8-TCDD (recommended $K_{oc}$ = 2.4X10$^7$ (Jackson, DR et al. (1986))) and its observed persistence in the environment indicates that the rate of 2,3,7,8-TCDD photolysis in the environment probably is not significant.

**· Max light absorption (nm):**　　　　No data
*Comment:*

**· Aq photol half-life:**　　　　　High:　　81 hours　　　　(3.4 days)
　　　　　　　　　　　　　　　　　　　Low:　　27 hours　　　　(1.1 days)
*Comment:* Scientific judgement based upon measured rate constant for photolysis in a 90:10 mixture of distilled water and acetonitrile under summer sunlight (low $t_{1/2}$) (Dulin, D et al. (1986)) and adjusted for relative winter sunlight intensity (high $t_{1/2}$) (Lyman, WJ et al. (1982)). Scientific judgement based upon a very measured high soil adsorption coefficient for 2,3,7,8-TCDD (recommended $K_{oc}$ = 2.4X10$^7$ (Jackson, DR et al. (1986))) and its observed persistence in the environment indicates that the rate of 2,3,7,8-TCDD photolysis in the environment probably is not significant.

**Photooxidation half-life:**
**· Water:**　　　　　　　　　　　　High:　　No data
　　　　　　　　　　　　　　　　　　　Low:

*Comment:*

**· Air:**　　　　　　　　　　　　　　High:　　223 hours　　　　(9.3 days)
　　　　　　　　　　　　　　　　　　　Low:　　22.3 hours
*Comment:* Scientific judgement based upon estimated rate constant for reaction with hydroxyl radicals (Atkinson, R (1987A)).

**Reduction half-life:**　　　　　　High:　　No data
　　　　　　　　　　　　　　　　　　　Low:
*Comment:*

**Hydrolysis:**
**· First-order hydr half-life:**
*Comment:*

**· Acid rate const (M(H+)-hr)$^{-1}$:**　　　　No hydrolyzable groups
*Comment:*

**· Base rate const (M(OH-)-hr)$^{-1}$:**
*Comment:*

# Dimethyl tetrachloroterephthalate

**CAS Registry Number:** 1861-32-1

**Half-lives:**

  · **Soil:**

| | High: | 2208 hours | (92 days) |
|---|---|---|---|
| | Low: | 432 hours | (18 days) |

*Comment:* Scientific judgement based upon unacclimated soil grab sample data (Choi, JS et al. (1988)).

  · **Air:**

| | High: | 1466 hours | (61.1 days) |
|---|---|---|---|
| | Low: | 147 hours | (6.1 days) |

*Comment:* Scientific judgement based upon estimated photooxidation half-life in air.

  · **Surface Water:**

| | High: | 2208 hours | (92 days) |
|---|---|---|---|
| | Low: | 432 hours | (18 days) |

*Comment:* Scientific judgement based upon estimated unacclimated aqueous aerobic biodegradation half-life.

  · **Ground Water:**

| | High: | 4416 hours | (184 days) |
|---|---|---|---|
| | Low: | 864 hours | (36 days) |

*Comment:* Scientific judgement based upon estimated unacclimated aqueous aerobic biodegradation half-life.

**Aqueous Biodegradation (unacclimated):**

  · **Aerobic half-life:**

| | High: | 2208 hours | (92 days) |
|---|---|---|---|
| | Low: | 432 hours | (18 days) |

*Comment:* Scientific judgement based upon unacclimated soil grab sample data (Choi, JS et al. (1988)).

  · **Anaerobic half-life:**

| | High: | 8832 hours | (368 days) |
|---|---|---|---|
| | Low: | 1728 hours | (72 days) |

*Comment:* Scientific judgement based upon estimated unacclimated aqueous aerobic biodegradation half-life.

  · **Removal/secondary treatment:**

| | High: | No data |
|---|---|---|
| | Low: | |

*Comment:*

**Photolysis:**

  · **Atmos photol half-life:**

| | High: | No data |
|---|---|---|
| | Low: | |

*Comment:*

· **Max light absorption (nm):**　　　　No data
*Comment:*

· **Aq photol half-life:**　　　　High:　　No data
　　　　　　　　　　　　　　　　　Low:

*Comment:*

**Photooxidation half-life:**
　· **Water:**　　　　　　　　　High:　　No data
　　　　　　　　　　　　　　　　　Low:

*Comment:*

　· **Air:**　　　　　　　　　　High:　　1466 hours　　　　(61.1 days)
　　　　　　　　　　　　　　　　　Low:　　147 hours　　　　(6.1 days)
*Comment:* Scientific judgement based upon an estimated rate constant for vapor phase reaction with hydroxyl radicals in air (Atkinson, R (1987A)).

**Reduction half-life:**　　　　　High:　　No data
　　　　　　　　　　　　　　　　　Low:

*Comment:*

**Hydrolysis:**
　· **First-order hydr half-life:**　　　No data
*Comment:*

　· **Acid rate const (M(H+)-hr)$^{-1}$:**
*Comment:*

　· **Base rate const (M(OH-)-hr)$^{-1}$:**
*Comment:*

# Benefin

**CAS Registry Number:** 1861-40-1

**Structure:**

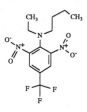

## Half-lives:

· **Soil:**　　　　　　　　　　High:　2880 hours　　　(120 days)

　　　　　　　　　　　　　　Low:　　504 hours　　　(21 days)

*Comment:* Based upon aerobic soil die-away test data for two different soils (low $t_{1/2}$: Zimdahl, RL and Gwynn, SM (1977); high $t_{1/2}$: Golab, T et al. (1970)).

· **Air:**　　　　　　　　　　High:　　7.82 hours

　　　　　　　　　　　　　　Low:　　0.782 hours

*Comment:* Scientific judgement based upon estimated photooxidation half-life in air.

· **Surface Water:**　　　　　High:　　864 hours　　　(36 days)

　　　　　　　　　　　　　　Low:　　288 hours　　　(12 days)

*Comment:* Scientific judgement based upon observed photolysis on soil TLC plates under summer sunlight (low $t_{1/2}$) (Helling, CS (1976)) adjusted for relative winter sunlight intensity (high $t_{1/2}$) (Lyman, WJ et al. (1982)).

· **Ground Water:**　　　　　High:　5760 hours　　　(240 days)

　　　　　　　　　　　　　　Low:　　144 hours　　　(6 days)

*Comment:* Scientific judgement based upon unacclimated aqueous aerobic (high $t_{1/2}$) and anaerobic (low $t_{1/2}$) biodegradation half-lives.

## Aqueous Biodegradation (unacclimated):

· **Aerobic half-life:**　　　High:　2880 hours　　　(120 days)

　　　　　　　　　　　　　　Low:　　504 hours　　　(21 days)

*Comment:* Based upon aerobic soil die-away test data for two different soils (low $t_{1/2}$: Zimdahl, RL and Gwynn, SM (1977); high $t_{1/2}$: Golab, T et al. (1970)).

· **Anaerobic half-life:**　　High:　　480 hours　　　(20 days)

　　　　　　　　　　　　　　Low:　　144 hours　　　(6 days)

*Comment:* Based upon anaerobic soil die-away test data (Golab, T et al. (1970)).

· **Removal/secondary treatment:**　　High:　　No data

661

Low:

*Comment:*

## Photolysis:
 · **Atmos photol half-life:**        High:    864 hours          (36 days)

                                      Low:    288 hours          (12 days)

*Comment:* Scientific judgement based upon observed photolysis on soil TLC plates under summer sunlight (low $t_{1/2}$) (Helling, CS (1976)) and adjusted for relative winter sunlight intensity (high $t_{1/2}$) (Lyman, WJ et al. (1982)).

 · **Max light absorption (nm):**        lambda max = 385, 395 nm (methanol)

*Comment:* Kennedy, JM and Talbert, RE (1977)).

 · **Aq photol half-life:**              High:    864 hours          (36 days)

                                      Low:    288 hours          (12 days)

*Comment:* Scientific judgement based upon observed photolysis on soil TLC plates under summer sunlight (low $t_{1/2}$) (Helling, CS (1976)) and adjusted for relative winter sunlight intensity (high $t_{1/2}$) (Lyman, WJ et al. (1982)).

## Photooxidation half-life:
 · **Water:**                           High:      No data

                                      Low:

*Comment:*

 · **Air:**                             High:    7.82 hours

                                      Low:    0.782 hours

*Comment:* Scientific judgement based upon estimated rate constant for reaction with hydroxyl radical in air (Atkinson, R (1987A)).

## Reduction half-life:                High:      No data

                                      Low:

*Comment:*

## Hydrolysis:
 · **First-order hydr half-life:**

*Comment:*

 · **Acid rate const (M(H+)-hr)$^{-1}$:**        No hydrolyzable groups

*Comment:*

 · **Base rate const (M(OH-)-hr)$^{-1}$:**

*Comment:*

# Octachloronaphthalene

CAS Registry Number: 2234-13-1

Half-lives:
  · Soil:                                     High:    8760 hours         (1 year)
                                          Low:    4320 hours       (6 months)
*Comment:* Scientific judgement based upon essentially no biodegradation observed for hexachlorobenzene in soil die-away tests (Griffin, RA and Chou, SJ (1981A)).

  · Air:                                     High:  16082 hours     (670 days)
                                          Low:    1608 hours       (67 days)
*Comment:* Scientific judgement based upon estimated photooxidation half-life in air.

  · Surface Water:                 High:    8760 hours         (1 year)
                                          Low:    4320 hours       (6 months)
*Comment:* Scientific judgement based upon essentially no biodegradation observed for hexachlorobenzene in soil die-away tests (Griffin, RA and Chou, SJ (1981A)).

  · Ground Water:               High:  17520 hours     (2 years)
                                          Low:    8640 hours     (12 months)
*Comment:* Scientific judgement based upon essentially no biodegradation observed for hexachlorobenzene in soil die-away tests (Griffin, RA and Chou, SJ (1981A)).

Aqueous Biodegradation (unacclimated):
  · Aerobic half-life:            High:    8760 hours         (1 year)
                                          Low:    4320 hours       (6 months)
*Comment:* Scientific judgement based upon essentially no biodegradation observed for hexachlorobenzene in soil die-away tests (Griffin, RA and Chou, SJ (1981A)).

  · Anaerobic half-life:        High:  35040 hours     (4 years)
                                        Low:    17280 hours     (24 months)
*Comment:* Scientific judgement based upon essentially no biodegradation observed for hexachlorobenzene in soil die-away tests (Griffin, RA and Chou, SJ (1981A)).

  · Removal/secondary treatment:    High:     No data
                                          Low:
*Comment:*

Photolysis:
  · Atmos photol half-life:       High:     No data
                                          Low:
*Comment:*

**· Max light absorption (nm):**      No data
*Comment:*

**· Aq photol half-life:**      High:      No data
                                          Low:
*Comment:*

**Photooxidation half-life:**
     **· Water:**      High:      No data
                                          Low:
*Comment:*

     **· Air:**      High:    16082 hours        (670 days)
                            Low:     1608 hours         (67 days)
*Comment:* Scientific judgement based upon estimated rate constant for reaction with hydroxyl radicals in air (Atkinson, R (1987A)).

**Reduction half-life:**      High: No reducible groups
                                       Low:
*Comment:*

**Hydrolysis:**
     **· First-order hydr half-life:**
*Comment:*

     **· Acid rate const (M(H+)-hr)$^{-1}$:**      No hydrolyzable groups
*Comment:*

     **· Base rate const (M(OH-)-hr)$^{-1}$:**
*Comment:*

# Diallate

**CAS Registry Number:** 2303-16-4

**Structure:**

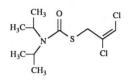

**Half-lives:**

· **Soil:** High: 2160 hours (3 months)
Low: 252 hours (0.35 months)
*Comment:* Based upon aerobic soil die-away test data (Anderson, JPE and Domsch, KH (1976); Smith, AE (1970)).

· **Air:** High: 5.8 hours
Low: 0.58 hours
*Comment:* Based upon estimated photooxidation half-life in air.

· **Surface Water:** High: 2160 hours (3 months)
Low: 252 hours (0.35 month)
*Comment:* Scientific judgement based upon aerobic soil die-away test data (Anderson, JPE and Domsch, KH (1976); Smith, AE (1970)).

· **Ground Water:** High: 4320 hours (6 months)
Low: 504 hours (0.70 month)
*Comment:* Scientific judgement based upon aerobic soil die-away test data (Anderson, JPE and Domsch, KH (1976); Smith, AE (1970)).

**Aqueous Biodegradation (unacclimated):**

· **Aerobic half-life:** High: 2160 hours (3 months)
Low: 252 hours (0.35 month)
*Comment:* Scientific judgement based upon aerobic soil die-away test data (Anderson, JPE and Domsch, KH (1976); Smith, AE (1970)).

· **Anaerobic half-life:** High: 8640 hours (12 months)
Low: 1008 hours (2.1 months)
*Comment:* Scientific judgement based upon aerobic soil die-away test data (Anderson, JPE and Domsch, KH (1976); Smith, AE (1970)).

· **Removal/secondary treatment:** High: No data
Low:

*Comment:*

## Photolysis:
· **Atmos photol half-life:**        High:   Not expected to directly photolyze
                                          Low:
*Comment:*

· **Max light absorption (nm):**
*Comment:*

· **Aq photol half-life:**           High:
                                          Low:
*Comment:*

## Photooxidation half-life:
· **Water:**                     High:   No data
                                          Low:
*Comment:*

· **Air:**                        High:   5.8 hours
                                        Low:    0.58 hours
*Comment:* Based upon estimated rate constant for reaction with hydroxyl radicals in air (Atkinson, R (1987A)).

## Reduction half-life:
                                       High:   Not expected to be significant
                                       Low:
*Comment:*

## Hydrolysis:
· **First-order hydr half-life:**      6.6 years                   ($t_{1/2}$ at pH 7)
*Comment:* Based upon measured first-order and base catalyzed hydrolysis rate constant (Ellington, JJ et al. (1987)).

· **Acid rate const $(M(H+)\text{-}hr)^{-1}$:**    No data        ($t_{1/2}$ = 6.6 years at pH 5)
*Comment:* Scientific judgement based upon measured first-order and base catalyzed hydrolysis rate constant (Ellington, JJ et al. (1987)).

· **Base rate const $(M(OH\text{-})\text{-}hr)^{-1}$:**    0.9 $M^{-1}$ $hr^{-1}$    ($t_{1/2}$ = 3.8 years at pH 9)
*Comment:* Based upon measured first-order and base catalyzed hydrolysis rate constant (Ellington, JJ et al. (1987)).

# N-Nitrosomethylvinylamine

CAS Registry Number: 4549-40-0

Half-lives:
· Soil:                         High:    4320 hours        (6 months)
                                Low:      672 hours        (4 weeks)
Comment: Based upon aerobic soil die-away data for N-nitrosodimethylamine and N-nitrosodipropylamine (high $t_{1/2}$: Tate, RL III and Alexander, M (1975); low $t_{1/2}$: Oliver, JE et al. (1979)).

· Air:                          High:    31.9 hours
                                Low:      3.39 hours
Comment: Scientific judgement based upon estimated photooxidation half-life in air.

· Surface Water:                High:    4320 hours        (6 months)
                                Low:      672 hours        (4 weeks)
Comment: Based upon aerobic soil die-away data for N-nitrosodimethylamine and N-nitrosodipropylamine (high $t_{1/2}$: Tate, RL III and Alexander, M (1975); low $t_{1/2}$: Oliver, JE et al. (1979)).

· Ground Water:                 High:    8640 hours        (12 months)
                                Low:     1344 hours        (8 weeks)
Comment: Based upon aerobic soil die-away data for N-nitrosodimethylamine and N-nitrosodipropylamine (high $t_{1/2}$: Tate, RL III and Alexander, M (1975); low $t_{1/2}$: Oliver, JE et al. (1979)).

Aqueous Biodegradation (unacclimated):
· Aerobic half-life:            High:    4320 hours        (6 months)
                                Low:      672 hours        (4 weeks)
Comment: Based upon aerobic soil die-away data for N-nitrosodimethylamine and N-nitrosodipropylamine (high $t_{1/2}$: Tate, RL III and Alexander, M (1975); low $t_{1/2}$: Oliver, JE et al. (1979)).

· Anaerobic half-life:          High:    17280 hours       (24 months)
                                Low:      2688 hours       (16 weeks)
Comment: Scientific judgement based upon estimated unacclimated aqueous aerobic biodegradation half-lives.

· Removal/secondary treatment:  High:    No data
                                Low:
Comment:

## Photolysis:

**· Atmos photol half-life:**         High:      No data

Low:

*Comment:*

**· Max light absorption (nm):**     lambda max = 200, 270, 376 nm

*Comment:* IARC (1978).

**· Aq photol half-life:**          High:      No data

Low:

*Comment:*

## Photooxidation half-life:

**· Water:**                   High:      No data

Low:

*Comment:*

**· Air:**                     High:     31.9 hours

Low:     3.39 hours

*Comment:* Scientific judgement based upon estimated rate constants for reaction with hydroxyl radical (Atkinson, R (1985)) and ozone (Atkinson, R and Carter, WPL (1984)).

## Reduction half-life:            High:      No data

Low:

*Comment:*

## Hydrolysis:

**· First-order hydr half-life:**      No data

*Comment:*

**· Acid rate const $(M(H+)-hr)^{-1}$:**      No data

*Comment:*

**· Base rate const $(M(OH-)-hr)^{-1}$:**      No data

*Comment:*

# Streptozotocin

**CAS Registry Number:** 18883-66-4

**Structure:**

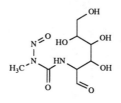

**Half-lives:**

  · **Soil:**

| | High: | 672 hours | (4 weeks) |
|---|---|---|---|
| | Low: | 168 hours | (7 days) |

*Comment:* Scientific judgement based upon unacclimated aqueous aerobic biodegradation half-life.

  · **Air:**

| | High: | 3.49 hours |
|---|---|---|
| | Low: | 0.349 hours |

*Comment:* Scientific judgement based upon estimated photooxidation half-life in air.

  · **Surface Water:**

| | High: | 672 hours | (4 weeks) |
|---|---|---|---|
| | Low: | 168 hours | (7 days) |

*Comment:* Scientific judgement based upon unacclimated aqueous aerobic biodegradation half-life.

  · **Ground Water:**

| | High: | 1344 hours | (8 weeks) |
|---|---|---|---|
| | Low: | 336 hours | (14 days) |

*Comment:* Scientific judgement based upon unacclimated aqueous aerobic biodegradation half-life.

**Aqueous Biodegradation (unacclimated):**

  · **Aerobic half-life:**

| | High: | 672 hours | (4 weeks) |
|---|---|---|---|
| | Low: | 168 hours | (7 days) |

*Comment:* Scientific judgement.

  · **Anaerobic half-life:**

| | High: | 2688 hours | (16 weeks) |
|---|---|---|---|
| | Low: | 672 hours | (28 days) |

*Comment:* Scientific judgement based upon unacclimated aqueous aerobic biodegradation half-life.

  · **Removal/secondary treatment:**

| | High: | No data |
|---|---|---|
| | Low: | |

669

*Comment:*

**Photolysis:**
　　**· Atmos photol half-life:**　　　　　High:
　　　　　　　　　　　　　　　　　　　　　　Low:
　　*Comment:*

　　**· Max light absorption (nm):**　　　No data
　　*Comment:*

　　**· Aq photol half-life:**　　　　　　High:
　　　　　　　　　　　　　　　　　　　　　　Low:
　　*Comment:*

**Photooxidation half-life:**
　　**· Water:**　　　　　　　　　　　　High:　　　No data
　　　　　　　　　　　　　　　　　　　　　　Low:
　　*Comment:*

　　**· Air:**　　　　　　　　　　　　　High:　　3.49 hours
　　　　　　　　　　　　　　　　　　　　　　Low:　　0.349 hours
　　*Comment:* Scientific judgement based upon estimated rate constant for reaction with hydroxyl radical in air (Atkinson, R (1987A)).

**Reduction half-life:**　　　　　　　　High:
　　　　　　　　　　　　　　　　　　　　　　Low:
　　*Comment:*

**Hydrolysis:**
　　**· First-order hydr half-life:**
　　*Comment:*

　　**· Acid rate const (M(H+)-hr)$^{-1}$:**　　No data
　　*Comment:*

　　**· Base rate const (M(OH-)-hr)$^{-1}$:**
　　*Comment:*

# Diaminotoluenes

CAS Registry Number: 25376-45-8

## Half-lives:
· Soil:                          High:    4320 hours        (6 months)
                                 Low:      672 hours        (4 weeks)
*Comment:* Scientific judgement based upon estimated unacclimated aqueous aerobic
biodegradation half-life.

· Air:                           High:     2.7 hours
                                 Low:      0.27 hours
*Comment:* Based upon estimated photooxidation half-life in air.

· Surface Water:                 High:    1740 hours        (72 days)
                                 Low:       31 hours        (1.3 days)
*Comment:* Based upon estimated photooxidation half-life in water.

· Ground Water:                  High:    8640 hours        (12 months)
                                 Low:     1344 hours        (8 weeks)
*Comment:* Scientific judgement based upon estimated unacclimated aqueous aerobic
biodegradation half-life.

## Aqueous Biodegradation (unacclimated):
· Aerobic half-life:             High:    4320 hours        (6 months)
                                 Low:      672 hours        (4 weeks)
*Comment:* Scientific judgement based upon unacclimated aerobic aqueous screening test data for
2,4-diaminotoluene which confirmed resistance to biodegradation (Sasaki, S (1978)).

· Anaerobic half-life:           High:   17,280 hours       (24 months)
                                 Low:     2688 hours        (16 weeks)
*Comment:* Scientific judgement based upon estimated unacclimated aerobic aqueous
biodegradation half-life.

· Removal/secondary treatment:   High:        34%
                                 Low:
*Comment:* Based upon activated sludge degradation results using a fill and draw method (Matsui,
S et al. (1975)).

## Photolysis:
· Atmos photol half-life:        High:
                                 Low:
*Comment:*

· **Max light absorption (nm):**   No data
*Comment:*

· **Aq photol half-life:**   High:
 Low:

*Comment:*

**Photooxidation half-life:**
 · **Water:**   High:   1740 hours   (72 days)
 Low:   31 hours   (1.3 days)
*Comment:* Scientific judgement based upon estimated half-life for reaction of aromatic amines with hydroxyl radicals in water (Mill, T (1989); Guesten, H et al. (1981)). It is assumed that diaminotoluenes react twice as fast as aniline.

· **Air:**   High:   2.7 hours
 Low:   0.27 hours
*Comment:* Based upon estimated rate constant for reaction with hydroxyl radicals in air (Atkinson, R (1987A)).

**Reduction half-life:**   High: Not expected to be significant
 Low:

*Comment:*

**Hydrolysis:**
 · **First-order hydr half-life:**
*Comment:*

· **Acid rate const (M(H+)-hr)$^{-1}$:**   No hydrolyzable groups
*Comment:*

· **Base rate const (M(OH-)-hr)$^{-1}$:**
*Comment:*

# Isopropalin

<u>CAS Registry Number:</u>  33820-53-0

<u>Structure:</u>

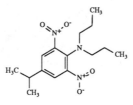

<u>Half-lives:</u>

· **Soil:**  High: 2520 hours (105 days)

Low: 408 hours (17 days)

*Comment:* Based upon aerobic soil die-away test data for one soil at 15°C (high $t_{1/2}$) and 30°C (low $t_{1/2}$) (Gingerich, LL and Zimdahl, RL (1976)).

· **Air:**  High: 7.43 hours

Low: 0.743 hours

*Comment:* Scientific judgement based upon estimated photooxidation half-life in air.

· **Surface Water:**  High: 864 hours (36 days)

Low: 288 hours (12 days)

*Comment:* Scientific judgement based upon observed photolysis on soil TLC plates under summer sunlight (low $t_{1/2}$) (Helling, CS (1976)) adjusted for relative winter sunlight intensity (high $t_{1/2}$) (Lyman, WJ et al. (1982)).

· **Ground Water:**  High: 5040 hours (210 days)

Low: 96 hours (4 days)

*Comment:* Scientific judgement based upon unacclimated aqueous aerobic (high $t_{1/2}$) and anaerobic (low $t_{1/2}$) biodegradation half-lives.

<u>Aqueous Biodegradation (unacclimated):</u>

· **Aerobic half-life:**  High: 2520 hours (105 days)

Low: 408 hours (17 days)

*Comment:* Based upon aerobic soil die-away test data for one soil at 15°C (high $t_{1/2}$) and 30°C (low $t_{1/2}$) (Gingerich, LL and Zimdahl, RL (1976)).

· **Anaerobic half-life:**  High: 360 hours (15 days)

Low: 96 hours (4 days)

*Comment:* Based upon data from an anaerobic soil die-away test which tested one soil (Gingerich, LL and Zimdahl, RL (1976)).

· **Removal/secondary treatment:**     High:     No data
                                      Low:
*Comment:*

## Photolysis:
· **Atmos photol half-life:**     High:     864 hours          (36 days)
                                 Low:     288 hours          (12 days)
*Comment:* Scientific judgement based upon observed photolysis on soil TLC plates under summer sunlight (low $t_{1/2}$) (Helling, CS (1976)) and adjusted for relative winter sunlight intensity (high $t_{1/2}$) (Lyman, WJ et al. (1982)).

· **Max light absorption (nm):**     lambda max = 390, 400 nm (methanol); approximately 320, 330, 360, 400 nm (acetonitrile)
*Comment:* Absorption in acetonitrile extends to >500 nm (methanol spectrum: Kennedy, JM and Talbert, RE (1977); acetonitrile spectrum: Draper, WM (1985)).

· **Aq photol half-life:**     High:     864 hours          (36 days)
                              Low:     288 hours          (12 days)
*Comment:* Scientific judgement based upon observed photolysis on soil TLC plates under summer sunlight (low $t_{1/2}$) (Helling, CS (1976)) and adjusted for relative winter sunlight intensity (high $t_{1/2}$) (Lyman, WJ et al. (1982)).

## Photooxidation half-life:

Water:     High:     No data
           Low:
*Comment:*

· **Air:**     High:     7.43 hours
              Low:     0.743 hours
*Comment:* Scientific judgement based upon estimated rate constant for reaction with hydroxyl radical in air (Atkinson, R (1987A)).

## Reduction half-life:     High:     No data
                          Low:
*Comment:*

## Hydrolysis:
· **First-order hydr half-life:**
*Comment:*

· **Acid rate const $(M(H+)-hr)^{-1}$:**     No hydrolyzable groups
*Comment:*

· **Base rate const (M(OH-)-hr)⁻¹:**
*Comment:*

# Propylene glycol, monoethyl ether

CAS Registry Number: 52125-53-8

Half-lives:
- Soil:                                High:     672 hours          (28 days)
                                       Low:      168 hours          (7 days)
Comment: Scientific judgement based upon estimated unacclimated aqueous aerobic biodegradation half-life.

- Air:                                 High:     31.9 hours
                                       Low:      3.19 hours
Comment: Scientific judgement based upon estimated photooxidation half-life in air.

- Surface Water:                       High:     672 hours          (28 days)
                                       Low:      168 hours          (7 days)
Comment: Scientific judgement based upon estimated unacclimated aqueous aerobic biodegradation half-life.

- Ground Water:                        High:     1344 hours         (8 weeks)
                                       Low:      336 hours          (14 days)
Comment: Scientific judgement based upon estimated unacclimated aqueous aerobic biodegradation half-life.

Aqueous Biodegradation (unacclimated):
- Aerobic half-life:                   High:     672 hours          (4 weeks)
                                       Low:      168 hours          (7 days)
Comment: Scientific judgement based upon unacclimated aqueous aerobic biodegradation screening test data for propylene glycol, monomethyl ether (Dow Chemical Company (1981)) and 2-ethoxyethanol (Bogan, RH and Sawyer, CN (1955)).

- Anaerobic half-life:                 High:     2688 hours         (16 weeks)
                                       Low:      672 hours          (28 days)
Comment: Scientific judgement based upon estimated unacclimated aqueous aerobic biodegradation half-life.

- Removal/secondary treatment:         High:     No data
                                       Low:
Comment:

Photolysis:
- Atmos photol half-life:              High:
                                       Low:

*Comment:*

**· Max light absorption (nm):**          No data
*Comment:*

**· Aq photol half-life:**          High:
                                    Low:
*Comment:*

**Photooxidation half-life:**
     **· Water:**          High:          No data
                           Low:
*Comment:*

     **· Air:**          High:          31.9 hours
                         Low:           3.19 hours
*Comment:* Scientific judgement based upon estimated rate constant for reaction with hydroxyl radical in air (Atkinson, R (1987A)).

**Reduction half-life:**          High:          No data
                                  Low:
*Comment:*

**Hydrolysis:**
     **· First-order hydr half-life:**
*Comment:*

     **· Acid rate const (M(H+)-hr)$^{-1}$:**          No hydrolyzable groups
*Comment:*

     **· Base rate const (M(OH-)-hr)$^{-1}$:**
*Comment:*

# Fluridone

<u>CAS Registry Number:</u>  59756-60-4

<u>Structure:</u>

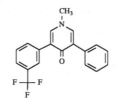

<u>Half-lives:</u>
- **· Soil:**   High:   4608 hours   (192 days)
  Low:   1056 hours   (44 days)

  *Comment:* Based upon soil die-away test data (low $t_{1/2}$) and field study soil persistence (high $t_{1/2}$) (Banks, PA et al. (1979)).

- **· Air:**   High:   3.20 hours
  Low:   0.359 hours

  *Comment:* Scientific judgement based upon estimated photooxidation half-life in air.

- **· Surface Water:**   High:   864 hours   (36 days)
  Low:   288 hours   (12 days)

  *Comment:* Scientific judgement based upon estimated photolysis half-life in water.

- **· Ground Water:**   High:   9216 hours   (1.05 years)
  Low:   2112 hours   (88 days)

  *Comment:* Scientific judgement based upon estimated unacclimated aqueous aerobic biodegradation half-life.

<u>Aqueous Biodegradation (unacclimated):</u>
- **· Aerobic half-life:**   High:   4608 hours   (192 days)
  Low:   1056 hours   (44 days)

  *Comment:* Based upon soil die-away test data (low $t_{1/2}$) and field study soil persistence (high $t_{1/2}$) (Banks, PA et al. (1979)).

- **· Anaerobic half-life:**   High:   18432 hours   (2.1 years)
  Low:   4224 hours   (176 days)

  *Comment:* Scientific judgement based upon estimated unacclimated aqueous aerobic biodegradation half-life.

- **· Removal/secondary treatment:**   High:   No data
  Low:

## Photolysis:
· **Atmos photol half-life:**        High:     864 hours       (36 days)

                                               Low:      288 hours       (12 days)

*Comment:* Scientific judgement based upon measured rate constant for summer sunlight photolysis in distilled water (low $t_{1/2}$) (Saunders, DG and Mosier, JW (1983)) and adjusted for relative winter sunlight intensity (high $t_{1/2}$) (Lyman, WJ et al. (1982)).

· **Max light absorption (nm):**       No data
*Comment:*

· **Aq photol half-life:**         High:     864 hours       (36 days)

                                               Low:      288 hours       (12 days)

*Comment:* Scientific judgement based upon measured rate constant for summer sunlight photolysis in distilled water (Saunders, DG and Mosier, JW (1983)) (low $t_{1/2}$) and adjusted for relative winter sunlight intensity (high $t_{1/2}$) (Lyman, WJ et al. (1982)).

## Photooxidation half-life:
· **Water:**                         High:     No data

                                               Low:

*Comment:*

· **Air:**                            High:     3.20 hours

                                               Low:      0.359 hours

*Comment:* Scientific judgement based upon estimated rate constant for reaction with hydroxyl radicals (Atkinson, R (1987A)) and ozone (Atkinson, R and Carter, WPL (1984)) in air.

## Reduction half-life:               High:     No data

                                               Low:

*Comment:*

## Hydrolysis:
· **First-order hydr half-life:**       No data
*Comment:*

· **Acid rate const $(M(H+)-hr)^{-1}$:**
*Comment:*

· **Base rate const $(M(OH-)-hr)^{-1}$:**
*Comment:*

# REFERENCES

Abrams, E.F., Derkics, C.V., Fong, D.K., et al. Identification of organic compounds in effluents from industrial sources. USEPA-560/3-75-002. Springfield, VA: Versar, Inc. 1975.

Adhya, T.K., Sudhakar-Barik, R., and Sethunathan, N. Fate of fenitrothion, methyl parathion and parathion in anoxic sulfur-containing soil systems. Pest. Biochem. Phys. 16: 14. 1981A.

Agnihotri, V.P. Persistence of captan and its effects on microflora, respiration, and nitrification of a forest nursery soil. Can. J. Microbiol. 17: 377-83. 1970.

Alexander, M. and Lustigman, B.K. Effect of chemical structure on microbial degradation of substituted benzenes. J. Agric. Food Chem. 14: 410-3. 1966.

Alexander, M. and Aleem, M.I.H. Effect of chemical structure on microbial decomposition of aromatic herbicides. J. Agric. Food Chem. 9: 44-7. 1961.

Alvarez, G.H., Page, S.W., and Ku, Y. Biodegradation of 14C-tris(2,3-dibromopropyl)phosphate in a laboratory activated sludge system. Bull. Environ. Contam. Toxicol. 28: 85-90. 1982.

Aly, D.M. and El-Dib, M.A. Studies on the persistence of some carbamate insecticides in the aquatic environment - I. Hydrolysis of Sevin, Baygon, Pyrolan, and Dimetilan in waters. Water Res. 5: 1191-205. 1971A.

Aly, O.M. and Faust, S.D. Studies on the fate of 2,4-D and ester derivatives in natural surface waters. J. Agric. Food Chem. 12: 451-6. 1964.

Anbar, M. and Neta, P. A compilation of specific bimolecular rate and hydroxyl radical with inorganic and organic compounds in aqueous solution. Int. J. Appl. Radiation Isotopes. 18: 493-523. 1967.

Anbar, M., Meyerstein, D., and Neta, P. The reactivity of aromatic compounds toward hydroxyl radicals. J Phys. Chem. 70: 2661-1. 1966.

Anderson, J.P.E. and Domsch, K.H. Microbial degradation of the thiocarbamate herbicide Diallate in soils and by pure cultures of soil microorganisms. Arch. Environ. Contam. Toxicol. 4: 1-7. 1976.

Apoteker, A. and Thevenot, D.R. Experimental simulation of biodegradation in rivers. Oxygen, organic matter, and biomass concentration changes. Water Res. 17: 1267-74. 1983.

Appleton, H., Banerjee, S., Pack, E., and Sikka, H. Fate of 3,3'-dichlorobenzidine in the aquatic environment. In: Pergamon Series in Environmental Science. 1: 473-4. 1978.

Atkinson, R. and Carter, W.P.L.  Kinetics and mechanisms of gas-phase ozone with organic compounds under atmospheric conditions.  Chem. Rev.  84: 437-70.  1984.

Atkinson, R., Aschmann, S.M., Winer, A.M., and Pitts, J.N., Jr.  Kinetics of the gas-phase reactions of $NO_3$ radicals with a series of dialkenes, cycloalkenes, and monoterpenes at 295 K.  Environ. Sci. Technol.  18: 370-5.  1984A.

Atkinson, R.  Kinetics and mechanisms of the gas-phase reactions of hydroxyl radical with organic compounds under atmospheric conditions.  Chem. Rev.  85: 69-201.  1985.

Atkinson, R., Aschmann, S.M., Winer, A.M., Jr., and Pitts, J.N.  Atmospheric gas phase loss process of chlorobenzene, benzotrifluoride, and 4-chlorobenzotrifluoride, and generalization of predictive techniques for atmospheric lifetimes of aromatic compounds.  Arch. Environ. Contam. Toxicol.  14: 417-25.  1985B.

Atkinson, R.  Estimation of OH radical rate constants and atmospheric lifetimes for polychlorobiphenyls, dibenzo-p-dioxins.  Environ. Sci. Technol.  21: 305-7.  1987.

Atkinson, R.  Structure-activity relationship for the estimation of rate constants for the gas-phase reactions of OH radicals with organic compounds.  Int. J. Chem. Kinetics.  19: 799-828.  1987A.

Bailey, R.E., Gonsier, S.J., and Rhinehart, W.L.  Biodegradation of monochlorobiphenyls and biphenyl in river water.  Environ. Sci. Technol.  17: 617-21.  1983.

Baird, R., Caroma, L., and Jenkins, R.L.  Behavior of benzidine and other aromatic amines in aerobic waste water treatment.  J. Water Pollut. Control Fed.  49: 1609-15.  1977.

Baker, M.D. and Mayfield, C.I.  Microbial and non-biological decomposition of chlorophenols and phenols in soil.  Water Air Soil Pollut.  13: 411.  1980.

Baker, M.D., Mayfield, C.I., and Inniss, W.E.  Degradation of chlorophenol in soil, sediment and water at low temperature.  Water Res.  14: 765-71.  1980.

Banerjee, S., Howard, P.H., Rosenberg, A., Dombrowski, A.E., Sikka, H.C., and Tullis, D.L.  Development of a general kinetic model for biodegradation and its application to chlorophenols and related compounds.  Environ. Sci. Technol.  18: 416-22.  1984.

Banerjee, S., Sikka, H.C., Gray, D.A., and Kelly, C.M.  Photodegradation of 3,3'-dichlorobenzidine.  Environ. Sci. Technol.  12: 1425-7.  1978A.

Banks, P.A., Ketchersid, M.L., and Merkle, M.G.  The persistence of fluridone in various soils under field and controlled conditions.  Weed Sci.  27: 631-3.  1979.

Barrio-Lage, G., Parson, F.Z., Nassar, R.S., and Lorenzo, P.A. Sequential dehalogenation of chlorinated ethenes. Environ. Sci. Technol. 20: 96-9. 1986.

Bartlett, P.D., Davis, J.W., Ross, S.D., and Swain, C.G. Kinetics and mechanisms of reactions of t-B-chloroethylamines in solution. II. Ethyl-bis-chloroethylamine. J. Amer. Chem. Soc. 69: 2977-82. 1947.

Battersby, N.S. and Wilson, V. Survey of the anaerobic biodegradation potential of organic chemicals in digesting sludge. Appl. Environ. Microbio. 55: 433-9. 1989.

Baur, J.R. and Bovey, R.W. Ultraviolet and volatility loss of herbicides. Arch. Environ. Contam. Toxicol. 2: 275-88. 1974.

Beck, J. and Hansen, K.E. The degradation of quintozene, pentachlorobenzene, hexachlorobenzene, and pentachloroaniline in soil. Pestic. Sci. 5: 41-8. 1974.

Belly, R.T. and Goodhue, C.T. A radiorespirometric technique for measuring the biodegradation of specific components in a complex effluent. In: Proc. Int. Biodegrad. Symposium. 3rd. pp. 1103-7. 1976.

Blades-Filmore, L.A., Clement, W.H., and Faust, S.D. The effect of sediment on the biodegradation of 2,4,6-trichlorophenol in Delaware River water. J. Environ. Sci. Health. A17: 797-818. 1982.

Blok, J. and Booy, M. Biodegradability test results related to quality and quantity of the inoculum. Ecotox. Environ. Saf. 8: 410-22. 1984.

Boethling, R.S. and Alexander, M. Effects of concentration of organic chemicals on their biodegradation by natural microbial communities. Appl. Environ. Microbiol. 37: 1211-6. 1979A.

Boethling, R.S. and Alexander, M. Microbial degradation of organic compounds at trace levels. Environ. Sci. Technol. 13: 989-91. 1979.

Bogan, R.H. and Sawyer, C.N. Biochemical degradation of synthetic detergents. II. Studies on the relation between chemical structure and biochemical oxidation. Sew. Ind. Wastes. 27: 917-28. 1955.

Bogyo, D.A., Lande, S.S., Meylan, W.M., Howard, P.H., and Santodonato, J. Investigation of selected environmental contaminants: Epoxides. EPA-650/11-80-005. U.S. EPA. Washington, DC. pp. 202. 1980.

Bollag, J.M., Blattmann, P., and Laanio, T. Adsorption and transformation of four substituted anilines in soil. J. Agric. Food Chem. 26: 1302-6. 1978.

Borighem, G. and Vereecken, J.  Study of the biodegradation of phenol in river water.  Ecological Modeling.  4: 51-9.  1978.

Bossert, I., Kachel, W.M., and Bartha, R.  Fate of hydrocarbons during oily sludge disposal in soil.  Appl. Environ. Microbiol.  47: 763-7.  1984.

Bourquin, A.W., Garnas, R.L., Pritchard, P.H., Wilkes, F.G., Cripe, C.R., and Rubinstein, N.I.  Interdependent microcosms for the assessment of pollutants in the marine environment.  Int. J. Environ. Studies.  13: 131-40.  1979.

Bourquin, A.W.  Biodegradation in the estuarine-marine environments and the genetically altered microbe.  U.S. EPA-600/D-84-051.  NTIS PB84-151 315.  U.S. EPA Environ. Res. Lab. Gulf Breeze, FL.  35 pp.  1984.

Bouwer, E.J., Rittman, B., and McCarty, P.L.  Anaerobic degradation of halogenated 1- and 2- organic compounds.  Environ. Sci. Technol.  15: 596-9.  1981.

Bouwer, E.J., McCarty, P.L., Bouwer, H., and Rice, R.C.  Organic contaminant behavior during rapid infiltration of secondary wastewater at the Phoenix 23rd Avenue Project.  Water Res.  18: 463-72.  1984.

Bouwer, E.J. and McCarty, P.L.  Transformations of 1- and 2-carbon halogenated aliphatic organic compounds under methanogenic conditions.  Appl. Environ. Microbiol.  45: 1286-94.  1983.

Bowman, M.C., Schechter, M.S., and Carter, R.L.  Behavior of chlorinated insecticides in a broad spectrum of soil types.  J. Agric. Food Chem.  13: 360-5.  1965.

Boyd, S.A., Shelton, D.R., Berry, D., and Tiedje, J.M.  Anaerobic biodegradation of phenolic compounds in digested sludge.  Appl. Environ. Microbiol.  46: 50-4.  1983.

Boyd, S.A., Kao, C.W., and Suflita, J.  Fate of 3,3'-dichlorobenzidine in soil: persistence and binding.  Environ. Tox. Chem.  3: 201-8.  1984.

Brahmaprakash, G.P., Reddy, B.R., and Sethunathan, N.  Persistence of hexachlorocyclohexane isomers in soil planted with rice and in rice rhizosphere soil suspensions.  Biol. Fertil. Soils.  1: 103-9.  1985.

Braun, B.A. and Zirrolli, J.A.  Environmental fate of hydrazine fuels in aqueous and soil environments.  ESL-TR-82-45.  NTIS AD-A125 813/6.  Tyndall AFB, FL: Air Force Eng. Serv. Ctr.  PP. 30.  1983.

Brice, K.A., Derwent, R.G., Eggleton, A.E.J., and Penkett, S.A.  Measurements of $CCl_3F$ and $CCl_4$ at Harwell over the period of January 1975 - June 1981 and the atmospheric lifetime of $CCL_3F$.  Atmos. Environ.  16: 2543-54.  1982.

Bridie, A.L., Wolff, C.J.M., and Winter, M. BOD and COD of some petrochemicals. Water Res. 13: 627-30. 1979.

Bringmann, G. and Kuehn, R. Biological decomposition of nitrotoluenes and nitrobenzenes by *Azotobacter agilis*. Gesundh.-Ing. 92: 273-6. 1971.

Brink, R.H., Jr. Studies with chlorophenols, acrolein, dithiocarbamates and dibromonitrilopropionamide in bench-scale biodegradation units. In: Proc. 3rd. Inter. Biodeg. Symp. pp. 785-91. 1976.

Bro-Rasmussen, F., Noddegaard, E., and Voldum-Clausen, K. Comparison of the disappearance of eight organophosphorus insecticides from soil in laboratory and in outdoor experiments. Pest. Sci. 1: 179-82. 1970.

Brochhagen, F.K. and Grieveson, B.M. Environmental aspects of isocyanates in water and soil. Cell Polym. 3: 11-7. 1984.

Brown, S.L., Chan, F.Y., Jones, J.L., Liu, D.H., and McCaleb, K.E. Research program on hazard priority ranking of manufactured chemicals (Chemical 61-79). NTIS PB-263 164. Menlo Park, CA. Stanford Res. Inst. 1975C.

Brown, J.A., Jr. and Weintraub, M. Biooxidation of paint process wastewater. J. Water Pollut. Control Fed. 54: 1127-30. 1982.

Brown, D. and Hamburger, B. The degradation of dyestuffs: Part III - Investigations of their ultimate degradability. Chemosphere. 16: 1539-53. 1987.

Brown, D. and Labourer, P. The aerobic biodegradability of primary aromatic amines. Chemosphere. 12: 405-14. 1983A.

Brunn, J., Peters, F., and Dethloff, M. UV spectra of alpha, beta-unsaturated esters and the effects of solvents and complex formation. J. Prakt. Chem. 318: 745-55. 1976.

Bunton, C.A., Fuller, N.A., Perry, S.G., and Shiner, V.J., Jr. The hydrolysis of carboxylic anhydrides. III. Reactions in initially neutral solution. J. Chem. Soc. 1963. pp. 2918-26. 1963.

Butler, L.C., Stauff, D.C., and Davis, R.L. Methyl parathion persistence in soil following simluated spillage. Arch. Environ. Contam. Toxicol. 10: 451-8. 1981.

Butler, A.R. and Gold, V. Kinetic solvent isotope effects in spontaneous hydrolysis of acyl derivatives. J. Chem. Soc. 1962. pp. 2212-7. 1962A.

Butz, R.G., Yu, C.C., and Atallah, Y.H. Photolysis of hexachlorocyclopentadiene in water. Ecotox. Environ. Safety. 6: 347-57. 1982.

Callahan, M.A., Slimak, M.W., Gabel, N.W., May, I.P., Fowler, C.F., Freed, J.R., Jennings, P., Durfee, R.L., Whitmore, F.C., Maestri, B., et al. Water-related environmental fate of 129 priority pollutants - Volume I. USEPA-440/4-79-029a. Washington, DC. U.S. Environ. Prot. Agency. 1979.

Callahan, M.A., Slimak, M.W., Gabel, N.W., May, I.P., Fowler, C.F., Freed, J.R., Jennings, P., Durfee, R.L., Whitmore, F.C., Maestri, B., et al. Water-related environmental fate of 129 priority pollutants - Volume II. USEPA-440/4-79-029b. Washington, DC. U.S. Environ. Prot. Agency. 1979A.

Calvert, J.G., Demeyan, K.L., Kerr, J.A., and McQuigg, R.D. Photolysis of formaldehyde as a hydrogen atom source in the lower atmosphere. Science. 175: 751-2. 1972.

Canalstuca, J. Biological degradation of formaldehyde. Pilot plant studies and industrial application. Ing. Quim. 15: 85-8. 1983.

Canton, J.H., Sloof, W., Kool, H.J., Struys, J., Pouw, T.J.M., Wegman, R.C.C., and Piet, G.J. Toxicity, biodegradability, and accumulation of a number of chlorine/nitrogen containing compounds for classification and establishing water quality criteria. Regul. Toxicol. Pharmacol. 5: 123-31. 1985.

Castro, E.A., Moodie, R.B., and Sanson, P.J. The kinetics of hydrolysis of methyl and phenyl isocyanates. J. Chem. Soc., Perkin Trans. 2. 5: 37-42. 1985.

Castro, T.F. and Yoshida, T. Effect of organic matter on the biodegradation of some organochlorine insecticides in submerged soils. Soil Sci., Plant Nutri. 20: 363-70. 1974.

Castro, T.F. and Yoshida, T. Degradation of organochlorine insecticides in flooded soils in the Phillipines. J. Agric. Food Chem. 19: 1168-70. 1971.

Castro, C.E. and Belser, N.O. Biodehalogenation. Reductive dehalogenation of the biocides ethylene dibromide, 1,2-dibromo-3-chloropropane, and 2,3-dibromobutane in soil. Environ. Sci. Technol. 2: 779-83. 1968.

Challis, B.C. and Li, B.F.L. Formation of N-nitrosamines and N-nitramines by photolysis. IARC Sci. Publ. 41: 31-40. 1982.

Chambers, C.W., Tabak, H.H., and Kabler, P.W. Degradation of aromatic compounds by phenol-adapted bacteria. J. Water Pollut. Control Fed. 35: 1517-28. 1963.

Chameides, W.L. and Davis, D.D. Aqueous-phase source of formic acid in clouds. Nature. 304: 427-9. 1983.

Chapman, R.A. and Cole, C.M. Observations on the influence of water and soil pH on the persistence of insecticides. J. Environ. Sci. Health. B17: 487. 1982.

Chau, A.S.Y. and Thomson, K. Investigation of the integrity of seven herbicidal acids in water samples. J. Assoc. Off. Anal. Chem. 61: 481-5. 1978.

Choi, J.S., Fermanina, T.W., Wehner, D.J., and Spomer, L.A. Effect of temperature, moisture and soil texture on DCPA degradation. Agron. J. 80: 108-13. 1988.

Chou, W.L., Speece, R.E., and Siddiqi, R.H. Acclimation and degradation of petrochemical wastewater components by methane fermentation. Biotechnol. Bioeng. Symp. 8: 391-414. 1979.

Chudoba, J., Prasil, M., and Emmerova, H. Residual organic matter in activated sludge process effluents. III. Degradation of amino acids and phenols under continuous conditions. Sb. Vys. Sk. Chem.-Technol. Praze, Technol. Vody. 13: 45-63. 1968.

Cohen, B., Van Artsdalen, E.R., and Harris, J. Reaction kinetics of aliphatic tertiary B-chloroethylamines in dilute aqueous solution 1. The cyclization process. J. Amer. Chem. Soc. 70: 281-5. 1948.

Collins, R.F. Perfusion studies with bromoxynil octanoates in soil. Pestic. Sci. 4: 181-92. 1973.

Conway, R.A., Waggy, G.T., Spiegel, M.H., and Berglund, R.L. Environmental fate and effects of ethylene oxide. Environ. Sci. Technol. 17: 107-12. 1983.

Coover, M.P. and Sims, R.C.C. The effects of temperature on polycyclic aromatic hydrocarbon persistence in an unacclimated agricultural soil. Haz. Waste Haz. Mat. 4: 69-82. 1987.

Corbin, F.T. and Upchurch, R.P. Influence of pH on detoxification of herbicides in soils. Weeds. 15: 370-7. 1967.

Cruickshank, P.A. and Jarrow, H.C. Ethylenethiourea degradation. J. Agric. Food Chem. 21: 333-5. 1973.

Crutzen, P.J., Isaksen, I.S.A., and McAfee, J.R. The impact of the chlorocarbon industry on the ozone layer. J. Geophys. Res. 83: 345-63. 1978.

Cupitt, L.T. Fate of toxic and hazardous materials in the air environment. EPA-600/3-80-084. U.S. EPA. Research Triangle Park, NC. 1980.

Dahlberg, J.A. The non-sensitized photo-oxidation of trichloroethylene in air. Acta Chemica Scandinavica. 23: 3081-90. 1969.

Davidson, J.M., Ou, L.T., and Rao, P.S.C. Adsorption, movement, and biological degradation of high concentrations of selected pesticides in soils. EPA-600/2-80-124. Cincinnati, OH: U.S. EPA. 124 pp. 1980.

Davidson, J.A., Schiff, H.I., Brown, T.J., and Howard, C.J. Temperature dependence of the rate constants for reactions of oxygen(1D) atoms with a number of hydrocarbons. J. Chem. Phys. 69: 4277-9. 1978.

Davis, E.M., Turley, J.E., Casserly, D.M., and Guthrie, R.K. Partitioning of selected organic pollutants in aquatic ecosystems. In: Biodeterioration 5. Oxley, T.A. and Barry, S., Eds. pp. 176-84. 1983.

Debont, J.A.M. Oxidation of ethylene by soil bacteria. Antonie van Leeuwenhoek. 42: 59-71. 1976.

DeLaune, R.D., Gambrell, R.P., and Reddy, K.S. Fate of pentachlorophenol in estuarine sediment. Environ. Pollut. Series B. 6: 297-308. 1983.

Digeronimo, M.J., Boethling, R.S. and Alexander, M. Effect of chemical structure and concentration on microbial degradation in model ecosystems. In: EPA-600/9-79-012. Microbial Degradation of Pollutants in Marine Environments. U.S. EPA. Gulf Breeze, FL. pp. 154-66. 1979.

Dilling, W.L., Tefertiller, N.B., and Kallos, G.J. Evaporation rates and reactivities of methylene chloride, chloroform, 1,1,1-trichloroethane, trichloroethylene, tetrachloroethylene and other chlorinated compounds in dilute aqueous solutions. Environ. Sci. Technol. 9: 833-8. 1975.

Dojlido, J.R. Investigations of biodegradability and toxicity of organic compounds. Final Report 1975-79. U.S. EPA-600/2-79-163. Cincinnati, OH: Municipal Environmental Research Lab. 118 pp. 1979.

Dojlido, J., Stodja, A., Gantz, E., and Kowalski, J. Biodegradation of methyl ethyl ketone, cyclohexanone, and cyclohexanol in sewage and surface water. Arch. Ochr. Srodowiska. 1: 115-23. 1984.

Domsch, K.H. Effects of pesticides and heavy metals on biological processes in soil. Plant Soil. 76: 367-78. 1984.

Dore, M., Brunet, N., and Legube, B. Participation of various organic compounds in the evaluation of global pollution criteria. Trib. Cebedeau. 28: 3-11. 1975.

Dorfman, L.M. and Adams, G.E. Reactivity of the hydroxyl radical in aqueous solution. NSRD-NDB-46. NTIS COM-73-50623. Washington, DC: National Bureau of Standards. 51 pp. 1973.

Dorn, P.B., Chou, C.S., and Gentempo, J.J. Degradation of Bisphenol A in natural waters. Chemosphere. 16: 1501-7. 1987.

Dow Chemical Company. The Glycol Ethers Handbook. Midland, MI. 1981.

Doyle, R.C., Kaufman, D.D., and Burt, G.W. Effect of dairy manure and sewage sludge on 14C-pesticide degradation in soil. J. Agric. Food Chem. 26: 987-9. 1978.

Drahanovsky, J. and Vacek, Z. Dissociation constants and column chromatography of chlorinated phenols. Collect. Czech. Chem. Commun. 36: 3431-40. 1971.

Draper, W.M. Determination of wavelength-averaged, near UV quantum yields for environmental chemicals. Chemosphere. 14: 1195-203. 1985.

Draper, W.M. and Crosby, D.G. Photochemical generation of superoxide radical anion in water. J. Agric. Food Chem. 31: 734-7. 1983A.

Duff, P.B. The fate of TDI in the environment. In: Polyurethane - New Paths to Progress-Marketing-Technology. Proc. SPI 6th Inter. Tech. Marketing Conf. pp. 408-12. 1983.

Dulin, D., Drossman, H., and Mill, T. Products and quantum yields for photolysis of chloroaromatics in water. Environ. Sci. Technol. 20: 72-7. 1986.

Earley, J.E., O'Rourke, C.E., Clapp, L.B., Edwards, J.O., and Lawes, B.C. Reactions of ethylenimines. IX. The mechanisms of ring openings of ethylenimines in acidic aqueous solutions. Anal. Chem. 27: 3458-62. 1958.

Edney, E.O. and Corse, E.W. Validation of OH radical reaction rate constant test protocol. NTIS PB86-166 758/as. Washington, DC: USEPA. 1986.

Edney, E.O., Kleindienst, T.E., and Corse, E.W. Room temperature rate constants for the reaction of OH with selected chlorinated and oxygenated hydrocarbons. Int. J. Chem. Kinet. 18: 1355-71. 1986.

Ehrenberg, L., Osterman-Golkar, S., Singh, D., and Lundqvist, U. On the reaction kinetics and mutagenic activity of methylating and beta-halogenoethylating gasoline additives. Radiat. Bot. 15: 185-94. 1974.

Eichelberger, J.W. and Lichtenberg, J.J. Persistence of pesticides in river water. Environ. Sci. Technol. 5: 541-4. 1971.

Ellington, J.J., Stancil, F.E., Payne, W.D., and Trusty, C.D. Measurement of hydrolysis rate constants for evaluation of hazardous waste land disposal: Volume 3. Data on 70 chemicals. EPA-600/3-88-028. NTIS PB88-234 042/AS. 1988A.

Ellington, J.J., Stancil, F.E., Payne, W.D., and Trusty, C.D. Measurement of hydrolysis rate constants for evaluation of hazardous waste land disposal: Volume 3. Data on 70 chemicals (preprint). EPA/600/S3-88/028. NTIS PB88-234042/AS. 1988.

Ellington, J.J., Stancil, F.E., and Payne, W.D. Measurement of hydrolysis rate constants for evaluation of hazardous waste land disposal. Volume 1. Data on 32 chemicals. U.S.EPA-600/3-86-043. NTIS PB87-140 349/GAR. 1987A.

Ellington, J.J., Stancil, F.E., Payne, W.D., and Trusty, C. Measurement of hydrolysis rate constants for evaluation of hazardous waste land disposal. Volume 2. Data on 54 chemicals. U.S. EPA-600/53-87/019. Washington, DC: USEPA. 1987.

Evans, W.H. and David, E.J. Biodegradation of mono-, di- and triethylene glycols in river water under controlled laboratory conditions. Water Res. 8: 97-100. 1974.

Feiler, H.D., Vernick, A.S., and Storch, P.J. Fate of priority pollutants in POTW'S. In: Proc. Natl. Conf. Munic. Sludge Manage. 8: 72-81. 1979.

Flathman, P.E. and Dahlgran, J.R. Correspondence on: Anaerobic Degradation of Halogenated 1- and 2-Carbon Organic Compounds. Environ. Sci. Technol. 16: 130. 1982.

Fochtman, E.G. Biodegradation and carbon adsorption of carcinogenic and hazardous organic compounds. U.S. EPA-600/S2-81-032. Cincinnati, OH: USEPA. 38 pp. 1981.

Fochtman, E.G. and Eisenberg, W. Treatability of carcinogenic and other hazardous organic compounds. U.S. EPA-600/S2-79-097. U.S.EPA. Cincinnati, OH. 38 pp. 1979.

Fogel, S., Lancione, R., Sewall, A., and Boethling, R.S. Enhanced biodegradation of methoxychlor in soil under enhanced environmental conditions. Appl. Environ. Microbiol. 44: 113-20. 1982.

Foschi, S., Cesari, A., Ponti, I., Bentivogli, P.G., and Bencivelli, A. Degradation and vertical movement of pesticides in the soil. Notiz. Mal. Piante. 82-83: 37-49. 1970.

Fournier, J.C. and Salle, J. Microbial degradation of 2,6-dichlorobenzamide in laboratory models. I. Degradation of 2,6-dichlorobenzamide in soil, comparison with the evolution of other benzamides with different substituents. Research of metabolic products. Chemosphere. 3: 77-82. 1974.

Freed, V.H. and Haque, R. Adsorption, movement, and distribution of pesticides in soils. Pestic. Formulations. pp. 441-59. 1973.

Freitag, D., Ballhorn, L., Geyer, H., and Korte, F. Environmental hazard profile of organic chemicals. Chemosphere. 14: 1589-616. 1985.

Freitag, D., Lay, J.P., and Korte, F. Environmental hazard profile - test rules as related to structures and translation into the environment. In: QSAR Environ. Toxicol. Proc. Workshop Quant. Struc.-Act. Relat. Kaiser, K.L.E., ed. Reidell Publ. Dordrecht. pp. 111-36. 1984A.

Freitag, D., Geyer, H., Kraus, A., Viswanathan, R., Kotzias, D., Attar, A., Klein, W., and Korte, F. Ecotoxicological profile analysis. VII. Screening chemicals for their environmental behavior by comparative evaluation. Ecotox. Environ. Safety. 6: 60-81. 1982.

Fujiwara, Y., Kinoshita, T., Sato, H., and Kojima, I. Biodegradation and bioconcentration of alkyl ethers. Yukagaku. 33: 111-4. 1984.

Fukuda, K., Inagaki, Y., Maruyama, T., Kojima, H.I., and Yoshida, T. On photolysis of naphthalenes in aquatic systems. Chemosphere. 17: 651-59. 1988.

Gaffney, P.E. Carpet and rug industry case study II: Biological effects. J. Water Pollut. Control Fed. 48: 2731-7. 1976.

Gardner, A.M., Alvarez, G.H. and Ku, Y. Microbial degradation 4C-diphenylamine in a laboratory model sewage sludge system. Bull. Environ. Cont. Toxicol. 28: 91-6. 1982.

Garraway, J. and Donovan, J. Gas-phase reaction of hydroxyl radical with alkyl iodides. J. Chem. Soc., Chem. Comm. 23: 1108. 1979.

Garrison, A.W. Analytical studies of textile wastes. Presented before the Division of Water, Air and Waste Chemistry. American Chemical Society. Unpublished work. 1969.

Gellman, I. and Heukelekian, H. Biological oxidation of formaldehyde. Sew. Indust. Wastes. 22: 13 21. 1950.

Gellman, I. and Heukelekian, H. Studies of biochemical oxidation by direct methods. V. Effect of various seed materials on rates of oxidation of industrial wastes and organic compounds. Sew. Indust. Wastes. 27: 793-801. 1955.

Gerhold, R.M. and Malaney, G.W. Structural determinants in the oxidation of aliphatic compounds by activated sludge. J. Water Pollut. Contr. Fed. 38: 562-79. 1966.

Gerike, P. and Fischer, W.K. A correlation study of biodegradability determinations with various chemicals in various tests. Ecotox. Environ. Safety. 3: 159-73. 1979.

Gibson, S.A. and Suflita, J.M. Extrapolation of biodegradation results to ground water aquifers: reductive dehalogenation of aromatic compounds. Appl. Environ. Microbiol. 52: 681-8. 1986.

Gingerich, L.L. and Zimdahl, R.L. Soil persistence of isopropalin and oryzalin. Weed Science. 24: 431-4. 1976.

Given, C.J. and Dierberg, F.E.  Effect of pH on the rate of aldicarb hydrolysis.  Bull. Environ. Contam. Toxicol.  34: 627-33.  1985.

Gledhill, W.E., Kaley, R.G., Adams, W.J., Hicks, O., Michael, P.R., Saeger, V.W., and LeBlanc, G.A. An environmental safety assessment of butyl benzyl phthalate.  Environ. Sci. Tech.  14: 301-5. 1980.

Going, J., Kuykendahl, P., Long, S., Onstol, J., and Thomas,K.  Environmental monitoring near industrial sites. Acrylonitrile. EPA-560/6-79-003. Washington, DC: U.S. EPA.  1979.

Golab, T., Herberg, R.J., Gramlich, J.V., Raun, A.P., and Probst, G.W.  Fate of benefin in soils, plants, artificial rumen fluid, and the ruminant animal.  J. Agric. Food Chem.  18: 838-44. 1970.

Gonsior, S.J., Bailey, R.E., Rhinehart, W.L., and Spence, M.W.  Biodegradation of o-phenylphenol in river water and activated sludge.  J. Agric. Food Chem.  32: 593-6.  1984.

Goodman, M.A., Tuazon, E.C., Atkinson, R. and Winer, A.M.  A study of the atmospheric reactions of chloroethenes with OH radicals.  In: ACS Div. Environ. Chem. 192nd Natl. Mtg.  26: 169-71.  1986.

Gore, R.C., Hannah, R.W., Pattacini, S.C., and Porro, T.J.  Infrared and ultraviolet spectra of seventy-six pesticides.  J. Assoc. Off. Analyt. Chem.  54: 1040-82.  1971.

Graham, P.R.  Phthalate ester plasticizers - why and how they are used.  Environ. Health Perspect.  3: 3-12  1973.

Grasselli, J. and Ritchey, W.  Atlas of Spectral Data and Physical Constants for Organic Compounds.  2nd Ed.  Volume 6.  The Chemical Rubber Co.  1975.

Graveel, J.G., Sommers, L.E., and Nelson, D.W.  Decomposition of benzidine, alpha-naphthylamine, and p-toluidine in soils.  J. Environ. Qual.  15: 53-9.  1986.

Greene, S., Alexander, M., and Leggett, D.  Formation of N-nitrosodimethylamine during treatment of municipal waste water by simulated land application.  J. Environ. Qual.  10: 416-21.  1981.

Griebel, G.E. and Owens, L.D.  Nature of the transient activation of soil microorganisms by ethanol or acetaldehyde.  Soil Biol. Biochem.  4: 1-8.  1972.

Griffin, R.A. and Chou, S.J.  Attenuation of Polybrominated Biphenyls and Hexachlorobenzene in Earth Materials. Final Report.  U.S. EPA-600/2-81-191.  Urbana, IL: Illinois State Geological Survey.  pp. 60.  1981A.

691

Groenewegen, D. and Stolp, H. Microbial breakdown of polycyclic aromatic hydrocarbons. Zentralbl. Bakteriol., Parasitenkd., Infektionskr. Hyg., Abt. 1: Orig., Reihe, B. 162: 225-32. 1976.

Guesten, H., Filby, W.G., and Schoop, S. Prediction of hydroxyl radical reaction rates with organic compounds in the gas phase. Atmos. Environ. 15: 1763-5. 1981.

Guirguis, M.W. and Shafik, M.T. Persistence of trichlorfon and dichlorvos in two different autoclaved and non-autoclaved soils. Bull. Entomol. Soc. Egypt, Econ. Ser. 8: 29-32. 1975.

Gummer, W.D. Pesticide monitoring in the prairies of western Canada. In: Water Qual. Interpretive report No. 4. Regina, Saskatchewan, Canada. Inland Waters Directorate. pp. 14. 1979.

Haag, W.R. and Mill, T. Effect of a subsurface sediment on hydrolysis of haloalkanes and epoxides. Environ. Sci. Technol. 22: 658-63. 1988.

Haag, W.R. and Mill, T. Direct and indirect photolysis of water soluble azodyes: kinetic measurements and structure-activity relationships. Environ. Toxicol. Chem. 6: 359-69. 1987.

Haag, W.R. and Mill, T. Direct and indirect photolysis of azodyes in water. Data generation and development of structure reactivity. Relationships for environmental fate process priorities. 68-02-3968. Menlo Park, CA: SRI International. pp. 45. 1985.

Haag, W.R. and Hoigne, J. Singlet oxygen in surface waters. 3. Photochemical formation and steady-state concentrations in various types of waters. Environ, Sci. Technol. 20: 341-8. 1986.

Haag, W.R. and Hoigne, J. Degradation of compounds in water by singlet oxygen. In: Water Chlorination: Chem. Environ. Impact Health Eff. Proc. Conf. Jolley, R.L.L., ed. pp. 1011-20. 1985A.

Haider, K., Jagnow, G., Kohnen, R., and Lim, S.U. Degradation of chlorinated benzenes, phenols, and cyclohexane derivatives by benzene- and phenol-utilizing soil bacteria under aerobic conditions. In: Decomposition of Toxic and Nontoxic Organic Compounds in Soil. Overcash, V.R. ed., Ann Arbor Sci, Publ., Ann Arbor, MI. pp. 207-23. 1981.

Haider, K., Jagnow, G., Kohnen, R., and Lim, S.U. Degradation of chlorinated benzenes, phenols, and cyclohexane derivatives by benzene and phenol utilizing soil bacteria under aerobic conditions. Arch. Microbiol. 96: 183-200. 1974.

Hallas, L.E. and Alexander, M. Microbial transformation of nitroaromatic compounds in sewage effluent. Appl. Environ. Microbiol. 45: 1234-41. 1983.

Hallen, R.T., Pyne, J.W., and Molton, P.M. Transformation of chlorinated ethers by anaerobic microorganisms. In: ACS Div. Environ. Chem. 192nd Nat'l Mtg. 26: 344-6. 1986.

Hamadmad, N. Photolysis of pentachloronitrobenzene, 2,3,5,6-tetrachloronitrobenzene, and pentachlorophenol. Diss. Abstr. B. 28: 1419. 1967.

Hambrick, G.A., DeLaune, R.D., and Patrick, W.H., Jr. Effects of estuarine sediment, pH and oxidation-reduction potential on microbial hydrocarbon degradation. Appl. Environ. Microbiol. 40: 365-9. 1980.

Hammerton, C. Observations on the decay of synthetic anionic detergents in natural water. J. Appl. Chem. 5: 517-24. 1955.

Hannah, S.A., Austern, B.M., Eralp, A.E., and Wise, R.H. Comparative removal of toxic pollutants by six wastewaters treatment processes. J. Water Pollut. Control Fed. 58: 27. 1986.

Hansen, J.L. and Spiegel, M.H. Hydrolysis studies of aldicarb, aldicarb sulfoxide and aldicarb sulfone. Environ. Toxicol. Chem. 2: 147-53. 1983.

Hanst, P.L., Spence, J.W., and Miller, M. Atmospheric chemistry of N-nitrosodimethylamine. Environ. Sci. Technol. 11: 403-5. 1977.

Harris, G.W., Atkinson, R. and Pitts, J.R., Jr. Kinetics of the reactions of the OH radical with hydrazine and methylhydrazine. J. Phys. Chem. 83: 2557-9. 1979.

Harrison, R.M. and Laxen, D.P.H. Sink processes for tetraalkylated compounds in the atmosphere. Environ. Sci. Technol. 12: 1384-92. 1978.

Hatfield, R. Biological oxidation of some organic compounds. Ind. Eng. Chem. 49: 192-6. 1957.

Hattori, Y., Kuge, Y., and Nakagawa, S. Microbial decomposition of phthalate esters in environmental water. Pollut. Control Cent. Osaka Perfect., Mizu Shori Gijutsu. 16: 951-4. 1975.

Hawkins, M.D. Hydrolysis of phthalic and 3,6-dimethylphthalic anhydrides. J. Chem. Soc., Perkin Trans. 2: 282-4. 1975.

Healy, J.B. and Young, L.Y. Catechol and phenol degradation by a methanogenic population of bacteria. Food Microbiol. Toxicol. 35: 216-8. 1978.

Heitkamp, M.A. Environmental and microbiological factors affecting the biodegradation and detoxification of polycyclic aromatic hydrocarbons. Diss. Abstr. Int. B. 48: 1926. 1988.

Helfgott, T.B., Hart, F.L., and Bedard, R.G. An Index of Refractory Organics. U.S. EPA-600/2-77-17 4. Ada, OK: USEPA. 1977.

Helling, C.S. Dintroaniline herbicides in soils. J. Env. Qual. 5: 1-15. 1976.

Hendry, D.G., Mill, T., Piszkiewicz, L., Howard, J.A., and Eigenmann, H.K. A critical review of H-atom transfer in the liquid phase: chlorine atom, alkyl, trichloromethyl, alkoxy and alkylperoxy radicals. J. Phys. Chem. Ref. Data. 3: 944-78. 1974.

Henson, J.M., Yates, M.V., and Cochran, J.W. Metabolism of chlorinated methanes, ethanes and ethylenes by a mixed bacterial culture growing on methane. J. Industrl. Microb. 4: 29-35. 1989.

Heukelekian, H. and Rand, M.C. Biochemical oxygen demand of pure organic compounds. J. Water Pollut. Control Assoc. 29: 1040-53. 1955.

Hewitt, C.N. and Harrison, R.M. Formation and decomposition of trialkyllead compounds in the atmosphere. Environ. Sci. Technol. 20: 797-802. 1986.

Hewitt, C.N. and Harrison, R.M. Tropospheric concentrations of the hydroxyl radical - a review. Atmos. Environ. 19: 545-54. 1985.

Horowitz, A., Shelton, D.R., Cornell, C.P., and Tiedje, J.M. Anaerobic degradation of aromatic compounds in sediment and digested sludge. Dev. Ind. Microbiol. 23: 435-44. 1982.

Hou, C.T., Patel, R., Laskin, A.I., Barnabe, N., and Barist, I. Production of methyl ketones from secondary alcohols by cell suspensions of C2 to C4 n-alkane-grown bacteria. Appl. Environ. Microbiol. 46: 178-84. 1983A.

Howard, P.H., Hueber, A.E., and Boethling, R.S. Biodegradation data evaluation for structure/biodegradability relations. Environ. Toxicol. Chem. 6:1-10. 1987.

Howard, P.H., Santodonato, J., Saxena, J., Malling, J., and Greninger, D. Investigation of selected potential environmental contaminants: Nitroaromatics. (Draft). U.S. EPA-560/2-76-010. Research Triangle Park, NC: USEPA. p. 600. 1976.

Howard, P.H. and Saxena, J. Persistence and degradability testing of benzidine and other carcinogenic compounds. U.S.EPA-560/5-76-005. U.S.EPA. Washington, DC. p. 23. 1976.

Howard, P.H. and Durkin, P.R. Sources of contamination, ambient levels and fate of benzene in the environment. U.S.EPA-560/5-75-005. U.S.EPA. Washington, DC. p. 8. 1975.

HSDB. Hazardous Substances Data Base. Report #2040. National Library of Medicine. 1988.

Hubrich, C. and Stuhl, F. The ultraviolet adsorption of some halogenated methanes and ethanes of atmospheric interest. J. Photochem. 12: 93-107. 1980.

Hungspreugs, M., Silpapat, S., Tonapong, C., Lee, R.F., Windom, H.L., and Tenore, K.R. Heavy metals and polycyclic hydrocarbon compounds in benthic organisms of the upper Gulf of Thailand. Mar. Pollut. Bull. 15: 213-8. 1984.

Hustert, K., Mansour, M., Parlar, H., and Korte, F. The EPA test - a method to determine the photochemical degradation of organic compounds in aquatic systems. Chemosphere. 10: 995-8. 1981.

Hutchins, S.R., Tomson, M.B., and Ward, C.H. Trace organic contamination of ground water from a rapid infiltration site: a laboratory-field coordinated study. Environ. Toxicol. Chem. 2: 195-216. 1983.

Hwang, H., Hodson, R.E., and Lee, R.F. Degradation of phenol and chlorophenols by sunlight and microbes in estuarine water. Environ. Sci. Technol. 20: 1002-7. 1986.

IARC. Monograph on the Evaluation of the Carcinogenic Risk of Chemicals to Humans. Vol. 20. Some Halogenated Hydrocarbons. Lyon, France: International Agency for Research on Cancer. 1979.

IARC. Monographs on the Evaluation of Carcinogenic Risk of Chemicals to Humans. Vol. 17. Some N-Nitroso Compounds. Lyon, France. International Agency for Research on Cancer. 1978.

IARC. Monographs on the Evaluation of Carcinogenic Risk to Man. Vol 9. Some Aziridines, N-, S- and O-Mustards and Selenium. Lyon, France: International Agency for Research on Cancer. 9: 235-41. 1975A.

Ide, A., Niki, Y., Sakamoto, F., and Watanabe, I. Decomposition of pentachlorophenol in paddy soil. Agric. Biol. Chem. 36: 1937-44. 1972.

Ilisescu, A. Behavior in the biological treatment plants of some pollutants removed with petrochemical wastes. In: Stud. Prot. Epurarea Apelor. pp. 249-66. 1971.

Inman, J.C., Strachan, S.D., Sommers, L.E., and Nelson, D.W. The decomposition of phthalate esters in soil. J. Environ. Sci. Health. B19: 245-57. 1984.

Ivie, G.W. and Casida, J.E. Photosensitizers for the accelerated degradation of chlorinated cyclodienes and other insecticide chemicals exposed to sunlight on bean leaves. J. Agric. Food. Chem. 19: 410-6. 1971.

Jackson, D.R., Roulier, M.W., Grotta, H.M., Rust, S.W., and Warner, J.S. Solubility of 2,3,7,8-TCDD in contaminated soils. Chlorinated Dioxins and Dibenzofurans in Perspective. Rappe et al., Eds. Lewis Publ. Co. pp. 185-200. 1986.

Jafvert, C.T. and Wolfe, N.L. Degradation of selected halogenated ethanes in anoxic sediment-water systems. Environ. Toxicol. Chem. 6: 827-37. 1987.

Jensen, S. and Rosenberg, R. Degradability of some chlorinated aliphatic hydrocarbons in sea water and sterilized water. Water Res. 9: 659-61. 1975.

Jensen-Korte, U., Anderson, C., and Spiteller, M. Photodegradation of pesticides in the presence of humic substances. Sci. Tot. Environ. 62: 335-40. 1987.

Johnson, B.T., Heitkamp, M.A., and Jones, J.R. Environmental and chemical factors influencing the biodegradation of phthalic acid esters in freshwater sediments. Environ. Pollut. B. 8: 101-18. 1984.

Johnson, B.T. and Lulves, W. Biodegradation of di-n-butyl phthalate and di-2-ethylhexyl phthalate in freshwater hydrosoil. J. Fish. Res. Board Can. 32: 333-9. 1975.

Kagiya, T., Takemoto, K., and Uyama, Y. Promotional oxidation degradation method for air pollutants using artificial photochemical processes, Paper 1036. Japan. Chem. Soc., Springterm Annl. Mtg. 32nd. Tokyo, Japan. 1975.

Kanazawa, J. Biodegradability of pesticides in water by microbes in activated sludge. Environ. Monit. Assess. 9: 57-70. 1987.

Kappeler, T. and Wuhrmann, K. Microbial degradation of the water soluble fraction of gas oil. I. Water Res. 12: 327-33. 1978.

Kato, S., Inoue, H., and Hida, M. The oxidative photolysis of anthraquinone derivatives. Nippon Kagaku Kaishi. pp. 1141-4. 1978.

Kaufman, D.D. and Doyle, R.D. Biodegradation of organics. Nat. Conf. Composting Municipal Residues Sludges. 75 pp. 1977.

Kawasaki, M. Experiences with the test scheme under the Chemical Control Law of Japan: An approach to structure-activity correlations. Ecotox. Environ. Safety. 4: 444-54. 1980.

Kearney, P.C., Isensee, A.R., Helling, C.S., Woolsen, E.A., and Plimmer,J.R. Environmental significance of chlorodioxins. Chlorodioxins - Origin and Fate. Adv. Chem. Ser. 120: 105-11. 1971.

Kearney, P.C., Nash, R.G., and Isensee, A.R. Persistence of pesticides in soil. Chemical Fallout: Current Research on Persistent Pesticides. Miller, M.W. & Berg, C.C., Eds. Springfield, IL: Charles C. Thomas. Chpt. 3., pp. 54-67. 1969C.

Kennedy, J.M. and Talbert, R.E. Comparative persistence of dintiroaniline type herbicides on the soil surface. Weed Sci. 25: 373-81. 1977.

Kincannon, D.F. and Lin, Y.S. Microbial degradation of hazardous wastes by land treatment. In: Proc. Indust. Waste Conf. 40: 607-19. 1985.

Kincannon, D.F., Weinert, A., Padorr, R., and Stover, E.L. Predicting treatability of multiple organic priority pollutant waste water from single-pollutant treatability studies. In: Proc. 37th Indust. Waste Conf. Bell, J.M., ed. Ann Arbor, MI: Ann Arbor Science Publ. pp. 641-50. 1983B.

Kincannon, D.F., Stover, E.L., Nichols, V., and Medley, D. Removal mechanisms for toxic priority pollutants. J. Water Pollut. Control Fed. 55: 157-63. 1983.

Kirkland, K. and Fryer, J.D. Degradation of several herbicides in a soil previously treated with MCPA. Weed Res. 12: 90-5. 1972.

Kitano, M. Biodegradation and bioaccumulation test on chemical substances. OECD Tokyo Meeting. Reference Book TSU-No. 3. 1978.

Klecka, G.M. Fate and effects of methylene chloride in activated sludge. Appl. Environ. Microbiol. 44: 701-7. 1982.

Kleopfer, R.D. and Fairless, B.J. Characterization of organic compounds in a municipal water supply. Environ. Sci. Technol. 6: 1036-7. 1972.

Kohnen, R., Haider, K., and Jagnore, G. Investigation on the microbial degradation of lindane in submerged and aerated moist soil. Environmental Quality and Safety Supplement Vol. III. pp. 223-6. 1975.

Kollig, H.P., Ellington, J.J., Hamrick, K.J., Jafverts, C.T., Weber, E.J., and Wolfe, N.L. Hydrolysis rate constants, partition coefficients, and water solubilities for 129 chemicals. A summary of fate constants provided for the concentration-based Listing Program, 1987. Athens, GA: USEPA. Environ. Res. Lab., Off. Res. Devel. Prepublication. 1987A.

Kollig, H.P., Parrish, R.S., and Holm, H.W. An estimate of the variability in biotransformation kinetics of xenobiotics in natural waters by aufwuchs communities. Chemosphere 16: 49-60. 1987.

Kolyada, T.I. Decomposition of thiourea in the soil. Vestsi Akad. Nauk. Belarus. 3: 36-40. 1969.

Kondo, M., Nishihara, T., Shimamoto, T., Watanabe, K., and Fujii, M. Screening test method for degradation of chemicals in water, a simple and rapid method for biodegradation test (cultivation method). Esei Kagaku. 34: 115-22. 1988A.

Kondo, M., Nishihara, T., Shimamoto, T., Koshikawa, T., Iio, T., Sawamura, R., and Tanaka, K. Biodegradation of test chemicals by cultivation method. Esei Kagaku. 34: 188-95. 1988.

Kondo, M. Simulation studies of degradation of chemicals in the environment: simulation studies of degradation of chemicals in the water and soil. Environment Agency, Office of Health Studies. Japan. 1978.

Kool, H.J. Influence of microbial biomass on the biodegradability of organic compounds. Chemosphere. 13: 751-61. 1984.

Kostovetskii, Y.I., Naishten, S.Y., Tolstopyatova, G.V., and Chegrinets, G.Y. Hygenic aspects of pesticide use in the catchment areas of water bodies. Vodn. Resur. 1: 67-72. 1976.

Krijgsheld, K.R. and Van der Gen, A. Assessment of the impact of the emission of certain organochlorine compounds of the aquatic environment. Part II: allylchloride, 1,3- and 2,3-dichloropropene. Chemosphere. 15: 861-80. 1986A.

Kuhn, E.P., Coldberg, P.J., Schnoor, J.L., Wanner, O., Zehnder, A.J.B., and Schwarzenbach, R.P. Microbial transformation of substituted benzenes during infiltration of river water to groundwater: laboratory column studies. Environ. Sci. Tech. 19: 961-8. 1985.

Lamb, C.B. and Jenkins, G.F. BOD of synthetic organic chemicals. Proc. 8th Indust. Waste Conf., Purdue Univ. pp. 326-9. 1952.

Lane, D.A. and Katz, M. The photomodification of benzo(a)pyrene, benzo(b)fluranthene, and benzo(k) fluranthene under simulated atmospheric conditions. Adv. Environ. Sci. Technol. 8 (Fate Pollut Air Water Environ): 137-54. 1977.

Larson, R.J. and Davidson, D.H. Acclimation to and biodegradation of nitrilotriacetate (NTA) at trace concentrations in natural waters. Water Res. 16: 1597-604. 1982.

Larson, R.J., Clinckemaillie, G.G., and VanBelle, L. Effect of temperature and dissolved oxygen on biodegradation of nitrilotriacetate. Water Res. 15: 615-20. 1981.

Laughton, P.M. and Robertson, R.E. Solvolysis in hydrogen and deuterium oxide. III. Alkyl halides. Can. J. Chem. 37: 1491-7. 1959.

Lee, R.F. and Ryan, C. Microbial degradation of organochlorine compounds in estuarine waters and sediments. In: EPA-600/9-79-012. Microbial Degradation of Pollutants in Marine Environments. Bourquin, A.W. and Pritchard, P.H., Eds. U.S. EPA. Gulf Breeze, FL. pp. 453-50. 1979.

Lee, R.F. and Ryan, C. Biodegradation of petroleum hydrocarbons by marine microbes. In: Proc. Int. Biodegradation Symp. 3rd. 1975. pp. 119-25. 1976.

698

Lee, M.L., Later, D.W., Rollins, D.K., Eatough, D.J., and Hansen, L.D. Dimethyl and monomethyl sulfate: presence in coal fly ash and airborne particulate matter. Science. 207: 186-8. 1980.

Lee, M.D., Wilson, J.T., and Ward, C.H. Microbial degradation of selected aromatics in a hazardous waste site. Devel. Indust. Microbiol. 25: 557-65. 1984.

Lemaire, J., Guth, J.A., Klais, O., Leahy, J., Merz, W., Philp, J., Wilmes, R., and Wolff, C.J.M. Ring test of a method for assessing the phototransformation of chemicals in water. Chemosphere. 14: 53-77. 1985.

Lichtenstein, E.P., Fuhremann, T.W., and Schulz, K.R. Persistence and vertical distribution of DDT, lindane and aldrin residues, ten and fifteen years after a single soil application. J. Agric. Food Chem. 19: 718-21. 1971.

Lichtenstein, E.P. and Schultz, K.R. Persistence of some chlorinated hydrocarbon insecticides influenced by soil types, rates of application and temperature. J. Econ. Entomol. 52: 124-31. 1959.

Liu, C.C.K., Tamrakar, N.K., and Green, R.E. Biodegradation and adsorption of DBCP and the mathematical simulation of its transport in tropical soils. Toxic. Asses. 2: 239-52. 1987.

Liu, D., Strachan, W.M.J., Thomson, K., and Kwasniewska, K. Determination of the biodegradability of organic compounds. Environ. Sci. Technol. 15: 788-93. 1981A.

Loekke, H. Degradation of 4-nitrophenol in two Danish soils. Environ. Pollut. Ser. A. 38: 171-81. 1985.

Lu, P.Y., Metcalf, R.L., Plummer, N., and Mandel, D. The environmental fate of 3 carcinogens: benzo(a)pyrene, benzidine, and vinyl chloride evaluated in laboratory model ecosystems. Arch. Environ. Contam. Toxicol. 6: 129-42. 1977.

Ludzack, F.J. and Ettinger, M.B. Chemical structures resistant to aerobic biochemical stabilization. J. Water Pollut. Control Fed. 32: 1173-200. 1960.

Ludzack, F.J. and Ettinger, M.B. Biodegradability of organic chemicals isolated from rivers. In: Purdue Univ. Eng. Bull., Ext. Ser. No. 115. pp. 278-82. 1963.

Ludzack, F.J., Schaffer, R.B., and Ettinger, M.B. Temperature and feed as variables in activated sludge performance. Water Pollut. Contr. J. 33: 141-56. 1961A.

Ludzack, F.J., Schaffer, R.B., and Bloomhuff, R.N. Experimental treatment of organic cyanides by conventional processes. J. Water Pollut. Control Fed. 33: 492-505. 1961.

Ludzack, F.J., Schaffer, R.B., Bloomhuff, R.N., and Ettinger, M.B.  Biochemical oxidation of some commercially important organic cyanides. I. River oxidation.  In: Proc. 13th Indust. Waste Conf. Eng. Bull. Purdue Univ., Eng. Ext. Ser.  pp. 297-312.  1958.

Lutin, P.A., Cibulka, J.J., and Malaney, G.W.  Oxidation of selected carcinogenic compounds by activated sludge.  In: Purdue Univ. Eng. Bull., Ext. Ser. No. 118.  pp. 131-45.  1965.

Lyman, W.J., Reehe, W.F., and Rosenblatt, D.H.  Handbook of Chemical Property Estimation Methods. Environmental Behavior of Organic Compounds.  McGraw-Hill. New York, NY.  pp. 960.  1982.

Mabey, W.R., Barich, V., and Mill, T.  Hydrolysis of polychlorinated alkanes.  In: Symp. Amer. Chem. Soc., Div. Environ. Chem. 186th Natl. Mtg. Washington, DC.  23:  359-61.  1983A.

Mabey, W.R., Smith, J.H., Podoll, R.T., Johnson, H.L., Mill, T., Chou, T.W., Gates, J., Partridge, I.W., and Van den Berg, D.  Aquatic Fate Process Data for Organic Priority Pollutants. U.S.EPA-440/4-81-014. USEPA. Washington, DC.  pp. 434.  1981.

Mabey, W. and Mill, T.  Critical review of hydrolysis of organic compounds in water under environmental conditions.  J. Phys. Chem. Ref. Data.  7: 383-415.  1978.

MacNaughton, M.G. and Farmwald, J.A.  Biological degradation of hydrazine.  ESL-TR-79-38. (NTIS AD-A084 426).  Tyndall Airforce Base, FL:  Engineering Service Lab.  1979.

Macrae, I.C., Yamaya, Y., and Yoshida, T.  Persistence of hexachlorocyclohexane isomers in soil suspensions.  Soil Biol. Biochem.  16: 285-6.  1984.

Malaney, G.W. and Gerhold, R.W.  Structural determinants in the oxidation of aliphatic compounds by activated sludge.  J. Water Pollut. Control Fed.  41: R18-R33.  1969.

Malaney, G.W., Lutin, P.A., Cibulka, J.J., and Hickerson, L.H.  Resistance of carcinogenic organic compounds to oxidation by activated sludge.  J. Water Pollut. Control Fed.  39: 2020-9.  1967.

Malaney, G.W. and Gerhold, R.W.  Structural determinants in the oxidative breakdown of aliphatic compounds by domestic activated sludge.  In: Proc. 17th Indust. Waste Conf., Purdue Univ., Ext. Ser.  112: 249-57.  1962.

Malaney, G.W.  Oxidative abilities of aniline-acclimated activated sludge.  J. Water Pollut. Control Fed.  32: 1300-11.  1960.

Mallik, M.A.B. and Tesfai, K.  Transformation of nitrosamines in soil and in vitro by soil organisms. Bull. Environ. Contam. Toxicol.  27: 115-21.  1981.

Marinucci, A.C. and Bartha, R. Biodegradation of 1,2,3- and 1,2,4-trichlorobenzene in soil and liquid enrichment culture. Appl. Environ. Microbiol. 38: 811-7. 1979.

Marion, C.V. and Malaney, G.W. Ability of activated sludge microorganisms to oxidize aromatic organic compounds. In: Proc. Ind. Waste Conf., Eng. Bull., Purdue Univ., Eng. Ext. Ser. pp. 297-308. 1964.

Matsui, S., Murakami, T., Sasaki, T., Hirose, Y., and Iguma, Y. Activated sludge degradability of organic substances in the waste water of Kashima petroleum and petrochemical industrial complex in Japan. Prog. Water Technol. 7: 645-59. 1975.

Maule, A., Plyte, S., and Quirk, A.V. Dehalogenation of organochlorine insecticides by mixed anaerobic microbial populations. Pestic. Biochem. Physiol. 277: 229-36. 1987.

McCall, P.J. Hydrolysis of 1,3-dichloropropene in dilute aqueous solution. Pestic. Sci. 19: 235-42. 1987.

Meallier, P. Kinetic study of the photochemical oxidation of simple aromatic amines. Ann. Chim. 4: 15-28. 1969.

Means, J.L. and Anderson, S.J. Comparison of five different methods for measuring biodegradability in aqueous environments. Water Air Soil Pollut. 16:301-15. 1981

Medvedev, V.A. and Davidov, V.D. The transformation of various coke industry products in Chernozem soil. In: Decomposition of Toxic and Nontoxic compounds in soil. Overcash, M.R., ed. Ann Arbor Sci. Publ. Ann Arbor, MI. pp. 245-54. 1981B.

Medvedev, V.A. and Davidov, V.D. Phenol and quinone degradation rates in Chernozem soil based on infrared spectroscopic data. In: Decomposition of Toxic and Nontoxic compounds in soil. Overcash, M.R., ed. Ann Arbor Sci. Publ. Ann Arbor, MI. pp. 193-9. 1981A.

Meylan, W., Papa, L., DeRosa, C.T., and Stara, J.F. Chemical of current interest propylene oxide: health and environmental effects profile. Tox. Indust. Hlth. 2: 219-57. 1986.

Miles, C.J. and Delfino, J.J. Fate of aldicarb, aldicarb sulfoxide, and aldicarb sulfone in Floridian groundwater. J. Agric. Food Chem. 33: 455-60. 1985.

Mill, T. Structure-activity relationships for photooxidation processes in the environment. Environ. Toxicol. Chem. 8: 31-43. 1989.

Mill, T., Winterle, J.S., Fischer, A., Tse, D., Mabey, W.R., Drossman, H., Liu, A., and Davenport, J.E. Toxic substances process data generation and protocol development. U.S.EPA Contract No. 68-03-2981. Washington, DC. 1985.

Mill, T. and Mabey, W. Photochemical transformations. Environ. Exposure Chem. 1: 175-216. 1985.

Mill, T., Winterle, J.S., Davenport, J.E., Lee, G.C., Mabey, W.R., Barich, V.P., Harris, W., Ingersoll, D., and Bawol, R. Validation of estimating techniques for predicting environmental transformation of chemicals. U.S.EPA Contract No. 68-01-6269. Washington, DC. 181 pp. 1982.

Mill, T. Hydrolysis and oxidation processes in the environment. Environ. Toxicol. Chem. 1: 135-41. 1982.

Mill, T., Mabey, W.R., Lan, B.Y., and Baraze, A. Photolysis of polycyclic aromatic hydrocarbons in water. Chemosphere. 10: 1281-90. 1981.

Mill, T., Hendry, D.G., and Richardson, H. Free-radical oxidants in natural waters. Science 207: 886-7. 1980.

Mills, E.J. Jr. and Stack, V.T. Jr. Biological oxidation of synthetic organic chemicals. In: Proc. 8th Indust. Waste Conf. Eng. Bull. Purdue Univ., Eng. Ext. Ser. pp. 492-517. 1954.

Mitchell, W.R. and Dennis, W.H., Jr. Biodegradation of 1,3-dinitrobenzene. J. Environ. Sci. Health. A17: 837-53. 1982.

Montgomery, C.W. and Rollefson, G.K. The quantum yields of the photochemical reactions of phosgene. J. Amer. Chem. Soc. 55: 4025-35. 1933.

Mookerjee, S.K. The environmental photodegradation of pesticides. Indian J. Agric. Chem. 18: 1-9. 1985.

Mudder, T.I. and Musterman, J.L. Development of empirical structure biodegradability relationships and biodegradability testing protocol for volatile and slightly soluble priority pollutants. In: Amer. Chem. Soc., Div. Environ. Chem., Mtg. Kansas City, MO. pp. 52-3. 1982.

Mudder, T. Development of empirical structure-biodegradability relationships and testing protocol for slightly soluble and volatile priority pollutants. Diss. Abstr. Int. B. 42: 1804. 1981.

Muel, B. and Saguem, S. Determination of 23 polycyclic aromatic hydrocarbons in atmospheric particulate matter of the Paris area and photolysis by sunlight. Inter. J. Environ. Anal. Chem. 19: 111-31. 1985.

Naik, M.N., Jackson, R.B., Stokes, J., and Swaby, R.J. Microbial degradation and phytotoxicity of picloram and other substituted pyridines. Soil Biol. Biochem. 4: 13-23. 1972.

Nakano, R. and Kuwatsuka, S. Formation and degradation of ethylene in flooded soil. III. Degradation of ethylene. Nippon Dojo-Hiryogaku Zasshi. 50: 55-60. 1979.

Nesbitt, H.J. and Watson, J.R. Degradation of the herbicide 2,4-D in river water-II. The Role of suspended sediment. Nutrients and Water Temperature. Water Res. 14: 1689-94 1980.

Norrish, R.G.W. and Searby, M.H. The photochemical decomposition of dicumyl peroxide and cumene hydroperoxide in solution. Proc. Roy. Soc. London., Ser. A. 237: 464-75. 1956.

Novak, J.T., Goldsmith, C.D., Benoit, R.E., and O'Brien, J.H. Biodegradation of methanol and tertiary butyl alcohol in subsurface systems. Water. Sci. Tech. 17: 71-85. 1985.

O'Grady, D.P., Howard, P.H., and Werner, A.F. Activated sludge biodegradation of 12 commercial phthalate esters. Appl. Environ. Microbiol. 49: 443-5. 1985.

Ohnishi, R., and Tanabe, K. A new method of solubility determination of hydrolyzing solute-solubility of benzyl chloride in water. Bull. Chem. Soc. Japan. 41: 2647-9. 1971.

Okey, R.W. and Bogan, R.H. Apparent involvement of electronic mechanisms in limiting microbial metabolism of pesticides. J. Water Pollut. Control Fed. 37: 692-712. 1965.

Oliver, J.E. Pesticide-derived nitrosamines. Occurrence and environmental fate. Agric. Symp. Ser. 1 74: 349-62. 1981.

Oliver, J.E., Kearney, P.C., and Kontson, A. Degradation of herbicide-related nitrosamines in aerobic soils. J. Agric. Food Chem. 27: 887-91. 1979.

Oremland, R.S., Marsh, L.M., and Polcin, S. Methane production and simultaneous sulphate reduction in anoxic salt marsh sediments. Nature. 296: 43-5. 1982.

Osborne, A.D., Pitts, J.N., Jr., and Darley, E.F. On the stability of acrolein toward photooxidation in the near ultra-violet. Inter. J. Air Water Pollut. 6: 1-3. 1962.

Osterman-Golkar, S. Reaction kinetics in water of chloroethylene oxide, chloroacetaldehyde and chloroacetone. Hereditas. 101: 65-8. 1984.

Ou, L.T. and Street, J.J. Hydrazine degradation and its effect on microbial activity in soil. Bull. Environ. Contam. Toxicol. 38: 179-83. 1987.

Ou, L.T., Sture, K., Edvardsson, V., Suresh, P., and Rao, C. Aerobic and anaerobic degradation of aldicarb in soils. J. Agric. Food Chem. 33: 72-8. 1985.

703

Overcash, M.R., Weber, J.B., and Miles, M.L.  Behavior of organic priority pollutants in the terrestrial system: di-n-butyl phthalate ester, toluene and 2,4-dinitrophenol.  NTIS PB82-224 544.  Report No. 171.  Water Resource Inst. Univ. NC, Raleigh, NC.  pp. 104.  1982.

Pahren, H.R. and Bloodgood, D.E.  Biological oxidation of several vinyl compounds.  Water Pollut. Control Fed. J.  33: 233-8.  1961.

Paris, D.F., Wolfe, N.L., Steen, W.C., and Baughman, G.L.  Effect of phenol molecular structure on bacterial transformation rate constants in pond and river samples.  Appl. Environ. Microbiol. 45: 1153-5.  1983.

Parsons, F., Lage, G.B., and Rice, R.  Biotransformation of chlorinated organic solvents in static microcosms.  Environ. Toxicol. Chem.  4: 739-42.  1985.

Patterson, J.W. and Kodukala, P.S.  Biodegradation of hazardous organic pollutants.  Chem. Eng. Prog.  77: 48-55.  1981.

PCFAP-USEPA.  GEMS. Graphic Exposure Modeling System.  U.S. Environmental Protection Agency.  1987.

PCHYDRO-USEPA.  GEMS. Graphic Exposure Modeling System.  U.S. Environmental Protection Agency.  1987.

Pearson, C.R. and McConnell, G.  Chlorinated C1 and C2 hydrocarbons in the marine environment. Proc. Roy. London Ser. B.  189: 305-32.  1975.

Penkett, S.A., Prosser, N.J.D., Rasmussen, R.A., and Khalil, M.A.K.  Measurements of $CHCl_2$ in background tropospheric air.  Nature.  286: 793-5.  1980.

Perbert, G., Filiol, C., Boule, P., and Lemaire, J.  Photolysis and photooxidation of diphenols in dilute aqueous solution.  J. Chim. Phys. Phys.-Chem. Biol.  76: 89-96.  1979.

Perdue, E.M. Association of organic pollutants with humic substances: Partitioning equilibria and hydrolysis kinetics In:  Aquatic and Terrestrial Materials. Christman, R.F. et al., (Eds.) Ann Arbor, MI. Ann Arbor Press. pp. 441-60.  1983.

Petrasek, A.C., Kugelman, I.J., Austern, B.M., Pressley, T.A., Winslow, L.A., and Wise, R.H.  Fate of toxic organic compounds in wastewater treatment plants.  J. Water Pollut. Control Fed.  55: 286-96.  1983.

Pettet, A.J. and Miles, E.V.  Biological treatment of cyanides with and without sewage.  J. Appl. Chem.  4: 434.  1954.

704

Pfaender, F.K. and Bartholomew, G.W. Measurement of aquatic biodegradation rates by determining heterotrophic uptake of radiolabled pollutants. Appl. Environ. Microbiol. 44: 159-64. 1982.

Pitter, P. Determination of biological degradability of organic substances. Water Res. 10: 231-5. 1976.

Pitts, J.N., Jr., Atkinson, R., Winer, A.M., Bierman,H.W., Cater, W.P.L., MacLeod, H., and Tuazon, E. C. Formation and rate of toxic chemicals in California's atmosphere. ARB/R-85/239. NTIS PB85-172 609. Air Resources Board. Sacramento, CA. 1984.

Pitts, J.N. Jr., Sandoval, H.L., and Atkinson, R. Relative rate constants for the reaction of oxygen(1D) atoms with fluorocarbons and nitrous oxide. Chem. Phys. Lett. 29: 31-4. 1974.

Portier, R.J. Comparison of environmental effect and biotransformation of toxicants on laboratory microcosm and field microbial communities. In: ASTM Spec. Tech. Publ. 865(Validation Predict. Lab. Methods Assess. Fate Eff. Contam. Aquat. Ecosyst.): 14-30. 1985.

Price, K.S., Waggy, G.T., and Conway, R.A. Brine shrimp bioassay and seawater BOD of petrochemicals. J. Water Pollut. Control Fed. 46: 63-77. 1974.

Queen, A. Kinetics of the hydrolysis of acyl chlorides in pure water. Can. J. Chem. 45: 1619-29. 1967.

Radding, S.B., Mill, T., Gould, C.W., Liu, D.H. and Johnson, H.L. The environmental fate of selected polynuclear aromatic hydrocarbons. EPA-560/5-75-009. Menlo Park,CA: Stanford Research Inst. pp. 131. 1976.

Reinert, K.H. and Rodgers, J.H. Fate and persistence of aquatic herbicides. Rev. Environ. Contam. Toxicol. 98: 69-91. 1987.

Rhodes, R.C. Studies with manganese(14C)ethylenebis(dithiocarbamate) ((14-C)maneb) fungicide and (14C)ethylenethiourea ((14-C)ETU) in plants, soil, and water. J. Agric. Food Chem. 25: 528-33. 1977.

Robbins, D.E. Photodissociation of methyl chloride and methyl bromide in the atmosphere. Geophys. Res. Lett. 3: 213-6. 1976.

Roberts, T.R. and Stoydin, G. The degradation of (Z)- and (E)-1,3-dichloropropenes and 1,2-dichloropropane in soil. Pestic. Sci. 7: 325-35. 1976.

Robertson, R.E. and Scott, J.M.W. The neutral hydrolysis of some alkyl and benzyl halides. J. Chem. Soc. 1961. pp. 1596-604. 1961.

Robertson, R.E. and Sugamori, S.E. The hydrolysis of dimethyl sulfate and diethyl sulfate in water. Can. J. Chem. 44: 1728-30. 1966.

Rogers, J.E., Li, S.W., and Felice, L.J. Microbial transformation kinetics of xenobiotics in aquatic environment. EPA-600/3-84-043. (NTIS PB84-162866). Richland, WA: Battelle Pacific Northwest Labs. 105 pp. 1984.

Rontani, J.F., Ranbeloarisoa, E., Bertrand, J.C., and Giuste, G. Favorable interaction between photooxidation and bacterial degradation of anthracene in sea water. Chemosphere. 14: 1909-12. 1985A.

Rosen, J.D., Strusz, R.F., and Hill, C.C. Photolysis of phenylurea herbicides. J. Agric. Chem. 17: 20 6-7. 1969.

Ross, R.D. and Crosby, D.G. Photooxidant activity in natural waters. Environ. Toxicol. Chem. 4: 773-8. 1985.

Ross, R.D. and Crosby, D.G. Photolysis of ethylenethiourea. J. Agric. Food Chem. 21: 335-7. 1973.

Roy, R.S. Spectrophotometric estimation of ester hydrolysis. Anal. Chem. 44: 2096-8. 1972.

Rubin, H.W. and Alexander, M. Effect of nutrients on the rates of mineralization of trace concentrations of phenol and p-nitrophenol. Environ. Sci. Technol. 17: 104-7. 1983.

Russell, D.J., McDuffie, B., and Fineberg, S. The effect of biodegradation on the determination of some chemodynamic properties of phthalate esters. J. Environ. Sci. Health A. 20: 927-41. 1985.

Sadtler. Sadtler Standard Spectra. Sadtler Research Lab. Philadelphia, PA.

Saeger, V.W. and Tucker, E.S. Biodegradation of phthalic acid esters in river water and activated sludge. Appl. Environ. Microbiol. 31: 29-34. 1976.

Saeger, V.W. and Tucker, E.S. Phthalate esters undergo ready biodegradation. Plast. Eng., Aug. 1973. pp. 46-9. 1973.

Saito, T., Widayat, Winiati, W., Hagiwara, K. and Murakami, Y. Study on organic pollution parameters for dyes and dyeing auxiliaries. Fresenius' Z. Anal. Chem. 319: 433-4. 1984.

Salkinoja-Salonen, M.S., Valo, R., Apajalahti, J., Hakulinen, R., Salakoski, L., and Jaakkola, T. Biodegradation of chlorophenolic compounds in wastes from wood processing industry. In: Current Perspectives in Microbiology. Klug, M.J. and Reddy, C.A., Eds. Nat. Acad. Sci. pp. 668-72. 1984.

706

Salkinoja-Salonen, M.S., Hakulenen, R., Valo, R., and Apajalohte, J. Biodegradation of recalcitrant organochlorine compounds in a fixed-film reactor. Water Sci. Technol. 15: 309-19. 1983.

Sanborn, J.R., Francis, B.M., and Metcalf, R.L. The degradation of selected pesticides in soil: A review of published literature. EPA-600/9-77-022. Cincinnati, OH: U.S. EPA. 616 pp. 1977.

Sasaki, S. The scientific aspects of the chemical substance control law in Japan. In: Aquatic Pollutants: Transformation and Biological Effects. Hutzinger, O. et. al., eds. Pergamon Press, Oxford, UK. pp. 283-98. 1978.

Saunders, D.G. and Mosier, J.W. Photolysis of the aquatic herbicide fluridone in aqueous solution. J. Agric. Food Chem. 31: 237-41. 1983.

Sawada, S., Hayakawa, K., Nakahata, K and Totsuka, T. A model calculation of ethylene removal capacity by soil acting as a natural sink in polluted atmospheres. Atmos. Environ. 20: 513-6. 1986.

Sawada, S. and Toksuka, T. Natural and anthropogenic sources and fate of atmospheric ethylene. Atmos. Environ. 20: 821-32. 1986.

Scheunert, I., Vockel, D., Schmitzer, J., and Korte, F. Biomineralization rates of 14C-labeled organic chemicals in aerobic and anaerobic suspended soil. Chemosphere. 16: 1031-41. 1987.

Schmidt-Bleek, F., Haberland, W., Klein, A.W., and Caroli, S. Steps toward environmental hazard assessment of new chemicals (including a hazard ranking scheme, based upon directive 79/831/EEC). Chemosphere. 11: 383-415. 1982.

Schneider, M. and Smith, G.W. Photochemical degradation of malathion. NTIS PB-286 115. College, AK: Inst. Water Resour. Univ. Alaska. 17 pp. 1978.

Schouten, M.J., Peereboom, J.W.C., and Brinkman, U.A.T. Liquid chromatographic analysis of phthalate esters in Dutch river water. Int. J. Analyt. Chem. 7: 13-23. 1979.

Schroeder, H.F. Chlorinated hydrocarbons in biological sewage purification - fate and difficulties in balancing. Water Sci. Technol. 19: 429-38. 1987.

Scully, F.E., Jr. and Hoigne, J. Rate constants for reactions of singlet oxygen with phenols and other compounds in water. Chemosphere. 16: 681-94. 1987.

Shanker, R., Ramakrishna, C., and Seth, P.K. Degradation of some phthalic acid esters in soil. Environ. Pollut. Ser. A. 39: 1-7. 1985.

Shashidar, M.A. Electronic adsorption spectra of some monosubstituted benzenes in the vapour phase. Spectrochemica. Acta. 27A: 2363-74. 1971.

Shelton, D.R., Boyd, S.A., and Tiedje, J.M. Anaerobic biodegradation of phthalic acid esters in sludge. Environ. Sci. Technol. 18: 93-7. 1984.

Shelton, D.R. and Tiedje, J.M. Development of tests for determining anaerobic biodegradation potential. EPA-560/5-81-013. NTIS PB84-166 495. Mich. State Univ., Dept. Crop Soil Sci., East Lansing, MI. 92 pp. 1981.

Sielicki, M., Focht, D.D., and Martin, J.P. Microbial transformations of styrene and 14C-styrene in soil and enrichment cultures. Appl. Environ. Microbiol. 35: 124-8. 1978.

Sikka, H.C., Appleton, H.T., and Banerjee, S. Fate of 3,3'-dichlorobenzidine in aquatic environments. EPA-600/3-78-068. Syracuse Research Corp. Syracuse, NY. 1978.

Silka, L.R. and Wallen, D.A. Observed rates of biotransformation of chlorinated aliphatics in groundwater. In: Superfund 88. Proc. 9th Natl. Conf., Haz Mat. Control Inst. pp. 138-41. 1988.

Silverstein, R.M. and Bassler, G.C. Spectrometric Identification of Organic Compounds. New York, NY: John Wiley & Sons, Inc. 1967.

Simmons, M.S. and Zepp, R.G. Influence of humic substances on photolysis of nitroaromatic compounds in aqueous systems. Water Res. 20: 899-904. 1986.

Sims, R.C. Fate of PAH compounds in soil: loss mechanisms. Accepted for Environ. Tox. Chem. 9: 1990.

Sims, G.K. and Sommers, L.E. Degradation of pyridine derivatives in soil. J. Environ. Qual. 14: 580-4. 1985.

Singh, H.B., Salas, L.J., Smith, J.A., and Shigeishi, H. Atmospheric measurements of selected toxic organic chemicals. EPA-600/3-80-072. U.S.EPA, Research Triangle Park, NC. 1980.

Skurlatov, Y.I., Zepp, R.G., and Baughman, G.L. Photolysis rates of (2,4,5-trichlorophenoxy)acetic acid and 4-amino-3,5,6-trichloropicolinic acid in natural waters. J. Agric. Food Chem. 31: 1065-71. 1983.

Small, M.J. Compounds formed from the chemical decontamination of HD, GB, and VX and their environmental fate. USAMBRDL-TR-8304. (NTIS AD-A149 515). Fort Detrick, MD: Army Med. Bioeng. Res. Devel. Lab. 92 pp. 1984.

Smith, A.E. and Hayden, B.J. Relative persistence of MCPA and mecoprop in Saskatchewan soils, and the identification of MCPA in MCPB-treated soil. Weed Sci. 21: 179-83. 1981.

Smith, A.E.  Soil persistence experiments with (14C) 2,4-D in  herbicidal mixtures and field persistence studies with tri-allate and trifluralin both singly and combined.  Weed Res.  19: 165-70.  1979A.

Smith, A.E.  Relative persistence of di- and tri-chlorophenoxyalkanoic acid herbicides in Saskatchewan soils.  Weed Res.  18: 275-9.  1978.

Smith, A.E.  Degradation, adsorption, and volatility of di-allate and tri-allate in prairie soils.  Weed Res.  10: 331-9.  1970.

Smith, J.H., Mabey, W.R., Bohonos, N., Holt, B.R., Lee, S.S., Chou, T.W., Venberger, D.C., and Mill, T.  Environmental pathways of selected chemicals in freshwater systems. Part II. Laboratory studies.  EPA-600/7-78-074.  U.S. EPA, Athens, GA.  1978.

Snelson, A., Butler, R., and Jarke, F.  A study of removal processes for halogenated air pollutants.  EPA-600/3-78-058.  U.S. EPA, Research Triangle Park, N.C.  1978.

Sokolova, L.P. and Kaplin, V.T.  Self-cleaning of natural waters. Elimination of alcohols of various structure under simulated conditions.  Gidrokhim. Mater.  51: 186-91.  1969.

Sonoda, Y. and Seiko, Y.  Degradation and toxicity of various compounds in anaerobic digestion. (I). Carbohydrates and alcohols.  J. Ferment. Technol. 46: 796-801.  1968.

Southworth, G.R.  Transport and transformations of anthracene in natural waters.  In: ASTM STP-667. Aquatic Toxicology.  Marking, L.L. & Kimberle, R.A., Eds.  Philadelphia, PA: American Society for Testing and Materials.  pp. 359-80.  1979.

Spain, J.C., Pritchard, P., and Bourquin, A.W.  Effects of adaptation on biodegradation rates in sediment water cores from estuarine and freshwater environments.  Appl. Environ. Microbiol. 40: 726-34.  1980.

Spanggord, R.J., Mabey, W.R., Mill, T., Chou, T.W., and Smith, J.H.  Environmental fate studies on certain munition wastewater constituents. Part 1. Model validation.  NTIS AD-A129 373/7. SRI Inter. Menlo Park, CA.  79 pp.  1981.

Spanggord, R.J., Mill, T., Chou, T.W., Mabey, W.R., Smith, J.H., and Lee, S.  Environmental fate studies on certain munitions wastewater constituents. Phase II - Laboratory studies.  NTIS AD A099 256.  U.S. Army Med. Res. Devel. Frederick, MD.  137 pp.  1980A.

Spanggord, R.J., Mill, T., Chou, T.W., Mabey, W.R., Smith, J.H., and Lee, S.  Environmental fate studies on certain munitions wastewater constituents. Final report. Phase I - Literature review. SRI Project No. LSU-7934.  Contract No. DAMD 17-78-C-8081.  U.S. Army Medical Res. Devel. Command. Fort Detrick, MD.  1980.

709

Speece, R.E. Anaerobic biotechnology for industrial wastewater treatment. Environ. Sci. Technol. 17: 416A-27A. 1983.

Steinhauser, K.G., Amann, W., and Polenz, A. Biodegradability testing of chlorinated compounds. Vom Wasser. 67: 147-54. 1986.

Stevens, B., Perez, S.R., and Ors, J.A. Photoperoxidation of unsaturated organic molecules O2'Delta G acceptor properties and reactivity. J. Amer. Chem. Soc. 96: 6846-50. 1974.

Stewart, D.K.R. and Chisholm, D. Long-term persistence of BHC, DDT and chlordane in a sandy loam clay. Can. J. Soil Sci. 61: 379-83. 1971.

Stover, E.L. and Kincannon, D.F. Biological treatability of specific organic compounds found in chemical industry wastewaters. J. Water Pollut. Control Fed. 55: 97-109. 1983.

Su, F., Calvert, J.G., and Shaw, J.H. Mechanism of the photoxidation of gaseous formaldehyde. J. Phys. Chem. 83: 3185-91. 1979.

Subba-Rao, R.V., Rubin, H.E., and Alexander, M. Kinetics and extent of mineralizatoin of organic chemicals at trace levels in freshwater and sewage. Appl. Environ. Microbiol. 43: 1139-50. 1982.

Sudhakar-Barik and Sethunathan, N. Metabolism of nitrophenols in flooded soils. J. Environ. Qual. 7: 349-52. 1978A.

Sugatt, R.H., O'Grady, D.P., Banerjee, S., Howard, P.H., and Gledhill, W.E. Shake flask biodegradation of 14 commercial phthalate esters. Appl. Environ. Microbiol. 47: 601-6. 1984.

Sugiura, K., Aoki, M., Kaneko, S., Daisaku, I., Komatsu, Y., Shibuya, H., Suzuki, H., and Goto, M. Fate of 2,4,6-trichlorophenol, pentachlorophenol, p-chlorobiphenyl, and hexachlorobenzene in an outdoor experimental pond: comparison between observations and predictions based on laboratory data. Arch. Environ. Contam. Toxicol. 13: 745-58. 1984.

Swain, H.M. and Somerville, H.J. Microbial metabolism of methanol in a model activated sludge system. J. Appl. Bacteriol. 45:147-51. 1978.

Swindoll, C.M., Aelion, C.M., and Pfaender, F.K. Inorganic and organic amendment effects of the biodegradation of organic pollutants by groundwater microorganisms. Amer. Soc. Microbiol. Abstr., 87th Annl. Mtg., Atlanta, GA. 298 pp. 1987.

Szeto, S.Y., Vernon, R.S., and Brown, M.J. Degradation of disulfoton in soil and its translocation into asparagus. J. Agric. Food Chem. 31: 217-20. 1983.

710

Tabak, H.H., Quave, S.A., Mashni, C.I., and Barth, E.F. Biodegradability studies for predicting the environmental fate of organic priority pollutants. In: Test Protocols for Environmental Fate and Movement of Toxicants. Proc. Symp. Assoc. Off. Analyt. Chem. 94th. Ann. Mtg., Washington, DC. pp. 267-328. 1981A.

Tabak, H.H., Quave, S.A., Mashni, C.I., and Barth, E.F. Biodegradability studies with organic priority pollutant compounds. J. Water Pollut. Control Fed. 53: 1503-18. 1981.

Tabak, H.H. and Barth, E.F. Biodegradability of benzidine in aerobic suspended growth reactors. J. Water Poll. Control Fed. 50: 552-8. 1978.

Tabatabai, M.A. and Bremner, J.M. Decomposition of nitrilotriacetate (NTA) in soils. Soil Biol. Biochem. 7: 103-6. 1975.

Takemoto, S., Kuge, Y., and Nakamoto, M. The measurement of BOD in seawater. Suishitsu Odaku Kenkyu. 4: 80-90. 1981.

Tate, R.L., III and Alexander, M. Microbial formation and degradation of dimethylamine. Appl. Environ. Microbiol. 31: 399-403. 1976.

Tate, R.L., III and Alexander, M. Stability of nitrosamines in samples of lake water, soil, and sewage. J. Natl. Cancer Inst. 54: 327-30. 1975.

Tester, D.J. and Harker, R.J. Groundwater pollution investigations in the Great Ouse Basin. Water Pollut. Control. 80: 614-31. 1981.

Thompson, J.E. and Duthie, J.R. The biodegradability and treatability of NTA. J. Water Pollut. Control Fed. 40: 306. 1968.

Tiravanti, G. and Boari, G. Potential pollution of a marine environment by lead alkyls: The Cavtat Incident. Environ. Sci. Technol. 13: 849-54. 1979.

Tou, J.C., Westover, L.B., and Soundabend, L.F. Kinetic studies of bis(chloromethyl)ether hydrolysis by mass spectrometry. J. Phys. Chem. 78: 1096-8. 1974.

Trzilova, B. and Horska, E. Biodegradation of amines and alkanes in aquatic environment. Biologia (Bratislava). 43: 209-18. 1988.

Tsao, C.W. and Root, J.W. A new primary process in the ultraviolet photolysis of methyl iodide. The direct photolysis to. J. Phys. Chem. 76: 308-11. 1972.

Tuazon, E.C., Winer, A.M., Graham, R.A., Schmid, J.P. and Pitts, J.N. Fuorier transform infrared detection of nitramines in irradiated amine-NOx systems. Environ. Sci. Technol. 12: 954-8. 1982.

U.S. EPA. Treatability manual- vol. 1. Treatability Data. EPA-600/8-80-042. U.S. EPA. Washington, DC. 1980.

U.S. EPA. PCHYDRO-PGEMS Graphical Exposure Modelling System. 1987.

Ugi, I. and Beck, F. Solvolysis of carboxylic acid derivatives. I. Reaction of carboxylic acid chlorides with water and amines. Chem. Ber. 94: 1839-50. 1961.

Urano, K. and Kato, Z. Evaluation of biodegradation ranks of priority organic compounds. J. Hazardous Materials. 13: 135-45. 1986.

Urushigawa, Y., Yonezawa, Y., Masunaga, S., Matsue, Y., Masunaga, H., and Yanagishi, T. Biodegradation of organic compounds by activated sludge. 13. Spectrophotometrical characteristics of mono-substituted phenols in biodegradation by activated sludge adapted to aniline, phenol, and m-cresol. Kogai Shigen Kenkyusho Iho. 14: 45-64. 1984B.

Urushigawa, Y., Yonezawa, Y., Masunaga, S., Tashiro, T., Hirai, M., and Tanaka, M. Biodegradability of mono-substituted phenols. I. Biodegradability by non-acclimated activated sludge microorganisms. Kogai Shigen Kenkyusho Iho. 12: 37-46. 1983.

Urushigawa, Y. and Yonezawa, Y. Chemico-biological interactions in biological purification system. II. Biodegradation of azocompounds by activated sludge. Bull. Environ. Contam. Toxicol. 17: 214-8. 1977.

Vaishnav, D.D. and Babeu, L. Comparison of occurrence and rates of chemical biodegradation in natural waters. Bull. Environ. Contam. Toxicol. 39: 237-44. 1987.

Van der Linden, A.C. Degradation of oil in the marine environment. Dev. Biodegrad. Hydrocarbons. 1: 165-200. 1978.

Van Veld, P.A. and Spain, J.C. Degradation of selected xenobiotic compounds in three types of aquatic test systems. Chemosphere. 12: 1291-305. 1983.

Vanluin, A.B. and Teurlinckx, L.V.M. The treatment of industrial wastewater in integrated activated sludge/powdered activated carbon systems. In: Manage. Hazard Toxic Wastes., Proc. Int. Congr. pp. 476-85. 1987.

Venkoba Rao, G. and Venkatasubramanian, N. Role of dimethyl sulphoxide as a solvent in hydrolysis of aliphatic dicarboxylic esters. Ind. J. Chem. 10: 178-81. 1972.

Versar, Inc. Development of POTW models for Estimating Removal of Chemical Substances During Wastewater Treatment. Draft Final Report. Versar Task No. 138. Prepared under contract no. EPA 68-02-4254 for U.S. EPA, Office of Toxic Substances, Exposure Evaluation Division, Washington, D.C. 1988.

Wagner, R. Investigation of the biodegradation of organic substances using the dilution bottle method. II. The degradation of test substances. Vom Wasser. 47: 241-65. 1976.

Wakeham, S.G., Davis, A.C., and Karas, J.L. Microcosm experiments to determine the fate and persistence of volatile organic compounds in coastal seawater. Environ. Sci. Technol. 17: 611-7. 1983.

Walker, A. and Zimdahl, R.L. Simulation of the persistence of atrazine, linuron and metolachlor in soil at different sites in the USA. Weed Res. 21: 255-65. 1981.

Walker, J.D. and Colwell, R.R. Measuring the potential activity of hydrocarbon-degrading bacteria. Appl. Environ. Microbiol. 31: 189-97. 1976.

Walker, W.W. Insecticide persistence in natural seawater as affected by salinity, temperature and sterility. EPA-600/3-78-044. Gulf Breeze, FL: U.S. EPA. 25 pp. 1978.

Walker, W.W. and Stojanovic, B.J. Microbial versus chemical degradation of malathion in soil. J. Environ. Qual. 2: 229-32. 1973.

Wallington, T.J., Dagaut, P., Liu, R., and Kurylo, M.J. Gas-phase reactions of hydroxyl radicals with the fuel additives methyl t-butyl ether and t-butyl alcohol over the temperature range 240-440K. Environ. Sci. Technol. 22: 842-44. 1988.

Ward, C.T. and Matsumura, F. Fate of 2,3,7,8-tetrachloro-p-dioxin (TCDD) in a model aquatic environment. Arch. Environ. Contam. Toxicol. 7: 349-57. 1978.

Ward, C.H., Tomson, M.B., Bedient, P.B., and Lee, M.D. Transport and fate processes in the subsurface. Water Resour. Symp. 13 (Land Treat.: Hazard. Waste Manag. Altern.): 19-39. 1986.

Weast, R.C., Astle, M.J., and Beyer, W.H. CRC Handbook of Chemistry and Physics. 54th Ed. CRC Press. Boca Raton, FL. 1962.

Weintraub, R.A., Jex, G.W., and Moye, H.A. Chemical and microbial degradation of 1,2-dibromoethane (EFB) in Florida ground water, soil, and sludge. ACS Symp. Ser., Garner, W.Y. et al., Eds. Washington, DC. 315(Eval. Pestic. Ground Water): 294-310. 1986.

Wendt, T.M., Cornell, J.H., and Kaplan, A.M. Microbial degradation of glycerol nitrates. Appl. Environ. Microbiol. 36: 693-9. 1978.

Wheeler, W.B., Rothwell, D.F., and Hubbell, D.H. Persistence and microbial effects of acrol and chlorobenzilate to two Florida soils. J. Environ. Qual. 2: 115-8. 1973.

Wilderer, P. A model river test to describe the various impacts of chemical substances on microbial biocommunities. AICHE Symp. Ser. 77: 205-13. 1981.

Wilson, B.H., Smith, G.B., and Rees, J.F. Biotransformations of selected alkylbenzenes and halogenated aliphatic hydrocarbons in methanogenic aquifer material: A Microcosm Study. Environ. Sci. Technol. 20: 997-1002. 1986.

Wilson, J.T., Cosby, R.L., and Smith, G.B. Potential for biodegradation of organo-chlorine compounds in ground water. R.S. Kerr Environmental Research Laboratory, Ada, OK. 1984.

Wilson, J.T., Enfield, C.G., Dunlap, W.J., et al. Transport and fate of selected organic pollutants in a sandy soil. J. Environ. Qual. 10: 501-506. 1981.

Wilson, J.T., McNabb, J.F., Balkwill, D.L., and Ghiorse, W.C. Enumeration and characterization of bacteria indigenous to a shallow water-table aquifer. Ground Water. 21: 134-42. 1983A.

Wilson, J.T., McNabb, J.F., Wilson, R.H., and Noonan, M.J. Biotransformation of selected organic pollutants in groundwater. Devel. Indust. Microbiol. 24: 225-33. 1983.

Wilson, W.G., Novak, J.T., and White, K.D. Enhancement of biodegradation of alcohols in groundwater systems. In: Toxic Hazard. Wastes, Proc. Mid-Atl. Ind. Waste Conf. 18: 421-30. 1986.

Wolfe, N.L., Kitchens, B.E., Macalady, D.L., and Grunde, T.J. Physical and chemical factors that influence the anaerobic degradation of methyl parathion in sediment systems. Environ. Toxicol. Chem. 5: 1019-26. 1986.

Wolfe, N.L., Zepp, R.G., Schlotzhauer, P., and Sink, M. Transformation pathways of hexachlorocyclopentadiene in the aquatic environment. Chemosphere. 11: 91-101. 1982.

Wolfe, N.L., Steen, W.C., and Burns, L.A. Phthalate ester hydrolysis: linear free energy relationships. Chemosphere. 9: 403-8. 1980B.

Wolfe, N.L., Burns, L.A., and Steen, W.C. Use of linear free energy relationships on an evaluative model to assess the fate and transport of phthalate esters in the aquatic environment. Chemosphere. 9: 393-502. 1980A.

Wolfe, N.L. Zepp, R.G., and Paris, D.F. Carbaryl, phospham and chloropropham: a comparison of the rates of hydrolysis and photolysis with the rate of biolysis. Water Res. 12: 565-71. 1978.

Wolfe, N.L., Zepp, R.G., Gordan, T.A., Baughman, G.L., and Cline, D.M. Kinetics of chemical degradation of malathion in water. Environ. Sci. Technol. 11: 88-93. 1977A.

Wolfe, N.L., Zepp, R.G., Paris, D.F., Baughman, G.L., and Hollis, R.C.  Methoxychlor and DDT degradation in water: rates and products.  Environ. Sci. Technol.  11: 1077-81.  1977

Wolfe, N.L., Zepp, R.G., Baughman, G.L., Fincher, R.C., and Gordon, J.A.  Chemical and photochemical transformation of selected pesticides in aquatic environments.  U.S. EPA-600/3-76-067.  U.S. EPA, Athens, GA.  153 pp.  1976.

Yordy, J.R. and Alexander, M.  Microbial metabolism of N-nitrosodiethanolamine in lake water and sewage.  Appl. Environ. Microbiol.  39: 559-65.  1980.

Zanella, E.F., McKelvey, R.D., and Joyce, T.W.  Effect of anthraquinone on toxicity and treatability of bleached kraft pulp mill effluents.  Tappi.  62: 65-7.  1979.

Zepp, R.G., Braun, A.M., Hoigne, J., and Leenheer, J.A.  Photoproduction of hydrated electrons from natural organic solutes in aquatic environments.  Environ. Sci. Technol.  21: 485-90.  1987.

Zepp, R.G. and Schlotzhauer, P.F.  Photoreactivity of selected aromatic hydrocarbons in water.  In: Polynuclear Aromatic Hydrocarbons.  Jones, P.W. and Leber, P., Eds.  Ann Arbor Sci. Publ. Inc., Ann Arbor, MI.  pp. 141-58.  1979.

Zepp, R.G., Wolfe, N.L., Azarraga, L.V., Cox, R.H., and Pape, C.W.  Photochemical transformation of DDT and methoxychlor degradation products, DDE and DMDE, by sunlight.  Arch. Environ. Contam. Tox.  2-3: 305-14.  1977A.

Zepp, R.G. and Cline, D.M.  Rates of direct photolysis in aquatic environment.  Environ. Sci. Technol.  11: 359-66.  1977.

Zepp, R.G., Wolfe, N.L., Gordon, J.A., and Fincher, R.C.  Light-induced transformation of methoxychlor in aquatic systems.  J. Agric. Food Chem.  24: 727-33.  1976.

Zeyer, J., Kuhn, E.P., and Schwarzenbach, R.P.  Rapid microbial mineralization of toluene and 1,3-dimethylbenzene in the absence of molecular oxygen.  Appl. Environ. Microbiol.  52: 944-7.  1986A.

Zhang, Z., Quan, W., Tian, D., and Zhang, J.  Phthalate esters in the atmospheric particulates of Beijing urban area.  Huanjing Huaxue.  6: 29-34.  1987.

Zhang, Z., Jiao, B., Wang, F., and Li, S.  Study on the reduction of carcinogenicity of nitrosamines.  Zhonhua Zhongliu Zashi.  5: 270-3.  1983.

Zhang, S., An, Q., Gu, Z., and Ma, X.  Degradation of BHC in soil. (CH).  Huanjing Kexue, 3: 1-3.  1982.

Zimdahl, R.L. and Gwynn, S.W.   Soil degradation of three dinitroanilines.   Weed Sci.   25:   247-51. 1977.

Zoeteman, B.C.J., Harmsen, K., Linders, J.B.H.J., Morra, C.F.H., and Sloof, W.   Persistent organic pollutants in river water and ground water of The Netherlands.   Chemosphere.   9: 231-49. 1980.

# CAS REGISTRY NUMBER INDEX

# CAS REGISTRY NUMBER INDEX

# CAS REGISTRY NUMBER INDEX

# CAS REGISTRY NUMBER INDEX

# CAS REGISTRY NUMBER INDEX

# CAS REGISTRY NUMBER INDEX

# CAS REGISTRY NUMBER INDEX